GREEK HOROSCOPES

O. NEUGEBAUER

AND

H. B. VAN HOESEN

THE AMERICAN PHILOSOPHICAL SOCIETY
INDEPENDENCE SQUARE
PHILADELPHIA
1987

Copyright 1959 by the American Philosophical Society
Reprinted 1987 by the American Philosophical Society

Memoirs of the

AMERICAN PHILOSOPHICAL SOCIETY

Held at Philadelphia

for Promoting Useful Knowledge

Volume 48

Library of Congress Catalog
Card Number: 59-11559
International Standard Book No.: 0-87169-048-9
US ISSN: 0065-9762

PRINTED IN THE UNITED STATES OF AMERICA
BY J. H. FURST COMPANY, BALTIMORE, MARYLAND

L'astrologie hellénistique est l'amalgame d'une doctrine philosophique séduisante, d'une mythologie absurde et de méthodes savantes employées à contre-temps.

A.-J. Festugière, O. P.
La Révélation d'Hermès Trismégiste I p. 89

PREFACE

What we know about Greek astronomy is dominated by Ptolemy's "Almagest" (written about 140 A.D.) and closely related works like Theon's "Handy Tables" (end of the fourth century) but we have only very little information about the practice of computing the positions of sun, moon, and planets during any period of antiquity. It seemed, therefore, desirable to make the Greek horoscopes available for study as a group which could be expected to reflect at least some of the current techniques of Greek astronomy.

About 60 horoscopes from the first five centuries of our era have been published since Young (1828) and Champollion-Figeac (1840) in the papyrological literature, which comprises about 18,000 texts. We have collected all horoscopes from this widely scattered material to the best of our knowledge and added a few unpublished pieces which were kindly put at our disposal. Wherever possible we have checked all readings in photographs or in the originals. Naturally the study of these texts as a class made it possible to progress occasionally beyond the original edition of the single documents. Instead of giving a list of such small improvements and referring the reader to the original publication for the main text we thought it simpler and more useful to give for all papyri, ostraca, and graffiti the Greek text in full, even if we had nothing to change in the original readings. We give a translation only, without the Greek text, in those cases where the original has been lost since its first publication.

A different situation prevailed in the "literary horoscopes," the majority of which are either contained in the *Anthology* of Vettius Valens (second century A.D.) or in the later astrological compilations made accessible through the *Catalogus Codicum Astrologorum Graecorum*. It would have taken us many more years, had we attempted to check the printed editions in all the variants of the original manuscripts. Consequently, for Vettius Valens we relied on Kroll's edition (1908); for the horoscopes published in the *CCAG*[1] we checked doubtful passages wherever possible and, as a rule, at least one of the manuscripts, but we did not reproduce the Greek text except for some notes on variant readings.

Since the primary objective of this study was astronomical, we have excluded from it purely fictitious horoscopes like the one given in the *Dionysiaca* of Nonnos whose author had so little concern with astronomy that he placed Venus almost in opposition to the sun. The time limits were given by the material itself: the earliest Greek horoscopes belong to the first century B.C. whereas the latest independent Greek texts are about contemporary with the beginning of the Islamic period.

Though we are dealing with texts of a comparatively well-defined character, we must admit that we are fully aware that we had to leave many a question unanswered. The discussion of texts of this type touches not only upon all the countless branches of papyrology, on legal, political, medical, and other questions, but also requires familiarity with an excessively complex and unwieldy astrological literature of which the twelve volumes of the *CCAG* are witness. All that we could hope to do is to present our material for further study in a form as accessible as possible. We hope that the indices and glossaries at the end will be found useful for quick orientation before details need to be consulted, e.g., in Bouché-Leclercq's unsurpassed *Astrologie grecque*.

Our sincere thanks are due to many scholars who helped us with advice and by collating doubtful passages or providing us with photographs. Our Egyptological colleagues, J. Černý and R. A. Parker, contributed a study of the Coptic sections of P. Lond. 98, based on a collation of the late Dr. P. Kahle. The important London papyri were repeatedly collated for us by Dr. T. C. Skeat to whom we are also indebted for several new readings. For the texts in Oxford we had the help of Dr. Colin H. Roberts. Dr. Bartoletti was helpful in checking papyri in Florence and Dr. Doresse made the parchment horoscope from the Louvre accessible to us (No. 326). Through Professor H. Youtie we obtained unpublished material from the Michigan collection. Dr. E. Boer in Berlin and the late F. H. Cramer were helpful in checking codices in Germany and France.

For permission to reproduce photographs of unpublished texts we are grateful to the authorities of the Cairo Museum (No. 46), Cornell University (No. 282), the Laurentian Library in Florence (No. L 497), the Louvre (No. 326), and the University of Michigan (Nos. 182 and 208).

We are greatly indebted to the American Philosophical Society for making the publication of this work possible.

---

[1] For the meaning of abbreviations consult p. 207 ff.

CONTENTS

	PAGE
Introduction	1
A. Numbering of the Texts	1
B. Symbols	1
C. Method of Dating	1
D. Glossary of Astrological and Technical Terms	2
Chapter I. The Original Documents	14
Chapter II. The Horoscopes from Literary Sources	76
Chapter III. General Comments	161
A. Chronological Distribution	161
B. The Horoscopes in General	162
C. Arrangement of the Basic Data	163
D. Calendaric Data	165
E. Astronomical Data	170
F. Astrological Data	173
Chapter IV. The Authors of the Literary Horoscopes	176
A. Vettius Valens	176
B. Critodemus	185
C. Antigonus of Nicaea	186
D. The Late Authors	187
1. Hephaestion	187
2. Palchus	187
3. Rhetorius	187
4. Eutocius	188
5. Stephanus	190
Glossary	191
A. Greek	191
1. General Greek Glossary	191
2. Greek Words, incomplete	200
3. Names of Persons and Places	200
4. Constellations and Star Names	201
B. Coptic	202
C. English	202
Concordances	205
A. Papyrus Publications	205
B. Vettius Valens	205
C. CCAG	206
D. Other Sources	206
References	207
Subject Index	211
Diagrams for Original Sources (pls. 1 to 9)	214
Diagrams for Literary Horoscopes (pls. 10 to 29)	219
Photographs (pls. 30 to 32)	229

INTRODUCTION

A. NUMBERING OF THE TEXTS

The texts are numbered, so far as possible, by their Julian dates. If L is prefixed we mean a "literary source" (Vettius Valens, etc.) as distinguished from an original text (papyri, etc.). For example No. 125 is a horoscope on papyrus for A.D. 125. If there are several horoscopes falling in the same year, they are differentiated by the addition of the month (or even day), e.g., L 114,V and L 114,VII. In case of duplication, the letters of the alphabet are used for distinction, e.g., 137,a and 137,b.

In case the date cannot be established accurately, the number is determined as follows. If definite limits are known (e.g., the reign of an emperor), these limits are used. Thus a horoscope dated by an unknown year of Antoninus is numbered 138/161. The estimate of a century to which a text may belong is indicated by using the fiftieth year of the century but always followed by 1, or 2, etc. For example, 250,1 and 250,2 mean two papyri tentatively assigned to the third century (whereas 250 alone means exactly the year 250). Similarly 300,1 means "from third to fourth century" 325,1 "first half of fourth century."

B. SYMBOLS

[ ]	restored passage
{ }	superfluous passage in text, e.g., duplication
( )	our addition
H	Horoscopos (i.e., rising sign)
M	Midheaven (upper culmination)
Δ	Dysis (setting sign)
IMC	Lower Midheaven (lower culmination)

⊙	sun; in texts: ♂︎	♈	Aries
☾	moon	♉	Taurus
☌	conjunction	♊	Gemini
☍	opposition	♋	Cancer
☊	ascending lunar node	♌	Leo
♄	Saturn — Phainon	♍	Virgo
♃	Jupiter — Phaeton	♎	Libra
♂	Mars — Pyroeis	♏	Scorpio
♀	Venus — Phosphoros	♐	Sagittarius
☿	Mercury — Stilbon	♑	Capricorn
		♒	Aquarius
		♓	Pisces

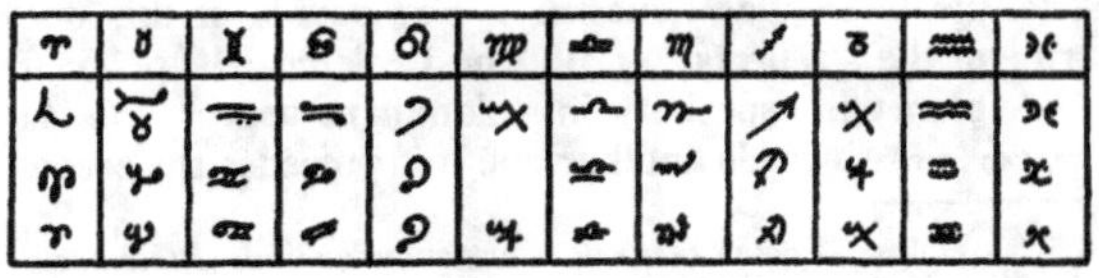

FIGURE 1.

In original documents no symbols occur except occasionally ♂︎ or ☾. The planetary and the zodiacal symbols appear only in Byzantine manuscripts, though written-out names occur at all periods. Some samples of zodiacal symbols from late manuscripts are shown in fig. 1 and fig. 20 p. 156. The symbol ⊙ for the sun does not seem to be attested before the Renaissance.

C. METHOD OF DATING

1. Only a small portion of horoscopes are explicitly dated. In the cases of papyri and ostraca the criteria for the date of such documents provide relatively narrow plausible limits, but in the case of literary horoscopes the possibility of fictitious compositions had to be faced. It is only because of the generally excellent agreement for the majority of horoscopes quoted by Vettius Valens with computed dates belonging to his period that we can be sure that the majority of the "literary" horoscopes are based on actual facts.

2. For the initial dating a graphical procedure was used. The positions of Saturn and Jupiter were computed in steps of 60 years from −60 to +600 (*cf.* table 1 given here). Then a long sheet of graph paper

Jan. 1	♃	♄	Jan. 1
−60	♌ 17	♉ 11	−60
0	♍ 9	♉ 27	0
+60	♎ 2	♊ 16	+60
120	♎ 21	♋ 2	120
180	♏ 11	♋ 20	180
240	♏ 29	♌ 7	240
300	♐ 18	♌ 22	300
360	♑ 7	♍ 8	360
420	♑ 27	♍ 23	420
480	♒ 14	♎ 6	480
540	♓ 7	♎ 19	540
600	♓ 28	♏ 2	600

TABLE 1.

was divided into 60 units of length, representing 60 years, and into 12 units of width, representing the twelve zodiacal signs. In the same scale two graphs of the mean motion of Saturn and Jupiter respectively were prepared on tracing paper.[1] To investigate, e.g., the interval from +60 to +120, one places the graph for Saturn at the left end on ♊ 16, on the right end on ♋ 2 (*cf.* the table) and similarly, Jupiter on ♎ 2 and ♎ 21 respectively. Then the mean places of both planets are

[1] If 1 cm represents one year in the x-direction and 30° longitude in the y-direction, then Saturn needs 29.4 cm, Jupiter 11.9 cm to cross 12 cm in width.

known for the whole interval. If, e.g., the horoscope requires a conjunction of Saturn and Jupiter, only those years are possible where the two graphs intersect, and the same holds, *mutatis mutandis*, for any other combination.

Thus one obtains very rapidly all possible dates within narrow limits. The position of the sun determines the month; the moon fixes the day within ± 1 day. Then the three remaining planets can be tested, Mars being most decisive for the first trial because of a motion of almost six signs per year.

3. The method of dating could be simplified considerably as work progressed. At first a wide range had to be tested in every single case. As more dates were established, a file of already known positions and dates became very helpful. All horoscopes were filed lexicographically according to planetary positions, beginning with Saturn, down to moon and sun. As more and more such positions were associated with definite dates, it became easier to estimate the proper date of a new horoscope from the date of a nearby position. Especially in the case of Vettius Valens did narrow limits appear to hold for the horoscopes of each chapter. This made it possible to date even incomplete horoscopes or to detect scribal errors.

4. All computations for the planets were carried out by means of P. V. Neugebauer's tables in *Astronomische Nachrichten* 248 (1932). The results are accurate to about one degree, which is much more than needed in the majority of cases where only zodiacal signs are given. For the moon volume 2 of the same author's *Tafeln zur astronomischen Chronologie* (1914) was used, but only for mean longitude and mean anomaly, ignoring latitude and all perturbations. This is again ample to determine the day with an accuracy of ± 1 in most cases. The date was usually chosen in such a way that the moon falls near the middle of the sign in question. In limiting cases, near the boundary of a sign, one has to take into account that ancient tables from the first centuries A.D. assign to longitudes values about 5° higher than modern tables (*cf.* below p. 180).

The hour was found in a simple schematic way by counting two hours for each zodiacal sign of solar distance from the horizon (which is known from the "Horoscopos"), disregarding all seasonal corrections. The error of this procedure will remain within ± 2 hours, in which time the moon moves about 1°. Again this is more than sufficient when nothing but zodiacal signs is given.

For the lunar nodes the tables of E. W. Brown (1919) were used though only by rounding off the results to ten minutes, which is amply sufficient for our purposes. About the same accuracy can be reached with P. V. Neugebauer's tables, quoted above, by computing the element Θ which equals $-\Omega$.

5. The determination of the culminating point (M) of the ecliptic from the rising point (H) is easy to carry out by means of Theon's Handy Tables [2] which give the solution of this problem on the basis of spherical trigonometry for the different climata (*cf.* p. 4 "Clima" and p. 9 "Midheaven"). During all periods, however, and particularly for the earlier horoscopes, the arithmetical methods developed in Babylonian astronomy were used. In such cases we followed the same method, using the rule of Paulus Alexandrinus (*cf.* p. 10).

D. GLOSSARY OF ASTROLOGICAL AND TECHNICAL TERMS

Hellenistic astrology developed a rich technical terminology. Except for rare cases all necessary information can be obtained from Bouché-Leclercq's masterful *Astrologie grecque* (1899). For any detailed study of the subsequently published horoscopes Bouché-Leclercq's work will be an indispensable tool. We found it useful, nevertheless, to have at hand a simple glossary of the technical terms and concepts which should suffice as a first orientation in reading the text of a horoscope.

Agathos Daimon ("Good Genius").

Locus 11 from Horoscopos, as "starting place" (τόπος ἀφετικός) inferior only to Midheaven and Horoscopos (*cf.* Bouché-Leclercq, *AG* p. 416). For its influence *cf.* Vettius Valens IV,12 (p. 179,26 Kroll). Similar influences were attributed to the locus 11 from the Lot of Fortune (Locus of Success, περιποιητικὸς τόπος) but the two were not identical; *cf.*, e.g., the horoscope No. L 50.

Alexandrian Calendar. *See* Calendar

Aphesis. *See* Starter

Apoklima (ἀπόκλιμα)

A sign which precedes, in the order of the signs of the zodiac, a center is called apoklima; thus ♈ is apoklima if, e.g., Midheaven is in ♉. *Cf.* also Loci, p. 8, note 22.

Ascensions. *See* Rising Times

Aspects

The *Tetrabiblos* I.13 distinguishes four combinations of symmetric arrangements of points of the ecliptic. "Opposition" or "diameter" (endpoints of a diameter), "trine" (equilateral triangle), "quartile" (square), "sextile" (regular hexagon). For details of this doctrine of the "aspects" *cf.* Bouché-Leclercq, *AG* p. 165 ff.

A planet is especially in "dominant aspect" (καθυπερτέρησις) towards another one if it precedes the second

---
[2] N. Halma, *Tables manuelles astronomiques de Ptolémée et de Théon*, Paris, 1823 (= vol. 2 of *Commentaire de Théon d'Alexandrie...*; vol. 1 p. 148 ff. contains the table for sphaera recta). We generally used a photocopy of cod. Vat. graec. 1291.

planet by 90° (" quartile ") in the sense of the daily rotation; *cf.* Bouché-Leclercq, *AG* p. 250, p. 252,4a. *Cf.* also the excerpts from Antiochus by Rhetorius, *CCAG* 1 p. 157 No. 26 and p. 155 No. 21. Serapion even defines " dominant " within one zodiacal sign, simply by precedence in the sense of daily rotation (*CCAG* 8,4 p. 226,30 to p. 227,5).

Augustus, Era of

Dates of the Alexandrian calendar can be converted into Julian dates by means of

Augustus 0 Thoth (I) 1 — — 30 August 29.

Basis (βάσις). *See* Lots

Calendar

The names of the months in the *Egyptian* (rotating) and *Alexandrian* (fixed) calendar are:

I Thoth	V Tybi	IX Pachon
II Phaophi	VI Mekheir	X Payni
III Hathyr	VII Phamenoth	XI Epiphi
IV Choiak	VIII Pharmouthi	XII Mesore
		epagomenai

In the Egyptian calendar the number of epagomenal days is constantly five. In the Alexandrian calendar a sixth epagomenal day is added every fourth year. Consequently the Alexandrian year remains in fixed relation with the Julian year, the first of Thoth being for three years in succession August 29, then once [a] August 30, etc.

If we wish to distinguish between these types of calendars we shall add in parenthesis " e " for the Egyptian, " a " for the Alexandrian and " j " for the Julian calendar. Thus I(a) 1 — VIII(j) 29.

In the horoscopes from Dura (Nos. 176, 219,I and 250,1) the names of the months of the *Macedonian* calendar appear. Their order in this late period is:

I Hyperberetaios	VII Xanthikos
II Dios	VIII Artemisios
III Apellaios	IX Daisios
IV Aydnaios	X Panemos
V Peritios	XI Loos
VI Dystros	XII Gorpiaios

with Hyperberetaios roughly corresponding to October.

Centers (κέντρα)

The " centers " are the four points which the ecliptic has in common with the horizon and the meridian at a given moment. Consequently these points are the rising and the setting points and the points of upper and lower culmination. The rising point is also called the " Horoscopos." The upper culmination is " Midheaven." *Cf.* Bouché-Leclercq, *AG* p. 258 ff.

---

[a] For the Julian year which precedes the Julian leap year.

In our diagrams the following abbreviations are used for the four centers

- H   Horoscopos or rising point
- Δ   δύσις or setting point
- M   Midheaven or upper culmination
- IMC   *imum coeli* or lower culmination
   — Lower Midheaven

Clima

In ancient geography the concept of " geographical latitude " (ϕ) plays only a secondary role beside the characterization of a locality by its longest daylight (M) or by the ratio of the longest to the shortest daylight ($M : m$). For example, $M = 14^h$ is always quoted as maximum for Alexandria, thus $m = 10^h$ and $M : m = 7 : 5$, but very rarely is the corresponding latitude of about $30\frac{1}{2}°$ mentioned. Similarly the clima of Babylon is defined by $M : m = 3 : 2$.

The longest daylight, M, is directly related to the rising times (*cf.* p. 11). Since the longest day occurs when the sun is located at the beginning of ♋ we can say that on that day (the summer solstice) ♋ rises with the sun. At sunset 180° of the ecliptic have risen, thus the six signs from ♋ to ♐. The total of the time required by their successive risings is M:

$$M = a_4 + a_5 + \cdots + a_9.$$

Since the a's are measured in degrees, M also is counted in the same units. Division by 15 changes degrees to hours.

The concept of " rising times " is of Babylonian origin though only applied for the geographical conditions of Babylon, i.e., for $M : m = 3 : 2$. These rising times were not found by exact trigonometric computation but approximately determined by arithmetical progressions. Two methods were known for this, called " System A " and " System B " respectively (*cf.* p. 12). The corresponding rising times are

System A	System B
$a_1 = a_{12} = 20°$	$a_1 = a_{12} = 21°$
$a_2 = a_{11} = 24$	$a_2 = a_{11} = 24$
$a_3 = a_{10} = 28$	$a_3 = a_{10} = 27$
$a_4 = a_9 = 32$	$a_4 = a_9 = 33$
$a_5 = a_8 = 36$	$a_5 = a_8 = 36$
$a_6 = a_7 = 40$	$a_6 = a_7 = 39.$

In both cases we have $M = a_4 + \cdots + a_9 = 3,36$, thus $m = 2,24$ (since $M + m = 6,0° = 24^h$) and $M : m = 3 : 2$.

With the transmission of Babylonian astronomical methods to Alexandria, similar schemes were adopted for this locality. In order to obtain $M : m = 7 : 5$ one must use

System A	System B
$a_1 = a_{12} = 21;40°$	$a_1 = a_{12} = 22;30°$
$a_2 = a_{11} = 25$	$a_2 = a_{11} = 25$

$$a_3 = a_{10} = 28;20 \qquad a_3 = a_{10} = 27;30$$
$$a_4 = a_9 = 31;40 \qquad a_4 = a_9 = 32;30$$
$$a_5 = a_8 = 35 \qquad a_5 = a_8 = 35$$
$$a_6 = a_7 = 38;20 \qquad a_6 = a_7 = 37;30$$

which give correctly $M = 3,30$, thus $m = 2,30$ and $M : m = 7 : 5$ for both systems.

By linear extrapolation from these two starting points a whole system of geographical zones, called "climata," was developed. It appears in two "Systems," A and B, and with two fundamental localities, (a) Alexandria and (b) Babylon, respectively. In all cases seven climata were distinguished, the "first" being Alexandria, the "second" Babylon. Actually, however, the two original lists do not occur in a common scheme. Starting from Alexandria one used the following values of M as characteristic for the climata:

$$3,30° \quad 3,34 \quad 3,38 \quad 3,42 \quad 3,46 \quad 3,50 \quad 3,54.$$

Thus Babylon, with $M = 3,36$, falls between the second and third value. Similarly Alexandria with $M = 3,30$ is omitted if one uses the sequence $3,32°$ $3,36...3,56$ which is based on Babylon. This holds for both systems, A and B, of rising times.

Vettius Valens (I,7) gives as definition of the climata, a scheme of rising times which are computed according to System A and based on Alexandria ("clima 1"). In fact, however, he also uses in his text the sequence based on Babylon (as "clima 2"); in both cases rising times of System B occur.[4] The same inconsistency runs through the whole ancient and mediaeval astrological and geographical literature. In order to be able to characterize a scheme exactly, we give in table 2 a complete list of all cases of rising times obtainable from Systems A and B, respectively, and with (a) Alexandria or (b) Babylon as basis.

With the development of spherical trigonometry, probably not much before the end of the first century A.D., an additional version of "seven climata" comes into existence. It is based on a sequence of geographical zones, beginning with the equator (this case is known as "sphaera recta"), such that the longest daylight M increases in steps of ½ hour. Ptolemy, in the Almagest II,8, gives a list of rising times for 11 such zones from $M = 12^h$ to $M = 17^h$. From this a selection of seven zones was made, beginning with $M = 13$ and ending with $M = 16$ (where $M : m = 2 : 1$). These are the "seven climata" used by many later authors[5] but known already to Ptolemy.[6] In this arrangement Alexandria (with $M = 14^h$) becomes the "third clima." Unfortunately, Byzantium with $M = 15;15^h$ falls between two climata and therefore later tables contain a special table for this latitude.

The relations between these different types of "climata," all of which remain in use during the Middle Ages, are represented in table 3.

Finally a remark is needed concerning the position of the equinoxes and solstices. If not explicitly stated to the contrary we always assume that these points coincide with the zero-point of their respective signs. In Babylonian astronomy, however, the vernal point was located either at ♈ 10 (in "System A") or at ♈ 8 (in "System B") and particularly the latter norm occurs quite frequently in Greek astrology. In this case one must define a_1 as the rising time of the ecliptic arc from ♈ 8° to ♉ 8° etc. and m and M refer to ♑ 8° and ♋ 8° respectively. It is characteristic for the syncretistic attitude of Hellenistic astrology that a table

TABLE 2.

---

[4] Cf. p. 184.

[5] Examples: Theon of Alexandria (about 360) Handy Tables (Halma III p. 34 f.) = P. Lond. 1278 and the closely related tables of Cod. Vat. Graec. 1291 (written about 816); Isidorus of Seville (about 600); al-Battānī (900); Marvazī (1050); Bar Hebraeus (1279), and *passim* in Islamic and Byzantine tables.

[6] Almagest II,13 (Heib. p. 174-187), VI,11,12 (Heib. p. 538, 10; 538,24/539,1); *Analemma* (Heib. p. 214, 10 ff.); *Planisphaerium* (Heib. p. 249,15); also *Geogr.* I,15 (Müller p. 41,1; 10 f.) with reference to Marinus of Tyre who wrote about A.D. 110.

for the length of daylight, computed with System A, may use the vernal point ♈ 8° of System B.[7]

TABLE 3.

Commanding and Obeying Signs

According to *Tetrabiblos* I,14 the zodiacal signs of the summer semicircle (from ♈ to ♍) are "commanding," the signs of the winter semicircle (from ♎ to ♓) "obeying" (*cf.* also Bouché-Leclercq, *AG* p. 163 f.).

A more complicated scheme is given by Vettius Valens I,8 (*cf.* fig. 2a) according to which ♓ obeys ♉, ♏ obeys ♍, while ♊ and ♌ and ♐ and ♒ "look" at each other (if we restore correctly the meaning of the somewhat confused text). The equinoctial and solstitial signs (♈ ♋ ♎ ♑) seem to be excluded from this scheme. The computation in Vettius Valens is based on the rising times for System A and the latitude of Babylon.

P. Mich. 149 distinguishes even three types of relations, to "see," to "hear" (*cf.* fig. 2b), and to "perceive."

Contact (συναφή)

There are several possibilities for the moon to be in "contact" with a planet according to a doctrine for which Vettius Valens VII,5 (p. 290,14-23 Kroll) quotes "the king," i.e. Nechepso. The moon might be in a sign next to the planet such that she will come soon in conjunction with the planet as her longitude increases; or the moon might reach the diameter of the planet (examples: Nos. L 120,IX, L 478, L 516) or the side of the quartile aspect [8] (example: No. L 475).

For a slightly different doctrine *cf.* Bouché-Leclercq, *AG* p. 245 ff.

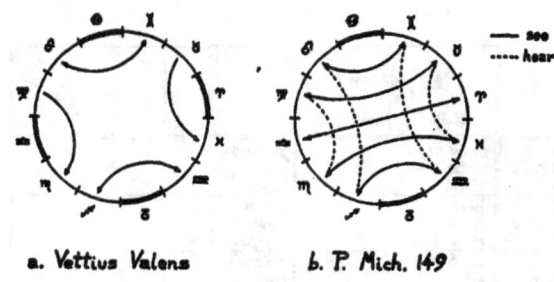

a. Vettius Valens   b. P. Mich. 149

FIGURE 2.

Culminating Point. *See* Midheaven

Decans

The "decans" of the Hellenistic and later periods are simply the thirds of the zodiacal signs, i.e., sections of the ecliptic of 10° lengths. Historically the decans go back to Egyptian lists of 36 constellations which were drawn up many centuries before the introduction of the zodiac.[9] Only when in the Ptolemaic period the zodiac was transplanted to Egypt did the decans become directly associated with this new system of reference for the solar motion through the year.

The Egyptian lists of decans experienced in the course of history many minor modifications, partly from intentional changes, partly by the accidents of tradition. Their latest form is represented by an astronomical frieze in the Temple of Edfu [10] (Ptolemy VII Euergetes II) and, with minor and obvious errors, by a frieze in the "Silver Room" of the Temple of Hathor in Dendera [11] (Augustus-Trajan). It is the list from Edfu which forms the basis for the second column in table 4 given below, except for the variant for the third decan in Gemini, which is taken from the Outer Hypostyle in Dendera [12] (Tiberius).

The Egyptian names of the decans are still reflected in several Greek and Latin versions though there were manufactured lists of decans which have nothing in common with the Egyptian originals (e.g., in the "Testa-

---

[7] *Cf.* the table in Vettius Valens VIII, discussed below in Ch. III,F (p. 174).

[8] Trine seems to be excluded.
[9] For the original positions of the decans *cf.* Neugebauer [9].
[10] Porter-Moss VI p. 134 (north wall).
[11] Porter-Moss VI p. 58 f. (Room XI).
[12] Porter-Moss VI p. 49 (ceiling).

mentum Solomonis"). The most frequently quoted Greek version is taken from Hephaistion of Thebes (fourth century A.D.), which is reproduced here in the third column but only in so far as it agrees with the Egyptian prototype. Professor R. A. Parker has recognized that the second half of Hephaistion's list is distorted by errors and duplications [13] which have not, however, affected to the same degree the Latin list preserved in Firmicus (fourth century). From this source are taken three names, adding one variant from Scaliger; cf. column (4).

(1)		(2) Egyptian	(3) Hephaistion	(4) Firmicus
		Edfu | Dendera		
♈	1	ḫntw ḫry	χονταρε	
	2	ḫntw ḫry	χονταχρε	
	3	sȝ kd	cικετ	
♉	1	bȝw	χωου	
	2	ꜥryt	ερω	
	3	rmn ḫry	ρομβρομαρε	
♊	1	ȝs ꜥrk	θοcολκ	
	2	wꜥrt	ουαρε	
	3	tpy-ꜥ spdt | pḥwy ḫry	φουορι	Tepis; Tepisatosoa
♋	1	spdt	cωθιc	
	2	ȝtw	cιτ	
	3	knmt	χνουμιc	
♌	1	ḫry bpd knmt	χηεχνουμιc	
	2	ḥȝt dȝt	ηπη	
	3	pḥwy dȝt	φουπη	
♍	1	tm	τωμ	
	2	wšȝty bkȝty	ουωcτευκωτι	
	3	ȝpst	αφοcο	
♎	1	sbḥs	couχωε	
	2	tpy-ꜥ ḫntt	ντηχουτ	
	3	ḫntt ḫrt	χονταρε	
♏	1	ḫntt ḥrt		Sentacer
	2	ṯms n ḫntt		Tepsisen(?)
	3	spty ḥnwy		
♐	1	ḥry ỉb wỉȝ	ρηουω	
	2	ȝsmw	cεcμε	
	3	knmw	κομμε	
♑	1	tpy-ꜥ smd		
	2	smd	cμaτ	
	3	srt	cεω	
♒	1	sȝ srt	ιερω	
	2	ḥry bpd srt		
	3	tpy-ꜥ ȝḥwy	πτιαυ	
♓	1	ȝḥwy	αευ	
	2	tpy-ꜥ bȝw	ντιβιον	
	3	bȝw	βιου	

TABLE 4.

In the Greek horoscopes the decans are mentioned by names only in P. Lond. 130 (No. 81) and P. Lond. 98 (No. 95). For these cases the commentaries to Nos. 81 and 95 should be consulted.

For the details of the astrological importance of the decans see Gundel, *DD* (1936) and the article "Dekane" in *RE* Suppl. VII, 116-123 (1940) by the same author. The reader should be cautioned, however, against Gundel's tendency to derive much too much from Egyptian sources (cf., e.g., Schott [1]).

Depression. *See* Exaltation

Diameter

Aspect in which two celestial bodies are located in opposite endpoints of a diameter of the ecliptic. *Cf.* also "Opposition."

Diocletian, Era of

The corresponding Julian dates are obtainable from the relation

Diocletian 0 Thoth (I) 1 = A.D. 283 Aug. 30

using the Alexandrian calendar for the years of the era Diocletian.

Disjunct (ἀσύνδετος). *See* Turning away

Diurnal. *See* Sect

Dodekatemoria

Assume a point A of a zodiacal sign being σ degrees from the beginning of the sign. With A is then associated as its "dodekatemorion" another point B whose distance from A is 12σ (and therefore from the beginning of the original sign 13σ). This relationship is one of the rare instances where a similar doctrine is also known from cuneiform sources.[14]

The following table gives the longitudes of B with respect to the beginning of the sign of A, measured in "signs" ($s = 30°$) and degrees.

A:	B:	A:	B:
$0^s 1$	$0^s 13°$	$0^s 16°$	$6^s 28°$
2	0 26	17	7 11
3	1 9	18	7 24
4	1 22	19	8 7
5	2 5	20	8 20
6	2 18	21	9 3
7	3 1	22	9 16
8	3 14	23	9 29
9	3 27	24	10 12
10	4 10	25	10 25
11	4 23	26	11 8
12	5 6	27	11 21
13	5 19	28	0 4
14	6 2	29	0 17
15	6 15	30	0 30

Example: original position: ♉ 21, thus A = 21 and B = ♉ $0° + 9^s 3 = $ ♒ $3°$ as dodekatemorion.

The dodekatemoria obviously increase the number of possibilities for the horoscopic predictions.

Dominant. *See* Aspects

---

[13] The decans No. 2 and 3 in ♑ and No. 1 of ♒ are taken by Hephaistion as the three decans of ♑. On the other hand the third decan of ♒ and the three decans of ♓ are given by Hephaistion two lines too high for the correspondence with the Egyptian names as shown in our table.

[14] *Cf.* Neugebauer-Sachs [1].

Dysis

Setting point of the ecliptic, exactly opposite the Horoscopos. Denoted in our diagrams by Δ. *Cf.* also "Centers."

Epanaphora (ἐπαναφορά)

A sign which follows, in the order of the signs of the zodiac, a center. Thus ♉ is epanaphora if ♈ is in Midheaven. *Cf.* also Loci, p. 8 note 21.

Equinox. *See* Clima

Era. *See* Augustus, Diocletian

Exaltation (ὕψωμα) and Depression (ταπείνωμα)

Zodiacal signs of special power and weakness respectively of the planets as indicated in fig. 3. (+ Exalt.,

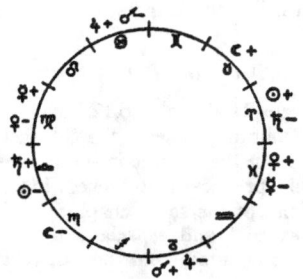

Exaltations and Depressions

FIGURE 3.

— Depr.). *Cf. Tetrabiblos* I,19 [20]. The arrangement of the planets on which this scheme is constructed follows the order of the geocentric distances and is therefore of Greek origin.

Occasionally exact longitudes are associated with the exaltations as follows:

☉ : ♈ 19	♂ : ♑ 28
☽ : ♉ 3	♀ : ♓ 27
♄ : ♎ 21	☿ : ♍ 15
♃ : ♋ 15	

The depressions are located 180° distant from the exaltations. *Cf.* Vettius Valens III,4 (p. 140, Kroll) and Bouché-Leclercq, *AG* p. 192 ff.

Exaltation (of the Nativity)

In Vettius Valens II,18 (p. 80 Kroll) the "exaltation" of a given nativity is defined by means of the position of the sun (in daytime) or of the moon (at night). This is done in the following way: one counts the distance from the sun to ♈ (which is the exaltation of the sun) or from the moon to ♉ (the lunar exaltation) respectively. The same distance is then assumed to separate the Horoscopos from the exaltation of the nativity.

Example:[15] The sun in ♌ is four signs ahead of ♈. If the Horoscopos is in ♎, then the exaltation of this nativity is ♓ because ♎ is four signs ahead of ♓.

For another example and for the definition *cf.* the text of No. L 50.[16]

Follow. *See* Precede

Horoscopos or Ascendant

The rising point of the ecliptic; *cf. Tetrabiblos* III,2. Denoted in our diagrams by H; see also "centers."

Houses (οἶκοι)

Each planet has two houses, symmetrically located to ♋ and ♌, which are assigned to moon and sun respectively (*cf.* fig. 4). *Cf. Tetrabiblos* I,17. The arrange-

Houses

FIGURE 4.

ment of the planets according to their distance from the earth speaks for a Greek origin of this doctrine.

In the horoscope No. 81 the name "sign" (ζώδιον) is used instead of "house."

Katarche (καταρχή)

Literally "beginning." As a branch of astrology it concerns the investigation of the influence of a momentary configuration of the celestial objects, e.g., at the beginning of a journey, upon the outcome of an enterprise. Thus the Katarche plays the same role with respect to a specific event as the horoscope does for the life as a whole. *Cf.* Bouché-Leclercq, *AG* p. 458 ff.

Loci (τόποι)

Astrological doctrine developed two other twelve-divisions of the ecliptic, beside the fixed division into the twelve zodiacal signs. These new twelfths are counted either from the Horoscopos (H) or, occasionally, from the Lot of Fortune. The single intervals are either of constant length, 30° each, or obtained by accurate tri-

---

[15] From No. L 188.
[16] Additional examples: Nos. L 65,X 95,V,18 97,XI 109 123,I 440 463 484.

section of the arcs between the "centers" (H, M, Δ, IMC) or rather points which precede the centers by 5 degrees.[17]

Whatever the procedure of division may be, all these doctrines have in common the idea of assigning each twelfth, called "locus," a special significance for a certain sphere of life, like parents, children, death etc. In details there exist many variants in associating influences to the loci. We quote in the following list the names most commonly used in astrological literature.[18] The variants—far from complete—are mostly taken from Vettius Valens, especially II,5 ff. (p. 62 ff.) and IV,12 (p. 179 Kroll[19]). It can be observed that the loci in Vettius Valens IV,12 are often diametrically opposite to the commonly accepted positions.[20]

The loci are always counted in the same order as the zodiacal signs, that is, with increasing longitude (cf. fig. 5), or opposite to the direction of the daily rotation which is indicated in figure 5 by an arrow.

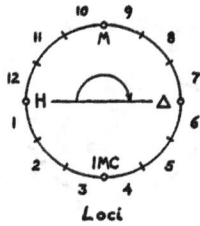

FIGURE 5.

A. From the Horoscopos

1. ὡροσκόπος (= H), life; also body, spirit or breath.
2. ἐπαναφορά,[21] livelihood, property; also "gate of Hades." It concerns livelihood, partnership, intercourse with women, business, profit from inheritance.
3. ἀπόκλιμα,[22] brothers; also "goddess" (moon); living abroad, royalty, wealth, friends, relatives, slaves.
4. ὑπογείον (= IMC), parents; spirits, life in temple; repute, children, etc.
5. ἐπαναφορά, children; also "good fortune." (ἀγαθὴ τύχη). Children, friendship, accomplishments; marriage; Venus.
6. ἀπόκλιμα, slaves; also "bad fortune." Illness, enmity, infirmity; Mars.
7. δύσις (= Δ), marriage.
8. ἐπαναφορά, death; trial, penalty, loss, weakness.
9. ἀπόκλιμα, travel; also "god" (sun). Friendship, travel, benefit from kings, revelations, manifestations of gods, soothsaying.
10. μεσουράνημα (= M), career and honors; accomplishment, reputation, children, wife.
11. ἐπαναφορά; also "good daemon."[23] Friends, hopes, gifts, children, freed persons; accomplishment.
12. ἀπόκλιμα; also "bad daemon." Enmity, foreign country, slaves, illness, dangers, court trials, infirmity, death.

B. From the Lot of Fortune

Vettius Valens II,40 (p. 126,12 Kroll) mentions the possibility of counting at least the 8th locus, the locus of death, not from the Horoscopos but from the Lot of Fortune. This practice is followed in the examples given at the end of the same chapter; cf. No. L 65,V. Also "Midheaven" and especially locus 11 "accomplishment" are counted from the Lot of Fortune.

Lots (κλῆροι)

From the longitudes of the Horoscopos, sun, moon, and planets, new points on the ecliptic are derived, which are considered significant for the nativity in question.

Lot of Fortune (κλῆρος τύχης)

According to some astrological doctrine the longitude λ_F of the "Lot of Fortune" is found from the longitude λ_H of the Horoscopos and from longitudes of sun and moon as follows

$$\Lambda_F = \lambda_H \pm (\pm \lambda_{\mathbb{C}} \mp \lambda_{\odot}) \quad \begin{cases} \text{daytime} \\ \text{night} \end{cases}$$

which is, of course, equivalent to

$$\Lambda_F = \lambda_H + \lambda_{\mathbb{C}} - \lambda_{\odot}$$

under all circumstances; cf. Bouché-Leclercq, AG p. 288 f. and p. 291 note 1. In No. 81 the use of

$$\Lambda_F = \lambda_H - (\lambda_{\mathbb{C}} - \lambda_{\odot})$$

is mentioned as a procedure of ignorant people. In practice, however, the rule

$$\Lambda_F = \lambda_H \pm (\lambda_{\mathbb{C}} - \lambda_{\odot}) \quad \begin{cases} \text{daytime} \\ \text{night} \end{cases}$$

---

[17] For a detailed discussion of this procedure cf., e.g., CCAG 8,1 p. 221 ff. (from Rhetorius). Cf. also Tetrabiblos III, 10 (= III,11 Boll-Boer) and P. Mich. 149 col. IX. As a medieval example can be quoted an anonymous Latin work on comets of A.D. 1238 (Thorndike, Comets, p. 60). There it is said that the 10th locus reaches from the point M at ♈ 28;37 over the whole sign of Taurus. Indeed, the Horoscopos being ♌ 4 a third of the arc from M to H is 31;48° long. On the other hand the loci between Δ and M cover only 28;12°.

[18] Cf. Bouché-Leclercq, AG p. 276 ff., also p. 415 ff.

[19] Cf. also CCAG 2 p. 158, 18 ff.

[20] Bouché-Leclercq, AG p. 280 f.

[21] All loci which immediately follow a center are called ἐπαναφορά "in process of rising" because they are next to become centers, e.g., Locus 2 is the next section to rise after the Horoscopos.

[22] All loci which immediately precede a center are called ἀπόκλιμα "declining" because they are past a center in the sense of the daily rotation of the sky.

[23] Cf. above sub "Agathos Daimon."

has been followed. It is stated in Vettius Valens II,22 p. 87,26 [24] and consistently applied in our material.

Lot of Daimon (κλῆρος δαίμονος)

This point is symmetric to the Lot of Fortune with respect to the Horoscopos; cf. Bouché-Leclercq, *AG* p. 294 ff. Hence

$$\Lambda_D = \lambda_H \mp (\lambda_{\mathbb{C}} - \lambda_{\odot})$$

if

$$\Lambda_F = \lambda_H \pm (\lambda_{\mathbb{C}} - \lambda_{\odot}).$$

Lot of Eros and Lot of Necessity (κλῆρος ἔρωτος and κλῆρος ἀνάγκης)

According to Paulus Alexandrinus (cf. Bouché-Leclercq, *AG* p. 307 f.) one should have

$$\Lambda_E = \lambda_H \pm (\lambda_{\venus} - \Lambda_D)^{25}$$
$$\Lambda_N = \lambda_H \pm (\lambda_{\mercury} - \Lambda_F).$$

No example of this procedure is preserved in the extant horoscopes. In Nos. 138/161 and 338 the following relations seem to be applied:

$$\Lambda_E = \lambda_H \pm (\Lambda_F - \Lambda_D)$$
$$\Lambda_N = \lambda_H \mp (\Lambda_F - \Lambda_D)$$

which is equivalent to saying that the Lots of Eros and of Necessity are symmetrically located with respect to the Horoscopos and accordingly at a distance from the Horoscopos of twice the elongation $\lambda_{\mathbb{C}} - \lambda_{\odot}$.

For additional usages of the term "Lot" cf. Bouché-Leclercq, *AG* p. 293 note 1 and the commentaries to Nos. 95 and 137.

Lot of Basis

In Vettius Valens II,22 (p. 87,29) a "κλῆρος τῆς βάσεως" is defined as follows: add to the Horoscopos the distance from the Lot of Fortune to the Lot of Daimon if the nativity falls in the day, otherwise the opposite direction is to be taken.

This principle is applied in Nos. L 78 and L 188.

Lower Midheaven

Point of the ecliptic in lower culmination, also called *imum coeli*. Denoted in our diagram by IMC; cf. also "Centers."

Macedonian Calendar. *See* Calendar

Magnitude of a Star

The star catalogue of the Almagest (books VII and VIII) contains, beside longitude and latitude of each star, also a number which measures its brightness, called μέγεθος "magnitude," ranging from 1 to 6, the sixth magnitude being assigned to the smallest visible brightness. These numbers agree quite well with modern photometric magnitudes; for details cf. Peters-Knobel, *PCS* p. 120 ff.

Male and Female Signs

According to *Tetrabiblos* I,12 the zodiacal signs are alternatingly male and female, beginning with Aries (cf. fig. 6). For other norms see Bouché-Leclercq, *AG* p. 155 f.

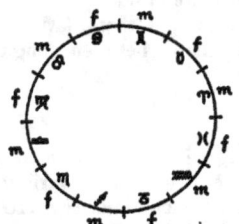

Male and female signs

FIGURE 6.

Midheaven (μεσουράνημα)

The culminating degree of the ecliptic, denoted by M in our diagrams; cf. also "centers."

In the horoscopes the problem often arises to determine M from the Horoscopos H at a given clima. The correct trigonometric solution to this problem is given by Ptolemy in the Almagest II,9 (Heib. p. 144,14). Let $\rho(H)$ denote the rising time, for the given clima, of the arc from ♈ 0° to H; then

(1) $$\alpha(M) = \rho(H) - 90$$

is the right ascension (= rising time for sphaera recta) of M. Thus the longitude λ of M can be found by using the tables for sphaera recta (Almagest II,8) in reverse direction.

The Handy Tables of Theon simplify this procedure still more by tabulating $\alpha(M) + 90$ directly as function of λ. Furthermore, the tabular interval is reduced from 10° steps in the Almagest to 1°. It is this type of table which was used in the late literary horoscopes.

The situation is different in the case of the arithmetical schemes, Systems A and B (cf. p. 3), for the rising times. If M denotes the longest daylight, measured in degrees, d the difference between rising times of consecutive zodiacal signs as explained on p. 12, then for all climata the following relations hold

System A: $d = \frac{1}{6}(M - 180)$
System B: $d = \frac{1}{12}(M - 180)$.

Since for sphaera recta $M = 180$, we see that the arithmetical schemes of rising times lead to the assumption

---

[24] For another variant see *CCAG* 1 p. 167 f.

[25] We owe to Miss E. Boer the text of Paulus Alexandrinus in which he gives an example according to which for daytime $\Lambda_E = (\lambda_{\venus} - \Lambda_D)$ + number of degrees of H within its sign. The omission of whole signs of H is, however, perhaps a mistake.

$d = 0$ for sphaera recta. This implies that the rising time of all zodiacal signs would be equal, namely, 30°, and no distinction made between ecliptic and equator.

We were informed by a kind communication of Miss E. Boer that in chapter 30 (p. 81 f. ed. Boer) of Paulus Alexandrinus (late fourth century A.D.) exactly this assumption is made for the computation of the longitude $\lambda(M)$ of Midheaven from the Horoscopos. Paulus' rule is

(2) $\qquad \lambda(M) = \rho(H) + ♑ 0°$.

Since ♑ 0° has the longitude $-90°$, we see from (1) that formula (2) implies that $\lambda(M) = a(M)$ or that no distinction is made between longitude and right ascension.

Monomoiria

Each single degree of the ecliptic is associated with a planet. There exists a great variety of systems to achieve this;[26] cf. Bouché-Leclercq, AG p. 216 note 3, also Gundel, HT p. 338 ff.

In our material only No. L 497 makes use of this concept, following a procedure described in Vettius Valens IV,26 (p. 202 Kroll). The subsequent table 5

Houseruler	♄	♃	♂	☉	♀	☿	☾
$d=$ 0° 7 14 21 28	♄	♃	♂	☉	♀	☿	☾
1 8 15 22 29	♃	♂	☉	♀	☿	☾	♄
2 9 16 23	♂	☉	♀	☿	☾	♄	♃
3 10 17 24	☉	♀	☿	☾	♄	♃	♂
4 11 18 25	♀	☿	☾	♄	♃	♂	☉
5 12 19 26	☿	☾	♄	♃	♂	☉	♀
6 13 20 27	☾	♄	♃	♂	☉	♀	☿

TABLE 5.

will serve to determine the monomoiria to a given longitude λ, expressed by a zodiacal sign s, degree d, and fractions f such that $0 < f \leq 0;60°$. Determine the house ruler, i.e., the planet whose house is s (cf. p. 7). Enter in the table the corresponding column and take the line in which d is found, ignoring f. The planet thus found is then the ruler of the degree in question.

Examples (from No. L 497):

1. The sun is in ♏ 4;22. Since ♏ is a house of ♂ we enter the table $d = 4$ under ♂ and find the moon as ruler.

2. The Lot of Fortune is in ♌ 16;0 = ♌ 15;60. Ruler of ♌ is the sun; thus for $d = 15$ we find ♀ as ruler of the 15th degree.

Month names of the Egyptian calendar. *See* Calendar

---

[26] See also Stoicheion.

Nativity or **Genesis** = horoscope

Nocturnal. *See* Sect

Oblique Ascension. *See* Rising Times

Opposition

Position of moon or planet diametrically opposite to the sun. Less accurately is this term applied to the case where moon or planet rises in the east when the sun sets in the west (hence ἀκρόνυκτος = beginning of the night). Cf. also "Diameter."

The outer planets (Saturn, Jupiter, Mars) are retrograde when near opposition.

Paradosis — Yielding

Transmission of rulership from one planet to another according to the sequence of periods of life; cf. Bouché-Leclercq, AG p. 493 f.

We also find the following procedure applied: if you investigate the nth year of life, reduce n modulo 12. The remainder m is then used to count the number of zodiacal signs from one planet A. If we then find that the mth sign is occupied by a planet B, the influence of A is "yielded" to B. For examples cf. Vettius Valens V,10 (No. L 102 and related horoscopes).

Phase

The concept of "phase" is probably derived from the observation of bright fixed stars. In the Greek calendars the characteristic risings and settings of stars are listed in their relation to the seasons. These first or last appearances in the morning or evening are called the "phases." Ptolemy's φάσεις ἀπλανῶν ἀστέρων καὶ συναγωγὴ ἐπισημασιῶν were edited by Heiberg in the *Opera astronomica minora* of Ptolemy and discussed by H. Vogt in part V of the *Griechische Kalender* edited by Boll. The corresponding "phases" of the planets are, of course, not in a fixed relation to the seasons. Their computation is discussed in the Almagest XIII,7-9. Tables are given in the Almagest XIII,10. More extensive tables were contained in the Handy Tables (Heiberg p. 174, Halma I p. 15). It follows from the horoscopes that this problem was already dealt with in the time before Ptolemy.

Periods

The following "periods" are often used:

☉ 19 [19 solar years = 235 lunar months: "Metonic cycle"]
☾ 25 [25 Egyptian years = 409 lunar months]
♄ 30 [30 years: approximate synodic period]
♃ 12 [12 years: approximate synodic period]
♂ 15 [15 years: approximate synodic period [27]]
♀ 8 [8 years: approximate synodic period = 99 lunar months]
☿ 20 [20 Egyptian years = 63 synodic periods [28]]

---

[27] Cf. *Tetrabiblos* IV,10 and P. Mich. 149 V,18 ff.
[28] Cf. *Almagest* IX,10 (II p. 293,4 f. Heiberg).

TECHNICAL TERMS: Periods — Sect

The astronomical arbitrariness of this collection of parameters is obvious. It is significant that the lunar period of 25 years is based on Egyptian years only,[29] whereas the 19-year intercalation cycle and the 8-year period of Venus are fundamental parameters of Mesopotamian practical astronomy.[30]

For astrological purposes these periods are used in order to associate definite numbers with the single planets; but these numbers may or may not signify years—they may also represent months or even days. For example: the total of these seven periods is 129. Counted as months 129 equals 10 years and 9 months and this interval has in this way become a new significant interval (cf. the title of Vettius Valens VI,5 " On the divisions in 10 years 9 months of favorable or unfavorable periods ").[31]

The above list of periods is not the only one used. In fact these periods are only the "minimal periods" beside which "maximal" and "mean" periods were assumed as follows:

	Maximal	Mean		Maximal	Mean
☉	120[32]	69½	♂	66	40½
☾	108[32]	66½	♀	82	45
♄	57[33]	43½[33]	☿	76	48
♃	79[33]	45½[33]			

The "mean" values are the arithmetical means between the extrema. Cf. also Bouché-Leclercq, AG p. 407 ff.

Planetary Symbols. Cf. Introduction p. 1.

Precede

A planet is said to "precede" a second one if it is farther ahead in the sense of the daily rotation (or of smaller longitude). The second planet "follows" the first one. The same sense of direction is meant in Babylonian astronomy when it is said that a planet is "in front of" (to the West) or "behind" (to the East) a fixed star.

Prosopa

The decans are associated as "countenances" or "faces" or "marks" with the planets according to the following list (table 6). Each column shows the sequence of the days of the planetary week. Cf. Bouché-Leclercq, AG p. 224 ff. and Gundel, DD p. 30 ff.

Right Ascension. *See* Rising Times and Sphaera recta

Rising Times

The rising times indicate how many degrees of the equator cross the horizon of a given locality simultaneously with the consecutive zodiacal signs. We denote these rising times by

$$a_1 a_2 \ldots a_{12}$$

following the order of the zodiacal signs from ♈ to ♓.

The exact determination of these rising times for a given geographical latitude is a problem of spherical trigonometry and is given in Book II of the Almagest.

	Decan 1	Decan 2	Decan 3	
♈	♂	☉	♀	♈
♉	☿	☾	♄	♉
♊	♃	♂	☉	♊
♋	♀	☿	☾	♋
♌	♄	♃	♂	♌
♍	☉	♀	☿	♍
♎	☾	♄	♃	♎
♏	♂	☉	♀	♏
♐	☿	☾	♄	♐
♑	♃	♂	☉	♑
♒	♀	☿	☾	♒
♓	♄	♃	♂	♓

TABLE 6.

Approximative arithmetical schemes were developed previously in Babylon, using constant differences ("System A") or a slightly refined scheme ("System B"). For details cf. s.v. Clima (p. 3) and Midheaven (p. 9).

In general the rising time of any given arc of the ecliptic at given geographical latitude is the length of the equatorial arc which crosses simultaneously the horizon. If in particular the arc in question is counted from ♈ 0° to a point P of longitude λ, then the corresponding rising time is also called the "oblique ascension" of P. In the special case of a geographical latitude $\phi = 0$ ("sphaera recta") one speaks of "right ascension."

Because 1° of the equator is the equivalent of 4 minutes of time, the rising times can be converted to hours and minutes by means of multiplication of the degree numbers by 4.

Sect ($\alpha \iota \rho \epsilon \sigma \iota \varsigma$)

Planets are said to belong either to the "diurnal" or "nocturnal sect" as follows [34]

diurnal:	nocturnal:
sun	moon
Jupiter	Venus
Mercury as morning star	Mercury as evening star
Saturn	Mars

---

[29] Cf. Neugebauer—Volten [1] on P. Carlsberg 9.
[30] The mathematical astronomy of the Seleucid period made use of much more accurate relations.
[31] For different examples cf., e.g., Nos. L 132 or L 158.
[32] Vettius Valens IV,6 p. 164,8-10.
[33] Vettius Valens III,16 p. 157,9-11.
[34] Cf. Tetrabiblos I,7; for finer details see Vettius Valens II,1 (p. 55,10 ff.) and III,5 (p. 141,17-26) or CCAG 1 p. 146, 1 ff. (from Rhetorius). Also Bouché-Leclercq, AG p. 103.

A similar classification is made for the signs of the zodiac:

diurnal:	nocturnal:
♈ ♊ ♌ ♎ ♐ ♒	♉ ♋ ♍ ♏ ♑ ♓

Solid Signs

According to *Tetrabiblos* I,11 the signs ♉, ♌, ♏, ♒, are "solid."

Sphaera recta

The phenomena of rising and setting, as seen at a point with geographical latitude $\phi = 0$, represent the case of "sphaera recta" in contrast to "sphaera obliqua" for latitudes $\phi \neq 0$.

The rising times of sphaera recta are called right ascensions.

Starter (ἀφέτης)

In some of the later horoscopes (e.g., No. L 401) a computation is carried out by which it is determined in which order and to which length the single planets rule the periods of life of the person in question. In this sequence of planetary chronocratorship, generally either the sun or the moon is the "starter" (for day- or night-birth respectively) while Saturn terminates the allotted span of life.

For details see Bouché-Leclercq, *AG* p. 411 ff., esp. p. 416 ff.; furthermore: Vettius Valens IV,2 (p. 158, Kroll) and V,11 (p. 232 ff., Kroll); Hephaistion II,11 (*CCAG* 8,2 p. 65 ff.); and Rhetorius (*CCAG* 8,1 p. 227,1-16).

Step (βάθμος)

This concept occurs in our material only in No. L 497. *Cf.* the commentary to this text (p. 154).

Stoicheion

Bouché-Leclercq (*AG* p. 291 note and p. 217 note) assumes that the στοιχεῖον is the same as μοῖρα "degree." He suggests that "l'expression στοιχείῳ Διός appliquée au ♈ 14 doit indiquer une répartition des planètes, degré par degré, dans l'order ♉ ♀ ♂ ♃ ♄ en commençant par ♉ = ♈ 1°." According to this scheme ♈ 14;6 would indeed fall in a degree assigned to ♃ as required by No. 81 line 60. One must wait for further evidence, however, before this scheme can be accepted. Equally possible would be an explanation, e.g., as a twelfth part of a sign (δωδεκατημόριον) counted in the same order as the "terms."

In *CCAG* 12, p. 161, 18 "ζῴδια καὶ στοιχεῖα of rising and setting" are mentioned without clear details but the translation "sign and degree" is at least plausible. Strabo, *Geogr.* II,16 (Loeb I p. 462) calls the geographical coordinates of a place its στοιχεῖα. For στοιχεῖα and decans see Gundel, *DD*. p. 227.

The zodiacal signs are sometimes represented by the letters (στοιχεῖα) of the alphabet; *cf.* Boll, *Sphaera*, p. 469. In this sense is to be understood Diog. Laert. VI,102 where he speaks about "an Arcadian hat with the twelve letters inwrought in it."

Systems A and B

Babylonian astronomy of the Seleucid period uses two different methods for the description of the motion of the sun and related phenomena. These two "Systems," called A and B respectively, find a continuation in Hellenistic astronomy. For our purposes the following facts are characteristic for the two Systems: (*a*) The vernal equinox is assumed to correspond to a solar longitude of ♈ 10° in System A, but ♈ 8° in System B (whereas the Almagest uses ♈ 0°). Especially the norm of System B is very frequently found in astrological writings far into the Middle Ages. (*b*) The rising-times of the zodiacal signs (*cf.* p. 11) are in both systems computed according to simple arithmetical schemes (in contrast to the tables in the Almagest II,8 which are based on spherical trigonometry). Let $a_1, a_2, \ldots, a_{12}$ denote the rising times of the 12 consecutive 30-degree sections of the ecliptic, beginning with the vernal point, then System A assumes $a_{i+1} = a_i + d$ for $i = 1$ to 5 and $a_{i+1} = a_i - d$ for $i = 7$ to 11 with constant difference d; in System B, however, we have $a_{i+1} = a_i + \delta$ for $i = 1, 2, 4, 5$, and $a_{i+1} = a_i - \delta$ for $i = 7, 8, 10, 11$ but $a_4 = a_3 + 2\delta$ and $a_9 = a_8 - 2\delta$. The values of d and δ depend on the geographical latitude or on the "clima"; *cf.* p. 4.

Terms (ὅρια)

The "terms" are sections of the signs associated with the planets. The *Tetrabiblos* I,20,21 mentions three different systems for the division of each zodiacal sign into five sectors, each of which is then associated with one of the five planets. Of these three methods, in our texts only the "Egyptian" system seems to be applied except in No. L 497 which also quotes the terms "according to Ptolemy" (Bouché-Leclercq, *AG* p. 211). Its scheme according to the *Tetrabiblos* is given in the subsequent table 7 (also Bouché-Leclercq, *AG* p. 207). Using it, e.g., for ♈, we see that Jupiter rules the first 6 degrees, then Venus from 6 to 12, Mercury from 12 to 20, etc. The same scheme is found in Vettius Valens I,3 with the exception only of the last three entries for ♎, which are ♃ 19, ♀ 26, ♂ 30.

A totally different scheme is given by Vettius Valens III,9 (Kroll p. 144 f.). It not only considers the planets but also sun and moon. The whole tables, one for daytime, one for night, are given by Bouché-Leclercq, *AG* p. 214, together with a summary of the motivation which Vettius Valens gives for the order and assignment of terms. This system of terms is used, at least partially, in Vettius Valens III,8 and in VIII,8.

Triangles

In *Tetrabiblos* I,18 the twelve zodiacal signs are

TECHNICAL TERMS: Triangles — Zodiacal Signs

grouped into four equilateral triangles (cf. fig. 7). Each triangle has a planet for its ruler during daytime and another for the night, except the triangle ♋ ♏ ♓ which has Mars as ruler both night and day with moon by night and Venus by day as co-rulers.

♈	♃	6	♀	12	☿	20	♂	25	♄	30	♈
♉	♀	8	☿	14	♃	22	♄	27	♂	30	♉
♊	☿	6	♃	12	♀	17	♂	24	♄	30	♊
♋	♂	7	♀	13	☿	19	♃	26	♄	30	♋
♌	♃	6	♀	11	♄	18	☿	24	♂	30	♌
♍	☿	7	♀	17	♃	21	♂	28	♄	30	♍
♎	♄	6	☿	14	♃	21	♀	28	♂	30	♎
♏	♂	7	♀	11	☿	19	♃	24	♄	30	♏
♐	♃	12	♀	17	☿	21	♄	26	♂	30	♐
♑	☿	7	♃	14	♀	22	♄	26	♂	30	♑
♒	☿	7	♀	13	♃	20	♂	25	♄	30	♒
♓	♀	12	♃	16	☿	19	♂	28	♄	30	♓

TABLE 7.

A more elaborate scheme is given by Vettius Valens II,1 (p. 55 Kroll) as follows:

Triangle	daytime	night
♈ ♌ ♐	☉ ♃ ♄	♃ ☉ ♄
♉ ♍ ♑	♀ ☾ ♂	☾ ♀ ♂
♊ ♎ ♒	♄ ☿ ♃	♄ ☿ ♃
♋ ♏ ♓	♀ ♂ ☾	♂ ♀ ☾

In general two planets are in the aspect of "triangle" or "trine" if their longitudes are 120° different. For details see Vettius Valens II,16 (Kroll p. 69 ff.).

Turning away (ἀπόστροφος)

Bouché-Leclercq, AG p. 166, considers ἀπόστροφος "turning away" and ἀσύνδετος "unconnected" or "disjunct" as equivalent expressions for "not in aspect" and indeed passages like CCAG 8,4 p. 226,6 speak about "the turning away of disjunct signs." The latter concept is then defined as signs which are contiguous to a sign or its diameter. These are indeed the signs which are neither in trine, nor in quartile nor in sextile to the given sign.[35]

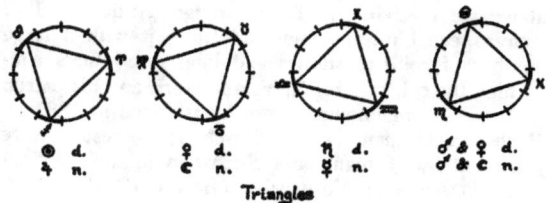

Triangles

FIGURE 7.

In our material the concept "turning away" occurs six times in literary horoscopes. Four times "Saturn, the ruler of the full moon" is ἀπόστροφος. In three of these cases (L 75 3rd version, L 87,VII, L 123,VII) Saturn is either one or five signs distant from the full moon, thus not in aspect. In the fourth case, however, ♄ and full moon are in quartile. Similarly, in No. L 101,III, Saturn "The ruler of Daimon" is in sextile to this Lot. Thus we have two cases in which ἀπόστροφος seems to mean something different from "not in aspect." Agreement between the two concepts is found again in L 114,V where Jupiter in ♉ is contiguous to the diameter ♈ (with ♄) — ♎.[36]

Vernal Point. See Clima

Yielding. See Paradosis

Zodiacal Signs

For the symbols and their arrangement cf. Introduction p. 1.

---

[35] Cf. also Tetrabiblos I,16 Robbins = I,17 Boll-Boer. It is characteristic for the flexibility of astrological interpretations that these "disjunct" signs, contiguous or 5 signs apart, can be considered as signs which show a special "affinity" (συμπάθεια) to one another (CCAG 1, p. 153,20).

[36] A similar example is Vettius Valens V,7 p. 216,1 f.: Mars and moon in ♑ are "turning away" from the sun in ♒.

I. THE ORIGINAL DOCUMENTS

No. —61

Nimrud Dagh

On the summit of the Nimrud Dagh, about 7,000 feet above sea level in the Taurus range, Antiochus I of Commagene built his tomb, adorning it with a large number of colossal statues and long inscriptions concerning the cult of himself, his ancestors, and his patron gods. These monuments were investigated in 1883 by Puchstein, Humann, and v. Luschan; the results were published by Humann and Puchstein in their *Reisen in Kleinasien und Nordsyrien* (Berlin, 1890) p. 232-253. The Western Terrace of this large complex contains, besides statues of Apollo-Helios-Hermes, the Goddess of Commagene, Zeus-Oromasdes, Antiochos, and Artagnes-Herakles-Ares, also a huge relief of a lion, covered with stars, and showing the moon and three planets:[1]

nearest to the head:	φαεθων διος	Jupiter
center:	ϲτιλβων απολλωνοϲ	Mercury
nearest to the tail:	πυροειϲ ηρακλ[εουϲ]	Mars

That the lion represents the constellation "Leo" cannot be doubted since the 19 stars which are depicted on and near the body of the lion agree very nearly with the positions of the 19 stars assigned to this constellation in the so-called Catasterisms of Eratosthenes.[2] The three planets above the back of the lion, however, are the very same planets which are associated with the three tutelary gods Zeus, Apollo, and Heracles who are represented by statues and repeatedly glorified in the inscriptions. Thus one may doubt whether we have here the representation of a real conjunction of three planets in Leo for horoscopic purposes or merely the planets of the patron gods represented beside Leo as symbol of royalty. The latter interpretation would, of course, destroy the interpretation of this monument as a horoscope and remove all possibilities of an astronomical dating. The strongest evidence in favor of a horoscope is the picture of the crescent on the chest of the lion because the combination of crescent and constellation Leo seems to have no independent significance. Thus it becomes more plausible that the three planets also be taken as representing an actual conjunction in Leo.

Humann and Puchstein did not question the horoscopic character of the monument. Their astronomical adviser, who attempted to determine its date, took the arrangement of the planets in the inscription more literally than seems historically permissible, by assuming that only a date is acceptable at which the arrangement of the planets followed exactly the order Jupiter-Mercury-Mars as indicated by the stars above the body of the lion. Following this principle the only date possible within the first century B.C. seemed to be —97 July 17[3] with the three planets in very close conjunction at the beginning of the constellation Leo and in the arrangement indicated, and also in conjunction with the sun, the moon, however, being beyond Leo in the middle of Virgo. Thus, this date does not account for the most significant feature of the horoscope, the crescent being in Leo. Furthermore, it leaves unexplained the absence in the monument of the sun, a fact which would be the more surprising as the sun would have stressed still more the "royal" character of the horoscope. Hence one must re-investigate the date under the condition of the moon being in Leo, but not the sun. The arrangement of the planets, however, plays a much more subordinate role. First of all, their position was in all probability computed and not observed: this is evident in the case of —97 July 17 when none of the planets was more than 3° distant from the sun and thus none was visible.[4] Secondly, in the enumeration of the planets in the contemporary Mesopotamian astronomy the strictly followed arrangement is Jupiter-Venus-Mercury-Saturn-Mars which agrees with the arrangement on the monument when Venus and Saturn are left out because they were not in Leo. Finally the arrangement of the three names in the narrow space

---

[1] Latest edition of the inscriptions: Jalabert-Mouterde *IGL SYR* I, 36 (p. 41; *cf.* also p. 18 and 825). Earlier edition: Dittenberger, *OGI* No. 383 (vol. 1 p. 591 ff.). For the general arrangement of the monument *cf.* Humann-Puchstein, *RKN* p. 328 and Goell [1]. *Cf.* also Cumont [10] p. 14-16.
The recent excavations of the whole monument, conducted since 1953 by F. K. Doerner and T. Goell (*cf.* e.g., *Illustrated London News*, July 2, 1955 p. 23-25) do not seem to have produced new material concerning the horoscope.

[2] *Cf.* Humann-Puchstein, *RKN* p. 329 note 1 and *Pseudo-Eratosthenis catasterismi*, ed. Olivieri, p. 15. In the often repeated reproduction of this monument (Bouché-Leclercq, *AG* p. 439 fig. 41) one star in the loop of the tail is omitted. It is barely visible on the photograph published in Humann-Puchstein *RKN* Tafel 40 or in F. Sarre, *Die Kunst des alten Persien*, Berlin 1923, pl. 55. This star is clearly shown on photographs taken in 1947 which we owe to the kindness of Miss Theresa Goell. The best published photograph is found in an otherwise valueless book: Philipp Metman, *Mythos und Schicksal*, Leipzig 1936, facing p. 128.
The deterioration of the monument has made great progress since it was uncovered in 1883.

[3] This date is often quoted erroneously as 97 B.C.

[4] What Puchstein says on p. 334 about the impressiveness of the consecutive appearance on the eastern horizon of Jupiter, Mercury, Mars, sun, and thereafter (!) Regulus, is astronomically absurd.

above the body of the lion may have been purely accidental.

A much more serious problem is the question whether the supposed conjunction of the three planets in Leo was computed for the constellation Leo as a whole or the zodiacal sign Leo of 30° length. In the first case (I) the longitudes of the planets should fall between about ♋ 15 and ♌ 25 of longitude, counted from the vernal point. For the zodiacal sign, on the other hand, four norms must be considered possible in the first century B.C.: (II) the scheme of Eudoxus according to which the sign of Leo would reach from ♋ 15 to ♌ 15; (III) the Mesopotamian System A with ♋ 20 and ♌ 20 as limits and the System B with ♋ 22 and ♌ 22; (IV) the norm of Hipparchus where Leo reaches from ♌ 0° to ♌ 30°. Each one of these combinations must be investigated separately, even if the first one seems to be rather implausible in the light of what we know about the methods of ancient astronomy.

Finally, there comes the question of the calendar. We know that the time of year must be about July because with Mercury in Leo the sun must be in or close to Leo. On the other hand we know that Antiochus was born on Aydnaios 16 and that his coronation fell on Loos 10.[5] As Puchstein remarked, the horoscope could not have been cast for the day of birth because the Aydnaios in the Roman calendar of the Province Asia corresponds roughly to the month of December.[6] Thus Puchstein took refuge in the "conjecture bien aventurée," to use Bouché-Leclercq's expression,[7] that Antiochus was a seven-month child whose conception was the basis of the horoscope for —97 July 17, the date of birth falling in the "beginning" (February) of the year —96. Obviously this "enchainement de postulats est bien fragile" and one must take the date of Loos 10th, the date of coronation, as the date of the horoscope because Loos corresponds roughly to the month of July as required by the horoscope. Thus, we can only hope to determine the date of coronation of Antiochus, not the date of his birth. We shall, nevertheless, extend our investigation over the whole interval which can possibly contain the life of Antiochus, that is from —119 to —35.

The first step consists in determining all the dates within this interval for which moon, Mercury, Mars, and Jupiter were between ♋ 15 and ♌ 30. A systematic investigation leads to five solutions the dates of which allow a small margin corresponding to the time interval for which the moon remains within the permitted limits. These cases are:

		☾	☿	♂	♃
(*a*)	—108 July 15	♌ 11	♌ 14	♌ 16	♌ 6
(*b*)	—97 July 16	♌ 25	♋ 16	♋ 17	♋ 15
(*c*)	—61 July 7	♌ 6	♌ 8	♌ 9	♋ 25
(*d*)	—61 Aug. 4	♌ 16	♌ 11	♌ 26	♌ 0
(*e*)	—48 July 13	♌ 17	♌ 16	♋ 27	♌ 24

The next step introduces the different possibilities for the interpretation of "Leo": (I) the constellation from ♋ 15 to ♌ 25; (II) the norm of Eudoxus for the sign Leo from ♋ 15 to ♌ 15; (III) the Babylonian norms from ♋ 20 or 22 to ♌ 20 or 22; (IV) the norm of Hipparchus which coincides with our norm ♌ 0 to ♌ 30.

At the same time one can make use of the omission of the sun, which is certainly astrologically significant, and the omission of Venus and Saturn. Combining this with the four cases (I) to (IV) we therefore require that ☾ ☿ ♂ ♃ be inside a given interval of one of these four types whereas ☉ ♀ ♄ remain outside.

With this requirement, (I) and (IV) yield no solution at all, and for (II) and (III) we have only the following possibilities:

(II) Leo from ♋ 15 to ♌ 15 (Eudoxus):

(*c*) —61 July 7:

☾	☿	♂	♃		☉	♀	♄
♌ 6	♌ 8	♌ 9	♋ 25	—	♋ 11	♊ 24	♉ 14

(III) Leo from ♋ 20/22 to ♌ 20/22 (Mesopotamia):

(*c*) —61 July 7:

☾	☿	♂	♃		☉	♀	♄
♌ 6	♌ 8	♌ 9	♋ 25	—	♋ 11	♊ 24	♉ 14

(*e*) —48 July 11:

☾	☿	♂	♃		☉	♀	♄
♋ 23	♌ 14	♋ 26	♌ 24	—	♋ 16	♋ 21	♎ 18

As the above table shows, —48 July 11 is only admissible when we are not too strict for the positions of Jupiter and Venus. We know furthermore that Antiochus died as an old man between —37 and —30.[8] Hence —48 is historically excluded both as date of birth and as date of coronation.

Consequently —61 July 7 remains the only[9] possibility. Astronomically it is the only date which satisfies all conditions excellently, showing furthermore that the positions must have been computed either with the

[5] *Cf.* Jalabert-Mouterde, *IGL SYR* I. 1 lines 83 to 85 and 102, 103. Jalabert-Mouterde (p. 18 note) speak erroneously about the 17th as the date of birth. The text says:

"I have sanctified the birthday of my flesh, the 16th of Aydnaios, and the day of my coronation, Loos 10th, to the manifestations of the great gods who were for me the guides of a fortunate reign and for the whole realm the source of common benefits" . . . "And for the rest of time, each single month I proclaimed that the days aforesaid—for my birth the 16th, and for my coronation the 10th—be forever honored by sacrifices."

The monthly celebration of the coronation is also mentioned in an inscription found in Selik near Samosata (Jalabert, *l.c.*, 51 line 5; *cf.* also Puchstein p. 369).

[6] *Cf.*, e.g., Ginzel, *Chronol.* III, p. 19.

[7] *AG* p. 373 note 2.

[8] *Cf.*, e.g., *RE* 1 col. 2489.

[9] Strictly speaking July 6 and 7 are equivalent dates.

Eudoxean or with the Mesopotamian norms for the sign "Leo," as seems historically plausible. It is also evident that —61 is excluded as date of birth of Antiochus; hence it must concern his coronation, again in very satisfactory agreement of the position of the month of Loos and July.[10] Unfortunately, —61 seems to be excluded as date of coronation because of the often quoted fact that Antiochus appeared as "king" before Lucullus in —68.[11] The solution of this difficulty can be found, however, in the relation of Antiochus to the Roman Empire. Calling himself repeatedly "Friend of the Romans" he must count his rule over Commagene as granted by the Romans during the final settlement of the organization of the Near East under Pompey, which happened indeed in the same year which was found by astronomical conditions, 62 B.C.[12] It must be this date to which our horoscope refers.[13]

No. —9

BGU 957

This fragmentary horoscope was excavated by Wilcken in Heracleopolis Magna in 1899 and destroyed by fire on a ship in Hamburg.[1] Fortunately, while still in Egypt, Wilcken had made a preliminary transcription which he then published in *BGU* III p. 281.

Since the date of the text is known (Augustus 20 = —10/—9), one can restore some of the lacunae on the basis of astronomical considerations. This leads to the following restoration of the text:

1. *ετουϲ κ̅ καιϲαροϛ μ[εϲορη τ]ηι κ̅δ̅*
2. *ειϲ την κ̅ε̅ κα[τ αρχαιουϲ ω]ρα γ νυκτο(ϲ̣)*
3. *ϲεληνη εν [υδρ]ηχωι και [κρονοϲ εν]*
4. *[υδρηχωι ηλιοϲ] εν λεοντι [και ερμηϲ]*
5. *εν λεοντι αρ[ηϲ εν ....]*

Rest destroyed.

TRANSLATION

1. Year 20 of Caesar (Augustus) M[esore] the 24th
2. to the 25th, according [to the old (calendar)] 3rd hour of the night.
3. The moon in Aquarius and [Saturn in]
4. [Aquarius. The sun] in Leo [and Mercury]
5. in Leo. Ma[rs in[2] ....]

Rest destroyed.

COMMENTARY

For the name of the month in line 1 one has only two possibilities: either Mekheir or Mesore. In line 2 one can restore a reference either to the old Egyptian calendar or to the calendar as reformed by Augustus. The 3rd hour of night from the 24th to the 25th corresponds then either to —9 February 15 or August 14 (Egyptian calendar) or to —9 February 18 or August 17 (Alexandrian calendar). On February 15 the moon was in ♋, on August 14 in ♒ as required by line 3. For February 18 and August 17 one obtains ♌ and ♓ respectively. Thus only Mesore 24/25 of the Egyptian calendar gives the correct lunar position. The phrase "*κατ' ἀρχαίουϲ*," without a preceding reference to the Alexandrian calendar is also found in No. 145.

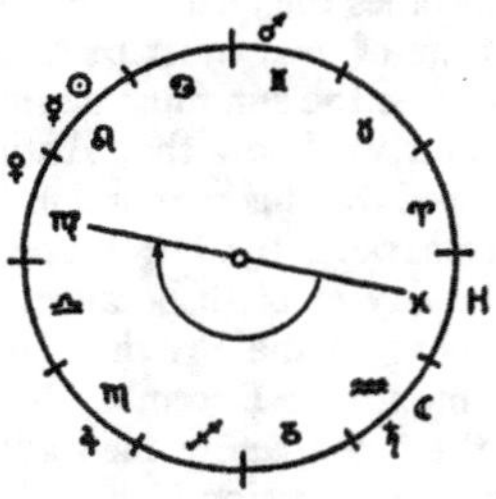

FIGURE 8.

Having now established the Julian date of this horoscope as —9 August 14, about 9 P.M., we can determine the position of all celestial bodies (*cf.* fig. 8):

☉	♌ 18	♂	♊ 23
☾	♒ 25	♀	♍ 2
♄	♒ 10	☿	♌ 22
♃	♏ 29	H	♓

This table shows that moon and Saturn are both in ♒, sun and Mercury both in ♌, and this determines the restorations in lines 3 and 4.

Though our date leads to a complete agreement with the preserved parts of the text, one must admit that the resulting arrangement in enumerating the planets is most unusual: instead of ☉ ☾ we begin here with ☾ (+ ♄) and ☉(+ ☿). We then should expect ♃ as next to be mentioned but the *αρ*[in line 5 can hardly be anything but *αρηϲ* = Mars. A glance at fig. 8 shows that the zodiacal order does not explain the adopted sequence

(☾ ♄)(☉ ☿) ♂ ....

No. —61

[10] Even the day numbers are very close. Ginzel, III p. 19 gives Loos 10 = July 4. *Cf.* also Johnson, *DS*, p. 6; Kubitscheck [1] p. 48.

[11] *Cf.*, *RE* 1, col. 2488, 20 following Cassius Dio XXXVI, 2. Also *RE* 13, col. 399, 22 and Suppl. 4, col. 978 ff.

[12] *Cf.*, e.g., Magie, *Roman Rule* I p. 377.

[13] We know of two more coronation horoscopes: No. L 484 (coronation of Leontius) and the Arabic horoscope for 531 August 18, the coronation of Khosro I, published Taqizadeh [1].

No. —9

[1] *Archiv* 1 (1901) p. 227 f.

[2] According to the planetary tables P (*cf.* p. 173) Mars is in ♊, entering ♋ on the 26th of Mesore.

No. —3

P. Oxy. 804

1. ετουϲ κζ και[ϲ]αροϲ
2. φαωφι ε̄ κα[τα] καιϲαρ
3. περι ωρ^L γ̄ τηϲ ημερ^L
4. ηλιοϲ εν ζυγω
5. ϲεληνη εν ιχθυϲι
6. κ[ρο]νοϲ εν τα[υρω]
7. ζ[ευ]ϲ εν κα[ρκινω]
8. [αρη]ϲ εν π[αρθενω]
9. [αφροδ]ε[ιτη εν ϲκορπιω]
10. [ερμηϲ εν παρθενω]
11. [ωροϲκοπει ϲκορπιοϲ]
12. [μεϲο]υρ[α]ν[ει λεων]
13. δυνει . τα[.]τ...
14. υπο γην υδροχοοϲ
15. εχει κινδυνουϲ
16. φυλαττου εωϲ ημερ^ω
17. μ̄ χαριν του αρεωϲ

1 κζ: *sic*, no bar over the number.

13 There are traces of writing between lines 13 and 14. In line 13 ταυροϲ must occur; perhaps one could read δυνει ειτα ταυροϲ.

TRANSLATION

1. Year 27 of Caesar (Augustus)
2. Phaophi 5 according to the Augustan calendar
3. about the 3rd hour of the day.
4. sun in Libra
5. moon in Pisces
6. Saturn in Taurus
7. Jupiter in Cancer
8. Mars in Virgo
9. [Venus in Scorpio]
10. [Mercury in Virgo]
11. [Scorpio is rising]
12. [Leo is at Midheaven]
13. [Taurus is then] setting.
14. Lower Midheaven Aquarius
15. There are dangers.
16. Take care for 40 days
17. because of Mars.

ASTRONOMICAL COMMENTARY

Diagram: pl. 1

For Augustus 27 II(a) 5 = —3 October 2[1] one finds:

Text		Computed	
☉	♎	♎	7
☾	♓	♓	2
♄	♉	♉	3
♃	♋	♋	10
♂	♍	♍	4
[♀	♏]	♏	22
[☿	♍]	♎	1
[H	♏]	about 9 A.M.	

[1] The editors write erroneously A.D. 4 instead of 4 B.C. Correct in Preisigke, *Berichtig.* p. 328.

We have here used the Alexandrian calendar. This is necessary, even without the explicit statement of the text (line 2), because the position of the moon agrees closely with the Alexandrian date and excludes the Egyptian date, which would be —3 September 27.

Our text can be compared with the Demotic planetary tables P,[2] which give the dates of the Egyptian calendar for the entry of the planets in the zodiacal signs. For our specific dates we find, changing Egyptian dates into Alexandrian, the following limits for the stay of the planets inside their respective signs

♄ in ♉	from year	27	I	10	to year	28	IX	28
♃ in ♋		26	XI	26		27	XII	24
♂ in ♍		27	I	10		27	II	9
♀ in ♏		27	I	13		27	II	10
☿ in ♍		27	I	19		27	II	8

The date of our horoscope, year 27 II 5, is contained within all these intervals. According to the last line in the above list Mercury must be at the end of Virgo whereas modern computation gives the first degree of Libra. In the case of Mars the second date of the Demotic text is certainly incorrect because Mars needs not 29 days but about 45 days to travel through Virgo.

ASTROLOGICAL COMMENTARY

At the end caution for 40 days is advised. At the moment of the horoscope, Mars, the evil planet, is in the ἀγαθὸϲ δαίμων,[3] he is in trine with the setting sign, in sextile with the Horoscopos, while the Lot of Fortune is in "bad fortune."[4] After 40 days Mars will be outside this zodiacal sign.

Strange as it may seem a reinvestigation of a horoscopic constellation at a later time is considered useful. Thus we read in the discussion of the horoscope of Hadrian (No. L 76) "not only this has to be considered but also the 3rd, the 7th, and the 40th day after birth"[5] and indeed all positions are recomputed for a date 40 days later.[6]

[2] *Cf.* below p. 173 or Neugebauer [4] p. 215.

[3] *Cf.* Bouché-Leclercq, *AG* p. 280 fig. 31. If the sign of the Horoscopos is counted as locus I, then the ἀγαθὸϲ δαίμων is locus XI, thus in our case Virgo.

[4] The elongation of the moon in Pisces from the sun in Libra is 5 signs. Because we are dealing with daylight the Lot of Fortune fell 5 signs behind the Horoscopos, thus in Aries. This is one sign below the setting sign, thus κακὴ τύχη (*cf.* Bouché-Leclercq, *AG* p. 280 fig. 31). The fact that Mars is in opposition to the moon also is dangerous; *cf.* Bouché-Leclercq, *AG* p. 488, note.

[5] *CCAG* 6 p. 71, 15 f.

[6] Similarly in Nos. L 40, L 113,IV, L 479. *Cf.* also Nos. 81 and 345.

No. 15/22
P. Oxy. 235

TRANSCRIPTION

1. αναγκαιον ηγησαμ[ενος ....]να ..[.......
2. γενεςεις παρα cου τριφων αγαπητε ε[......
3. πειραcομαι προς τους δοθεντας η[μιν ....
4. χρονους τυν[χ]α[ν]ουcι δε ουτοι κατα [το ....
5. ετος τιβεριου μηνι φαωφι ā κατ[α δε τους]
6. αρχαιους χρονους φαωφι ιā εις [ιβ̄]
7. ωρα τεταρτη της νυκτος τυνχανει [ηλιος]
8. εν ζυγω ζωδιω αρcενικω οικω αφ[ροδιτης]
9. cεληνη εν ταυρω ζωδιω θηλυκω οικω [αφροδιτης]
10. κρονος ζευς εν τοξοτη [ζω]διω αρcεν[ικω οικω]
11. διος αρης εν ζυγω οικω αφροδιτης [ερμης αφρο]
12. διτη εν ςκορπιω ζωδιω αρcενικω [οικω αρεως]
13. ωροςκοπει ταυρος .... οικος αφροδιτ[ης μεςουρανει]
14. υδροχοω ζωδιον αρςενικον οικητη[ριον κρονου]
15. δυνει ςκορπιος οικος αρεως υπο[γην εν λεοντι]
16. οικος ηλιου οικοδεςποτει αφροδ[ιτη]

The following diagram (fig. 9) gives an approximate representation of the figure at the end of the text.

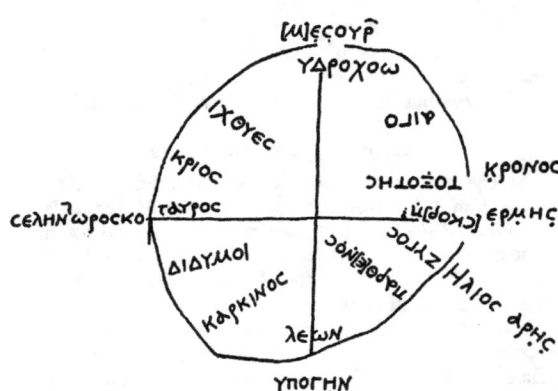

Figure 9.

TRANSLATION

1. Thinking it proper [to ....]
2. your descendants(?), dear Tryphon [......]
3. I shall try to [set forth] for the dates [which you have given us ....]
4. These happen to be: [the .....]
5. year of Tiberius, Phaophi 1, according to
6. the old calendar Phaophi 11 to [12],
7. at the fourth hour of the night. [The sun] happens to be
8. in Libra, a male sign, house of Venus;
9. the moon in Taurus, a female sign, house [of Venus];
10. Saturn (and) Jupiter in Sagittarius, a male sign, [house]
11. of Jupiter; Mars in Libra, house of Venus; [Mercury]
12. (and) Venus in Scorpio, a male sign, house [of Mars;]
13. Taurus is the Horoscopos, house of Venus; [Midheaven]
14. in Aquarius, a male sign, house [of Saturn;]
15. Scorpio is setting, house of Mars; Lower [Midheaven in Leo,]
16. house of the sun. Venus is the house ruler (of the Horoscopos).

COMMENTARY

Diagram: pl. 1

This text is dated by a year of Tiberius in which Phaophi (II) 1 of the Alexandrian calendar corresponds to Phaophi 11/12 of the Egyptian calendar. We now must investigate two cases (cf. fig. 10 a and b respec-

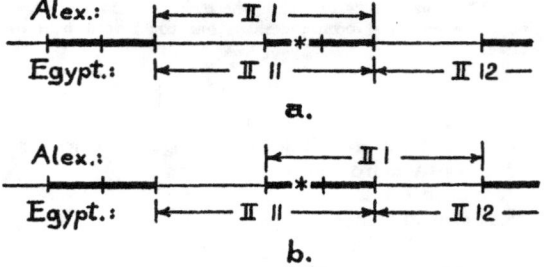

Figure 10.

tively): (a) both Egyptian and Alexandrian calendar use morning epoch. In this case we have simply Alexandrian II,1 = Egyptian II,11, a relation which holds for the years Tiberius 2 to 5, i.e., for A.D. 15 to 18.[1] In case (b), however, we assume evening epoch for the Alexandrian calendar, morning epoch for the Egyptian calendar, following the norm which is suggested by the double dates in No. 81.[2] Then the daytime of Alexandrian II,1 corresponds to the daytime of Egyptian II,12 which is true for Tiberius 6 to 9, i.e., for A.D. 19 to 22.[3] It would be easy to decide between these possibilities if the astronomical data were correct. Saturn and Jupiter are both said to be in ♐ as is indeed the case in A.D. 15 and therefore excluded for A.D. 16 to 22. But Mars in ♎ is only correct for A.D. 18 and 20 but entirely wrong for the year 15. To make the confusion complete, a lunar position in ♉ on September 28 or 29 requires

---

[1] September 29 for A.D. 15, September 28 for the next three years.
[2] For a detailed discussion of these possibilities cf. p. 168.
[3] September 29 for A.D. 19, September 28 for the next three years.

the year 17. Venus is in ♏ in the years 16 and 18 and only the sun is always at the beginning of Libra at the end of September. This excludes a position of Mercury in ♏ since it would require the maximum possible elongation of this planet; it only reaches ♎ 23 in the year 19. Obviously none of the basic astronomical data can inspire any confidence.[4]

The astrological data are correct except for denoting Scorpio a "male" sign, since it is "female," at least according to the *Tetrabiblos* and to Vettius Valens. Perhaps we are dealing here with a different system according to which six consecutive signs are male and female respectively; *cf.*, for such possibilities Bouché-Leclercq, *AG* p. 154 ff.

BIBLIOGRAPHICAL NOTES

After the original publication in P. Oxy. II (1899) this text was discussed by Smyly [2] (1901), Nicklin [1] (1902), Laudien, *GPO* p. 29 f. and p. 56 f. (1912), Macnaughton, *SEC* p. 330-332, and David-Van Groningen, *PL* p. 130 f. (1940).[5] The last quoted work and Laudien give the Greek text.

Nicklin [1] p. 119 says "the horoscope is entirely correct" for the year 14, but this must be a mistake. Macnaughton, *SEC* p. 330, prefers the year 16. His remark that the symbols ♍ and ♏ can be easily confused makes no sense because the use of these symbols is not attested before the Middle Ages. Smyly [2] p. 86 investigated only the calendaric conditions.

No. 32
PSI 1276 verso

TRANSCRIPTION

1. ετους ιβ τιβερ[ιου
2. παυνι ᾱ ω[ρα θ ημερας]
3. ηλιος α[φροδιτη ερμης]
4. [ε]ν διδυ[μοις]
5. σεληνη [ζεις εν ταυρω]
6. κρονος ε[ν καρκ]ινω
7. αρης [εν ιχθυσι]
8. ωροσκοπο[ς ε]ν ζυ[γω]

TRANSLATION

1. Year 12 of Tiber[ius
2. Payni 1 [9th hour of the day]
3. Sun, V[enus (and) Mercury]
4. in Gemi[ni]
5. Moon [(and) Jupiter in Taurus]
6. Saturn i[n Can]cer
7. Mars [in Pisces]
8. Horoscopos in Li[bra]

---

[4] Since H = ♉ and ☉ = beginning of ♎ the time is correctly about the 4th hour of the night and M = ♒.
[5] English edition: PP p. 142 f. (1946).

COMMENTARY
Diagram: pl. 9

This horoscope was first published by Bartoletti in *Aegyptus* 19 (1939) p. 177. The recto contains a paraphrase of Homeric verses.

For the date given in the text, Tiberius 18, Payni 1 = A.D. 32 May 26, one finds full agreement between text and computation:

Text		Computed
☉		♊ 2
[♀]	} ♊	♊ 0
[☿]		♊ 15
☾	} ♉	♉ 23
[♃]		♉ 28
♄	♋	♋ 8
♂	[♓]	♓ 5
H	♎	about 3 P.M.

The restorations are quite certain since they are confirmed by the arrangement of the planets.

No. 46
P. Oxy. 307
(unpublished)
Photo: pl. 30

1. [ε]υρισκομεν τον ηλιον τον κοσμο[κρα]τω[ρα]
2. [εν] αιγογερωι μοιρων ια ∠ οριοις διος
3. την σεληνηι μητηρ του παντος
4. [ευρισκομε]ν εν κριωι μοιρων ια
5. [οριοις] αφροδιτηι
6. [φαι]νων ο του κρονου εν τοξοτη[ι]
7. [μοιρω]ν λ ενγιστωι οριοις αρεως
8. [φαε]θων ο του διος εν καρκινωι
9. [μοι]ρων ιθ εν αναποδισμωι
10. οριοι[ς] κρ[ον]ου πυροις
11. ο του αρεως εν υγρη[χ]ωι μοιρ[ω] ιδ ∠
12. οριοις διος εν ιδιωι οικωι
13. φωσφορος ο της αφροδιτηι
14. εν αιγογερωι μοιρων ιθ
15. οριοις αφροδιτηι στιλ
16. [σ]τιλβωι ο του ερμου
17. εν τοξοτηι μοιρω λ
18. οριοις αρεως οροσκοπι παρ[θ]ενωι
19. με[σο]υρανι διδυμ[οις] δυν[ι] ιχθυσι
20. υπογην τοξ[οτηι]

11 μοιρ[ω] ιδ: one could also read μοιρ[ω]ν δ but the terms of Jupiter (line 12) exclude this interpretation. For μοιρω instead of μοιρων cf. line 17.

15 στιλ: probably left incomplete and then resumed in the next line.

18 αρεως: ε corrected from ω.

TRANSLATION

1. We find the sun, the ruler of the world,
2. in Capricorn 11½ degrees, terms of Jupiter.
3. The moon, the mother of all,
4. we find in Aries 11 degrees,
5. terms of Venus.
6. Phainon, the (star) of Saturn, in Sagittarius
7. 30 degrees approximately, terms of Mars.
8. Phaethon, the (star) of Jupiter, in Cancer,
9. 19 degrees, in retrogradation,
10. terms of Saturn. Pyroeis,
11. the (star) of Mars, in Aquarius, 14½ degrees,
12. terms of Jupiter, in the house of the same.
13. Phosphoros, the (star) of Venus,
14. in Capricorn, 19 degrees.
15. terms of Venus. Stil(bon)
16. Stilbon, the (star) of Mercury,
17. in Sagittarius 30 degrees,
18. terms of Mars. The Horoscopos in Virgo,
19. Midheaven in Gemini, setting (point) in Pisces,
20. Lower Midheaven in Sagittarius.

COMMENTARY

Diagram: pl. 1

Within more than one century in each direction the only possible date is A.D. 46 January 3. For this date one obtains the following results:

	Text		Computed	Text—comp.
☉	♑	11½	♑ 12	0
☾	♈	11	♈ 8	+3
♄	♐	30	♐ 22	+8
♃	♋	19	♋ 23	—4
♂	♒	14½	♓(!) 18	[—4]
♀	♑	19	♑ 14	+5
☿	♐	30	♑ 6	—6
H	♍		about 10 P.M.	

Jupiter being close to opposition is retrograde as stated in the text. The only serious discrepancy is found in the case of Mars where the text gives the wrong zodiacal sign for the position; in the reference to the "house of Jupiter," however, the correct sign ♓, is used. The rulers of the terms as given in the text require longitudes which are given in the following list:

☉	terms of	♃	thus	♑	7 to 14
☾		♀		♈	6 to 12
♄		♂		♐	26 to 30
♃		♄		♋	19 to 26
♂		♃		♒	13 to 20
♀		♀		♑	14 to 22
☿		♂		♐	26 to 30

It is seen that they all agree with the longitudes found in the text although in the case of Jupiter the lower boundary is taken as determining the ruler, while in the case of Saturn and Mercury the upper boundary was chosen. Another case of the use of the lower boundary is found in No. 138/161 (P. Princ. 75) lines 2 and 3; for the upper boundaries cf. No. 137 a and b (p. 41) and No. L 482 (p. 146 notes 3 and 6).[1]

The text was found in Behnesa (= Oxyrhynchos) by Grenfell and Hunt (1897).[2]

Nos. 50,1, 50,2, and 50,3

P. Abd. 126

This little fragment seems to belong to a text which had collected several horoscopes. On the recto we suggest separating into two columns what the editor took as one text.

On the left hand we see, following a dividing line, No. 50,1:

1. ....]ηρης
2. ...]
3. ..]
4. .]αρης

remaining lines destroyed.

Traces of writing in the space between line 1 and 4 were interpreted as ηλιος by the editor, continued by αι[γοκερω which we read αρ[ης belonging to another column.

Toward the right the beginning of five lines occupies about the same space as four lines in the first column. We read No. 50,2:

Beginning destroyed
1. κρο[νος
2. ζευς [
3. αρ[ης
4. αφρ[οδειτη ερμης]
5. ιχθυ[ςιν

remainder destroyed.

Assuming the correctness of our restoration in line 4, one can conclude that also the sun must be near ♓, thus the time late winter or early spring.

The verso, in a different hand, gives

No. 50,3: 1. .]..υδρ[οχοω
           2. .]..αφροδ[ιτη

beginning and end being destroyed.

---

[1] Cf. also No. 138/161, note to line 9.
[2] Cairo. Cat. gén. X p. 3 No. 10012.

No. 81

P. Lond. 130

TRANSCRIPTION

I

1. [Τ]ων αιγυπτιων οι το
2. παλαιον αν[δρ]ες γε
3. νομενοι [γ]νησιως
4. τε περι τα ουρανι
5. α φιλοπονησαν
6. τες και επιγνον
7. τες την των επτα
8. θεων κεινησιν
9. τα ολα ϲ[υ]νεχου
10. σαν τε και διοικου
11. σαν αφ[θ]ονως η
12. μειν δια κανο
13. νων αιωνιων
14. απελειπον την
15. περι αυτων γνω
16. ϲιν οθεν εκα
17. ϲτον ψηφισας α
18. κριβως τεταχα
19. ϲοι προς τε μοι
20. ραν και λεπτον
21. ϲχημα τ[ε] και φα
22. ϲιν και τα προς

II

23. επισκεψιν απλως
24. τεινοντα ινα μη
25. επιλεγομενος ε
26. καστον ευχρονιζω·
27. Ουτως γαρ ο εν αστρο
28. λογια προρητικος
29. τροπος αναμφι
30. βολος τουτεστιν
31. ομολογος κατορ
32. θουται ερρωσο
33. μοι φιλτατε ερ
34. μον

35.   χρονος
36.   [τη]ς τροπης
37. Ετους τριτου θεου
38. τιτου φαρμου
39. θι τη επιφωϲκου
40. ϲη εκτηι επι
41. τριτης της νυ
42. κτος ωρας: ως
43. δε ρωμαιοι αγουϲι
44. καλανδαις απρι
45. λειας: κατ αρχαι
46. ους δε παχων νεο

III

47. μηνια εις την δευ
48. τεραν: οθεν ο
49. μεν μεγιστοτατος
50. ηλιος και των ολων
51. δυναστης· απο εα
52. ρινης ισημεριας φε
53. ρομενος κριου μοι
54. ραϲ επειχεν δεκα
55. τεϲϲαρας· και προς
56. ετι λεπτα εξ· του
57. τεϲτιν μερος δεκα
58. τον μοιρης: ζωδι
59. ωι αρεως: οριοις ερ
60. μου: στοιχειωι δι
61. ος: αρρενικωι και
62. προς βορρωι ζωδι
63. ωι υψουμενος: ε
64. π[ι] της λαγονος του
65. κριου σελαγιζων:
66. δεκανωι δευτε
67. ρωι καλουμενωι
68. ϲενταχορ: ου το δω
69. δεκατημοριον
70. διηυγαζεν σκορ
71. πιου περι τον πρω

IV

72.   τον σφονδυλον:
73. Η δε θεα και σελασφο
74. ρος σεληνη απ α
75. νατολης φερομε
76. νη οξυγωνιος ταυ
77. ρου μοιρας ετρε
78. χε δεκατρεις· και
79. ετι μερος χιλιοστ°
80. μοιρης: ζωδιωι
81. Αφροδιτης: υψω
82. ματι ιδιωι: ορι
83. οις ερμου: θηλυ
84. κωι και στερεωι
85. ζωδιωι: χρυσωι
86. ομοια: επι της
87. ραχεως αναβιβα
88. ζουσα του ταυρου:
89. δεκανωι δευτε
90. ρωι καλουμενω
91. αρωθ: ης το δωδε
92. κατημοριον σκορ
93. πιου παλιν περι
94. τον αυτον τοπον
95. εξελαμπεν:
96. Φαινων δ ο του κρο
97. νου αστηρ ιχθυ

V

98. ων μοιρας επληρου
99. εξ· λειπουσας μερος
100. εξηκοστον μοιρης:
101. Ζωδιωι διος: οριοις
102. και υψωματι αφροδι
103. της: εωος ανατολι
104. κος: επι του χελει
105. δονιαιου ιχθυος κα
106. ταβιβαζων: επι
107. τριτος ογκωι:
108. Φαεθων δ ο του διος
109. αστηρ υψος ανατρε
110. χων επειχε καρκι
111. νου μοιρας εξ· και
112. ετι λεπτα τριτα δε
113. κ[α] α δη μερος μοι
114. ρης εστιν δισμιρο
115. χιλεξακοσιοστον:
116. Ζωδιωι σεληνης: ο
117. ριοις και ταπει
118. νωματι αρεως:
119. το δευτερον στη
120. ριζων: του εν τωι
121. νωιτωι λαμπρου
122. αστερος βορειοτε
123. ρος δακτυλο[ι]ς δυει:
124. ημιολιος τε ογ
125. κωι και προστασσων

VI

126. πυροε[ι]ς δ ο του αρε
127. ως αστηρ υδροχοου
128. μοιρας ανεβιβα
129. ζεν δε[κα] εξ· εικο
130. στον: ζω[δ]ιον κρο
131. νου: τριγωνον ερ
132. μου: οριον διος: α
133. στερα τον εν τηι χλα
134. μυδι καλουμενον
135. γανυμηδην ομω
136. νυμως τωι ολωι ει
137. δωλωι: εωος αγαν:
138. Φωσφορος δ ο της α
139. φροδιτης αστηρ
140. ιχθυων μοιρας ε
141. πληρου δεκα εξ·
142. και λεπτα τεσσαρα·
143. α δη πεντεκαιδε
144. κατον μοιρης με
145. ρος εστιν: ζωδι
146. ωι διος: υψωματι ι
147. διωι: εωα ανατο
148. λικη: επι του νο

149. τειου ιχθυος: ο
150. μοια κρυσταλλωι
151. οριοις ερμου: του
152. εν τοις αποτετα
153. μενοις λινοις
154. λαμπρου αστερος

VII

155. απεχουσα σεληνια[κ]
156. α μεγεθη δυο:
157. στιλβων δ ο του ερ
158. μου αστηρ κριου ε
159. πετρεχε μοιρας στε
160. ρεας δεκα: περιγει
161. ος: προ επτα φασιν
162. πεποιημενος: δι
163. ο οικοδεσποτησει
164. το διαθεμα·
165. Ο δ οιαξ των ολων ωρο
166. σκοπος σκορπιου
167. απετεμνε μοιρας
168. δεκα οκτωι: οριον
169. ερμου: ζωδιον αρε
170. ως: τριγωνον σε
171. ληνης: δεκανον
172. θουμουθ:
173. ο δε προς ορθας τωι δε
174. μεσημβρινος λεον
175. τος την ραχιν εκρου
176. εν:
177. Ο κληρος δε της τυχης
178. εις διος και ζωδιον
179. και τριγωνον ανα
180. παλιν εσται: ον τινες
181. ζυγωι δωρησονται α
182. γνοια: ομως αφρο
183. διτης παλιν α κλη
184. ρος: διευτυχει·

VIII space of 8 lines left blank

185. Τιτος
186. Πιτη
187. νιος
188. εψηφι
189. σα ως
190. προκει
191. ται

rest of column not inscribed

Detached fragment:

IX

192. α' ερμ'
193. β' η[λιος]

ORIGINAL DOCUMENTS: No. 81   23

194.   α' κρ[ονος]
195.   β' αφρο⁸
196.   ινα και της
197.   ταξεως
198.   μη διασφα
199.   [λ]ης των δι^π
200.   εψηφισθη
201.   εν ερμου
202.   πολει ενθα
203.   λογον ε[χ]ει
204.   ο οριζω[ν]
205.   τον επτ[α]
206.   προς πεν
207.   τε
           ———
208.   ο σποριμ[ος]
209.   χρονος
210.   ημερω[ν]
211.   ς' ο' ς'
rest of column blank

X, in bold letters beside lines 202 and 205:

212.   αγαθηι
213.   τυχηι

33-36 margin: cf. note on lines 182-184.
45-46 κατ αρχαιους: supply χρονους; cf., e.g., No. 15/22 (= P. Oxy. 235) line 6.
74-76 απ ανατολης φερομενη: "waxing" not "rising."
76 οξυγωνιος: "crescent," literally "acute-angled."
112-113 λεπτα τριτα δεκ[α]: not "three or thirteen minutes" (Bouché-Leclercq, *AG* p. 291 n. 1), but "ten parts of the third order."
159-160 μοιρας στερεας: "full degrees"; similar: στερεαι in P. Mich. 149 III,20.
161 προ επτα: "before the seventh," cardinal instead of ordinal numeral, as commonly in papyri and N. T. Greek.
172 θουμουθ: the reading of the final θ seems certain (communication of Dr. T. C. Skeat).
182-184: At the end of Col. VII one finds on the margin the same elaborate coronis which marks the end of the introduction in Col. II beside lines 33 to 36.
188,189 εψηφιςα: we owe this reading to Dr. T. C. Skeat.
199 δι^π: Dr. Skeat drew our attention to the fact that δι is characterized as an abbreviation by a raised sign or letter (π or c?) above the ι.
202 ενθα: reading of Dr. Skeat.

TRANSLATION

Col. I

The Egyptian men of old who had faithfully studied the heavenly bodies and had learned the motions of the seven gods, compiled and arranged everything in perpetual tables and generously left to us their knowledge of these things. From these I have accurately calculated and arranged for

Col. II

each one (of the seven gods) according to degree and minute, aspect and phase, and, simply, not to waste time in enumerating each item, whatever concerns its investigation. For thus the way of astrological prediction is made straight, unambiguous, that is, consistent. Farewell, dearest Hermon.

Time of the (equinoctial) tropic of the third year of the Divine Titus, the sixth day of Pharmouthi, the third hour of the night; as the Romans reckon, the Kalends of April; according to the old (calendar), Pachon

Col. III

the first to the second.

Accordingly, the sun, the mightiest and ruler of all, moving from the spring equinox had attained in Aries fourteen degrees and six minutes, that is, the tenth part of a degree; in the sign of Mars; in the terms of Mercury; in the stoicheion of Jupiter; in exaltation in a male and northerly sign; shining upon the Flank of Aries; in the second decan, called Sentachor; the dodekatemorion was shining upon the First Joint (of the tail) of Scorpio.

Col. IV

And the divine and light-bringing moon, waxing in crescent had advanced in Taurus thirteen degrees, and a thousandth part of a degree; in the sign of Venus; in its own exaltation; in the terms of Mercury; in a female and solid sign; like gold; mounting the Back of Taurus; in the second decan called Aroth; its dodekatemorion again was shining on about the same place in Scorpio.

Col. V

And Phainon, star of Saturn, had completed six degrees in Pisces, lacking a sixtieth of a degree; in the sign of Jupiter; in the terms and the exaltation of Venus; at its morning rising; descending from the Swallow-Fish; of magnitude four thirds.

Phaeton, the star of Jupiter, traversing its exaltation, in Cancer, had attained six degrees and ten sixtieths of the third order, which are one twenty-one-thousand-six-hundredth part of a degree; in the sign of the moon; in the terms and depression of Mars; at its second station; two fingers north of the Bright Star at the Back; of the magnitude three halves and commanding.

Col. VI

And Pyroeis, the star of Mars, has mounted in Aquarius sixteen degrees and a twentieth; (and has mounted) the sign of Saturn; the triangle of Mercury; the term of Jupiter; the Star in the Cloak, called Ganymede, homonymous with the whole constellation, far to the east.

And Phosphoros, the star of Venus, had completed in Pisces sixteen degrees and four minutes, which is the fifteenth part of a degree; in the sign of Jupiter; in its own exaltation; rising at dawn; at the Southern Fish; like crystal; in the terms of Mercury;

Col. VII

distant two lunar diameters from the Star in the Connecting Cords.

And Stilbon, the star of Mercury, had mounted in Aries ten full degrees; in perigee; having completed its phase before the seventh (day of the month); wherefore it will dominate the theme (of the nativity).

And the rudder of them all, the Horoscopos, has cut off eighteen degrees in Scorpio; the terms of Mercury; the sign of Mars; the triangle of the moon; the decan Thoumouth.

And the meridian at right angles to this had struck the Back of the Lion.

And the Lot of Fortune, (counting) inversely, will be in the sign of Jupiter and in its triangle; this (Lot) some will ignorantly assign to Libra; thus the Lot would be back (in the house) of Venus.

Good luck!

Col. VIII

Titus Pitenius computed it as is set forth.

Fragment, Col. IX

1. Mercury
2. sun
1. Saturn
2. Venus

in order that you may not make a mistake in the arrangement of the . . .

Computed in Hermopolis, where the horizon has the ratio seven to five.

The time of pregnancy: 276 days.

Col. X

With good fortune.

ASTRONOMICAL COMMENTARY

Diagram: pl. 1

Col. I

"*Perpetual tables.*" The existence of tables with this name is known from Ptolemy's introduction to his planetary theory (Almagest IX,2). In criticizing earlier attempts to develop a consistent theory of planetary motion by means of the so-called "perpetual tables" (διὰ τῆς καλουμένης αἰωνίου κανονοποιίας [1]) Ptolemy tells us that these tables were built on the assumption of eccentric or epicyclic models, though of so primitive a structure that the resultant retrogradations were of constant amount. The fact that our text mentions the perigee of Mercury is at least in principle in agreement with Ptolemy's description.

From our present text it seems furthermore to follow that these tables were computed to at least three sexagesimal places. This is not surprising in itself because the fundamental period relations can only be guaranteed if the mean motions are computed accurately enough. Three places are given, e.g., for the moon in P. Lund. 35a and four places in P. Ryl. 27.[2] It is no contradiction that often no fractions at all appear in the final results where a higher accuracy is of no practical value. The statement of our text (Col. II) that the horoscope was computed "according to degree and minute" is not to be taken literally, as the higher accuracy with the moon and with Jupiter shows. The remaining longitudes, however, do not go beyond minutes.

"*Time of the (equinoctial) tropic.*" The longitude of the sun being ♈ 14;6 (*cf.* also Col. III) shows that the expression "tropic" here is meant in the general sense in which τροπή is used not only for the solstices but also for the equinoxes; *cf.*, e.g., *CCAG* 7, p. 194,16.

"*Third year of the Divine Titus.*" Our present horoscope was cast for A.D. 81 March 31 but was written after the death of Titus (September 81) as the appellative "divine" shows. Though Titus ruled only little over two years he is represented with three years in the so-called "Ptolemaic Canon."[3]

The planetary tables P. Tebt. II, 274 quote not only the regnal years of Trajan but also the years of Titus from 29 to 37, continued until Titus 42 by P. Lund 35 b.[4] An era Titus is otherwise unknown.

Col. II

"*The sixth day, the third hour of night.*" This translation of τη επιφωσκουςη εκτηι επι τριτης της νυκτος ωρας is astronomically necessary. We know that the sun is in the middle of Aries while the middle of Taurus is setting. Consequently, only about two hours can have elapsed since sunset and certainly not about nine hours as would be the case if επιφωςκουςη meant the time between midnight and sunrise. The same result has been reached from the discussion of passages in the N.T.[5]

---

[1] *Opera* II p. 211,5 ed. Heiberg.

[2] Knudtzon-Neugebauer [1] and Neugebauer [8].

[3] *Cf.*, e.g. Kubitschek, *Zeitr.* p. 60. For occurrences of a year "Titus 3" in non-astronomical context *cf.* Hohmann [1].

[4] Neugebauer [4] and Knudtzon-Neugebauer [1] p. 86 to 88.

[5] *Cf.* Smyly [2] p. 83 note 1 or Burkitt [1]. Sethe, *Zeitr.* p. 133 note 7 assumed a meaning "the night leading to the 6th" in order to rescue the identity of epoch for Egyptian and Alexandrian calendar.

Figure 11 illustrates the relative position of the three different calendar dates which can be reconciled only on the assumption of an evening epoch for the Alexandrian and for the Roman date. For the discussion of this problem within a wider frame *cf.* below p. 167 ff.

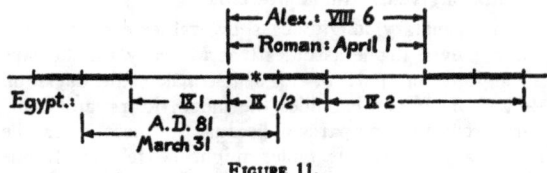

FIGURE 11.

Col. III

"*Moving from the spring equinox.*" The sun being in Aries 14 would mean 6° beyond the spring equinox if the vernal equinox is assumed to be in Aries 8 as is often the norm in astrological literature and in the Babylonian "System B" of the solar-lunar theory. Modern computation gives about Aries 9 for the position of the sun. *Cf.* below p. 27 the comparison between text and modern computation.

"*Second decan,* cενταχορ." Aries 14 belongs to the second decan whose name is given by Hephaistion as χονταχρε, by Firmicus as Senacher. The Egyptian form is ḫntw ḥry "the lower ḫntw (jugs)." *Cf.* also p. 6.

"*First Joint (of the tail) of Scorpio*": *Cf.* note under Col. IV "Dodekatemorion in Scorpio."

Col. IV

"*Mounting the Back of Taurus*" αναβιβαζουσα was taken by Bouché-Leclercq (*AG* p. 122 note 1) in a technical sense. Referring to the present passage and to a similar passage in Col. V "*Descending from the Swallow-Fish*" he says "ces mots 'monter' et 'descendre' signifient être au N. ou au S. de l'équateur." Such a usage is not only unknown elsewhere but explicitly contradicted here by the computation which shows that both moon and Saturn had southern latitude, the moon 4;40°, Saturn 1;49°. These values furthermore demonstrate that both moon and Saturn were far from their nodes, thus excluding definitely the ordinary interpretation of the two words in question as referring to the ascending or descending nodes.

"*Dodekatemorion in Scorpio.*" The dodekatemorion of the place of the sun, ♈ 14, is ♎ 2, the dodekatemorion of the place of the moon, ♉ 13, is ♎ 19. Though the Scorpion extended originally over Libra and Scorpio, Libra being called the "Claws," we know from Ptolemy's catalogue of stars that the star in the First Joint of the Tail of Scorpio had a longitude of ♏ 18;30 (Heiberg II p. 110/111,11). Thus one has to assume a gross error in the computation of the dodekatemorion. The result would be correct if the longitude of the sun had been about ♈ 18.

"*Waxing.*" This is correct because the conjunction fell in the early hours of March 29.

"*A thousandth part of a degree.*" We have no tables which give decimal fractions of degrees. Probably the corresponding sexagesimal number was erroneously omitted; *cf.* the cases of sun, Saturn, Jupiter, and Venus. For Mars, however, we find only "one twentieth."

"*Second decan,* αρωθ." Correct for ♉ 13; Egyptian: 'ryt.

Col. V

"*Lacking a sixtieth of a degree.*" Very unusual terminology for 59 minutes.

"*Descending.*" *Cf.* Col. IV sub "Mounting."

"*Morning rising.*" Saturn, being about 38° ahead of the sun, rises indeed in the morning before the sun.

"*Swallow Fish.*" Boll, *Sphaera*, p. 389 considers this constellation as belonging to the "sphaera barbarica"; *cf.* also below p. 27.

"*Magnitude 4/3.*" The translation of ὄγκος "weight" by "magnitude" is a mere guess. Nowhere else in ancient astronomy is this terminology attested nor do we have any proof for ascribing variable numerical values to the magnitude of planets. Computation, however, shows that Saturn had at the given moment the magnitude 1.3 in excellent agreement with the text. On the other hand, we find for Jupiter a magnitude of —1.8 whereas the text says "weight 3/2" i.e., +1.5. Of course there is no trace in Greek astronomy of negative stellar magnitudes. Both Sirius and Canopus are listed in the Almagest and in Arabic astronomy as stars of first magnitude.[a]

"*21600th part of a degree.*" Indeed $\frac{1}{21600} = \frac{10}{60^3}$.

"*Second station.*" For the appearances this remark is correct because Jupiter is still very close to its second stationary point which occurred about February 22. The stationary points are always given in Babylonian ephemerides of the outer planets.

"*Two fingers,*" probably corresponding to 0;5° each. This is the norm both in Babylonian and in Greek astronomy (*cf.* Kugler, *SSB* II p. 547 ff. and Manitius II, p. 396).

"*Magnitude 3/2.*" *Cf.* above.

"*Rising at dawn.*" Venus is morning star; elongation 28°.

---

[a] *Cf.* Peters-Knobel, *PCS* p. 139, 140.

Col. VII

"*Two lunar diameters*," probably corresponding to $\frac{1}{2}°$ each. This is the norm used in the Almagest (*cf.* Manitius II, p. 396). The earliest observation which uses lunar diameters and which is quoted in Almagest IX,10 (Heiberg II p. 288) is from the year —264 (Era Dionysius 21). The latest observation was made by Ptolemy himself +140 (Almagest X,1 Heiberg II p. 297).

"*Perigee*." Aries 10 is exactly the perigee of the eccenter of Mercury according to Almagest IX,7 (Heiberg II p. 264).

"*Completed its phase before the 7th.*" Mercury is practically in upper conjunction with the sun, thus completing its phase as morning star. Obviously the text assumes that the upper conjunction will be reached before the 7th, i.e., in less than 24 hours. Because the elongation at the beginning of the 6th is 4°, from ♈ 10° to ♈ 14°, a daily motion of about 5° seems to be assumed. Actually Mercury travels at most about 2° per day, as was, of course, known to the ancients; consequently, it would take the planet more than two days to reach the sun. Hence the statement of the text would only be correct if the longitude of Mercury were about ♈ 12 or 13. This would also bring Mercury in line with the outer planets in their comparison with modern computations (*cf.* the summary below).

"*Decan* θουμονθ." The corresponding decan (2nd in ♏) in the Egyptian lists is tms (n ḥntt).

"*Meridian at right angles*" is only qualitatively correct as ♌ 23;40° culminates when ♏ 18° rises. *Cf.* Bouché-Leclercq, *AG* p. 259 note 2.

Fragm. Col. IX

"*1. Mercury.*" The purpose of the enumeration of four planets is not clear. Perhaps it refers to Mercury and sun both being in ♈ (10 and 14;6 respectively) and Saturn and Venus both in ♓ (5;59 and 16;4 respectively). The subsequent sentence "in order that you may not make a mistake in the arrangement of the pairs(?)" perhaps gave the explanation. Unfortunately, the last word is written by an abbreviation beginning with δι, which might be διπλόων.

"*The horizon has the ratio of 7 to 5.*" This is the standard ratio of longest to shortest day for Lower Egypt (*cf.* p. 5 table 3).

The planetary positions were computed with P. V. Neugebauer's *TACh* and *ACh*. For the Horoscopos and Midheaven more accurate computation would be without interest in view of the approximate data of the text. Hence these positions were computed (for the moment when 3 seasonal hours of night were completed) according to the rules given by Ptolemy, Almagest II,9 (Heiberg p. 143 f.), and by Theon in his introduction to the Handy Tables (Halma p. 40 ff.). This was furthermore combined with different schemes for the length of daylight, computed with System A and System B as well. The results vary only inconsiderably around the values given in the above table.

The planetary longitudes show rather irregular deviations even if one assumes that Mercury should have a longitude of ♈ 12 or 13 which makes the deviation + 1 or 2. If one excludes the moon, Mars, and Mercury from the comparison because in these cases the inaccuracy of ancient tables might be felt most, one obtains the deviations + 5, + 5, + 5$\frac{1}{2}$, + 8$\frac{2}{3}$. It seems doubtful whether this can be taken as an indication that the vernal point was called Aries 8°. A vernal point in Aries 5° would be obtained, however, if we follow the theory of "Trepidation" mentioned by Theon as the theory of the old astrologers. According to this theory the vernal point recedes 1° in 80 years from ♈ 8° beginning in 159 B.C. Our horoscope being just three 80-year periods later would thus consider ♈ 5° as the vernal point.

Beside the longitudes, relations to stars in constellations are also given. This reminds one of the Babylonian "Almanacs" and "Diaries" where not only longitudes but also relations to "normal stars" are entered.[7]

In the case of the sun, the moon, and the meridian, the description is undetermined enough so as to make agreement possible, though it is not easy to see how the moon can be "mounting the Back of Taurus" when its latitude is about —4$\frac{2}{3}$°. Jupiter is said to be "2 fingers (= 0;10°) north of the bright star in the Back" of Cancer. According to computation, this would be ♋ 0;24 β = +0;20 or ♋ 6 according to the text. A star which seems to fit this description is ζ Canc which is called λαμπρός by Ptolemy (Heiberg II p. 94,12) and whose coordinates are given as ♋ 5;40 β = —2;40 magn. 4 in the Almagest, actually corresponding to ♋ 4;52 β = —2;27 in A.D. 100, and to ♋ 1;42 in 130 B.C.[8] Another possibility is μ Canc ♋ 2;40 β = +1 magn. 5 in the Almagest, ♋ 3;3 β = +1;8 in A.D. 100, and ♊ 29;53 in 130 B.C.[9] Neither one of these identifications is very convincing.

Saturn, in ♓ 6°, is said to be "descending from the Swallow-Fish." Because the "Southern Fish" is mentioned in connection with Venus, Bouché-Leclercq and, following him, Boll[10] have assumed that the "Swallow-Fish" is the same as the "Northern Fish." This is explicitly confirmed by the scholion to Aratus 242[11]

---

[7] *Cf.* Sachs [1].
[8] Peters-Knobel, *PCS* p. 61 and p. 83 No. 448.
[9] Peters-Knobel, *PCS* No. 456.
[10] Bouché-Leclercq, *AG* p. 148 note 2; Boll in Roscher VI, 979, 46 ff.; Boll, *Sphaera*, p. 196 f.
[11] Maass, *CAR* p. 382, 27 ff.

where this name is declared to be of Babylonian origin. According to the Almagest the Northern Fish belongs more to the first half of the zodiacal sign, in agreement with our computed longitude. The latitude, however, of almost −2° does not agree with the picture to be derived from the Almagest.

Venus, in ♓ 16, is said to be in the "Southern Fish, two lunar diameters from the star in the Connecting Cords." The star in the Cords is usually taken to be α Pisc,[12] whose coordinates in the Almagest are ♈ 2;30

treatise of "Hermes Trismegistos," discovered by Gundel, with a longitude of ♓ 14;26.[14]

Finally Mars is found in ♒ 16, at the "Star in the Cloak, called Ganymede, homonymous with the whole constellation, far to the East." That Aquarius is represented as Ganymede is well known.[15] In the Cod. Vat. Graec. 1291[16] the cloak reaches from Ganymede's right shoulder to his left hand which is stretched out towards Capricorn. Thus the most conspicuous part of the cloak is seen at the western end of the constellation. Hence

SUMMARY OF COMPUTATION

	Text		Computation		Text − Comput. Longitudes	
	Titus 3 III(a) 6 3ʰ night = April 1 = IX(e) 1/2		+ 81 March 31 21ʰ Alexandria time			
☉	♈ 14;6	Flank of Aries	♈ 9;14		+ 4;52 ≈	+ 5°
☾	♉ 13,0,3,36	Back of Taurus	♉ 10;44	β = −4;40	+ 2;16	+ 2
♄	♓ 5;59	Swallow-Fish	♓ 0;49	− 1;49	+ 5;10	+ 5
♃	♋ 6,0,0,10	Bright star at the Back of Cancer	♋ 0;24	+ 0;28	+ 5;36	+ 5½
♂	♒ 16;3	Ganymede, star in the Cloak, far to the East	♒ 13;8	− 1;22	+ 2;55	+ 3
♀	♓ 16;4	Southern Fish, star in the Connecting Cords	♓ 7;27	+ 3;48	+ 8;37	+ 8⅔
☿	♈ 10		♈ 10;56	− 0;16	− 0;56	− 1
H	♏ 18		♏ 19;30			
M	at right angle		♌ 25;30			

TABLE 8.

β = −8;30 magn. 3, actually ♈ 2;55 β = −9;10 in A.D. 100 and ♓ 29;45 in 130 B.C.[13] This is in very bad agreement with the longitude ♓ 16 of the text and with the computed position ♓ 7;25 β = +3;48 even if one adds or subtracts 2 lunar diameters = 1°. Perhaps the star d Pisc is meant, near the tail of the Southern Fish. The Almagest gives for this star ♓ 11 β = + 5;45 magn. 6, actually ♓ 11;34 β = +5;27 in A.D. 100 and ♓ 8;24 in 130 B.C. It also appears in the star catalogue of the

our text must refer to the end of the cloak near the body, slightly east of the middle of the sign.

ASTROLOGICAL COMMENTARY

Col. I

"*Phase.*" This term is used here in a much looser sense

---

[12] Aratus 244; Almagest, Heiberg II p. 126, 10.
[13] Peters-Knobel, PCS No. 692.
[14] Peters-Knobel, PCS No. 682. Gundel, HT, p. 159 No. 67.
[15] Cf. Bouché-Leclercq, AG p. 146 note 3 or Roscher VI, 975, 4 ff.; 976, 50 ff. Boll, *Sphaera*, p. 389 considers "Ganymede" as a borrowing from the "Sphaera Barbarica."
[16] Boll [4].

than ordinarily (*cf.*, Introduction p. 10). Mercury at superior conjunction is said to have "completed its phase." This can only mean that its "phase" as morning star is ended. This is confirmed by the other cases which obviously concern the "phases" mentioned here in the introduction of the horoscope. Saturn is said to rise before dawn and Venus is mentioned as morning star but neither one of these planets is in a specific position of a "phase" in the strict sense of this term. For the astrological significance of the phases *cf.* Bouché-Leclercq, *AG* p. 309, IV; also p. 113 note 1.

Col. III

"*Sign of Mars.*" "Sign" (ζῳδιον) is here used for "house" (οἶκος).

"*Northerly sign.*" This classification seems to be unknown, even in lists like *CCAG* 1 p. 164 f. or *CCAG* 4 p. 152. See, however, the "northerly triangle" ♈ ♌ ♐, *Tetrabiblos* I, 18. *Cf.* also Bouché-Leclercq, *AG* p. 196 note 2.

Col. IV

"*Like gold.*" Bouché-Leclercq, *AG* p. 314 note 2; "accidentellement peut-être"; the ordinary metal associated with the moon is silver. *Cf.* Col. VI. *CCAG* 1 p. 173,2 and 2 p. 82,7 the moon is called χρυσάμπυξ. For the importance of conjunctions between the moon and fixed stars *cf. CCAG* 6, p. 69 1 f.

Col. V

"*Commanding.*" According to the *Tetrabiblos* Cancer is a commanding sign, but not so according to Vettius Valens.

"*Second station.*" For the astrological significance of this point *cf.* Bouché-Leclercq, *AG* p. 429 note 1.

Col. VI

"*Like crystal.*" This seems to be an unknown association with Venus. In al-Bīrūnī's *Astrology* p. 243 crystal is mentioned (among others) with the moon, gold with Venus, exactly opposite to our text. The moon representing crystal is again mentioned as a Hindu doctrine in al-Bīrūnī, India LXXX (Sachau II p. 214). *Cf.* also the table in Boll, *FS* p. 20.

Col. VII

"*It will dominate the theme.*" Obviously the fact is decisive that Mercury is in superior conjunction with the sun. According to *Tetrab.* I,24 a planet is most powerful if its true motion is greatest and this is the case for superior conjunction. *Cf.* also Bouché-Leclercq, *AG* p. 113 note 1 and p. 406 note 2. It looks as if the bad influences (e.g., Horoscopos in Scorpio) were purposely ignored.

"*The rudder of them all.*" *Cf.* Bouché-Leclercq, *AG* p. 258 note 4.

"*Lot of Fortune.*" Sagittarius is the "house" (or "sign") of Jupiter and the "triangle" ♐ ♈ ♌ is at night assigned to Jupiter. The elongation of the moon is ♉ 13 — ♈ 14 — 29°. Adding this to the Horoscopos ♏ 18 we obtain ♐ 17. Subtraction would lead to ♎ 19 but only "ignorant" people do not know that addition is proper for the night, subtraction for daytime. *Cf.* Bouché-Leclercq, *AG* p. 291 note 1.

"*(counting) inversely.*" This translation of ἀνάπαλιν is based on Bouché-Leclercq's remark (*AG* p. 291 note 1) which relates our passage to "Nechepso's" rule for the computation of the Lot of Fortune. *Cf.* Introduction p. 8.

"*back (in the sign) of Venus*" not in the "Lot of Venus" as Bouché-Leclercq, *AG* p. 291 note 1 has it.

Fragm. Col. IX

"*Pregnancy 276 days.*" Vettius Valens I, 23 and 24 gives an extensive method for finding the moment of conception from the date of birth. As the mean value of the duration of pregnancy, 273 days is assumed. The accurate duration is computed by means of the elongation of the moon from the Horoscopos. Neither the mean value nor the accurate value of Vettius Valens' theory agrees with our text. There exists, however, a doctrine which considers the third day after birth as the decisive day (*cf.* Bouché-Leclercq, *AG* p. 488 note).

The Egyptians assumed a duration of pregnancy of 275 days (*cf.* Jonckheere, [1] p. 44 f.).

BIBLIOGRAPHICAL NOTES

Photographs of columns I to VIII: P. Lond. Facsimiles Pl. 77-79; of columns IX and X: Macnaughton, *SEC* pl. XVII (facing p. 332) taken from J. C. S. Thompson, *The Mystery and Romance of Astrology*, New York, Frederick A. Stokes Company, 1930; of column IV, 72-85: W. Schubart, *Griechische Palaeographie* (*Handb. d. Altertumswiss.* 1,4,1 München 1925) p. 122 fig. 81.

This text is very frequently referred to in the literature, e.g., by Bouché-Leclercq, *AG* p. 196,3 213,e *et passim*; Wilcken, *Ostr.* II p. 792 f.; Gundel, *DD*. p. 48, p. 409, etc.

No. 95

P. Lond. 98

INTRODUCTION

This papyrus became famous through the fact that its verso bears the Funeral Oration of Hyperides. Its recto contains an elaborate Greek horoscope cast for the year

A.D. 95, followed by an astrological treatise in Greek and early Coptic.

The papyrus originally belonged to the Stobart collection and was purchased in Egypt in 1853/1854, probably in Thebes. Griffith suggested that it was found together with the wooden tablets which contain Demotic planetary tables for at least the years A.D. 71 to 132.[1]

manuscripts of normal Coptic are datable so soon after." We are greatly indebted to our Egyptological colleagues, Professor J. Černý and Professor R. A. Parker, for putting at our disposal their results in studying this extremely difficult Coptic document. The late Dr. P. Kahle, Jr., made a careful copy of the original in the British Museum and Dr. T. C. Skeat, Keeper of the

SUMMARY OF ASTROLOGICAL DATA FROM NO. 81

	Longitude	Phase	House of	Term of	Triangle of	ϲτοιχ. of	Exalt.(+) Depr.(−) of	male or female	solid	material	north.	constell.	Decan	Dodecat.	magn.(?)	command.	Conclusion
☉	♈ 14,6	equin.	♂	☿		♃	+ ☉	m.			n.	×	×	♍			
☾	♉ 13 $\frac{1}{1000}$	cresc.	♀	☿			+ ☾	f.	s.	gold		×	×	♍			
♄	♓ 5;59	morn. rising	♃	♀			+ ♀					×			4/3		
♃	♋ 6;0,0,10	2nd station	☾	♂			+ ♃ − ♂					×			3/2	c.	
♂	♒ 16;3		♄	♃	☿							×					
♀	♓ 16;4	rising at dawn	♃	☿			+ ♀			cryst.		×					
☿	♈ 10	perigee															ruler of the theme
H	♍ 18		♂	☿	☾								×				
M	♌																
Lot of Fort.	♐ or ♎		♃ ♀		♃												

TABLE 9.

The horoscope began with an introduction, now almost completely destroyed, followed by the enumeration of positions and astrological characterization of sun, moon, and the five planets. On several occasions Demotic signs are used in the spelling of names of decans. The end of the horoscope proper gives the four cardinal points and four "Lots." This brings us to the end of line 82.

What now follows (lines 83 to 101) is almost entirely illegible because there are breaks in the papyrus and overlaid patches, and all the lines have been erased or cancelled. A few less conjectural readings, however, indicate that we have to do with the periods of life and their subdivisions.

With line 102 seems to begin a detailed discussion of the single periods of life, first in Greek, then, from line 118 on, in Old-Coptic with some Greek insertions. Griffith [2] p. 81 note 1 quotes the following remark by Crum: "What strikes me in these texts is not their likeness to Coptic, but their extreme unlikeness, while

Department of Manuscripts, provided us with infrared and ultraviolet photographs. Professor Eric G. Turner of the University of London collated once more the whole text. In spite of this combined effort much remains enigmatic in this last section.

A. TRANSCRIPTION OF LINES 1 TO 82

Col. I

1. ....]ϲ̣ηϲ
2. ....].ϲ
3. ....]...
4. ....]...
5. ....]ϲ̣το
6. ....].ωνα
7. ....].θεου
8. ....]ε̣ιϲ
9. ....].χο

1]ϲηϲ.: followed by one more letter(?) ending in a long stroke slanting down to the right.

[1] *Cf.* below p. 173, tables S.

10. ....].λο
11. ....]...
12. ....]....
13. ......]..
14. ....]. ωι
15. .. της ημε]ρας
16. [ηλιος ευρισκομεν την κε]ινησιν ποι
17. [ουμενον εν ....]
18. [.................]
19. [....... ου το δωδεκατη]μοριον εν λεοντι
20. [μοιρων .. λεπτων .. οικωι ιδιωι] οριοις ερμους
21. [των δε λϛ λαμπρων] ωροσκοπων χεν
22. [σεληνη ευρ:σκομεν τ]ην κει
23. [νησιν ποιουμενην εν .....] μοιρ
24. [..............]πλατει
25. [..........β]αθμοις
26. [των δε λϛ λαμπρων ωροσκοπω]ν ου ονο^μ
27. [.... ου το δωδεκατημοριον εν c]κορπιωι
28. [μοιρ]ων η̄ λεπτων νβ̄ οι[κωι αρεως ταπιν]ωματι σελη
29. [οριοι]ς αφροδειτης δεκ[ανος ......]τι

Col. II

30. φαινων τον του κρονου ευρισκομεν την κεινησιν
31. ποιουμενον εν παρθενωι μοιρης ᾱ οικου ερμους
32. υψωματι ερμους ταπινωματι αφροδειτης οριοις
33. ερμους ου το δωδεκατημοριον εν παρθενωι οικωι
34. ερμους των δε λϛ ωροσκοπων ου ονομα ετdε
35. των δεκανος β̄ ου ονομα [ονες]dη βικωτ ——
36. φαεθων τον του διος ε[υρισκομεν] την κεινησιν ποι
37. ουμενον εν παρθενωι [μοιρων .. λεπ]των κ̄ οικωι
38. ερμους υψωματι ερμου[ς ταπινωματ]ι αφροδιτη οριοις
39. αφροδειτης ου το δωδε[κατημοριον εν] τοξοτηι μοιρων ιγ
40. οικωι ιδιωι οριοις ερμο[υς των δε λϛ λα]μπρων ωροσκοπων
41. ου ονομα φουτdε τω[ν δεκανος .....] αρωι αρωι
42. αρης πυροις ευρ[ισκομεν την κεινη]σιν ποιουμενον
43. εν διδυμοις μοι[ρων .. λεπτων.]ā οικωι ερμους ορι[οις
44. οριοις ερμους . [......μηκει(?) πρ]οστιθωι τοις αριθμοις
45. των δε λϛ λαμπ[ρων ωροσκοπ]ων ου ονομα αρου >
46. [ου το δωδεκατημοριον] εν λεοντι μοιρων κ̄
47. οικωι ηλιου οριοις ε[ρμου των δε]κανος ā ρεμενααρε
48. φωσφορους τον [της] αφροδειτης ευρισκομεν την κειν
49. σιν ποιουμενον ε[ν ταυρ]ωι [μοιρων ..] οικωι ιδιωι υψωματι
50. σεληνηι οριοις δ[ιος ου το δωδεκατημ]οριον ενληγει εν
51. υδρηχωι μοιρων γ̄ οικωι [κρονου οριοις ερ]μους των λϛ λαμπ
52. ωροσκοπων ου ονομα χω[...... δεκαν]ος ᾱ ου ονομα αρου >
53. στιλβων τον του ερμου ευ[ρισκομεν την] κεινησιν ποιουμενον
54. τον του ερμου εν κριωι περι [μοιρων .. οι]κωι αρεως
        υψωματι
55. ηλιου ταπινωματι κρονου ορι[οις ....]των λϛ λαμπρων
56. ωροσκοπων ου ονομα χεντα[.. ου] το δωδεκατημοριον εν
57. ληγει εν τοξοτηι μοιρων κ̄ οικ[ωι διο]ς οριοις αφροδειτης των
58. δεκανος β̄ κατ/ κονατ/

Col. III

59. τον δε ωρονομον ωροσκοπον προσδει δι οληϊ
60. την [αν]αλογιαν εκ του αναφορικου προς κλεψυ
61. δραν εκπειπτει εν καρκινωι περι μοιρων κ̄ε οκωι
62. σεληνης οριοις διος ου το δωδεκατημοριον εν
63. ταυρω[ι] μοιρων κ̄ε οικωι αφροδειτης υψωματι
64. σεληνης των [λ]ϛ λαμπρων ωροσκοπων ου ονο^μ
65. ρεμεναχ.[....] των δεκανος β̄ κνουμε ḥρακνου^μ
66. μεσουρ[ανημα ε]ν κριωι μοιρων ῑ λεπτων λ̄
67. οικωι αρε[ως οριοις αφ]ροδειτης ου ονομα ϲρωι
68. υπογην [εν ζυγωι οικ]ωι αφροδειτηι υψωματι κρονος
69. ταπινωμ[ατι ηλιου οριο]ις ερμους ου ονο^μ στωμα
70. δισικον [εν αιγοκερωι] μοιρων κ̄ε οικωι κρονου οριοις
71. ιδιοις [.... ου ονο]μα σιϲρωι
72. ο πρ[ωτος κληρος της γ]ενεσεως ο εστιν αγαθος δαιμων
73. [εν ταυρωι μοιρων .. λε]πτων λδ οικωι αφροδειτης υψ
74. υψω[ματι σεληνης οριοις] ερμου και αρεως
75. β̄ κλη[ρος ....]εν κριωι μοιρων κγ λεπτων [..]
76. οικωι [και ο]ρ[ιοι]ς αρε[ως]
77. ο τριτο[ς κ]ληρος περι [θα]νατων εν υδρηχωι μοιρων
        [..]λε..
78. οικωι κ[ρον]ου ταπινω[μα]τι ερμου οριοις διος και αρε[ω]ς
79. ο φυσ[ικος ο]ικοδες[ποτης] της γενεσεως αφροδειτης και
        ερμους
80. εν τω[ι κεν]τρωι αφρο[δειτ]ης αναφερεται εν τωι κεντρω[ι]
81. μεσ[ουρανημ]ατ]ος
82.   [αποτελεςματι]κα των ε αστερων περι ζοης

34 ετdε: d Demotic sign. *Cf.* Griffith [1] p. 74, and Crum [1].
35 [ονες]dη βικωτ: Professor Parker suggested this restoration as a possible rendering of the corresponding Egyptian name wšȝty bkȝty.
41 φουτdε: d Demotic sign. *Cf.* Griffith [1] p. 74.
45 and 52 αρου >: Last sign is perhaps Demotic though we did not succeed in identifying it. It is not mentioned in Crum [1] though in his text (about A.D. 150) an unknown sign <· also occurs, written often in the inner part of words as well as at the end.
56 το: τ written over δ.
61 οκωι: sic for οικωι.
65 ḥρακνουμ: initial letter Demotic ḥ.
69 στωμα: μα doubtful; perhaps Demotic?
80 αναφερεται: ν written over φ.
82]κα: αποτελεςματι]κα or χρηματιςτι]κα.

B. TRANSLATION OF LINES 1 TO 82

Col. I

Lines 1 to 27. Only the last letters of these lines are visible. The beginnings are now covered by brown paper of the mounting of the papyrus.[2] Only in the lines 28 and 29 has transparent paper been used and hence also

---
[2] This detached part of the text is not shown in the facsimile reproduction of P. Lond. 98. We owe all information concerning this fragment to Dr. T. C. Skeat.

parts of the beginning can be read. The middle part of the whole first column is broken away.

The last line of the introduction (line 15) can probably be read as ending in "... hour of the d]ay." The two following sections concern the sun and the moon and can be restored to some extent by means of the pattern of the subsequent sections for the planets.

16. [The sun we find] moving
17. [in ....]
18. [..............]
19. [... whose dodekate]morion is in Leo
20. [.. degrees .. minutes, his own house,] terms of Mercury,
21. [and of the 36 bright] Horoscopes khen.
22. [The moon we find] mo-
23. [ving in ....] degrees
24. [...................] latitude
25. [.............] steps;
26. [and of the 36 bright Horoscopes] the one named
27. [......, whose dodekatemorion is in] Scorpio
28. 8 degrees 52 minutes, in the [house of Mars,] depression of the moon,
29. [terms] of Venus, decan [......]ti.

Col. II

30. Phainon, the (star) of Saturn we find moving
31. in Virgo 1 degree, house of Mercury,
32. exaltation of Mercury, depression of Venus, terms
33. of Mercury; whose dodekatemorion is in Virgo, house
34. of Mercury, and of the 36 Horoscopes the one named Etdje;
35. whose decan is the 2nd, named [Usdj]e-Bikot.
36. Phaeton, the (star) of Jupiter we find moving
37. in Virgo [.. degrees] 20 minutes, house
38. of Mercury, exaltation of Mercury, depression of Venus, terms
39. of Venus; whose dodekatemorion is in Sagittarius 13 degrees,
40. his own house, terms of Mercury, and of the 36 bright Horoscopes
41. the one named Phoutdje; whose decan is the [2nd], Aroi Aroi.
42. Mars Pyroeis we find moving
43. in Gemini [... degrees ... minutes]; house of Mercury, {terms}
44. terms of Mercury; [... in longitude] he is adding to (his) numbers;
45. and of the 36 bright Horoscopes the one named Arou;
46. [whose dodekatemorion] is in Leo 20 degrees,
47. house of the sun, terms [of Mercury; whose] decan is the 1st, Remenaare.
48. Phosphorus the (star) of Venus we find
49. moving in Taurus [... degrees], her own house, exaltation
50. of the moon, terms of Jupiter; whose dodekatemorion ends in
51. Aquarius 3 degrees, house of [Saturn, terms] of Mercury, of the 36 bright
52. Horoscopes the one named Cho[...]; whose decan is the first, named Arou.
53. Stilbon, the (star) of Mercury we find moving
54. {the (star) of Mercury} in Aries about [... degrees], house of Mars, exaltation
55. of the sun, depression of Saturn, terms of [.......]; of the 36 bright
56. Horoscopes the one named Chenta[chori]; his dodekatemorion
57. ends in Sagittarius 20 degrees, house of [Jupiter]; terms of Venus; whose
58. decan is the 2nd, Kat/ Kouat/.

Col. III

59. And the governing Horoscopos is fixed in complete
60. conformity (with the rules) of the rising-times (as well as) with the water clock;
61. it falls in Cancer about 25 degrees, house
62. of the moon, terms of Jupiter; whose dodekatemorion is in
63. Taurus 25 degrees, house of Venus, exaltation
64. of the moon; of the 36 bright Horoscopes the one named
65. Remenach [......]; whose decan is the 2nd, Knoum Hraknoum.
66. Midheaven in Aries 10 degrees 30 minutes,
67. house of Mars, [terms] of Venus; (the bright Horoscopos?) named Sroi.
68. Lower Midheaven [in Libra,] house (of) Venus, exaltation (of) Saturn,
69. depression [of the sun], terms of Mercury; (the bright Horoscopos?) named Stoma.
70. Setting [in Capricorn] 25 degrees, house of Saturn,
71. his own (i.e. Saturn's) terms; [....] (the bright Horoscopos?) [nam]ed Sisroi.
72. The first Lot of the nativity, that is Agathos Daimon,
73. [in Taurus .... degrees] 34 minutes, house of Venus, {exaltation}
74. exaltation [of the moon, terms] of Mercury and Mars.
75. The second [Lot ....] in Aries 23 degrees [..] minutes,
76. house [and terms] of Mars.
77. The third Lot, concerning death, in Aquarius [...] degrees [..] minutes,
78. house of Saturn, depression of Mercury, terms of Jupiter and Mars.

79. The [natural] ruler of the nativity (is) (the star) of Venus and, (the star) of Mercury being
80. in the center, (the star) of Venus is rising (toward³) the center
81. of Mid[heaven ....]
82. [the influen]ces of the five stars concerning life.

C. LINES 83 TO 117

The lines 83 to 104 are almost completely obliterated by erasures. The following indicates what words still can be read with some degree of certainty.

83.⁴ πρω[τος .... αρ]ροδει[τ]η ....
84. (ετους) ς̄ [................ κρονο[υ ....
85. μερ[ιζει ....
86. μερ[ιζει ....

Col. IV

90. ετ[ους ...... εως] ετους κ̄ε μηνας β̄ ημερας κ̄ε
93. ....αφροδειτης και ερμου μηνας [..]
95. .... σεληνης
96. .... αφ.ερουμε
98. ... χρηματ[ιζει ....
99. ....]η̄ μηνης ᾱ
100. .... απο ....

In spite of the bad condition of the text it would seem clear that this section is dealing with the periods of life (cf. particularly line 90) and similar subdivisions. This is supported by the occurrence of frequent overlines, marking numbers, though the numbers themselves are usually obliterated.

In line 102 the first period of life is mentioned. The second period begins with line 116 to be continued in the Coptic text (lines 118 ff.). There are erasures and breaks in the Greek lines, and where the reading is clear, it is rather incoherent grammatically. Our text of these lines is as nearly complete as we can make it but our translation is a free version.

TRANSCRIPTION

102. [..............] πρωτος χρηματι[ζει
103. απο [ετους ....] .... ς̄ .... [....]..
         αφ[ροδ]ειτης
104. κλ[ηρος ....
105. ε..ρ... τερπνων εαν δε δουλος ην [ελευθερωθησεται] ..
106. πλουσι ...[.....] εαν δε πλουσιος π[λουσιωτε]ρος
         γενησεται και σω

---

³ The literal translation of the text "Venus rises *in* the center" makes no sense since Venus is in the sign which precedes Midheaven.

⁴ Near the beginning of the lines 83 to 89 a small strip from another papyrus is pasted over the original (perhaps an ancient repair). This narrow strip contains one letter or two in each of five lines which have nothing to do with the original text which shows in all columns a narrower spacing of lines.

107. μ[ατων .... οπο]ιων κυριευσι επιθη[σει] οσα και
         η γενεσις σημαι
         γενωνται
108. νει φυτειαν δε οποιουν..α καλως κα[ι τε]κνων
         ποιουνται υπο τινος
109. κοιτης κα..κα...α φυσις γονεισιος γονεις και
         συστασεσιν υπερεχων
110. γειρησεται εισχ[ωρ]ουσιν οπου κοσ⟨μος⟩μου κρυεμος
         η χρωματων
111. η πα[θων ....τ]αυτα δε αποτελει αρης γενηται
         επαναφερη
112. [αφροδε]ιτης και κρονος τριγωνιζει ψυχρος περι των
         γυναικων
113. [.......] υπο την αρχην ψογους αοιμους κινδυνευ[σε]ι
         ου ονειρος
114. [....ο]υς κακοπαθησεται και ξενιτευει κα[......]αν
         ο γαρ πρωτος
115. [κληρος] απολυς η δευτερα ομοιως απολυς κ[αι η τριτη
         ο]μοιας απολυς

FREE TRANSLATION

The first period Venus rules up to 6 [years 9(?) months 25 days. Its position in Lot 1, Agathos Daimon, indicates well being and abundance] of pleasure. If he is a slave [he will be set free; if he is poor he will become] rich; if he is rich he will become richer and [whatever slaves] he is master of he will increase, as the nativity indicates; and [his crops(?) of whatever sort they are] will do well; and the children from whatever union born will have inborn in them a fine nature like their parents. And he will have excellent associations extending everywhere. He will suffer from cold or fevers or other illnesses; such are the things that Mars portends, . . . Venus is in Epanaphora. And Saturn is in trine to Venus (and portends that) he will be cold as regards women. He will be tried before the magistrate for unspeakable accusations . . . he will suffer miserably and will live a miserable life abroad. For the first lot is a separation(?), the second likewise a separation(?) and the third likewise a separation(?)

D. THE COPTIC TEXT

by

J. Černý and R. A. Parker

The discussion of the second period of life probably began in line 116 with the statement that Mars(?) decides the second period, which, according to the next line,⁵ extends "from year 6 month 9 (or 5) day 25 to year 25 [month 2 day 25]."⁶ In the next line the Coptic text begins. The occasional Greek sentences which are

---

⁵ One can read with some security in line 117: ...].. (ετους) ς̄ μηνας θ̄ [ημ]ερας κ̄ε εως ....[....].... χρηματιζει.
⁶ Cf. below line 126.

inserted in the Coptic are also written by a different hand; here they are printed in italics.[7]

The following translation has been published by us in an article on "The Old Coptic Horoscope" in *JEA* 43 (1957) p. 86 to 100 to which we refer the reader for the text itself[8] and our commentary.

118. [If] he had the Sun as friendly star (and) Mercury, and they were at enmity to this st[ar . . .], it shall not be
119. . . . (effective?) . . . as hostile star. The two gods . . . (are . . .) during his life.
120. [If] (they were) in akhĕlŏ, they shall make twenty-four evils against him . . . of his life (and) his occupation,
121. . . . after thirty-five (evils) from it. If he had M[ercury itse]lf as his krahtouŏt,

Col. V

122. his growth(?) shall be established in its likeness, whether(?) it be against his good or his evil.
123. If there was an evil star as his krahtouŏt, he shall . go to . . . man(?)
124. or god, ⟨or⟩ he shall go into misfortune, or (he shall) see a man's death.
125. It shall come through him or through his snake. *Third period: Jupiter decides.*
126. *From year 25, month 2, day 25 until year 34, month 5, day 24*
127. *Jupiter decides.* If Jupiter (was) a hostile star on the ⟨day⟩ of his birth,
128. perhaps he shall pass by (— neglect) his wife, or be at enmity to her, or his children shall
129. misbehave in turn, or he shall part (from them) in mercantile business. If there was an evil star
130. . . . , a wife shall not make a term for him ⟨for ever⟩, a child shall not come to him
131. for ever. Evil shall happen after him. They shall . . . hireling
132. . . . and [t]hey shall cause the S[un] to come again, it shining forth there opposite Jupiter,
133. and a wife shall burn unto (the) hair. If a good star gave to him an omen, you should set
134. it against me. *Fourth period: [Mars] decides from year 34, month 5, day 24*
135. *until year 54, month 10, day 4.* If the Moon was a friendly star for him on the day
136. of his [birth], . . . shall . . . upward. After a closing as friendly star for him on the d[ay]
137. of his birth, pe[rhaps it was](?) a friendly star for him again. The said man, the half of his . . .

---

[7] ⟨ ⟩ indicate emendations.
[8] On pl. XI (which is printed upside down) read Col. IV and Col. VI instead of Col. 1 and Col. 3 respectively and on pl. XII read Col. V instead of Col. 2.

138. shall . . . against him. His evils shall multiply. A woman of . . .
139. . . . [shall] . . . him, and his heart shall ⟨become good⟩ through her. He shall take a name(?) . . .
140. . . . service of kingship . . . his kindred, he causing them to appear as those who
141. take a wreath of gold to their head(s) in the years named, and he shall speak and they shall act at his voice
142. and he shall see a law which is great to his heart. If an evil star was coming up
143. after them, against his good and for his evil, a woman shall cause him to take shame, or he shall be
144. hunted(?) from year 42 upwards. A woman shall be to him. He shall take one
145. until year 94. He shall see a [wife's] death or shall be parted from her and ⟨they⟩ shall examine him
146. while he . . . and . . . of his shall make his evils or . . . of his.
147. If a good star was . . . , he shall escape his place of confusion . . .
148. . . . to them . . . and they shall cast 31 winters upon them(selves).
149. He shall take counsel from a woman and [perhaps she] shall take counsel from him. If he had
150. a good star take closing on the d[ay] of his birth, perhaps he shall see (the) death
151. of a man who is near to him ⟨or⟩ (the) death of a relative (and) he shall not make new profit with him
152. (but) his heart shall become good for a wife (and) send him to a child from year 42 {from year
153. 42} again and according to the way with him which . . . before it is decided for his good (or)
154. his evil, and his ⟨heart⟩ shall become good for his life and he shall not have died in a past hour (but shall)
155. see . . . child . . . All these which are below,

Col. VI

156. you shall set the[se . . .
157. . . . year . . . life. He shall . . .
158. death in his presence, They shall make . . .
159. for six months. He shall see . . .
160. going away to a town . . .
161. God said: one (woman) and . . .
162. become good for one (woman). His heart shall l[ove . . .
163. six. His life shall come to . . .
164. through him. His ⟨heart⟩ shall become good . . .
165. [pi]ty . . . in his heart . . .
166. He shall set . . .

167. year 30 . . .
168. Jupiter . . . day . . .
169. he has . . . in flesh. His . . .
170 to 172. . . .
173. . . . to his hand . . .
174. his life, and he shall fly . . .
175. . . . and he shall drink his . . .
176. and he shall take a wife and she shall leave . . .
177. . . . against him. She shall . . .
178. year 42 . . .
179. the number. He shall make . . .
180. West. If there was . . .
181. he shall make . . .
182. come up . . .
183. pity . . .
184. someone . . . to him . . .
185. his dream. He . . .
186. again. I shall become . . .
187. a scorpion was not dumb . . .
188. shall live . . .
189. He shall . . .

E. ASTRONOMICAL COMMENTARY

Diagram: pl. 1

The first section undoubtedly contained, beside the general introduction, the date of the horoscope. Line 15 ends in ρας, to be restored to τῆς ἡμέρας, showing us that we are dealing with a daytime birth. This, however, could have been deduced from the subsequent sections. We know that Mercury was in Aries, Venus in Taurus. Consequently the sun must have been in or near Aries. We know furthermore that Aries was in the meridian. Combining this with the previous information we know that the time was about noon.[9] Thus the accessible fragments of the first section do not bring us any farther toward the date than we would be without them.

Consequently, we must try to date the text on purely astronomical grounds within the historically plausible limits of the first or second century A.D. To this end we must first of all determine the longitudes of the superior planets.

Saturn is said to be in ♍ 1. This is in agreement with two other pieces of information, namely (a) that Saturn is in the terms of Mercury, which extend from ♍ 0 to ♍ 7, and (b) that the dodekatemorion lies also in ♍ which limits the original position to the first two degrees of ♍. Consequently this element is not subject to any doubt.

Jupiter is also in ♍, the number of degrees being lost. The terms are the terms of Venus; thus Jupiter must have a longitude between ♍ 7 and 17. The corresponding dodekatemorion would then fall between ♐ 1 and ♈ 11. The text gives ♐ 13 which would require an initial position of ♍ 7;25,42, . . . whereas the text preserves only the fractional part 0;20. Thus it is tempting to restore ♍ [7;]20. Unfortunately, however, the value ♐ 13 for the dodekatemorion is not fully reliable, because it is said to fall in the terms of Mercury and these extend from ♐ 17 to 21, thus excluding ♐ 13. Consequently the initial position might also have been ♍ 8;20 which would lead to ♐ 18 as dodekatemorion.

Similar discrepancies are observable in the case of Mars which is located in Gemini, the degrees again being lost. The terms of Mercury restrict us to the first 6 degrees of ♊; consequently the extremal value for the dodekatemorion is ♌ 18. But the text gives ♌ 20, which would correspond to an initial position of ♊ 6;9,13, . . .

To sum up we might say that we have for the outer planets the following positions:

♄ :   ♍ 1°
♃ :   ♍ about 7;20° or 8;20°
♂ :   ♊ about 6°

This suffices to restrict the possible dates of our horoscope to the years 95 and 155, other possibilities being about 59 or 60 years distant from these two and being ruled out on paleographical grounds and on the basis of the Coptic linguistic features. We shall return presently to the details when we have the additional elements of sun and moon and the inferior planets at our disposal.

Venus must be in Taurus, as appears from the fact that she is in the exaltation of the moon (♉) and in her own house (♉ or ♎). The terms of Jupiter limit us furthermore to the interval from ♉ 14 to 22. The dodekatemorion is ♒ 3 to which ♉ 21 belongs as original position. There is no reason to doubt the correctness of the value for the dodekatemorion since it falls in the terms of Mercury which cover the first 7 degrees of ♒. Thus ♉ [21] seems to be a safe restoration for the longitude of Venus.

Mercury is in ♈; both longitude and terms are lost. The dodekatemorion ♐ 20 requires ♈ 20 for the initial longitude. Unfortunately the dodekatemorion is said to be in the terms of Venus which extend only from ♐ 12 to 17. To ♐ 12 corresponds ♈ 19;23, . . . as original longitude. Thus ♈ [20] might be slightly too high a value for the longitude of Mercury.

For the moon we only know that its dodekatemorion is ♏ 8;52. This is confirmed by the further information that it falls in the depression of the moon (♏) and in the terms of Venus which extend from ♏ 7 to 17. The original position which leads exactly to this dodekatemorion is ♐ 26;4. This is a basic element for the dating of our text.

Finally we know that the sun had its dodekatemorion

---

[9] This conclusion is confirmed by more accurate computation. As we shall see presently, the sun must have had a longitude of about ♈ 11. The culminating point is given as ♈ 10;40. Thus the time must be very close to noon.

in Leo, the number of degrees being lost. Traces, however, seem to indicate that the terms were the terms of Mercury, thus limiting us to the arc from ♌ 18 to 24. This alone does not suffice to determine the original longitude. Because of the positions of Mercury (in ♈ [20]) and of Venus (in ♉ [21]) the sun must be located somewhere in the interval from about ♈ 5 to ♉ 15. The corresponding dodekatemoria range from ♊ 5 to ♎ 15 and only an original position very close to ♈ 11 leads to a dodekatemorion between ♌ 18 and ♌ 24, namely ♌ 23.

Thus we have to reckon with the following elements

| ☉: | about ♈ 11° | ☾: | ♐ 26;4 |
| ☿: | ♈ 20 or slightly less | ♀: | ♉ 21 |

These elements should determine the date accurately within the year 95 or 155 required for the outer planets.

The sun being in Aries requires a date in or near April. The moon is in the end of Sagittarius during this period in 95 March 17 and April 13 and in 155 March 14 and April 10. The dates in March give solar longitudes in the end of Pisces, thus about 20° less than expected. Hence there remains only 95 April 13 and 155 April 10 for nearer investigation.

For these two dates one finds the following results

Text		95 April 13:	155 April 10:
☉	[♈ 11]	♈ 21	♈ 19
☾	[♐ 26]	♐ 26	♐ 26
♄	♍ 1	♍ 1	♍ 14(!)
♃	♍ [8]	♍ 5	♍ 30(!)
♂	♊ [6]	♊ 6	♉ 14(!)
♀	♉ [21]	♊ 4(!)	♓ 25(!)
☿	♈ [20]	♉ 10(!)	♉ 3(!)

It is clear that 155 is ruled out. For 95 the agreement is excellent for the outer planets [10] but sun, Venus, and Mercury show deviations. All planets and the sun would be very well represented for 95 April 3. Sun, Venus, and Mercury have then practically exactly the longitudes we restored and only Mars is slightly worse (♊ 1) whereas Jupiter and Saturn agree exactly with the text. The moon, however, is completely out of place, in Leo, instead of Sagittarius. In view of this fact 95 April 13 seems the only possible date.[11]

The position of the Horoscopos, ♋ 25, is perfectly possible for an hour close to noon. Given this rising point one obtains ♈ 14;10 as culminating point, using the tables of Almagest II,8 and assuming the latitude of Alexandria.[12] A somewhat better agreement with the text (♈ 10;30) is reached (♈ 11;23,20) if one uses the procedure of Paulus Alexandrinus [13] and the rising times of System A instead of Ptolemy's more accurate tables.

The "governing Horoscopos" is said to be "fixed in complete conformity with the rising-times and the water clock." This remark is very puzzling indeed. Evidently the elements of a horoscope had to be found by computation, and not by observation, especially for a day birth when the planets were all invisible. The rising-times were needed for determining the culminating degree as well as for the rising and setting points when the hour was given. The role of a water clock, however, remains obscure. Our text has a parallel in *Tetrabiblos* III,2 (Robbins) where also instrumental methods for the determination of the Horoscopos and the rising-times are compared. Unfortunately the details are equally obscure in Ptolemy's text, obviously because he assumes them to be well known.

F. ASTROLOGICAL COMMENTARY

Our text represents a quite unusual type among the earlier horoscopes in the detailed character of its data as well as in the addition of long excursus (Greek in col. IV, Coptic in cols. V and VI) on astrological doctrine. In both respects it resembles much more the treatises of the fifth century than the simple factual statements in the bulk of horoscopes on papyrus or even the samples quoted by Vettius Valens. On the other hand the great number of discrepancies between our text and the orthodox doctrine as well as unexplained features might well be due to the fact that we know very little about the details of astrological theory before the second century (Ptolemy and Vettius Valens).

For all seven celestial bodies and for the Horoscopos not only the positions but also the corresponding dodekatemoria are given. Table 10 p. 36 shows these data together with the additional information concerning houses, exaltations and depressions. Discrepancies are indicated by (!) in the explanatory remarks.

No dodekatemoria are given with the remaining three centers nor with three "lots":

M	♈ 10;30	house:	♂
		[terms]:	♀ (♈ 6 to 12)
IMC	♎ (10;30)	house:	♀
		exalt.:	♄
		depr.:	[☉]
		terms:	☿ (♎ 6 to 14)
Δ	♑ [25]	house:	♄
		terms:	♄ (♑ 22 to 26)

---

[10] It is said in II,44 that Mars is "adding to his numbers." Indeed, Mars is moving forward in both cases listed here, and the same holds for 95 March 17, April 1 and 155 March 14.

[11] Blass, *Hyp. or.*⁽²⁾ p. XVIII f. also came to the conclusion that the year 95 is preferable to the year 155. He based this result on planetary positions computed for 95 April 1 and 155 May 15 respectively without taking into consideration what is known about sun and moon.

[12] The tables of Theon would give very slightly less than ♈ 15 but such accurate tables were probably not in existence in the first century A.D.

[13] Cf. p. 10.

36 ORIGINAL DOCUMENTS: No. 95

κλῆρος 1: ἀγαθὸς δαίμων ´[....];34
    house: ♀ (either ♎ or ♉)
    exalt.: [☽]
    terms: ♉ and ♂ (♉ 30)

κλῆρος 2: [..........][14] ♈ 23;[..]
    house: ♂
    [terms]: ♂ (♈ 20 to 25)

κλῆρος 3: περὶ θανάτων ♒ [....]
    house: ♄
    depr.: ☿ (no depr. or exalt.
            in ♒; depr. of
            ☿ in ♓.)
    terms: ♃ and ♂ (♒ 20)

The most serious difficulties arise, however, in connection with the "decans." Apparently in connection with the dodekatemoria decans 1, 2, and [3] are mentioned and names quoted which agree generally well with the Egyptian names of decans. Beside the "decans," however, "36 Horoscopes" are also listed (sometimes with the positions, sometimes with the dodekatemoria) which again carry names obviously related to the Egyptian decans.[14a] Finally these Egyptian names are quoted without further specification for the positions of M, IMC, and Δ.

The hypothesis which at first suggests itself, that the "decans" are the decans of the dodekatemoria, the "36

	Position		Dodekatemorion	
	Text	Remarks	Text	Remarks
☉	[♈ 11]		♌ [.....] house: [☉] [terms]: ♀	♈ 11 has ♌ 23 as Dodek. ♌ 18 to 23
☽	[♐ 26;4]		♏ 8,52 house:[♂] depr.: ☽ terms: ♀	♐ 26,4 has ♏ 8,52 as Dodek. ♏ 7 to 17
♄	♍ 1 house: ☿ exalt.: ☿ depr.: ♀ terms: ♀	♍ 0 to 7	♍ house: ☿	♍ 1 has ♍ 13 as Dodek.
♃	♍ [...];20 house: ☿ exalt.: ☿ depr.: ♀ terms: ♀	♍ 7 to 17	♐ 13 house: ♃ terms: ♀	♍ [8];20 has ♐ 18[(1)] as Dodek. ♐ 17[(1)] to 21
♂	♊ [......] house: ☿ terms: ♀	♊ 0 to 6	♌ 20 house: ☉ terms: [♀]	♊ 0 to 6 has ♊ 0 to ♌ 18[(1)] as Dodek. ♌ 18 to 24
♀	[♉ 21;..] house: ♀ exalt.: ☽ terms: ♃	♉ 14 to 22	♒ 3 house: [♄] terms: ♀	[♉ 21] has ♒ 3 as Dodek. ♒ 0 to 7
☿	♈ [20;..] house: ♂ exalt.: ☉ depr.: ♄ terms: [...]	♀ : ♈ 12 to 20 ♂ : ♈ 20 to 25	♐ 20 house: [♃] terms: ♀	♈ [20] has ♐ 20 as Dodek. ♐ 12 to 17[(1)]
H	♋ 25 house: ☽ terms: ♃	♋ 19 to 26	♉ 25 house: ♀ exalt.: ☽	♋ 25 has ♉ 25 as Dodek.

TABLE 10.

In spite of some minor discrepancies the concepts mentioned so far belong to standard astrological repertoire. The only strange element is the numbering and the names of the "lots." They agree neither with the Lot of Fortune (about ♈ 10) nor with the Lot of Daimon (about ♍ 10) but must be found according to different principles which are unknown to us.

Horoscopes" the decans of the longitudes themselves, does not lead to satisfactory arrangement of the decans whose order is known from Egyptian sources or from Hephaistion (end of fourth century A.D.). As R. A. Parker has realized one obtains a fair agreement if one

---
[14] Perhaps πράξεως καὶ δόξης or ἀξιωματικός.

[14a] Also the Corpus Hermeticum (Asclepius) mentions "the 36, which are called Horoscopes," but no further explanation is given; cf. Nock-Festugière, *Corpus Herm.* II p. 319.

considers both the "36 Horoscopes" and the "decans" as referring to the original positions and not to the dodekatemoria. This is shown in table 11 in which the numbering of the Egyptian decans follows the arrangement on the zodiac in Edfu (cf. table 4 p. 6). Perhaps one may suggest as a possible explanation for the multiplicity of decanal names a survival of the distinction between constellations in the original Egyptian meaning of the decans and the Hellenistic usage of the same names for the thirds of the zodiacal signs.[15]

would fit much better a general astrological treatise than an individual horoscope. Nevertheless, also in the Coptic text definite dates are mentioned for the single periods.

The following information can be gathered for the lengths of the single periods and their rulers:

lines 83 and 103: I. Venus. (From birth) until year 6, month 9 (or 5), day 25.

lines 90 and 117: II. Mars(?). From year 6, month 9 (or 5), day 25 until year 25, month 2, day 25.

		"36 Horoscopes"	"Decan"	?
M	♈ 10,30			
☉	[♈ 11]	χεν[.ι..] = ḫntw (♈ Decan 1 or 2)		
☿	♈ [20]	χεντα[...] = ḫntw (♈ Decan 1 or 2)	2: κατ′ κουατ′ = kd(?) (♈ Decan 3)	
♀	♉ [21]	χω[....] = ḥ³w (♉ Decan 1)	1: αρου> = wʿrt (♊ Decan 2)	
♂	♊ 6	αρου> = wʿrt (♊ Decan 2)	1: ϩεμεναααρε = rmn ḥry (♉ Decan 3)	
H	♋ 25	ϩεμεναχ[...] = rmn ḥry (one or two Decans after ♉ Decan 3)	2: κυουμε ϩρακυουμ = knmt and ḥry ḫpd knmt (♋ Decan 3 and ♌ Decan 1)	
♄	♍ 1	ετδε = ḥ³t d̲³t (♌ Decan 2)	2: [....]δη βικωτ = wš³ty bk³ty (♍ Decan 2)	
♃	♍ 8	φουτδε = pḥwy d̲³t (♌ Decan 3)	[..]: αρωι αρωι = ḥry ib wi³ (?) (= ρηουω) (♐ Decan 1)	
IMC	♎ 10,30			ϲτωμα = tm³ (♍ Decan 1)
☾	[♐ 26]		[..]: [.....]τι	
Δ	♑ 25			ϲιϲρωι = s³ srt (♒ Decan 1)

Table 11.

The end of our text (col. III, 83 ff.) is concerned with the periods of life. The specific numbers assigned to these intervals require specific data for their computation and it is natural to assume they are the data of the preceding horoscope. The remarks in line 111 f. (♀ in Epanaphora and ♄ in trine with ♀) refer to the conditions of this horoscope. The predictions, however, are held in an extremely vague form and particularly in the Coptic part many conditional sentences occur ("if he had the sun as a friendly star" etc.) which

lines 125 ff.: III. Jupiter. From year 25, month 2, day 25 until year 34, month 5, day 24.

lines 134 f.: IV. Ruler? From year 34, month 5, day 24 until year 54, month 10, day 4.

This major division is apparently partitioned still further. In line 167 year 30 is mentioned and in line 144 f. an interval from year 42 to year 92 occurs. The year 42 is referred to again in lines 152 f. and 178.

We do not know how these data were determined. Similar detailed divisions of lifetimes appear again in

---
[15] Cf. for the history of the decans Neugebauer [9].

later literary horoscopes, e.g. in the compilation of Rhetorius (Nos. L 401, 440, 488, 516).

G. BIBLIOGRAPHICAL NOTES

Photographs of columns II to IV: P. Lond. Facsimiles, pls. 72 to 74; of columns II and V in Cat. of Anc. Ms. I, pls. 4 and 5.

This papyrus, which contains on the reverse the funeral oration of Hyperides, was brought to London by H. Stobart in 1856. Lines 30 to 82 (i.e. column II and the greater part of column III) were published and discussed by Goodwin [1] (1864). Wessely [1] p. 150 ff. (1888) extended his investigation over the whole Greek text, which he transcribed as far as possible up to the Coptic text (line 117). *Cf.* furthermore F. Blass, *Hyperidis orationes sex* (3) Leipzig, Teubner, 1894 p. XXIV f.; [16] Wilcken in *GGA* 1894 p. 731 f. and in *Archiv* 2 (1903) p. 175; Gundel, *Dek.* p. 46 ff., p. 408 f. and many other references in the literature to this horoscope, e.g., Bouché-Leclercq, *AG*, index p. 647 *s.v.* Papyrus astrologiques.[17]

The Coptic section was first discussed by Goodwin [2] (1868) and then by Griffith [1] and [2] (1900/1901).

No. 125
P. Fouad 6

TRANSCRIPTION	TRANSLATION
Beginning destroyed	....
1. του.......	1. the .... [.....
2. φαρμου^θ β[	2. Pharmouthi 2(?) [....]
3. ϕ γ̄	3. 3rd hour
4. της νυκτ[ος]	4. of the night.
5. κρονος ωροσκ[οπος]	5. Saturn (and) Horoscopos
6. παρθε[νω]	6. in Virgo
7. ηλις ζευς	7. sun (and) Jupiter
8. ιχθυσι	8. in Pisces
9. αρης κριου	9. Mars in Aries
10. αφροτι^{τ}ης	10. Venus
11. ηγροχ[ο]ω	11. in Aquarius
12. σελην^η λεων[τι]	12. moon in Leo
13. ερ^{μ}ου ιχθυσι	13. Mercury in Pisces

1 του ..... : The editors read του κυρι[ο]υ. As shown in our commentary the calendar used must be the Egyptian. In this case one should expect a phrase like κατα δε] τους αρχαιους. Neither space nor traces exclude this restoration.

[16] Also Blass, *Hyp. or.* $^{(2)}$ p. XVIII f.
[17] Goodwin [2] p. 23 f. quotes as "a text of the same kind" a P. Anastasi 1073. This text is now Bibl. Nat. suppl. gr. 574 and published by Preisendanz in his *Griechische Zauberpapyri* I p. 64-181. It is a magical text and not a horoscope.

COMMENTARY

Diagram: pl. 1

The sun is in Pisces from about February 19 to March 20. The month Pharmouthi (VIII) of the Alexandrian calendar begins March 27. Thus our horoscope must be based on the Egyptian calendar. For the beginning of our era VIII(e) 1 falls on March 20 and by A.D. 240 the last day of VIII(e) has moved to February 19. Thus our horoscope must belong to the period later than 0 but earlier than 240. Combining the positions of Jupiter and Saturn with the moon one obtains three possible dates within this interval, namely 65, 125, and 185. The earliest date is ruled out by Mars and the latest by Jupiter. Thus only 125 remains, for which one finds for March 4 = Pharmouthi 15

Text		Computed
♄ } H }	♍	♍ 10 about 7 P.M.
☉ } ♃ }	♓	♓ 13 ♓ 21
♂	♈	♈ 30
♀	♒	♓ 15 (!)
☾	♌	♌ 10
☿	♓	♒ 29

The only bad deviation occurs in the case of Venus, where the text gives Aquarius instead of Pisces.

It is, of course, the position of the moon which determines the day number as Pharmouthi 15 (or 16). This would require in line 2 the reading ι[ε (or ι[ς). In fact, however, only β seems compatible with the remainder of the letter the right half of which is broken away. For this date, Pharmouthi 2 = February 19 one would obtain

Text		Computed
♄ } H }	♍	♍ 10 about 8 P.M.
☉ } ♃ }	♓	♓ 0 ♓ 12
♂	♈	♈ 22
♀	♒	♒ 29
☾	♌	♒ 14 (!!)
☿	♓	♓ 9

Though all planetary positions are now correct the moon is 180° wrong. This almost looks as if full moons and new moons were interchanged in some table of syzygies.

As time of the horoscope is given the third hour. This is barely possible if the Horoscopos is at the very end of ♍ and the sun at the beginning of ♓. This again speaks in favor of the date Pharmouthi 2 as against Pharmouthi 15.

The order of enumeration of the planets is very unusual since it follows neither the customary arrangement according to planets, nor the order of the zodiacal signs.

Nos. 137a, b, and c
P. Paris. 19, P. Lond. 110, P. Paris. 19 bis

INTRODUCTION

Nos. 137a and b provide essentially the same text. Kenyon, the editor of No. 137b, considers No. 137a the draft from which the final copy was made. No. 137c concerns the same horoscopic moment (137 December 4) but gives a different and much more elaborate text; the numerical data also seem to be slightly different. Perhaps Anoubion, the owner of the two first horoscopes, went to another astrologer who cast an independent horoscope. Unfortunately No. 137c is so badly damaged that a comparison of details is made impossible.

TRANSCRIPTION OF NO. 137a
(Louvre N 2342)

Col. I
1. α[γα]θ[η τ]υχη γενεcιc ανουβιωνοc ε̄
2. ψανcνωτοc
3. (ετουc) α αντωνινου καιcαροc του κυριου
4. μηνοc αδριανου η̄ κατα δε τουc αρχαι°
5. τυβι ιη̄ ωραc ᾱ τηc ημεραc αρχ
6. ηλιοc εν τοξοτηι μοιρων ιγ̄ λεπτ° κ̄γ
7. οικω διοc οριοιc αφρ[οδε]ιτηc
8. cεληνηc υδροχω μ[οιρων] γ̄ λεπτ° ϛ
9. ανατολικοc [οικω κρονου] οριοιc ερμου
10. κρονοc εν υδρο[χ]ω μοιρ[ων] γʃ λεπτ° η̄
11. εν τω β̄ cτηριγμω ιδιω οικω οριοιc
11a. ερμου
12. ζευc εν κρειω μοιρων [ι]β̄ λεπτ° μ̄δ
13. εν τω β̄ cτηριγμω [οικ]ω αρεωc
14. υψωματι ηλιου ταπι[ν]⁻ κρονου
15. οριοιc αφροδειτηc
16. αρηc εcχατα αιγοκερω μοιρων λ̄
17. λεπ[τ]ον ουδεν εν τω β̄ cτεριγμω
18. οικω κρονου ιδιω υψωματι
19. ταπινωματι διοc οριοιc ιδιοιc
20. αφροδειτηc εν τοξοτη μοι[ρ]ων θ̄
21. λεπτον νδ̄ εωα ανατολαc
22. οικω διοc οριοιc διοc

Col. II
23. ερμηc εν τοξοτη [μοιρ]ων ιε̄
24. λεπτον β̄ εν εcπερ[ια] οικω διοc
25. οριοιc αφροδειτη[c]
26. οροcκοποc εν τοξοτη μοιρων ιε̄
27. οικω διοc οριοι[c α]φροδειτηc
28. γαμοcτολοc εν διδυμο[ι]c μοιρων ιε̄
29. οικω ερμου οριοιc ερμου
30. μεcουρανημα εν παρθενω μοιρων
31. η̄ οικω ερμου υψωματ[ι] ερμου
32. ταπινωματι αφροδει⁻ οριοι[c] αφροδ[ειτηc]
33. υπογην εν ιχθυcι μοιρων [η̄]
34. οικω διοc υψωματι αφροδ[ειτηc]
35. ταπινωματι ερμ[ο]υ ορ[ιοι]c αφρ[οδ]ει'
36. ο ᾱ κληροc τηc τυχηc εν α[ι]γοκερω
37. μοιρων ιθ̄ οικω [κ]ρονο[υ] υψω[μ]ατι
38. αρεωc ταπινωματι διοc ο[ρ]ιοιc αφροδ
39. ο β̄ κληροc τηc τυχηc εν κα[ρ]κειν[ω]
40. μοιρων ιᾱ οικω c[ελ]ηνηc υψω[μα]τι
41. διοc ταπινωματι αρεωc οριοιc αφ[ρ]ο̣δ
42. ο οικοδεcποτηc τηc γενεcεωc αυτου
43. ο τηc αφροδειτηc αcτηρ

The facsimile in Young, *Hierogl.* II pl. 52 shows a much better state of preservation of the text, all words and letters which we enclose in [ ] in column I were still preserved. In line 9 Young gives κρονοc where we restore κρονου.

TRANSLATION OF NO. 137a

Col. I
1. Good Fortune. Nativity of Anoubion 5(th? son)
2. of Psansnos.
3. Year 1 of Antoninus Caesar the Lord,
4. month of Hadrian 8, according to the old (calendar)
5. Tybi 18, hour 1 of the day, at the beginning.
6. sun in Sagittarius 13 degrees, 23 minutes,
7. house of Jupiter, terms of Venus.
8. moon in Aquarius, 3 degrees, 6 minutes,
9. rising, [house of Saturn,] terms of Mercury.
10. Saturn in Aquarius, 3¼ degrees, 8 minutes,
11. in his 2nd station, his own house, terms
11a. of Mercury.
12. Jupiter in Aries, 12 degrees, 44 minutes,
13. in his 2nd station, house of Mars,
14. exaltation of sun, depression of Saturn,
15. terms of Venus.
16. Mars at the end of Capricorn, 30 degrees,
17. no minutes, in his 2nd station,
18. house of Saturn, his own exaltation,
19. depression of Jupiter, his own terms.
20. Venus in Sagittarius, 9 degrees,
21. 54 minutes, morning rising,
22. house of Jupiter, terms of Jupiter.

Col. II
23. Mercury in Sagittarius, 15 degrees,
24. 2 minutes, in the evening, house of Jupiter,
25. terms of Venus.
26. Horoscopos in Sagittarius, 15 degrees,
27. house of Jupiter, terms of Venus,
28. Marriage-maker in Gemini, 15 degrees,
29. house of Mercury, terms of Mercury.

30. Midheaven in Virgo, 8 degrees,
31. house of Mercury, exaltation of Mercury,
32. depression of Venus, terms of Venus.
33. Lower Midheaven in Pisces, [8] degrees,
34. house of Jupiter, exaltation of Venus,
35. depression of Mercury, terms of Venus.
36. The 1st Lot of Fortune in Capricorn,
37. 19 degrees, house of Saturn, exaltation
38. of Mars, depression of Jupiter, terms of Venus.
39. The 2nd Lot of Fortune in Cancer,
40. 11 degrees, house of moon, exaltation
41. of Jupiter, depression of Mars, terms of Venus.
42. The ruler of his nativity
43. the star of Venus.

TRANSCRIPTION OF NO. 137b

(P. Lond. 110)

Col. I

1. [γενεcιc ανουβι]ωνοc ε̄ ψανcνωιτοc
2. [(ετουc) ᾱ αντωνιν]ου καιcαροc του κυριου
3. [μηνοc αδριαν]ου η̄ κατα δε τουc αρχαι
4. [ουc τυβι ιη̄ ω]ραc ᾱ τηc ημεραc αρχαc
5. [ηλιοc εν τοξοτ]ηι μοιρων ιγ̄ λεπτων
6. [κγ̄ οικωι διοc] οριοιc αφροδειτηc
7. [cεληνη εν υδροχ]ωι μοιρων γ̄ λεπτων
8. [ξ̄ ανατολικοc] οικωι κ[ρ]ονου οριοιc [ε]ρμου
9. [κρονοc εν υδροχωι μοιρ]ων γ𐅵 λε[πτων]
10. [η̄ εν τωι β̄ cτηριγμωι ι]διωι οικω[ι οριοιc]

Col. II

11. ερμου
12. ζευc εν κρ[ιω]ι μοιρων ιβ̄ λε[πτ]ων
13. μδ̄ εν τωι β̄ [cτ]ηριγμωι οικωι αρεωc
14. υψωματι ηλ[ιο]υ ταπινωματι κρονου
15. οριοιc αφρο[δε]ιτηc
16. αρηc εcχατ[α] αιγοκερωι μοιρων λ̄
17. λεπτων ουδ[ε]ν εν τω β̄ cτηριγμωι
18. οικωι κρονου [ι]διωι υψωματι ταπι
19. νωματι δι[οc ορ]ιοιc ιδιοιc
20. αφροδει[τη εν τοξ]οτηι μοιρων θ̄ λ[ε]πτων
21. δ̄ εωια [ανατολαc] οικωι διοc [ορ]ιοιc

Col. III

22. ιδιοιc
23. ερμηc εν τοξοτηι μοιρων ιε̄ λεπτων β̄
24. εν εcπερια οικωι διοc οριοιc αφροδει
25. τηc
26. ωροcκοποc εν τοξοτηι μοιρων ιε̄ οικωι
27. διοc οριοιc αφροδειτηc
28. γαμοcτολοc εν διδυμοιc μοιρων ιε̄ οικωι
29. ερμου οριοιc ερμου
30. μεcουρανημα εν παρθενωι μοιρων η̄
31. οικωι ερμου υψωματι ερμου ταπινω
32. ματι αφροδειτηc οριοιc αφροδειτηc

Col. IV

33. υπογην εν ἰχθυcι μο[ιρ]ων η̄ οικωι διοc υψωματι
34. αφροδειτηc ταπινωματι ερμου οριοι[c] αφ[ρ]οδει[τηc]
35. ο ᾱ κληροc τηc τυχηc εν αιγοκερωι μ[οι]ρων ιθ̄ οι[κω]ι
36. κρονου υψωματι αρεωc ταπινωματι διοc οριοι[c]
37. αφροδειτηc
38. ο β̄ κληροc τηc τυχηc εν καρκινωι μοιρων [ιᾱ] [οι]κωι
39. cεληνηc υψωματι διοc ταπινωματι αρ[εωc] ορ[ιοιc]
40. αφροδειτηc
41. οικοδεcποτηc τηc γενε[c]εωc αυτου ο τηc α[φρ]οδ[ειτηc]
42. αcτηρ
43.   αγαθη τυχηι   ευτυχ[ε]ι

Column I is written on the reverse of a small sheet. The three following columns are now mounted separately but seem to have been written originally on one piece of papyrus of about 47 cm length.

To the left of the last lines of col. IV a large coronis is written.

TRANSLATION OF NO. 137b

Col. I

1. [Nativity of Anoubi]on 5(th? son) of Psansnois,
2. [year 1 of Antonin]us, Caesar the Lord,
3. [month Hadrian] 8, according to the old (calendar)
4. [Tybi 18,] hour 1 of the day, at the beginning.
5. [sun in Sagittarius,] 13 degrees, [23] minutes,
6. [house of Jupiter,] terms of Venus.
7. [moon in Aquari]us, 3 degrees, [6] minutes,
8. [rising,] house of Saturn, terms of Mercury.
9. [Saturn in Aquarius,] 3½ degrees, [8] minutes,
10. [in his 2nd station,] his own house, [terms]

Col. II

11. of Mercury.
12. Jupiter in Aries, 12 degrees, 44 minutes,
13. in his 2nd station, house of Mars,
14. exaltation of sun, depression of Saturn,
15. terms of Venus.
16. Mars at the end of Capricorn, 30 degrees,
17. no minutes, in his 2nd station,
18. house of Saturn, his own exaltation, de-
19. pression of Jupiter, his own terms.
20. Venus in Sagittarius, 9 degrees, 4 minutes,
21, 22. morning [rising,] house of Jupiter, his own terms.

Col. III

23. Mercury in Sagittarius, 15 degrees, 2 minutes,
24, 25. in the evening, house of Jupiter, terms of Venus.
26. Horoscopos in Sagittarius, 15 degrees, house of
27. Jupiter, terms of Venus,
28. Marriage-maker in Gemini, 15 degrees, house of
29. Mercury, terms of Mercury.
30. Midheaven in Virgo, 8 degrees,
31. house of Mercury, exaltation of Mercury, depres-
32. sion of Venus, terms of Venus.

Col. IV

33. Lower Midheaven in Pisces, 8 degrees, house of Jupiter, exaltation
34. of Venus, depression of Mercury, terms of Venus.
35. The 1st Lot of Fortune in Capricorn, 19 degrees, house of
36. Saturn, exaltation of Mars, depression of Jupiter, terms of
37. Venus.
38. The 2nd Lot of Fortune in Cancer, [11] degrees, [house of]
39. moon, exaltation of Jupiter, depression of M[ars, terms of]
40. Venus.
41, 42. The ruler of his nativity the star of [Venus].
43. Good fortune. Farewell!

ASTRONOMICAL COMMENTARY TO NOS. 137a AND b

Diagram: pl. 2

The date in question is Antoninus 1, month Hadrian (= IV(a)) 8 = V(e) 18 = 137 December 4, about 8 A.M. For this date one finds:

	Text	Computed	Text—comp.
☉	♐ 13;23	♐ 12	+1;30
☾	♒ 3;6 rising	♑ 29;30	+3;30
♄	♒ 3;38 2nd station	♑ 27	+6;30
♃	♈ 12;44 2nd station	♈ 11	+1;30
♂	♑ 30;0 2nd station	♒ 3	—3
♀	♐ 9;54[1] morn. rising	♐ 12	—2
☿	♐ 15;2 evening	♍ 20 (!)	[—5]
H	♐ 15	about 8 A.M.	
M	♍ 8	♍ 27;50 (!)[2]	

The agreement of the longitudes is fair with exception of Mercury where one should expect ♏ instead of ♐.[3] The culminating point also is essentially wrong; if ♐ 15 rises, then the culminating point lies beyond ♍ 15 and not before it. The text says furthermore that all three outer planets are in the second stationary point. This is correct for Jupiter but Saturn is about 2½ months beyond it and Mars is about 4½ months past his second station. The moon is about 3 days past conjunction and therefore again clearly visible or "rising" (ἀνατολικός). Similarly Venus is reappearing as morning star. Mercury, accepting the longitude given in the text, would be invisible, though setting after the sun, which can be considered as a reason for the remark "in the evening" (ἐν ἑσπέρᾳ). Actually Mercury is morning star near to greatest elongation.

[1] Or 9;4 in No. 137b.

[2] Computed for Alexandria, using Almagest II,8. The procedure of Paulus Alexandrinus (*cf.* p. 10) would lead to ♍ 29;10.

[3] Bouché-Leclercq, *AG* p. 283 note 2, assumes that Mercury was purposely placed in the same sign and even degree with the Horoscopos. Our general experience does not support the assumption of an arbitrary modification of astronomical data.

ASTROLOGICAL COMMENTARY TO NOS. 137a AND b

The following information for the celestial bodies is given in this version:

		Phase	House	Exalt.	Depr.	Terms
☉	♐ 13;23		♃			♀
☾	♒ 3;6	rising	♄			☿
♄	♒ 3;38	2nd station	♄			☿
♃	♈ 12;44	2nd station	♂	☉	♄	♀
♂	♑ 30;0	2nd station	♄	♂	♃	♂
♀	♐ 9;4	morn. rising	♃			♃
☿	♐ 15;2	evening	♃			♀

Of interest are the terms of Jupiter and Mars. Jupiter's longitude is ♈ 12;44 and the terms of Venus extend from ♈ 6 to ♈ 12, followed by the terms of Mercury. Our text gives the terms of Venus, which shows that the whole 12th degree is still assigned to the terms which end in ♈ 12. Similarly in the case of Mars ♑ 30 is counted as belonging to the terms of Mars which extend from ♑ 26 to 30.[4]

After the positions of the planets six more points are given: Horoscopos and setting point, here called γαμοστόλος because this point is the locus of marriage;[5] then the points of upper and lower culmination; and finally a "1st" and a "2nd" Lot of Fortune. With these points the following data are associated:

		House	Exalt.	Depr.	Terms
H	♐ 15	♃			♀
Δ	♊ 15	☿			☿
M	♍ 8	☿	☿	♀	♀
IMC	♓ 8	♃	♀	☿	♀
1st L. of F.	♑ 19	♄	♂	♃	♀
2nd L. of F.	♋ 11	☾	♃	♂	♀

In the case of the setting point (♊ 15) the terms are erroneously given as of Mercury instead of Venus. The astrologer thus overlooked the peculiar accident that all six points fall in the terms of Venus.

How the two "Lots of Fortune" are computed is not clear.[6] The first one lies 34° beyond the Horoscopos, the second one 26° beyond the setting point. Neither one of these numbers equals the elongation of 50° of the moon from the sun. *Cf.* Bouché-Leclercq p. 293 note 1 who considers the possibility that the astrologer who had cast this horoscope "prétend avoir fait des calculs auxquels il n'entendait rien." If one takes, however, the "Lots" as equivalent with the "Loci" then

[4] For similar instances *cf.* the commentary to No. 46 (p. 20).

[5] *Cf.* Bouché-Leclercq *AG* p. 281. Vettius Valens IX,2 (p. 335, 11 Kroll) says: τὸ ζʹ γαμοστόλος τῆς γενέσεως .... τόπος.

[6] Cumont [9] p. 79 note 2 assumes that these two "Lots" represent δαίμων and τύχη respectively.

♑ 19 falls in the Locus of "Possession" or "Gain" and ♋ 11 in the Locus of Marriage. For a possible identification of "Lots" and "Topoi" cf. No. 95.

Venus is the ruler of the nativity since she is in the same sign with the sun [†] in the Horoscopos.

TRANSCRIPTION OF NO. 137c

(Louvre 2342 bis)

Col. I

1.   επτα θε[οι]
2. σκεψαμενος απο πολλων βιβλων ως παρεδοθη
3. ημειν απο σοφων αρχαιων τουτεστιν χαλδαικων
4. και [π]ετοσιρις μαλιστα δε και ο βασιλευς νεχευς
5. ωσπερ και αυτοι συνΰδρευσαν απο του κυριου ημων
6. ερμου και ασκληπιου ο εστιν ιμουθου υιος ηφηστου
7. κατα τον δοθεντα μοι χρονον επι α (ετους) αντωνινου
8. καισαρος του κυριου μηνος αδρια[ν]ου η
9. κατα των ηλληνων κατα δε τους αιγυπτιους
10. τυβι ι̅η̅ ωρας ᾱ της ημερας
11. θεον ηλ[ιο]ν ψηφεισθης κατα την ελαχιστην καὶ μεγεθου
12. τ[αχ...]ς ευρεθη ζωδιω τοξοτη μοιρων ι̅ς̅
13. λε[πτων τ̅]ης εξηκονταδος λε ο εστιν [με]ρ[ος]
14. [μοιρας] ημ[ις]ι δωδεκατοψ οικω διος τριγω[νω]
15. [διος μεγιστω] δρ[ο]μω συ[ναν]α̣τελλεται δε αυτω
16. [...........................]ω και μοιρα ο εστ[ιν]
17. [...........................]ρω εν τω ουρανε[ισκω]
18. [...........................]τω βελι στ[
19. [...........................]πο αιγοκερ[ω
20. [...........................]χιμεριν .[
21. [...........................]πλατος ε[

I,1 επτα θε[οι]: this obvious restoration was made by Seyffarth (Beitr. 4 p. 212, 1833) on the basis of the facsimile in Young, Hierogl. II pl. 52, which still shows a better stage of preservation; at present only επτ[is left. Cf. also Boll, SCP p. 238 and Lauth [2] p. 96. Brunet de Presle in the edition of 1865 gives " επ' α[γαθω] ? "

I,14 to 26: Wilcken, Ostr. I p. 792 note 2 assumed that the fragment which begins with οικω διος in line 14 is incorrectly placed because of disagreement on the verso. Contexts as well as fibers and coloring, however, leave no doubt that Wilcken was wrong except for a very minor adjustment of the lower fragment toward the right by about the space of one letter. On the verso the main text ends on line 11 whereas the fragment shows the endings of a few lines of entries made further to the left.

I,15 δρ[ο]μω: traces of δ certain, ρ also read by Wessely though both letters were ignored in the facsimile in Notices et Extraits. Thus there is very little doubt that one had to read δρ[..]μω but the space between ρ and μ would rather suggest two missing letters.

I,18 to 21: A loose fragment is incorrectly mounted here instead of in front of lines 23 to 26. This was not recognized by Wessely in his transcription Mit. P. Erzh. Rainer 2,3 p. 6 and concerns the letters which he read νβ, αν, and νχ in lines 17 to 20. The correct position is still shown on pl. XIX (fragment d) of Notices et Extraits 18,2.

[†] Bouché-Leclercq, AG p. 406 note 2.

22. [...........................ε]ν ζωδ[ι]ω[
23. [...........................]λβ κιν[
24. [...........................]λλ ...[
25. [...........................]ν..ιου[...
26. [...........................]...[....

Col. II   Detached fragment:

1. φαινω[ν...   1. .....]το..ν.[....
2. ζωνη[....   2. ...]βιος και τυχ[η ...
3. κατα[....   3. ...]. αναγκη τερο[ς ...
4. μοιρας[....   4. ....]αιτη ποιει το δε [...
5. μερος [....   5. ...]ετα εαν τις αυτον[...
6. ερμου[....   6. ...]τερον ει ετασθεις ρ[...
7. [α]στερες[....   7. ....]ονται στησαι.[....
8. στηριγ[μω ....   8. ....τ]ων γονεων κ[...
9. ανατελ[λεται ....   9. ...τη]ς τυχης βιου τ[...
10. αστερ.[....   10. ....]...[....
11. μοιρω[ν ....
12. δρομ[ω ....
13. λεπ]των ....
14. αποτ[....

II,4: This line was omitted by Wessely.

Detached fragment: This fragment (called c in Notices et Extraits 18,2 P. XIX) is now broken in two, the parts being mounted separately. The facsimile shows occasionally more than is now preserved. In line 1 one might perhaps restore: το δε υπ[ογειον.

TRANSLATION OF NO. 137c

1.   The seven gods
2. After examination of many books as it has been handed down
3. to us from ancient wise men, that is the Chaldeans,
4. and Petosiris and especially also King Necheus
5. just as they themselves took counsel from our lord
6. Hermes and Asclepius, that is Imouthes, son of Hephaestus.
7. For the time given to me, in the 1st (year) of Antoninus
8. Caesar the Lord, month Hadrian
9. according to the Greeks, and according to the Egyptians
10. Tybi [1]8, 1st hour of the day.
11. The divine sun calculated according to the lowest and the greatest
12. [of velocity] was found in the sign Sagittarius, 16 degrees
13. 35 parts of sixty, which is the
14. half (and) twelfth [part of a degree], house of Jupiter, triangle
15. [of Jupiter, greatest] mo[tion]. There rises simultaneously with it . . . . .

For the continuation see the commentary.

COMMENTARY TO NO. 137c

Though the data and the hour of this horoscope are exactly the same as in Nos. 137a and b we find for the sun a different longitude. In Nos. 137a and b the sun was said to be in ♐ 13;23 whereas we find now ♐ 16;35. In lines 11 and 12 we are told that the sun was computed according to the method of smallest and greatest (?) velocity (?) which perhaps means a method based on a linear zigzag function [8] and not on an epicyclic or eccentric model. Perhaps in line 15 was mentioned the "[greatest] velocity," which would be close to the fact because the solar perigee lies in Sagittarius.

The remaining parts are too badly preserved for a consistent translation. It is interesting to see that "Synanatellonta" are mentioned,[9] that is simultaneously rising stars. No other horoscope is known which mentions this doctrine.

In I,16 one can perhaps restore εν τω αυτω ζωδι]ω και μοιρα " in the same zodiacal sign] and degree" followed by " that is [♐ 16;35]." If our restoration ουρανε[ισκω (line 17) is correct then the constellation "southern crown" would be meant which according to Geminus (*Isag.* III, 11 Manitius p. 40, 17 f.) is also called οὐρανίσκος (similarly Schol. 400 to Aratus, Maass p. 418, 7). It also occurs in Teucros (Boll, *Sphaera*, p. 263). It must be admitted, however, that a reading ουρανω[is not excluded.

In line 18 the "arrow" of Sagittarius seems to be mentioned. Lines 19 and 20 contain the words "capricorn" and "winter [solstice]" and probably indicate the fact that the sun is approaching the winter solstice in Capricorn.

In line 21 we read "latitude" while in line 22 a "zodiacal sign" is mentioned; in line 23 a number 32 [minutes] occurs, perhaps followed by κιν[.... "motion." All this points to the description of the motion of the moon. The two λs in line 24 belong perhaps to "synanatellonta."

Column II begins with "Phainon," that is Saturn. "Degree" and "part" in lines 4 and 5 respectively point to a description of the longitude in degrees and fractions of a degree (cf. I,12 to 14) followed in the next line by "[terms] of Mercury" (cf. No. 137a and b). The "stars" in line 7 should again be "synanatellonta." In lines 8 to 10 we find a "station" mentioned, then again "anatellonta" and "stars." Thus it is plausible to assume here a description of Jupiter's position. The "degrees" and "motion" in lines 11 and 12 may concern Mars, the "minutes" in line 13 Venus.

The fragment mentions Tyche twice and seems concerned with general predictions or doctrine. For "Asclepius, that is Imouthes, son of Hephaestus" cf. Edelstein, *Asclepius* I, Text 331, II p. 252.

BIBLIOGRAPHICAL NOTES

We are dealing here with three papyri:

No. 137a: P. Par. 19 — Louvre N 2342 } duplicates
No. 137b: P. Lond. 110
No. 137c: P. Par. 19 bis — Louvre 2342 bis; variant

The extensive literature on these texts is made particularly difficult to follow since the numbers 19 and 19 bis were interchanged by several authors.

No. 137a. Young, *Hierogl.* II Pl. 52 (1828) gives a facsimile which still shows several passages now destroyed. The present state is shown in the facsimile on pl. 22 of the *Notices et Extraits* 18,2 (Planches) while a transcription by Brunet de Presle was given on p. 237 f. of the corresponding text volume (1865).

The text comes supposedly from Thebes [10] and was purchased in Livorno with the Salt collection by the Louvre in 1826.[11] The date of the horoscope was verified in an anonymous note published in the *Quarterly Journal of Science, Literature and Art*, January to June 1830 p. 353 f. Then, based on Young's facsimile, Seyffarth *Beitr.* 4 p. 231 ff. gave a transcription of the Greek text, a Latin translation and commentary. Seyffarth apparently assumed that No. 137a represented col. II and III of No. 137c of which he knew only the main part of column I from Young. The Greek text and a French translation by Champollion-Figeac [12] was given in C. L. E. Panckoucke, *Bibliothèque Latin-Française* vol. 31 p. 380-385 as an appendix to Cicero's *De divinatione* (edited by DeGolbéry).[13]

The date of the horoscope was misread by Franz *CIG* 3 p. 375 (1853), then discussed by Browne [1] (1855), by Hincks [1] (1865) and finally in detail by Schram in *Mit. P. Ersh. Rainer* 2/3 (1887) p. 37 to 40, following a short note by Wessely on p. 5 in the same volume, where Wessely seems to have been the first one to confuse the numbers 19 and 19 bis of P. Par., in which mistake he is then followed by Wilcken, *Ost.* II p. 792.

The first and last lines of this horoscope are also cited in *CCAG* 8,4 (p. 96) as No. 121 B but with a wrong reading α' for ε' at the end of line one and giving an incorrect date (138 instead of 137).

No. 137b. The Greek text of this horoscope was first

---

[8] *Cf.* the ιση πρεσθες.ς και αφαιρεςις " constant increase and decrease " in P. Mich. 149, XVII 24/25.

[9] For this concept *cf.* Boll, *Sphaera*, p. 75 ff. or Gundel, Art. "Paranatellonta" in *RE* 18,3 cols. 1214-1275.

[10] Young, *Hierogl.* II pl. 52. The anonymous article, quoted below, has, however, the title "Astrological MS. from Saccara."

[11] Brunet de Presle in P. Par. p. 235.

[12] With slight variations repeated in Champollion-Fig.. *Eg.* p. 101 f.

[13] We have only seen the edition of 1840 which is said to repeat the first edition of 1837.

given by Wessely [14] [1] p. 152 f. (1888) though without making use of the duplicate No. 137a, as was done by Kenyon in P. Lond. 110 (1893).[15] This horoscope is often quoted, e.g. by Macnaughton, *SEC* p. 333. A photographic reproduction is given in the volume of British Museum Facsimiles pl. 75 f. A photograph of col. III is given by Roberts, *GLH* Pl. 18.

No. 137c. This text belonged to the same collection as No. 137a and was published simultaneously with it by Young, *Hierogl.* II Pl. 52, though without the fragments of column I and II which were added by Brunet de Presle on Pl. 19 of the *Notices et Extraits* 18,2. The corresponding transcription (p. 236), however, again ignores the fragments.

A transcription and Latin translation of the first 14 lines is contained in Seyffarth, *Beitr.* 4 p. 212, combined with No. 137a. More than fifty years elapsed before a transcription was published which also included the fragments. This was done by Wessely [1] p. 5 f. (1887), who, however, erroneously called the text "Papyrus XIX." Wilcken, *Ostr.* I p. 792 refers to this horoscope as P. Paris. 19 in his text but as P. Paris. 19 bis in the notes.

Concerning the two horoscopes Nos. 137a and c, the appendix to Cicero, *De Divinatione* in the *Bibl. Lat.-Franc.* vol. 31 [16] p. 379 says "rapportés d'Égypte par Champollion le Jeune, expliqués par MM. Champollion-Figeac et de Golbéry." Then follows a transcription [17] and translation of the main part of column I and comments (p. 380-388).

In *CCAG* 8,4 p. 95 the horoscope No. 137c is listed as No. 121,A. The beginning lines have been quoted repeatedly with more or less accuracy in reproducing the actual spellings or readings. e.g. Lauth [2] p. 96 (lines 2 to 10) or Riess [1] p. 331 (lines 2 to 6 — Testim. 6).

No. 138/161

P. Princeton 75

TRANSCRIPTION

Preceding column (or columns) lost

1. [μοιρων] κα̅ οικω κρονου οριοις αφροδειτης
2. [ο κ]ληρος της τυχης ευρεθη καρκινω μοιρων ι̅γ̅ οικω σε
3. [ληντ]ς οριο]ις ερμου ουτος σημαινει μετρον της τυχης
4. [ο κλ]ηρος ο του δαιμονος ευρεθη αιγογερω μοιρων κ̅ζ̅ οικω

5. [κρονου οριοις αρεω]ς ουτος σημαινει τον περι ηθους και αγωγης τροπον
6. [ο κληρος ε]ρωτος ευρεθη ταυρω μοιρων δ̅ οικω και οριοις α
7. [φροδειτης ουτος σημαιν]ει τον περι φιλιας και συστασεως
8. [ο κληρος] της αναγκης ευρεθη κριω μοιρων ς̅ οικω αρεως
9. [οριοις διος ου]τος σημαινει τον περι εκχθρων και παντοδαπου
10. [δυστυχημα]τος ..ω τε απας περι του οικοδεσπο
11. [του των] αστερων ψηφον επικρατουντας παντων δε
12. [..........]και αστερων τον του ερμου αστερα εσχηκοτα
13. [.......]..γ.[....]ος̣κο[.] το της οικοδεσποτειας βραβειον εσχατα
14. [.....]δεδει μηνας ι̅ζ̅ μεθ ους παρηλθε τον σινωτικον
15. [κλιμακτ]ηρικον ορον και γεινεται η υποστασις απο του οικοδε
16. [σποτου] δι[ευτυχ]ει

second hand:

17. [γενεσις ο]φελλιου [(ετους)..] αντωνινου [κ]αισαρος του κυριου
18. [..........]τος ς̅ [εις ζ̅] ωρας ζ̅ νυ[κτ]ος

7 φιλιας: the editors read θεσεως. For φιλία and σύστασις see Vettius Valens p. 42, 15/16.

9 παντοδαπου: the editors read παντο[ς] λαττ θου.

9 [οριοις διος: since ♈ 6 is the boundary between the terms of ♃ and ♀ (in this order) one might restore the name of either one of these planets; cf. the commentary to No. 46. The available space favors the shorter word.

10 ω τε απας: the editors read εξετε παν.

14 ι̅ζ̅: or ι̅γ̅?

TRANSLATION

.... in Capricorn]
1. 21 [degrees], house of Saturn, terms of Venus.
2. [The] lot of Fortune was found in Cancer, 13 degrees, house of the m-
3. [oon], terms of Mercury; this (lot) indicates the measure of fortune.
4. [The] lot of Daimon was found in Capricorn, 27 degrees, house of
5. [Saturn, terms of Mars;] this (lot) indicates the way of life, as regards character and conduct.
6. [The lot] of Eros was found in Taurus, 4 degrees, house and terms of V-
7. [enus; this (lot) indicates] the [place(?)] regarding friendship and association.
8. [The lot] of Necessity was found in Aries, 6 degrees, house of Mars.
9. [terms of .....;] this (lot) indicates the [place(?)] regarding enemies and every kind of
10. [misfortune.] And then the whole [.......] regarding the rul[ing] of the nativity
11. [by those] of the stars which have prevailing influence and of all

---

[14] Previously mentioned by Wessely, *Mit. P. Erzh. Rainer* 2/3 (188) p. 6. Also: Catalogue of additions to the Manuscripts in the British Museum in the years 1854-(1875). Vol. 2 p. 833 f. (1877).

[15] The date is given incorrectly as A.D. 138.

[16] Cf. above p. 43.

[17] Also Lauth [2] p. 96 (1875); Boll, *SCP* p. 238 (1894); Festugière, *Rev.* p. 430 and p. 126 (1950) etc.

12. [the five(?)] stars; the star of Mercury which had
13. [....................] control of rulership last
14. [....................] beware 17(?) months after which it has passed the baneful
15. climactic term and there comes the promise from the ruler of the nati-
16. [vity.] Good luck!
17. [Nativity of O]phellios, [year ..] of Antoninus Caesar the Lord,
18. [.......................] ... 6 [to 7,] 7th hour of night.

COMMENTARY

The first preserved line assigns a 21st degree to the house of Saturn and the terms of Venus. The houses of Saturn are Capricorn and Aquarius. In the first case the terms of Venus extend from 14° to 22°, in the second from 7° to 13°. Hence we can only restore [Capricorn] 21. We furthermore find that the four "Lots" are symmetrically located with respect to the point ♈ 20°:

Lot of Fortune:	♋ 13 = ♈ 20 + 83°
Lot of Daimon:	♑ 27 = ♈ 20 − 83°
Lot of Eros:	♉ 4 = ♈ 20 + 14°
Lot of Necessity:	♈ 6 = ♈ 20 − 14°

Consequently it is plausible to consider ♈ 20 either as the Horoscopos or as the setting point. In the second case the Horoscopos would be ♎ 20. In order to distinguish between these two possibilities we compute the culminating points (for the latitude of Alexandria). The result is:

If H = ♈ 20° then M = ♑ 12;49   IMC = ♋ 12;49
If H = ♎ 20° then M = ♋ 12;24   IMC = ♑ 21;24

This shows that only the second assumption explains the number "[♑]21" with which our text begins. Thus ♎ 20 is the Horoscopos.

According to the final line, which was added by another hand, the hour of the birth was the 7th hour of night; therefore the sun must be located not more than about 15° beyond lower culmination. Thus we obtain at least ♑ 6 for the longitude of the sun. Unfortunately we do not know, however, which part of the 7th hour is referred to and we must therefore consider a range of about 15° for the solar longitude. Consequently, we can conclude that the date of our text falls between December 27 and January 11 in the Julian calendar or Tybi (V) 1 to 16 in the Alexandrian calendar. During the reign of Antoninus the corresponding Egyptian dates would range between Mekheir (VI) 11 and Phamenoth (VII) 2.

In line 18 the best reading seems to be "...]..6 [to 7]." As the above discussion shows, the Egyptian dates do not include the days 6 or 7. In the Alexandrian calendar, however, we would have Tybi 6 to 7 at our disposal. Unfortunately this does not explain the traces]τος which precede the day number.

There exist additional relations between the four "Lots" mentioned in our text. Let Λ_F to Λ_N denote the longitudes of the Lots of Fortune, Daimon, Eros, and Necessity respectively. Then one finds for $\Lambda_H = ♎ 20 = ♈ 200$ the following relations satisfied

$$\Lambda_E = \lambda_H - (\Lambda_F - \Lambda_D) = 200 + 194 = ♉ 4$$
$$\Lambda_N = \lambda_H + (\Lambda_F - \Lambda_D) = 200 - 194 = ♈ 6$$

and

$$\Lambda_F = \lambda_H - 97 = 103 = ♋ 13$$
$$\Lambda_D = \lambda_H + 97 = 297 = ♑ 27$$

where the 97° can be interpreted as the elongation between sun and moon. Similar relations seem to hold in No. 338.

The advice of line 14 "beware 17(?) months" might mean that the person for whom the horoscope has been cast will have passed a critical age; otherwise one might think of Saturn having left a critical zodiacal sign.

No. 139

P. Oxy. 596, Col. II

(Photo: Roberts *GLH* Pl. 16)

TRANSCRIPTION

1. $\overline{\beta}$ αντωνινου κ[αιcαροc]
2. του κοριου φαμ[ενωθ]
3. ·$\overline{\lambda}$· εις $\overline{α}$ ωρας $\overline{δ}$ ν[υκτος]
4. ηλιος ιχθυςι [
5. ςεληνη διδυ[μοις]
6. κρονος ερμης [
7. υδροχω· αρη[ς
8. ζευς ταυρω [
9. [ωροςκ]ο[π]ος [

TRANSLATION

1. (Year) 2 of Antoninus C[aesar]
2. the Lord. Pham[enoth (from)]
3. 30 to 1; 4th hour of the n[ight]
4. sun in Pisces [....]
5. moon in Gem[ini ...
6. Saturn (and) Mercury [...
7. in Aquarius. Mars [in ...]
8. Jupiter in Taurus [...
9. Horoscopos [....

COMMENTARY

Diagram: pl. 2

As far as preserved the papyrus shows a blank space at the left in the width of about a third of a column; then the text of No. 149 follows and after it, in the next column, the horoscope No. 139. Possibly No. 139 was written first (but by the same hand as No. 149) since line 5 of No. 149 was too short to write the word ιχθυcι in a straight line, thus forcing the scribe to bend his line down and to raise the last two letters in small script above the υ.

Both horoscopes are dated, the one in the left hand column to Antoninus 13 Mekheir 8/9, the other one to Antoninus 2 Phamenoth 30/1. "Antoninus" would naturally be interpreted as Antoninus Pius. This would lead to the dates 149 December 21 and 139 February 14, respectively. Since, however, the agreement with modern computation is rather poor, we have not only investigated the possibility of the use of the Egyptian and of the Alexandrian calendar, but also the case that "Antoninus" could mean Marcus Aurelius Antoninus.[1] The result is that the smallest number of disagreements is reached if we interpret the horoscope in the left column as referring to Marcus Aurelius 13 Alexandrian Mekheir 8 (= 173 February 2) and the second one (column II) to Antoninus Pius 2 Egyptian Phamenoth 30 (= 139 February 14). However, the shift of meaning of "Antoninus" combined with a shift in calendaric notation makes this solution very implausible, not to mention the time interval of 34 years. Therefore we interpret Antoninus 13 as A.D. 149.

The numerical data for the present horoscope (column II) are as follows:

Text	Antoninus Pius 2		Marcus Aurelius 2	
Phamenoth 30/1	Eg. cal. 139 Feb. 14	Alex. cal. Mar. 25	Eg. cal. 162 Feb. 8	Alex. cal. Mar. 26
☉ ♓	♒ 25(!)	♈ 3(!)	♒ 19(!!)	♈ 4(!)
☾ ♊	♑ (!!)	♋ (!)	♉ (!)	♑ (!!)
♄ } ♒	♒ 17	♒ 21	♐ (!!)	♐ (!!)
☿ } ♒	♒ 17	♓ 10(!)	♒	♈ (!!)
♂ [...]	[♏ 18]	[♐ 4]	[♑]	[♒]
[♃] ♉	♉ 18	♉ 24	♈ (!)	♉
	♀: ♑ 18	♀: ♓ 4	♀: ♓	♀: ♉

The case of 139 February 14 shows the smallest number of discrepancies, but the moon is entirely wrong and it is also suspicious that the Egyptian calendar should have been used without explicitly saying so.[2]

Complete agreement between the preserved text of No. 139 and computation would be obtained if one could interpret the date "Phamenoth (VII) 30 to 1" as meaning "Mekheir (VI) 30/Phamenoth 1." By a peculiar accident exactly the same phrase "Phamenoth 30 to the 1st" occurs in the Almagest.[3] There, as all interpreters since Petavius have realized, it is necessary, because of the astronomical data, to emend Ptolemy's date to "Mekheir 30/Phamenoth 1." This coincidence suggests that we have here a specific terminology for dates which involve two different month names, but we have no additional evidence to support or to disprove such a hypothesis.[4]

Whatever the case may be, the date Antoninus Pius 2 Alexandrian calendar Mekheir (VI) 30/Phamenoth (VII) 1 = 139 February 24 10 P.M. leads to the following comparison:[5]

Text	Computed
☉ ♓	♓ 5
☾ ♊	♊ 20
♄ } ♒	♒ 18
☿ } ♒	♒ 30
♂ [...]	♏ 23
[♃] ♉	♉ 20

The only objection one can raise against this emendation (or interpretation) of the given date consists in the fact that the other horoscope of the same text also shows serious discrepancies between text and computation without any corresponding emendation which would eliminate these differences.

The interpretation of the traces in line 8 as ζευc is supported by computation. Venus seems to have been omitted unless we assume mention in line 6 after Mercury. However, this would place the planet in ♒ in contradiction to computation in all cases.

[1] It is easy to see that the dates of other rulers of the Antonine house are a priori excluded by the positions of ♃ and ♄.

[2] *Cf.* p. 166.

[3] Heiberg II p. 265,16: Φαμενὼθ λ' εἰς τὴν α'.

[4] *Cf.* perhaps No. 184 (below p. 51) where no irregularity appears in the counting of a last day of a year.

[5] Venus would be in ♑ 30.

No. 145

P. Hamb. 96

TRANSCRIPTION

1. ογδοον αντωνεινου
2. φαρμουτι κατ αρχαιους
3. ι̅δ̅ ωρα τριτη νυκτος

2 αρχαιους: writing cramped toward the end, the final ς coming very close to the edge of the papyrus. Consequently the present edge of the papyrus seems to be the original one and the restoration of a double day number ι̅γ̅ εις ι̅δ̅ seems to be excluded though it is the rule for an hour of night.

TRANSLATION

1. Eighth (year) of Antoninus
2. Pharmouthi according to the ancients,
3. 14, third hour of the night.

COMMENTARY

Faint traces of erased writings suggest interpreting the above lines as the beginning of a horoscope for 145 February 25.

No. 149

P. Oxy. 596, Col. I

(Photo: Roberts *GLH* pl. 16)

TRANSCRIPTION

1. ι̅γ̅ αντωνινου καισαρος
2. μεχειρ ·η̣· εις θ̣ ωρας
3. ε̣· νυκτος ·
4. ηλιος ερμης αφροδι̤τη
5. αιγοκερο σεληνη ιχ[θ]υ^σι
6. κρονος ζευς κριω·
7. αρης τοξοτη·
8. ωροσκοπος ζυγω·

TRANSLATION

1. (Year) 13 of Antoninus Caesar
2. Mekheir 8 to 9.
3. 5th hour of the night.
4. Sun, Mercury, (and) Venus
5. in Capricorn. Moon in Pisces.
6. Saturn (and) Jupiter in Aries.
7. Mars in Sagittarius
8. Horoscopos in Libra.

COMMENTARY

Diagram: pl. 2

This horoscope is written in the column which precedes the horoscope No. 139. The problem of interpretation of "Antoninus" and that of the calendar date are discussed in the commentary to No. 139 (*cf.* p. 46). The table given at the bottom of this page shows the comparison of the text with the figures in the four possibilities under consideration.

The position in Aries of both Saturn and Jupiter is satisfied for 173/174 and therefore again around 113/114 but certainly not in the intermediate sixty years which contain the reign of Antoninus Pius.

No. 150

P. Oslo 6

TRANSCRIPTION

1. γενεσις φιλοης ι^γ (ετους) αντωνινο[υ]
2. καισαρος του κυριου φαμε^θ ι̅ε̅
3. εις ι̅ς̅ ωρας δ̅ νυκτος ηλιος εν
4. ιχθυσι ζευς ερμης εν κριω
5. [κρον]ο̤ς̤ ε̤ν̤ καρκινω αρης λεον^τ
6. αφρο̣δ σεληνη υδρη̣[χο]ω
7. ωροσκοπ̣ει σκωρ'

7 The editors read: ωροσκω̣π̣ο̤ς̤ σκωρπ(ιος).

TRANSLATION

1. Nativity of Philoe. Year 13 of Antoninus
2. Caesar the Lord. Phamenoth 15
3. to 16, 4th hour of the night, sun in

Text		Antoninus Pius 13		Marcus Aurelius 13	
		Eg. cal.	Alex. cal.	Eg. cal.	Alex. cal.
Mekheir 8/9		149 Dec. 21	150 Feb. 2	172 Dec. 15	173 Feb. 2[1]
☉	♑	♑ 0°	♒ 13(!!)	♐ 22(!)	♒ 14(!!)
☿	♑	♐ 6(!)	♒ (!)	♑	♑ 23
♀	♑	♐ 3(!)	♑	♐ (!)	♒ 13(!)
☾	♓	♒ 18(!)	♍ (!!)	♉ (!!)	♓ 20
♄	♈	♋ 12(!!)	♋ (!!)	♈	♈ 9
♃	♈	♈ 17	♈	♓ (!)	♈ 0
♂	♐	♌ 27(!!)	♍ (!!)	♏ (!)	♐ 8
H	♎				

[1] No. L 173 is one day later.

4. Pisces, Jupiter and Mercury in Aries,
5. Saturn in Cancer, Mars in Leo,
6. Venus (and) moon in Aquarius.
7. Scorpio is the Horoscopos

COMMENTARY

Diagram: pl. 2

This horoscope was first published in a preliminary form in Pap. Osloenses I p. 21 and a corrected reading in P. Osloenses II p. 146-151 (with photo on pl. I) where Poul Heegaard explained the method by which one arrives at the date of a text of this type. He found complete agreement between text and computation with exception only of the moon which is in the second half of Capricorn instead of Aquarius.

For 150 March 11 which is the equivalent of Phamenoth 15 of the Alexandrian calendar one obtains

Text		Computed
☉	♓	♓ 20
♃ }	♈	♈ 28
☿ }		♈ 4
♄	♋	♋ 8
♂	♌	♌ 8
♀ }	♒	♒ 6
☾ }		♑ 19(!)
H	♏	about 10 P.M.

The position of the moon would be correct one day later.[1]

BIBLIOGRAPHICAL NOTE

This horoscope has been incorporated in the "Select Papyri" by Hunt and Edgar (vol. I p. 44). The Greek text reproduces the final readings from Pap. Osloenses II p. 151 whereas the translation is based on the preliminary text from Pap. Osloenses I p. 21, adding the error "10th year" (instead of 13th year) but giving the correct date A.D. 150.

A free translation is given in Winter, LLP p. 263 f. Macnaughton, SEC p. 331 note * also discusses the date of this horoscope.

No. 161

P. Fay. 139, verso

TRANSCRIPTION

1. αρποκρατιων [
2. α (ετους) αντωνινου [και]
3. ουηρου των κυριων
4. ϲεβαϲτων καθ ελληναϲ
5. μεϲορη ε ωρα ζ ημερα
6. κατα δε τους αρχαιους θωθ ις

---

[1] Use of the Egyptian calendar would lead to a position of the moon in ♌, thus to total disagreement.

TRANSLATION

1. Harpocration [......
2. Year 1 of Antoninus [and]
3. Verus, the Lords
4. Augusti; according to the Greeks
5. Mesore 5, 7(th) hour (of the) day,
6. according to the old (calendar) Thoth 16

COMMENTARY

The date XII(a) 5 = I(e) 16 corresponds to 161 July 29. The text is written near the upper edge of the verso of a large sheet, the lower part of which was left blank. Bundles of fibres from the recto of the papyrus show that the sheet extended to the left. Thus it is possible that our text is the end of a horoscope, the main part of which is lost with the preceding column.

The text was found in Qasr el Banat in the Faiyūm.[1] Its date is also discussed by Macnaughton, SEC p. 333.

No. 171

P. Harris 52

TRANSCRIPTION

Beginning destroyed
1. [............ προ ιδ καλανδων]
2. [μαι]ων κοτυλα [.] της [νυκτος]
3. [κρο]νος ιχθυϲι ζευς υδ[ρο]
4. [χοω] ηλις ερμης κριω ϲε[ληνη]
5. [αφρο]διτη ιχυϲι ωρα [......]

TRANSLATION

1. [...... 14 days before the calends of]
2. [Ma]y, [..] cotyla of the [night].
3. Saturn in Pisces, Jupiter in Aqua-
4. rius, sun (and) Mercury in Aries, moon
5. (and) Venus in Pisces. Horoscopos [.....]

COMMENTARY

Diagram: pl. 2

Between the first and the sixth century there is only the year A.D. 171 which gives agreement between the data of the text and our computation. One finds for 171 April 18

Text		Computed
♄	♓	♓ 23
♃	♒	♒ 19
☉ }	♈	♈ 27
☿ }		♈ 11
☾ }	♓	♓ 7
♀ }		♈ 7(!)

---

[1] Cairo, Cat. gén. X p. 102 No. 10812.

The discrepancy in the case of Venus is probably due to an inaccuracy in the ancient computation. Mars is omitted;[1] the arrangement of the planets is unusual.

The beginning of the text is unfortunately destroyed. We assume that the first preserved line refers to a number of " cotylai " of water (one cotyle is about 0.2 liter[2]) which measured the time by means of a water clock. The use of these units of capacity in connection with water clocks is attested from Theophrastus by Athenaeus.[3]

The editor of P. Harris, J. E. Powell, assumed for this and for No. 245 a date in the fourth century. It seems to us, however, that the style of writing is not incompatible with the astronomically determined date in the late second century. *Cf.* Schubart, *Pal.* p. 66 ff. and *Pap. Graec. Berol.* plate 26 ff.

No. 176

Dura, Graffito in Private House

Dura, Rep. II p. 161 ff. and pl. 51 (photograph and copy) and Johnson [1] p. 76 f.

Above diagram:

ζπυ παν θ
τ ρ(?)

Diagram: crude circle, with horizon and meridian as diameters. Beginning with the left endpoint of the horizon (= H) we read

ιχθ	*παρθ*
κρι	*ζυγ*
ταυ	[.....]
φαιν διδ	*τοξ*
καρκ φαε̣[...[1]	*αιγ*
λεων φωcφ cτιλβ̣ πυ	*[υδ]ρ̣*

Because ♀ and ☿ are in ♌ the sun must have been mentioned after ♃ in ♋, since no other space is available for a restoration near *καρκ*. Finally the moon must have been placed either in ♏ or in ♒. On this basis D. Brouwer gave the following alternatives:

176 July 3 to 5 if moon in ♏
July 10 to 12 if moon in ♒.

J. Johnson decided for the first possibility because the date above the diagram

(Seleucid Era) 487 (= A.D. 175/6)
Panemos 9[2]

excludes the second date. The nearest conjunction was June 24 about 10 A.M.; assuming that Panemos 1 began on June 25 at sunset, about 33 hours after conjunction one finds that Panemos 9 began at sunset of July 3. Since the sun was in ♋ and IMC = ♊ the hour was about 10 P.M. Computation for this date shows that the conditions of the horoscope can be satisfied for three consecutive days; the results for the middle one, July 4, are as follows (*cf.* the diagram on pl. 2):

	Text	Computed
♄	♊	♊ 4
♃	♋	♋ 22
♂	♌	♌ 11
[☉	♋]	♋ 11
♀	♌	♌ 2
☿	♌	♌ 9
[☾	♏]	♏ 15

Counting backwards from Panemos, approximately July 176, one finds that the preceding Hyperberetaios began after the conjunction of 175 October 2. This is in perfect agreement with the calendar still in use in Dura at the time of the horoscope of 219 January 9 (*cf.* below p. 55).

No. 177

P. Oslo 164

TRANSCRIPTION

1. *γενεcιc πρωταρχ̣ο̣υ [*
2. *ιζ (ετουc) φαρμουθ̣ι η [ειc θ ωραc .. τηc]*
3. *νυκτοc κρονοc ε[ν δ]ι̣δυ[μ]ο̣[ιc] ζ̣ε̣[υc ωροc]*
4. *κοποc λεοντι αρη[c αι]γ̣ο̣κ̣ε̣ρ̣ω̣ α̣[φροδιτη]*
5. *ερμηc η̣λ̣ι̣ο̣c ε̣ν̣ υ̣[δροχοω*

3 the editors read after *κρονοc*: *c[κ]ορ[π]ι̣ω̣ ...[.*
4 editors: *τ]ο̣ξοτη αφ[ροδιτη.*

TRANSLATION

1. Nativity of Protarchos [........
2. (year) 17, Pharmouthi 8 [to 9, ... hour of the]
3. night. Saturn in [Gem]ini, [Jupiter (and) Horos-]
4. copos in Leo, Mar[s in Capri]co[rn], V[enus,]
5. Mercury (and) sun in A[quarius ....

COMMENTARY

Diagram: pl. 3

Before discussing the date of this horoscope we first have to explain the restoration of *ζευc* in line 3. Without this we would have the arrangement

♄ in ♊; Horoscopos in ♌; ♂ in ♑

No. 171

[1] Computation: end of Sagittarius.
[2] *Cf. RE* 11, 1546 f.
[3] *Cf.* C. P. Schmidt, *EAW* p. 90, X.

No. 176

[1] The photograph clearly shows that J. Johnson's reading is correct, whereas *SEG* 7 No. 364 adopted *καρκιν*, which is in itself an unlikely abbreviation.

[2] The two letters(?) below the date are probably not numbers.

which would follow neither planetary nor zodiacal order beginning with the Horoscopos. Furthermore Jupiter would have been omitted or, at best, mentioned in the last place. If, however, we restore "[Jupiter and] Horoscopos in ♌" we obtain the ordinary planetary arrangement with the often attested modification of combining all data concerning the same zodiacal sign.

Heegaard, in discussing this horoscope in *Symb. Osl.* 15/16 p. 100 objected to the combination of an hour of the night with Leo as rising sign for the month Pharmouthi (VIII) because the sun would then be in Aries and thus above the horizon. This argument, however, is only correct for the Alexandrian calendar. In the Egyptian calendar Pharmouthi corresponds to a solar longitude of ♓/♈ in A.D. 100, to ♒/♓ in A.D. 200 and to ♑/♒ in A.D. 300. Consequently the date of our horoscope may fall anywhere after about A.D. 100. It must be admitted, however, that a month name, without further specification, normally is to be understood as referring to the Alexandrian calendar.[1]

The preserved year number 17 limits us to a date before Diocletian because, after 284, without exception, all horoscopes on papyrus give their date in the Diocletian era, the name Diocletian following the year number. And even if this text were an exception, the year Diocletian 17 is astronomically impossible, as can be shown by computation.

Now the year number 17 can be utilized. For the period in question only Antoninus Pius, Marcus Aurelius, and Severus are possible. The corresponding dates for Pharmouthi (VIII) 8 of the Egyptian calendar are

+154 February 18	with ♄ in ♌	♃ in ♍
+177 February 12	♊	♌
+209 February 14	♋	♈

the first and third of which are excluded by the given positions. For +177 February 12 one finds

Text		Computed
♄	♊	♊ 3
[♃]	♌	♌ 9
H	♌	about 7 P.M.
♂	♑	♑ 20
♀	♒	♒ 15
☿	♒	♓ 10 (!)
☉	♒	♒ 24

in good agreement with the text except for Mercury.

[1] *Cf.* p. 166.

No. 182

P. Mich. Inv. 6329

(unpublished)

Photo: pl. 31

Beginning destroyed

1. ζευς α̣γογερω μ̣(οιρων) [
2. αρεως [λ̇]εοντι μ̄(οιρων) [
3. αφροδιτης αγογερω̣ [
4. ερμης ιχθυς μ(οιρων)[
5. ηλιος ιχθυς μ̄̇(οιρων)[
6. σεληνη ιχθυς μ̣(οιρων)[
7. φ α̣γωγερω μ̄̇(οιρων) [
8. κληρους τυχης [
9. ευτυχηαι

TRANSLATION

1. Jupiter in Capricorn [....] degrees
2. Mars in Leo [....] degrees
3. Venus in Capricorn [....] degrees
4. Mercury in Pisces [....] degrees
5. sun in Pisces [....] degrees
6. moon in Pisces [....] degrees
7. Horoscopos in Capricorn [....] degrees
8. Lot of Fortune [.....]
9. Farewell

COMMENTARY

Diagram: pl. 3

The date of this horoscope (from Karanis in the Faiyûm) is found to be 182 February 20 to 22. The date could be determined exactly if the degrees of the moon were preserved. The hour must be about three hours before sunrise. The positions are for February 21:

Text		Computed
♃	♑	♑ 13
♂	♌	♌ 22
♀	♑	♑ 16
☿	♓	♓ 4
☉	♓	♓ 2
☾	♓	♓ 14
H	♑	about 3 A.M.

No. 184

P. Mich. 152

TRANSCRIPTION

1. (ετους) κδ επαγο/ ε̄ της
2. εις την ᾱ νεου ε̣τους
3. ωρα η̄ νυκτος
4. ηλις πα̣ρ̣θ[εν]ω̣

4 ηλις: not ηλιος as given in the edition.

5. *κρονος παρθενω*
6. *ερμης παρθενω*
7. *ζε[υς] κρειω*
8. *αφρ[οδ]ειτ ζυγω*
9. *σεληνη σκορπ̣ιω*

Remainder destroyed

TRANSLATION

1. Year 24, epagomene 5
2. to the first of the new year,
3. 8th hour of the night
4. sun in Virgo
5. Saturn in Virgo
6. Mercury in Virgo
7. Jupiter in Aries
8. Venus in Libra
9. moon in Scorpio

....

COMMENTARY

Diagram: pl. 3

As the editors have remarked, the year is the 24th year of Commodus, the last day of the Alexandrian year, corresponding to 184 August 28. The computed positions confirm the data of the text:[1]

	Text	Computed
☉	♍	♍ 4
♄	♍	♍ 16
☿	♍	♍ 12
♃	♈	♈ 2
♀	♎	♎ 9
☾	♏	♏ 3
		♂ : ♏ 17

No. 187

P. Aberd. 13

TRANSCRIPTION

1. *ε̣τ̣[ους εικοσι]*
2. *επτ[α ωρα δ της]*
3. *ημ[ερας μεσορη ς]*
4. *κα[θ] ελλ̣̣ηνων*
5. *κατα̣ δε [α]ι̣γυπτι[ω]ν*
6. *(ετους) κ̣η̣ θωθ̣ κ̣γ*
7. *κρο[ν]ος εν ζυ[γω]*
8. *ζευς εν καρ[κι]ν̣ο̣υ*
9. *αρη[ς] εν καρκ[ινω]*
10. *αφ[ρο]δ̣ε̣ι̣τη [ε]ν̣ λ[εοντι]*
11. *ερμ̣η̣ς̣ νε λεο[ντι]*
12. *ηλις̣ νε λεον[τι]*
13. *σεληνη νε δοξ[οτη]*
14. *ωρο[ς]κ̣οπος ζ[υγω]*

TRANSLATION

1. Ye[ar twenty-]
2. seven, [4th hour of the]
3. day, [Mesore 6]
4. according to the Greeks;
5. according to the Egyptians
6. year 28, Thoth 23.
7. Saturn in Libra
8. Jupiter in Cancer
9. Mars in Cancer
10. Venus in Leo
11. Mercury in Leo
12. sun in Leo
13. moon in Sagittarius
14. Horoscopos in L[ibra]

COMMENTARY

Diagram: pl. 3

The date of this text is Commodus 27 (= A.D. 187) Mesore 6 in the Alexandrian calendar, Commodus 28 Thoth 23 in the Egyptian calendar. The editors read in line 6 *θωθ ις*, but the last number is certainly *γ* and not *ς*; the first number is hardly *ι* and *κ* is a possible reading. Thoth 13 = July 20 is astronomically excluded. The sun is in Cancer 26 (instead of Leo) and the moon in Gemini (instead of Sagittarius). Ten days later, however, the sun is correctly in Leo and the moon is in Scorpio, i.e., not too far from the position given in the text. All other positions are correct:

	Text	Computed
♄	♎	♎ 14
♃	♋	♋ 6
♂	♋	♋ 28
♀	♌	♌ 14
☿	♌	♌ 29
☉	♌	♌ 5
☾	♐	♏ 9(!)
H	♎	about 10 A.M.

Since the sun is in ♌ and the hour is an hour of daylight, only a sign between ♌ and ♑ can be rising. In line 14 just enough is visible of the first letter of the Horoscopos to exclude all signs but *ζυγος*. The hour can then be restored as *δ* (line 2).

No. 190

P. Oslo 163

TRANSCRIPTION

1. *[λ (ετους)] κ̣ομ[μοδο]υ τυβι λ̇*
2. *φ αρχουσης ενvατη[ς ημερας]*
3. *ηλις ερμης υδρη̣[χοω]*

3 *ηλις*: *sic*, not *ηλιος* as in edition.

[1] Also in the case of Saturn, contrary to the statements made in the edition.

4. ϲεληνη   ιχθυϲ[ι]
5. κρονοϲ   ϲκορπι[ω]
6. ζευϲ   παρθ[ενω]
7. αρ[η]ϲ αφροδειτη τ[οξοτη]
8. οροϲκοποϲ διδυμ[οιϲ]

TRANSLATION

1. [Year 30] of Commodus, Tybi 30,
2. beginning of the ninth hour [of the day.]
3. sun (and) Mercury in Aquarius
4. moon in Pisces
5. Saturn in Scorpio
6. Jupiter in Virgo
7. Mars (and) Venus in Sagittarius
8. Horoscopos in Gemini

COMMENTARY

Diagram: pl. 3

As P. Heegaard [1] p. 98 has shown, the computation agrees fully with the data of the text for 190 January 25:

Text		Computed
☉ }	♒	♒ 5
☿ }		♒ 4
☾	♓	in ♓
♄	♏	♏ 22
♃	♍	♍ 18
♂ }	♐	♐ 12
♀ }		♐ 18
H	♊	about 2 P.M.

The calendar used is the Alexandrian.

If the sun is in ♒, the Horoscopos in ♊, then the time must be at least 4^h after sunrise and at most 8^h after sunrise in Alexandria. Since the text gives the "beginning of the 9th hour" it is clear that seasonal hours must be meant. In this case one finds as extremal value (for ☉ = ♒ 0, H = ♊ 30) 9;20 seasonal hours after sunrise.[1] This agrees well with our result of a solar position at the beginning of ♒.

---

[1] Accurately computed. Using the rising times of System A or B the limit becomes 9;30 seasonal hours.

No. 207

Bodleian Greek Ostracon 244

Wilcken, Ostr. 1602

TRANSCRIPTION

1. κρονοϲ διδυμ [....
2. ζευϲ υδριχω μ[....
3. αρηϲ αιγοϲαιρω [....
4. αφροδιτη κριω [....
5. ερμηϲ υδριχω [....
6. ϲεληνη ταυρω [....
7. ηλιοϲ ιχθυϲι λ[επ ....
8. φ υδριχω μιρ / [....
9. παρα αιγυπτιοιϲ φ[αρμουθι ....
10. κγ̄

7 λ[επ(των)]: suggested by Bilabel, *Berichtig.* II,1 p. 124.
8 φ : reading certain. μιρ / : sic.
9 φ[αρμουθι : perhaps followed by κβ̄ εις ; cf. note 2.

TRANSLATION

1. Saturn in Gemini [....
2. Jupiter in Aquarius [..] degrees [...
3. Mars in Capricorn [....
4. Venus in Aries [....
5. Mercury in Aquarius [....
6. Moon in Taurus [....
7. Sun in Pisces [..] mi[nutes]
8. Horoscopos in Aquarius [..] degrees [...
9. According to the Egyptians [Pharmouthi [1]
10. 23

COMMENTARY

Diagram: pl. 3

Computation leads to an excellent agreement with the data of the text for 207 February 20 Pharmouthi 23 (Egyptian calendar). One finds

Text		Computed
♄	♊ [...]	♊ 11
♃	♒ [...]	♒ 19
♂	♑ [...]	♑ 11
♀	♈ [...]	♈ 11
☿	♒ [...]	♒ 12
☾	♉ [...]	about ♉ 4
☉	♓ (beg. of)	♓ 1
H	♒ [...]	about 5 A.M.[2]

The same date was found by Fotheringham according to Bilabel, *Berichtig.* II,1 p. 124. The accuracy of his hour "between 4.55 and 6.25 A.M., if observed at

---

[1] *Cf.* note 2.
[2] For an hour of night one should perhaps restore in lines 9/10 [Pharmouthi 22 to] 23. The positions one day earlier would still be in agreement with the text, except for the moon.

Thebes," is, of course, purely fictitious since H may fall anywhere in ♒ thus leaving us with a margin of about 2 hours. The uncertainty is increased by the difference of several degrees between ancient and modern longitudes and the fact that the horoscope was computed and not "observed."

No. 208

P. Mich. Inv. 1461

(unpublished)

Photo: pl. 31

TRANSCRIPTION

1. (ετους) ιε // μεχ̣ε̣ιρ ι̅.[
2. κρονος καρκ̣(ινω) β̣[
3. ο ενιαυτος τ[
4. ζωδιον επ̣.[

Subsequent lines destroyed

We are indebted to Professor Herbert C. Youtie for the reading of the last line. One might restore επ̣ο̣[μενον; *cf.* Bouché-Leclercq, *AG* p. 418 ff.[1]

TRANSLATION

1. Year 15, Mekheir 10[+ x
2. Saturn in Cancer ..[
3. The ..[........] year [...
4. the f[ollowing (?)] zodiacal sign [....

COMMENTARY

The date of this horoscope is very uncertain. We know from the first line that it falls in the 15th year of some ruler, month Mekheir, i.e. roughly February. This restricts us for the time from Augustus to Diocletian to very few possibilities which all can be ruled out by the condition that Saturn should be in Cancer. The closest approach to a situation in which both conditions are satisfied is the year Septimius Severus 16 = 207/8 since Saturn is found in February 208 at the end of ♊ or entering ♋ if one takes the difference of norming longitudes into consideration. Thus it seems as if our text counted the year 193/4 as year 1 of Septimius Severus and not 192/3 as is usually assumed.

[1] Or ἐπαναφερόμενον ζῴδιον as in Bouché-Leclercq, *AG* p. 270 note 1.

No. 217

P. Warren 21

Verso Col. I, 67-72

TRANSCRIPTION

1. διδ̣ υμ κρονος ζ[υ]γω
2. ζευς ☾ αιγοκ(ερω)
3. αφροδιτη κρ[ι]ω
4. ☉ ταυρω
5. ερμης αρης διδυμοις
6. ⏀ λεοντι

TRANSLATION

1. Didymos.[1] Saturn in Libra
2. Jupiter (and) moon in Capricorn
3. Venus in Aries
4. sun in Taurus
5. Mercury (and) Mars in Gemini
6. Horoscopos in Leo

COMMENTARY

Diagram: pl. 4

Text and computation agree for 217 May 12 and 13 very well. For May 12 one obtains

Text		Computed
♄	♎	♎ 21
♃	♑	♑ 2
☾	♑	♑ 12
♀	♈	♈ 7
☉	♉	♉ 20
☿	♊	♊ 1
♂	♊	♉ 28
H	♌	about noon

and practically the same holds for one day later, the moon being in about ♑ 26.

BIBLIOGRAPHICAL NOTE

The *Magical Papyrus Warren* was first published by Hunt [1] (1932) and then again in *The Warren Papyri* (1941). The first column of the verso contains four horoscopes in the following order:

No. 219,II,1
No. 219,II,12
No. 217
No. 244.

[1] Or twins?

No. 219,I

Dura, House of Nebuchelus
Graffiti on South Wall of West Room

On a wall of an important house[1] in the center of Dura[2] six more or less complete versions of a horoscope were found, all scratched in a crude fashion into the plaster of the wall. Its date is, as will be shown presently, 219 January 9.

As the editors stated[3] it was probably the author of these drawings who wrote on the same wall

1. εποιςα cφηρον δαcιου ā
2. κατα cεληνην λ̄ μνηcθ[η] ο γραψας
3. ημερα cεληνης

"I have made the sphere on Daisios 1, according to the moon 30. May the writer be blessed.[3a] Monday."

The use of σφαίρα for a circular diagram is of interest and probably stems from the meaning "circular orbit" of σφαίρα in the astronomical literature.[4]

The terminology "Daisios 1, according to the moon 30" must mean that the 1st of Daisios in the civil calendar corresponded to the 30th of the preceding accurately taken lunar month (Artemisios). This coexistence of a somewhat inaccurate or schematic civil calendar and a strict lunar calendar is reminiscent of the Athenian calendar of the Hellenistic period.

Since the date of the horoscope in the lunar calendar is Aydnaios 5, corresponding to A.D. 219 January 9, one finds for the conjunction preceding the next month Daisios the date 219, about midnight from May 30 to May 31. Since May 30 is a Monday[5] the horoscope was actually cast, in all probability, when the child was five months old.

The seven versions of the horoscope itself[6] are written in six groups of graffiti, henceforth denoted by (a) to (g) as follows:[7]

(a) Dura, *Rep.* IV No. 232 upper half
(b)   lower half
(c)   No. 235
(d)   No. 236
(e)   No. 237
(f)   No. 238
(g)   No. 239

---

[1] *Cf.* Dura, *Rep.* IV p. 79 and fig. 1 p. 92.
[2] Also called "House of the Archives."
[3] *Rep.* IV p. 95 f., No. 220.
[3a] *Cf.* for this translation of μνησθῇ Rehm [1] p. 16.
[4] *Passim*, e.g., in the Almagest. The editors refer to the σφαίρα ἀστρολογική of Plutarch, Mor. 838c (*Rep.* IV p. 95 note 22) but in this passage a celestial globe, not a horoscope, is meant. For context and correct interpretation *cf.*, e.g., Schlachter, *Globus*, p. 58.
[5] The editors suggested Monday 235 May 5 but their procedure is influenced by the incorrect dating of the horoscope.
[6] We ignore No. 231 (*Rep.* IV, p. 105): "A circle, 0.065 m in diameter drawn for a horoscope never completed."
[7] Also *SEG* 7 Nos. 364 to 369.

The last copy was never completed.

The originals of the numbers 232, 235, and 236 are now at Yale University and were kindly put at our disposal, together with all relevant notes and photographs, by Professor C. B. Welles.

Version (a). Copy: *Rep.* IV p. 106[8] No. 232. Crude circle with horizon and meridian as diameters. Inscriptions, beginning with the horoscopos (left end of horizon):

υδρο	λεω
ιχθ	παρθ
κρι	ζυγ
ταυ	κρονος cκρο (sic)
διδ	αφροδ τοξ
καρ	ζευς αιγ

Version (b) written below (a). Same diagram as in (a). Above it the following text:

1. ετους λφ μηνος αυδιναιου θ
2. κατα cεληνην ε ημερα κρονου περι γ̄
3. φ ημεριυης εγενηθη αλεξαν μακεδων
4. απολλωνικου εν ιχθυδιω φ υδροχους

"Year 530 month Aydnaios 9, according to the moon 5, Saturday, about 3rd hour of the day, was born Alexander Macedonius son of Apollonicus ....; Horoscopos: Aquarius."

Around the diagram we read:

ζευς υδρ		λεω
αρης ☾ ιχθ		παρ
κριο		ζυγ
ταυ		cκροπιον (sic)
διδ		αφροδ τοξ
καρ		η^λ ερμης αιγ

The last line of the text is translated by the editors[9] as "son of Apollonicus; in the constellation of Pisces; Aquarius was rising on the horizon." Grave objections must be raised, however, against this interpretation. If the Horoscopos is ♒ then it makes no sense to say "(born?) in Pisces," in addition to the fact that "(born) in such and such a sign" is attested in no other horoscope. Furthermore, it is extremely unlikely that the constellation ἰχθύες is renamed ἰχθύδιον, even if there existed a preference for diminutives in Dura. Consequently, one must assume that the word in question belongs to the proper name, giving perhaps a locality or a profession or a nickname.[10]

---

[8] A photograph was published by Johnson [1] p. 79, but no details are legible.
[9] *Rep.* IV p. 107.
[10] Gagé [1] based far-reaching conclusions on this horoscope, using as support an astronomically meaningless emendation of a corrupt passage in the Alexander romance.

In the lower half of the diagram two lines are written for which the editors suggest a reading ἀρχὴ(ν) ἔχει. The reading itself is by no means clear and the significance extremely doubtful. What one would expect astrologically is κλῆρος τύχης which would be located in Aries,[11] right where the lower line was written. Nevertheless, this conjecture is not supported by the few letters which can be read: ... χη (or λη?) preceded by φρ or φ or ψ in the upper line, ετ or ευ[...] γλ or γχ in the lower line (*cf.* fig. 12 which indicates roughly what seems to be still visible on the original).

FIGURE 12.

The astronomical data from the two versions agree, except for Jupiter, which is supposed to be in ♑ in (*a*), in ♒ in (*b*). The latter version is supported also by (*c*) and (*d*). The editors suggested 218 December 11 (Saturday) as the corresponding date. Since the date of the horoscope is given as (Seleucid Era) 530 (= A.D. 218/219) Aydnaios 9 (civil) 5 (lunar), the month Aydnaios had to be equated with December instead of January, as expected. But computation reveals that Mars, sun, Venus, and Mercury were also one sign short, and it is very simple to repair all these discrepancies by going one month ahead. This is shown in the following comparison:

	Text	218 Dec. 11	219 Jan. 9
♄	♏	♏ 15	♏ 17
♃	♑ or ♒	♒ 4	♒ 15
♂	♓	♒ 24 (!)	♓ 16
☉	♑	♐ 18 (!)	♑ 19
♀	♐	♏ 7 (!)	♐ 10
☿	♑	♐ 25 (!)	♒ 7 (!)
☾	♓	♓ 1	♓ 27
	Saturday	Saturday	Saturday
		conj.: Dec. 5	conj.: Jan. 4

It is obvious that the second date is preferable. There remains only a small discrepancy for Mercury as is very common for this difficult planet. All other conditions are satisfied and the calendar leads to the expected position of Hyperberetaios beginning about 218, October 7.[12] Finally Daisios coincides with June, thus leading to the most natural date for the writer's inscription which is without a year number: 219 May 30 Monday for Artemisios (lunar) 30 = Daisios (civil) 1.

Since the sun is located in ♑, the Horoscopos in ♒, it is clear that the hour must be between one and three of daylight. The text above the diagram of (*b*) says "about 3rd hour" whereas (*c*) says "first hour." It seems as if the exact hour was unknown.

Version (*c*). No copy published; *cf. Rep.* IV p. 115 No. 235. Professor Welles kindly put a tracing of the text at our disposal; the upper part of the original is now broken away.

Above the circle two lines of text:

1. ετους λφ [μηνος] αυδυνεου θ
2. φ πρωτης ημερινης

Around the circle which is divided by two perpendicular diameters one reads (beginning at H)

ζευς υδρ	λ[εω]
☾ ιχθ	παρ
αρης	ζυγ
κρι	κρονος σκροπι (*sic*)
ταυρ	αφρο τοξ
διδ	η^λ ερμης αιγ
καρ	

These positions agree with version (*b*).

Version (*d*). Circle with horizon and meridian as in (*a*) to (*c*). Above it inscription (copy: *Rep.* IV p. 115 No. 236):

1. ετους λφ μηνος αυδενεου θ̄
2. κατα σεληνην ε περι φ α ημερινης
3. εγενηθη οκκας

For the last word the editors say "if it is a proper name, it could be the Semitic name of Alexander, or of Alexander's otherwise unattested (elder) twin brother. More probably ... something like ὁ παῖς."

Round the diagram we have

ζευς υδρ	λεω
☾ ιχθ	παρθ
αρης	ζυγ
κρι	κ̣ρονος σκορπ
ταυ	αφροδ τοξ
διδ	η^λ ερμης αιγ
κ̣α̣ρκ	

Version (*e*). Very crude circle, divided as before (copy: *Rep.* IV p. 116 No. 237). Outside:

ζευς υδρο	λ̣[εω] (?)
☾ ιχθ̣υ	παρ (?)

---

[11] Because $A_F = \lambda_H + (\lambda_{☾} - \lambda_{☉}) = ♒ + 2^s = ♈$.

[12] The famous hemerologia, discussed by Kubitschek [1] identify Hyperberetaios and October for the "Hellenes" (i.e. Antioch and Syria; *cf.* Kubitschek [2] p. 101). *Cf.* also Ginzel, *Chron.* III p. 29 ff.

αρηc	....
κρι	ζυγ
ταυρ	κρονοc cκορ
διδ	αφρο τοξ
καρκ	ηλ ε[ρμ]ηc αιγ

Inside the circle, in the lower half, is written what the editors interpret as ἀρχὴν ἔχεις and where we would expect the Lot of Fortune; *cf.* version (b).

Version (f). No copy published. *Rep.* IV p. 119 No. 238: " A copy of a horoscope of which little can be deciphered. The legend is better preserved."

1. [ἔτους λφ′] μηνὸς Αὐδυνέου θ′
2. ὥρ(ας) πρώτης ἡμερινῆς, ἐγενήθη
3. OKIΣ

Version (*g*). No copy published. *Rep.* IV p. 119 No. 239: " An unfinished copy of the horoscope, without the quarterings or the names of constellations and planets. At the left of the circle occurs the word ἔτους, at the right the beginning of the legend, ἔτους λφ′ μηνὸς ..... Below occurs the numeral ρμδ′." We have no explanation to offer for the last mentioned number (144).

No. 219,II,1

P. Warren 21

Verso Col. I, 52-58

TRANSCRIPTION

1. (ετουc) β αντωνι$^{\nu}$ κορυ'
2. μεχειρ ̅ϛ ειc την ̅ζ
3. ϕ ̅ζ νυκτοc
4. κρο′ ϕ cκορπιω
5. ζευ/ ☉ υδροχω^{o}
6. αρηc κριω
7. αφρο$^{\delta}$ ☾ ερμη^{c} αιγο$^{\kappa}$

TRANSLATION

1. Year 2 of Antoninus,[1] the Lord (?)
2. Mekheir 6 to 7
3. 7th hour of the night
4. Saturn (and) Horoscopos in Scorpio
5. Jupiter (and) sun in Aquarius
6. Mars in Aries
7. Venus, moon, (and) Mercury in Capricorn

1 κορυ' or κυρυο': one would expect του κυριου or καιcαροc.

COMMENTARY

Diagram: pl. 4

Cf. the bibliographical note to No. 217.

The data of the text agree with the results obtained by computation for 219 February 1, which is the equivalent of VI(a) 6 after midnight.

[1] Elagabalus.

Planet	Text	Computed
♄	♏	♏ 19
H	♏	after midnight
♃	♒	♒ 19
☉	♒	♒ 12
♂	♈	♈ 4
♀	♑	♑ 8
☾	♑	♑ 20
☿	♑	♑ 23

Fotheringham based his computations on the assumption " that the place of observation was Arsinoe " (*The Warren Papyri*, p. 63). Actually this assumption plays no role whatsoever because the data of the horoscope would fit equally well if he had computed for Tripoli or for Basra.

No. 219,II,12

P. Warren 21

Verso Col. I, 59-66

TRANSCRIPTION

1. (ετουc) β του αυτου
2. μεχειρ ̅ι̅ζ ειc ̅ι̅η
3. ϕ ια νυκτοc
4. κρο′ cκορπιω
5. ζευc ☉ ερ$^{\mu}$ υδροχω[
6. αρηc κριω
7. αφρο$^{\delta}$ ϕ ερ$^{\mu}$ αιγοκερω
8. cεληνη διδυμοιc

5 ερ$^{\mu}$: erased.

TRANSLATION

1. Year 2 of the same,[1]
2. Mekheir 17 to 18
3. 11th hour of the night
4. Saturn in Scorpio
5. Jupiter, sun {Mercury} in Aquarius
6. Mars in Aries
7. Venus, Horoscopos, (and) Mercury in Capricorn
8. moon in Gemini

COMMENTARY

Diagram: pl. 4

The date in the first line refers to the preceding horoscope[2] which is dated " year 2 of Antoninus, Mekheir 6 to 7." In the fifth line " Mercury " was erroneously mentioned with the sun but afterwards erased and correctly listed with Venus.

Because the present horoscope falls only 11 days later than the preceding one it is evident that the sun and the outer planets remain within their respective signs. One finds for 219 February 12

[1] Elagabalus; *cf.* No. 219,II,1.
[2] *Cf.* the bibliographical note to No. 217.

Text		Computed
♄	♏	♏ 19
♃	♒	♒ 20
☉		♒ 23
♂	♈	♈ 12
♀		♑ 21
H	♑	about 4 A.M.
☿		♑ 26
☾	♊	♊ 27

No. 220

P. Oxy. 585

TRANSCRIPTION

1. ετ[ουc .. μεχει]ρ [..]
2. ωραc θ ημεραc
3. ηλιc ερμηc υδροχω
4. κρονοc cκορπιω
5. ζευc ιχθυcι αρηc
6. τοξοτη αφροδειτη
7. κριω cεληνη αιγοκερω
8. ωροcκοποc εcχατα
9. διδυμων

TRANSLATION

1. Year [.. Mekhei]r [23]
2. 9th hour of the day
3. sun (and) Mercury in Aquarius
4. Saturn in Scorpio
5. Jupiter in Pisces, Mars in
6. Sagittarius, Venus in
7. Aries, moon in Capricorn
8. Horoscopos at the end of
9. Gemini

COMMENTARY

Diagram: pl. 4

Computation shows perfect agreement with the date of the text for 220 February 18 and excludes any other date within many centuries earlier and later.

Text		Computed
☉	♒	♒ 29
☿		♒ 7
♄	♏	♏ 30
♃	♓	♓ 20
♂	♐	♐ 14
♀	♈	♈ 13
☾	♑	♑ 22
H	end of ♊	about 3 A.M.

With H ≈ ♊ 30 the 9th (seasonal) hour requires a solar position toward the end of ♒;[1] this confirms our result.

[1] For $\lambda_{\odot}$ = ♒ 30 the time is about 8;30 seasonal hours after sunrise but 9;10 for $\lambda_{\odot}$ = ♒ 0.

No. 227

P. Iand. 89

TRANSCRIPTION

Beginning destroyed
1. α̣γαθ̣[η τυ]χ̣η
2. [γε]νεcιc πτολ̣[ε]μαιδοc
3. ετουc ζ̄ cε̣ουηρου αλεξανδρου
4. χυακ ϛ̄ ω̣ρα αρχη β̄ ημεραc

TRANSLATION

....
1. Good Fortune
2. Nativity of Ptolemais
3. Year 7 of Severus Alexander
4. Choiak 6, beginning of the second hour of the day.

COMMENTARY

The preserved lines are followed by blank space and probably belong to the end of a horoscope (*cf.*, e.g., No. 138/161). No positions are preserved. The date is 227 December 4, assuming Alexandrian calendar.

No. 244

P. Warren 21

Verso Col. I, 73-75

TRANSCRIPTION

1. διονυcια (ετουc) α φι[λι]ππ[ο]υ επεφ η̄ Φ β ημε^ρ
2. κρο/ αρη[c π̣αρθ]/ ζευ αφροδ τα[υρω]
3. ερμηc δ̣[ιδυμοιc Φ] ☿̣ καρκινω [

Remainder destroyed

TRANSLATION

Dionysia. Year 1 of Philippus, Epiphi 8, 2nd hour of the day. Saturn (and) Mars in [Virgo]. Jupiter (and) Venus in Taurus. Mercury in Gemini. Horoscopos (and) sun in Cancer. [.....

COMMENTARY

Diagram: pl. 4

Computation shows good agreement for 244 July 2 = Philippus Julius Verus 1 XI(a) 8:

Text		Computed
♄	[♍]	♍ 23
♂		♍ 13
♃	♉	♈ 27
♀		♉ 23
☿	♊	♊ 24
H	♋	after sunrise
☉		♋ 9

Cf. also the bibliographical note to No. 217.

No. 245

P. Harris 53

TRANSCRIPTION

Beginning destroyed
1. [ⓗ] ηλιος υ[δροχοω]
2. [ζευς] κριω
3. [ερμη]ς ταυρω
4. [κρο]νος ζυγω
5. [σε]ληνη αρ^χ σκορπιω
6. αφροδειτη υδροχοω
7. αρης ϊχθυσι

TRANSLATION

....
1. [Horoscopos,] sun in A[quarius]
2. [Jupiter] in Aries
3. [Mercury] in Taurus
4. Saturn in Libra
5. moon in the beginning of Scorpio
6. Venus in Aquarius
7. Mars in Pisces

COMMENTARY

Diagram: pl. 5

The restoration [ζευς] in line 2 and [ερμη]ς in line 3 instead of the inverse arrangement is based on the relative length of the two words in comparison with the available space. In both cases Mercury has too great an elongation from the sun.

Saturn in ♎ and Jupiter in ♈ would be correct for the years 186, 245, 305, 364. Mars, however, eliminates all but one of these dates if we take into account that only the months from January to March are possible because of the solar longitude. The moon is in the beginning of ♏ on February 6 or one month earlier or later but only in February is the sun in ♒. Thus we obtain for 245 February 6:

Text		Computed sunrise
[H]	♒	
☉	♒	♒ 18
[♃]	♈	♈ 25
[☿]	♉	♑ 30 (!!)
♄	♎	♎ 9
☾	beg. of ♏	♏ 3
♀	♒	♒ 23
♂	♓	♒ 24 (!)

No. 250,1

Dura, Graffito

"On a piece of fallen plaster from Wall Street near Tower 19." *Dura, Rep.* VI No. 736, p. 246 f. with copy. Above the crude diagram

[μ]ηνος περιτιου γ

This line is continued by some six or seven letters which, according to the editors, may either contain a personal name or be read as (κα)τα σελη(νην) α. The last letters look to us like δεια.

In the second line one can read γρα followed by the traces of two or three letters, perhaps [ε]γραψε. The editors suggest either [μνηςθῇ ὁ] γράψα[ς or [ἐγεννήθη ὁ] γράψα[ς.

Below these two lines a circle is drawn with two diameters roughly at 45° and 135° with respect to the horizontal direction. To the right of the circle, about the level of the center we read

αιγ ηλ[ι]ς

whereas the editors suggested αιγοκ[ερω]ς. Our reading seems not only to fit the traces very well but is supported by the fact that the month Peritios corresponds roughly to January, when the sun is in Capricorn.

To the left of the diagram one reads διδ and above it ζευ or (much less likely) ταυ. Jupiter in Gemini during January would suggest, e.g., the years 258, 270, 282, 294, etc. and, less likely, 246 or even earlier.

Above the diagram perhaps the moon, thus near ♓ or ♈.

No. 250,2

P. Ryl. 524

TRANSCRIPTION

Beginning and end of both columns destroyed

I		II	
1.	[ ]. λε	16.	αφ[ροδειτη]
2.	[τριγωνω α]φροδει	17.	φωςφ[ορος την]
3.	[της]	18.	πεμπτ[ην ταξιν ε]
4.	[..........]	19.	τυγ'χαν[εν κινου]
5.	[...... την π]ρωτην	20.	μενος ε[.......]
6.	[ταξιν ετυγ']χανεν	21.	αιγοκερ[...... ας]
7.	[κινουμεν]ος εν	22.	τερων λ[...... τρι]
8.	[ιχθυσι μ]οιρων	23.	γωνω ι[διω αφαιρει]
9.	[... οικ]ω διος	24.	τοις αρι[θμοις]
10.	[τριγω]νω α	25.	εωαν δυ[σιν]
11.	[ρεως ....]ν	26.	[ερμης]
12.	[......]	27.	ςτ[ιλβων εν τω]
13.	[......]	28.	ζω[διω ....]
14.	[την δευ]τεραν	29.	κα[.....
15.	[ταξιν ετυγ'χανεν]	30.	[......

The reverse contains in a second hand:

....]α' ημερ
....]ιτη

1 λε: or perhaps λ only.
6 ταξιν: chosen among other possibilities because this seems to have the right number of letters for the length of line; similarly for κινουμενος in 7 and in 19/20.
11: perhaps δυνω]ν.
18/19 ετυγ'χανεν: possibly one should read ν instead of γ'

ORIGINAL DOCUMENTS: No. 250,2 — No. 258

since γ and ν are interchangeable before γ, κ, and χ (cf. No. 14, line 7) and ν appears in a form similar to this, with horizontal transverse, at least in the Ptolemaic period (cf. Schubart, *Papyri Graecae Berolinenses* pl. 4c).

20-23: A possible restoration would be

  κινου]μενος ε[πι των εν]
  αιγοκερ[ου ριπι ας]
  τερων λ[αμπρων τρι]γωνω

TRANSLATION

I

1. [.......]...
2. [triangle of] Ven-
3. [us]
4. [........]
5. [.... had] the first
6. [place,]
7. [moving] in
8. [Pisces ...] degrees
9. [house] of Jupiter
10. [triangle] of M-
11. [ars .....] ..
12. [......]
13. [.......]
14. [had the se]cond
15. [place,]

II

16. Ve[nus]
17. Phosph[oros, had the]
18. fifth [place,]
19. [mo-]
20. ving [.......]
21. Capricorn [....]
22. ....[.... tri-]
23. angle of its[elf, subtracting]
24. from its num[bers (of longitude)]
25. evening set[ting]
26. [Mercury,]
27. St[ilbon in the ...]
28. zo[diacal sign ....]
29. ..[......
30. [......

COMMENTARY[1]

The text is divided into sections, one for each planet. These sections are partly designated by ordinal numbers. The word "first" appears in the second section (lines 4-11) the word "second" occurs in the next section (lines 12 ff.). In the second column the section of Venus is associated with "fifth."

From line 25 we know that Venus was morning star in ♑ and near its last visibility. Consequently the sun

---
[1] Previously published: Neugebauer—Van Hoesen [1].

must have about 15° more longitude than Venus, that is to say it must be in ♑ or ♒. From the note on the reverse we know that we are dealing with the first hour of the day. Thus the Horoscopos must be about ♒ or ♓. Mercury is always near the sun and in our case must be between Venus and the sun.

We can furthermore determine the longitude of the object No. 1. From line 9, "House of Jupiter," we have only ♐ or ♓ at our disposal. Lines 10/11 can only be restored "Triangle of Mars" or "of Venus." Neither Mars nor Venus rules a triangle which contains ♐, but Mars and Venus in daytime govern the triangle ♋ ♏ ♓. Thus ♓ is the only possible restoration in line 8. We furthermore restore Mars in line 11 and not Aphrodite, because of the length of the line. The blank space above the section of Venus in column II suggests that we are dealing with the beginning of a column. Thus column I must concern Nos. 1 to 4, column II No. 5 (♀), No. 6 (♂), and No. 7. It is furthermore clear that at least one more column must have preceded column I.

These facts suggest the following restoration of the structure of the text. It began with an introduction, now completely lost, and with the description of the Horoscopos, the last two lines of which are partly preserved in lines 1 to 3. We restore "[triangle] of Venus" which includes for daytime the sign of ♓. Then begins the enumeration of the celestial bodies following the order of the zodiac: beginning with No. 1 in ♓ and ending with No. 5 — Venus in ♑, No. 6 — Mercury in ♑ and finally the sun as No. 7 in ♑ or ♒. Unfortunately these elements do not suffice for an astronomical dating. Paleographically the early third century seems likely.

No. 258

P. Oxy. 1563

TRANSCRIPTION

1. ϛ (ετους) των κυριων η[μ]ων ουαλεριανου [και
  γαλλιηνο]υ και κορνηλιου σαλωνεινου
2. σε[βα]στων θωθ κθ [ωρα ϛ της] ημερας
3. κρονος ιχθυσι μοι(ρων) β(?)′ ....
4. αρης ζευς καρκινω μ[οι(ρων) .... επ]ι[κεντ]ροι
5. αφ[ρ]οδειτη λεοντι μοι(ρων) ....
6. ηλ[ι]ς ζυγω μοι(ρων) γ′
7. ερμης ζυ[γω] μοι(ρων) γ′
8. σε[λ]ηνη [υδρο]χοω μοι(ρων) κ[...].. μγ
9. ωροσκοπ[ος αιγο]κερω [........]

2: The date 29 (κθ) is certain. The editors read 27 (κζ) but the apparent ζ is only the result of a prolonged stroke of the κ, which thus connects with the lower part of the θ.

3-8: All around our text are traces of writing. It seems to us very doubtful whether minutes of longitude were indicated. A number 43 at the end of line 8 is certain, but it is written somewhat out of line and it is therefore doubtful whether it belongs to our text.

4 επ]ι̣[κεντ]ροι: For the importance of the centers *cf.*, e.g., Vettius Valens p. 22,27 or p. 49,33. We restore a similar passage in No. 260 line 7.

8 [υδρο]χο̣ω: this reading agrees better with the traces than [αιγο]κ̣ε̣ρω and is astronomically necessary.

The editors transcribe one more line where they read the proper name Eunopha. In the present state of preservation (as it appears in the photograph) we see no trace of this name.

TRANSLATION

1. (Year) 6 of our lords, Valerian [and Gallienus] and Cornelius Saloninus,
2. Augusti. Thoth 29 [6th hour of] the day.
3. Saturn in Pisces 2 (?) degrees ......
4. Mars (and) Jupiter in Cancer .. degrees, [in cen]ter
5. Venus in Leo [19 (?)] degrees ....
6. sun in Libra 3 degrees
7. Mercury in Libra 3 degrees
8. moon in [Aquarius] 20[+ x] degrees ...
9. Horoscopos in Capricorn [.......

COMMENTARY

Diagram: pl. 5

The chronological interest of this and related horoscopes is discussed in *Oxyrhynchos Papyri* vol. XII p. 229 ff. The astronomical data were checked by Fotheringham but for a date which is two days too early (258 September 24 instead of September 26) because of an erroneous reading of the date.

The high accuracy of Fotheringham's computations is in any case fictitious. It makes no sense to operate with local Oxyrhynchos time when the moment for which the computation has to be made shows a margin of uncertainty of about one hour either way. It is furthermore evident that the ancient computations were based on tables computed for "Lower Egypt" without any distinction between Alexandria and Oxyrhynchos or any other locality in northern Egypt. The longitude of Mars, as given by Fotheringham, is the result of a computing error of about 10°.

The agreement between text and computation for 258 September 26 is very good, showing Jupiter and Mars in almost exact conjunction and setting:

	Text	Computed
♄	♓ 2(?)	♓ 6
♂ }	♋ [. . .]	♋ 13
♃ }		♋ 12
♀	♌ 19(?)	♌ 20
☉	♎ 3	♎ 3
☿	♎ 3	♎ 12
☾	[♒] 20 [+ x]	♒ 21

The discrepancy in the case of Mercury is undoubtedly caused by the inaccuracy of the ancient tables for this troublesome planet.

The present horoscope was found together with No. 260 and was written by the same hand.

No. 260

P. Oxy. 1476

TRANSCRIPTION

1. α (ετους) μακρινιανου και κυιητου
2. cεβαcτων φαωφι β/ ημεραc
3. φ ι/ πε̣πληρωμενη μοι(ρων) β/
4. ωροcκοποc υδρηχο̣υ μοι(ρων) κη
5. κρονον κρειω ι̣α/ λε̣[π](των) λ̣β/ ακρωνυ
6. κτον αφαιρετι[κο]c
7. αρηc λεοντι επ̣[ικε]ν̣[τ]ροc
8. ζευc παρ̣θενω γ λεπτων
9. αφροδειτη παρθενω η′ ιϛ̄
10. ερμηc παρθενω κγ̄ μδ/
11. ηλιοc ζυγω η/
12. cε̣λ̣[η](νη) α̣[ιγοκ(ερω)] μ̣ο̣ι(ρων) η/ λεπ(των) λβ/
13. cαραπ[α]μμ[ω]νοc

5/6 κρονον and ακρωνυκτον: *sic*, not -οc.

7 επ̣[ικε]ν̣[τ]ροc: Fotheringham suggested "that the mutilated word means 'at the very beginning'." Our restoration assumes similarity with No. 258 line 4.

TRANSLATION

1. Year 1 of Macrinianus and Quietus,
2. Augusti, Phaophi 2, the 10th hour of the day
3. completed (and) 2 degrees.
4. Horoscopos in Aquarius 28 degrees
5. Saturn in Aries 11 (and) 32 minutes, in
6. opposition and retrograde
7. Mars in Leo in center
8. Jupiter in Virgo 3 (and ..) minutes
9. Venus in Virgo 8;16
10. Mercury in Virgo 23;44
11. sun in Libra 8
12. moon [in Capricorn] 8 degrees (and) 32 minutes
13. (Nativity) of Sarapammon

COMMENTARY

Diagram: pl. 5

For the chronological importance of this horoscope and its relation to No. 258 *cf. Oxyrh. Pap.* vol. XII p. 229 ff. and A. Stein in *Archiv* 7 (1924) p. 34 f. The accurate date, 260, September 29, was established by Fotheringham, who found good agreement between text and computation.

	Text	Computed	Text—comp.
H	♒ 28	♒ 26;32	+ 1;30
♄	♈ 11;32	♈ 5;46	+ 6
♂	♌ (?)	♌ 0;8	
♃	♍ 3	♍ 3;35	— 0;30
♀	♍ 8;16	♍ 11;1	— 3
☿	♍ 23;44	♍ 27;23	— 4
☉	♎ 8	♎ 6;19	+ 2
☾	[♑] 8;32	♑ 9;58	— 1

The position of Saturn is far different from the

correct value. This has, of course, no influence on the fact that Saturn is practically in opposition to the sun and thus retrograde.

The accuracy in the indication of time, 10 hours and 2 degrees (≈ 8 minutes), is most unusual and obviously senseless. Computation with Theon's tables lead for the latitude of Alexandria to 9;53,36 equinoctial hours after sunrise which is equivalent to 10;6,24,... seasonal hours or 10 seasonal hours plus 1;34 ≈ 2 time degrees, in excellent agreement with the text.

In line 8 the scribe forgot to enter the minutes of the longitude of Jupiter.

The text is also published, with a translation, in Hunt-Edgar, *Sel. P.*, p. 444 f. (No. 200).

No. 277
PSI 764

TRANSCRIPTION

1. αγαθη τυχη γενεςις θεωνι [νου]
2. β (ετους) προβου τυβι κϛ
3. φ β/ ημερας
4. κρονος σκορπιω
5. ζευς αρης σεληνη
6. αιγοκερω
7. αφροδιτη ιχθυσι
8. ερμης ηλιος υδρηχοω
9. ωρα υδρηχοω
10. διευτυχει

9 ωρα: sic for ωροςκοπος.

TRANSLATION

1. With good fortune. Nativity of Theoninus
2. 2nd year of Probus, Tybi 26,
3. 2nd hour of day
4. Saturn in Scorpio
5. Jupiter, Mars, (and) moon
6.   in Capricorn
7. Venus in Pisces
8. Mercury (and) sun in Aquarius
9. Horoscopos in Aquarius.
10.   Good luck!

COMMENTARY

Diagram: pl. 5

Except for Mars and Venus, computation for 277 January 21 confirms the data of the text (Alexandrian calendar):

Text		Computed
♄	♏	♏ 10
♃ }		♑ 10
♂ }	♑	♒ 16 (!)
☾		♑ 30

Text		Computed
♀	♓	♒ 6 (!)
☿ }		♒ 19
☉ }	♒	♒ 2
H	♒	about 7 A.M.

No. 282
P. Cornell, Inv. No. 78
(unpublished [1])

Photo: pl. 31

TRANSCRIPTION

1. αγαθη τυχη
2. γενεσις α (ετους) καρινου θωθ ιϛ
3. φ αρχη θ φ υδρηχοω
4. ζευς σεληνη καρκινω
5. αφροδιτη λεοντι
6. ηλιος ερμης παρθενω
7. κρονος το[ξο]τη
remaining lines destroyed

TRANSLATION

1. Good fortune.
2. Nativity, year 1 of Carinus, Thoth 16,
3. beginning of the 9th hour (of the day); Horoscopos in Aquarius
4. Jupiter and moon in Cancer
5. Venus in Leo
6. sun and Mercury in Virgo
7. Saturn in Sagittarius
[8. Mars in Capricorn]

COMMENTARY

Diagram: pl. 5

The agreement between text and computation is excellent for 282 September 13:

Text		Computed
H	♒	about 3 P.M.
♃ }	♋	♋ 18
☾ }		♋ 3
♀	♌	♌ 10
☉ }	♍	♍ 20
☿ }		♍ 29
♄	♐	♐ 29

The entry for Mars is destroyed, but the arrangement according to increasing longitudes places Mars between Saturn in Sagittarius and the Horoscopos in Aquarius, exactly as expected from computation (♑ 14).

---

[1] We are indebted to the late Professor W. L. Westermann, who identified this horoscope in the Cornell collection, for permission to publish the text in the present volume.

No. 283

P. Oxy. 1564

TRANSLATION [1]

1. Nativity of Pichime
2. Year 1 of Carinus, Phamenoth
3. 27, 1st hour of the day
4. Horoscopos (and) sun in Aries,
5. Mars in Taurus
6. moon (and) Jupiter in Cancer
7. Saturn in Capricorn
8. Mercury (and) Venus in Pisces.
9. Good luck!

COMMENTARY

Diagram: pl. 5

Computation for 283 March 23 confirms the text

	Text	Computed
H	♈	about 6 A.M.
☉	♈	♈ 2
♂	♉	♉ 15
☾	♋	♋ 0
♃	♋	♋ 14
♄	♑	♑ 16
☿	♓	♓ 4
♀	♓	♓ 13

No. 284

PSI 765

TRANSCRIPTION

Recto

1. αγαθη τυχη κατα τον̣
2. αναδοθεντα μοι χρονον̣
3. υπο cη φιλτατη εγεν̣ηθηc
4. τηc ευτυχηcτατηc
5. διοκλ̣ητιανου βαcιλια̣c
6. μινι κατ' ελληνα̣c α̣θυρ
7. ιδ∫// ⏀ ς∫// νυκτοc
8. ουτωc εχουcι οι
9. αcτεραιc εcτι δε
10. ηλιοc cκορπιω μοιρων ι̣θ̣
11. cεληνη ταυρω μοιρων ῑ.
12. κρονοc αιγοκερω μο:ρων λ̣̄
13. ζευc παρθενω μοιρων η̄
14. αρηc ιχθυcι μοιρων ιθ̄
15. αφροδιτη cκορπιω μοιρων [...]
16. ερμηc c̣κ̣ο̣ρπ̣ιω̣ μοιρων [...]
17. ωρ[οc]κ̣[οποc λεοντι μοιρων ...]
18. .. [....
19. .[......

remainder destroyed

Verso. The editors read

1. ∠ λ̣α ∫ Διο(κλητιανοῦ)
2. Ἀθὺρ ιδ ∫ //
3. ὥρ(α) ς νυκτός

Except for the last line we see only faint traces which might belong to three lines (not only two) and the whole possibly might be restored as follows

1a. [(ετουc)] α̣ διοκ̣

1b. [λ]η̣τ̣ι̣α̣ν̣ο̣υ̣

2. [αθ]υρ ι̣δ̣

3. ⏀ ς νυκτοc

4: Traces at the end might be the expected α (ετουc).

11 ῑ.: traces of a letter following ι.

12 λ̣̄: or κ[ ].

13 η̄: ιη̄ not impossible.

TRANSLATION

1. With good fortune. According to the
2. date given me
3. by you, dearest (lady), you were born
4. in the blessed
5. reign of Diocletian, (year one)
6. according to the Greek (calendar) the month Athyr
7. the 14th, 6th hour of night.
8. Thus stand the
9. stars, that is:
10. sun in Scorpio 19 degrees
11. moon in Taurus 10 [+ x] degrees
12. Saturn in Capricorn 30 degrees
13. Jupiter in Virgo 8 degrees
14. Mars in Pisces 19 degrees
15. Venus in Scorpio [..] degrees
16. Mercury in Scorpio [..] degrees
17. Horoscopos [in Leo ..] de[grees]

....

Verso

1. (year) 1 of Diocletian
2. Athyr 14
3. 6th hour of night

COMMENTARY

Diagram: pl. 6

The editors interpreted the traces of the date on the reverse as "year 31 of Diocletian" (A.D. 314). Computation of the planetary positions for this date, however, shows total disagreement. The only possible date turns out to be the first year of Diocletian, as is shown by the following comparison for 284 November 10. Diocletian 1 III(a) 14:

	Text	Computed	Text—comp.
☉	♏ 19	♏ 19	0
☾	♉ 10[+ x]	♉ 19	
♄	♑ 30	♑ 24	+6

[1] The original which should be in Washington University, St. Louis, is now lost.

♃	♍ 8(?)	♍ 18	
♂	♓ 19	♓ 21	—2
♀	♏ [ ]	♏ 6	
☿	♏ [ ]	♏ 7	

For the 6th hour of night the Horoscopos must be somewhere between about ♌ 3 and ♌ 23.

No. 293,VIII

P. Oxy. 1565

TRANSCRIPTION

1. θ (ετουc) και η (ετουc) και α (ετουc) πληρọυμ[ε]
2. νου επ̣α̣γ̣ο̣μ̣ε̣ν̣ω̣ν̣ ε ωρα α΄
3. ημεραιc̣
4. ω̣ρ̣α̣ ηλιοc παρθ̣ε̣νω
5. αφροδιτη ζυγω
6. αρηc cκορπιω
7. cεληṿ[η] τοξοṭη πολλ[α]ν
8. κρονοc [δ]ι̣[δ]υμοιc αρχαc
9. ζ[ευ]c̣ δ̣ι̣δ̣υμοιc
10. [ερ]μ̣ηc λ̣εοντι ε̣ξ[.].....

9 ζ̣[ευ]c̣: the traces agree very badly with the only plausible restoration.

10: The editors read ερμηc δι ολου του ξ[ο]ανου λ̣εο̣[ν] [το]c assuming "an unprecedented use of the word (ξόανον 'image') for a sign of the zodiac." We should expect a term for "going out" or "leaving."

TRANSLATION

1. 9th (year) and 8th (year) and 1st (year) completed
2. 5th (day) of the epagomenai, 1st hour
3. of the day.
4. Horoscopos (and) sun in Virgo
5. Venus in Libra
6. Mars in Scorpio
7. moon in Sagittarius far advanced(?)
8. Saturn in the beginning of Gemini
9. [Jupiter] in Gemini
10. Mercury in Leo, going out (?)

COMMENTARY

Diagram: pl. 6

The initial date refers to the 9th year of Diocletian, and the 8th year of Maximianus and the first year of the Caesars Constantius and Galerius. Computation for Diocletian 8 Epag. 5 = 293 August 28 shows good agreement with the text:

	Text	Computed
H	♍	about 6 A.M.
☉	♍	♍ 5
♀	♎	♎ 21
♂	♏	♏ 5
☾	♐	♐ 26
♄	beg. of ♊	♉ 29
♃	♊	♊ 22
☿	leaving ♌	♌ 30

The reading of the last word in line 7 is very doubtful. Perhaps it means "many (degrees)" in Sagittarius. The moon is in first quarter.

No. 293,XII

P. Fouad 7

TRANSCRIPTION		TRANSLATION
Beginning destroyed		Beginning destroyed
1. [ηλιοc]	εν τοξοτηι	1. [sun] in Sagittarius
2. [cελην]η	εν παρθενωι	2. [moo]n in Virgo
3. [κρονοc]	εν ταυρωι	3. [Saturn] in Taurus
4. [ζευc]	εν διδυμοιc	4. [Jupiter] in Gemini
5. [αρηc]	ε̣ν̣ αιγοκερω	5. [Mars] in Capricorn
remainder destroyed		....

Between lines 1 and 2 and between lines 4 and 5 is a free space of about half a line in height.

COMMENTARY

Diagram: pl. 6

In line 2 a final η suggests the restoration of either cεληνη or αφροδιτη. The arrangement of the zodiacal signs does not follow their natural order. Therefore, we may assume that the text follows the order of the planets. The mention of Venus in line 2 would necessitate supplying in lines 4 and 5 the Horoscopos and another center but Gemini and Capricorn are neither diametrically opposite to each other nor 90° distant. Thus Venus is excluded and we must restore the moon in line 2.

Under these assumptions a possible date would be 293 December 8 because the sun is at this date in ♐ 17, the moon about ♍ 20, Saturn in ♉ 23, Jupiter in ♊ 20, and Mars in ♑ 21.

No. 316

Grenfell [1]

TRANSLATION [1]

1. Good fortune. Nativity
2. of Heracleides.
3. 33rd year of Diocletian
4. ⟨Phaophi⟩ Thoth 28
5. 5th hour of day
6. sun and Mercury in Libra
7. moon in Cancer
8. Saturn in Aquarius

4 ⟨Phaophi⟩: crossed out. Phaophi is not the equivalent in the Egyptian calendar of Alexandrian Thoth.

6 Sun and Mercury: και written above the line.

[1] We were not able to learn the present location of the text.

9. Jupiter in Gemini
10. Mars in Aries
11. Venus in Virgo
12. Horoscopos
13. in Sagittarius

COMMENTARY

Diagram: pl. 6

One obtains exact agreement for 316 September 25 which is the date corresponding to Diocletian 33 I(a) 28:

	Text	Computed
☉		♎ 3
☿	♎	♎ 27
☾	♋	♋ 0
♄	♒	♒ 27
♃	♊	♊ 1
♂	♈	♈ 11
♀	♍	♍ 10
H	♐	about 11 A.M.

No. 320

P. Iand. 88

TRANSCRIPTION

1. αγαθη τυχη γ[ενεcιc ....
2. ετοc λη[....
3. λη διοκλητιαν[ου]
4. τυβι δ ω[ρα ....
5. ωρα ερμ[ηc ....
6. κρονοc [....
7. ζευc ζ[υγω ....
8. αρηc κα[ρκινω]
9. c ηλιοc αφ[ροδιτη ....
10. cεληνη [....
followed by empty space and
11. γενεcιc [.....
remainder destroyed

2 λη: probably the same correction of λζ to λη as in the next line.

3 λη: η written over ζ.

4 ω[ρα: or ⲫ; *cf.*, e.g., Nos. 277 and 282.

9 ηλιοc: corrected from ηλιον. The c in the margin of the line corresponds to the c written above the ν of ηλιον.

Traces of previous writing are visible at different points.

TRANSLATION

1. Good Fortune. N[ativity of ...
2. year 3[8 ....
3. 38 of Diocletian
4. Tybi 4 [11th] hour [of the night]
5. Horoscopos (and) Mercury [in Sagittarius]
6. Saturn [in Aries]
7. Jupiter in L[ibra]
8. Mars in Can[cer]
9. sun (and) Ven[us in Capricorn]
10. moon [in Gemini]
11. Nativity [of ....]

COMMENTARY

Diagram: pl. 6

The date of this horoscope was changed from Diocletian 37 Tybi 4 (= 320 December 30) to Diocletian 38 Tybi 4 (= 321 December 30). Computing for both dates one obtains:

	Text	320 Dec. 30	321 Dec. 30
H		about 4 A.M.	about 4 A.M.
☿	[....]	♐ 17	♐ 27
♄	[....]	♈ 17	♉ 2
♃	♎	♎ 6	♏ 4 (!)
♂	♋	♋ 4	♑ 10 (!)
☉		♑ 10	♑ 10
♀	[....]	♑ 1	♑ 30
☾	[....]	♊ 19	♏ 6

From this comparison it is evident that only the original earlier date corresponds to the longitudes of Jupiter and Mars given in the text. In neither case is the moon in the same sign with sun and Venus, thus we have to assume a greater length for line 9 than for the ones immediately preceding.

No. 326

Louvre 10.390
(unpublished)

Photo: pl. 30

Parchment folded in zigzag, the ends pasted to stiff cover. Size of the pages about 51 × 58 mm. The first three pages contain some legal statements, followed by signatures, but without a date. Then follow three empty pages, and finally two pages inscribed by another hand and upside down as compared with the text at the beginning. This obviously later text is a horoscope, published here for the first time. We are obliged to Dr. J. Doresse for drawing our attention to this horoscope and for providing us with a photograph.

TRANSCRIPTION

I

1. αγαθη τυχη
2. γενεcιc κοπριου
3. μβ (ετουc) απο διοκλητιανου
4. μεχειρ ιδ ⲫ ᾱ ημεραc
5. ωραcκοποc υδρηχοω
6. ⲫ υδρηχοω

II

7. κρονοc διδυμοιc

8. ζευϲ αρηϲ ϊχθυϲι
9. ηλιοϲ ερμηϲ υδρηχο
10. αφροδιτη αιγοκερω
11. κρονοϲ διδυμοιϲ
12. ϲεληνη ζυγω

5 *ωραϲκοποϲ*: *sic*; *cf.* No. 338.
7: followed by signs written on the left part of page III, perhaps μοι ϛ (or μ̣ο ιϛ).
8: followed perhaps by a κ written on page III.
9: followed by ιϛ (or ιϲ ?) written on page III.
12: followed by a sign which is either ϛ̣ or ϛ, written on page III.

TRANSLATION

I

1. Good fortune
2. Nativity of Koprios
3. year 42 of Diocletian,
4. Mekheir 14, hour 1 of the day.
5. Horoscopos in Aquarius
6. Horoscopos in Aquarius

II

7. Saturn in Gemini
8. Jupiter (and) Mars in Pisces
9. sun (and) Mercury in Aquarius
10. Venus in Capricorn
11. Saturn in Gemini
12. moon in Libra

COMMENTARY

Diagram: pl. 6

Computation for the given date, Diocletian 42, Mekheir (VI) 14, (−326 February 8) leads to the following comparison

	Text		Computed
H	♒		about 6 A.M.
♄	♊ 6(?) or 16(?)		♊ 29
♃	♓	20(?)	♒ 29
♂			♓ 13
☉	♒	16(?)	♒ 20
☿			♒ 1
♀	♑		♑ 8
☾	♎	6(?)	♎ 10

The agreement is good except for the rather doubtful numbers of degrees which the text seems to add.

No. 338

PSI 23,a

TRANSCRIPTION

1. αγαθη τυχη
2. γενεϲιϲ ερμηϲιω̣[νοϲ]
3. ν̅ε̅ (ετουϲ) διοκλητιανου
4. χοιακ κ̅η̅ ειϲ κ̅θ̅ γ̄ νυκτοϲ
5. ωραϲκοποϲ λεοντι μυρ/ ιγ
6. ηλιοϲ αιγογερω: μυρ/ ν
7. ϲεληνη ϲκορπιω: μυρ/ λε
8. κρονοϲ τοξοτη μυρ/ θ λʹ α
9. ζευϲ ιχθυϲι: μυρ/ κβ
10. αρηϲ αιγογερω μυρ/ ιδ ʹλ ϛ̣
11. αφροτιδη: ϲκορπιω μυρ/ κθ
12. ερμηϲ αιγογερω μυρ/ ιθ
13. κληροϲ τυχηϲ π̣αρθενω μυρ/ κθ̣
14. κλ(ηροϲ) δαιμονοϲ διδ[υ]μοιϲ μυρ/ κ
15. κλ(ηροϲ) ερωτοϲ τοξ[οτ]η μυρ/ ιε
16. κλ(ηροϲ) αναγκηϲ τ̣[αυρω] μυρ/ ι λε θ̣
17. η προγεναμενη αποκρο̣υ̣ϲιϲ
18. ε̣γε̣νετο διδυμοιϲ μυρ/ β λε/ μθ
19. οικου κλιου οριοιϲ θ κρονου διμ/ β
20. οικου κρονου ορ̣ι̣ο̣ι̣ϲ̣ ε̣ρμο̣υ̣ διμ/ α
21. οικου αρ̣ε̣ω̣ϲ ορ̣ι̣ο̣ι̣ϲ̣ ερμο̣υ̣ διμ/ β̣
22. οικου διοϲ οριοιϲ διοϲ διμ/ α
23. οικου δ̣ι̣ο̣ϲ̣ οριοιϲ αρεωϲ διμ/ γ
24. οικου κ̣ρ̣ονου οριοιϲ αφροτ̣ι̣δ̣η διμ/ β
25. οικου αρεωϲ οριοιϲ κρονου διμ/ γ
26. οικου κρονου οριοιϲ αφροτιδη διμ/ β
27. ο̣ι̣κ̣ο̣υ̣ ε̣ρ̣μ̣ου οριοιϲ κρονου διμ/ γ
28. οικου διοϲ οριοιϲ αφροτιδη διμ/ β
29. οικου αρεωϲ οριοιϲ ερμου διμ/ β
30. <u>οικου ερμου οριοιϲ αρεωϲ διμ/ γ</u>

5 *ωραϲκοποϲ*: *sic*; *cf.* No. 326.
13 *κθ̣*: the editors read *κε* but *κθ* is equally possible and is required by the fact that the *κλῆρος τύχης* falls in the terms of Saturn (line 27) which extend only from ♍ 28 to ♍ 30.
16 *τ̣[αυρω]*: Dr. V. Bartoletti kindly writes us: "le tracce sono molto incerte, tuttavia a me pare *κ̣[ριῷ]*, meno probabile *τ̣[αύρῳ]*." Computation favors Taurus.
19 *κλιου*: the editors read *ἡλίου* but *κ* seems certain.
19 *θ κρονου*: the editors read *θ̄ κρόνου* but we see no bar over the *θ* nor is there any number expected in this context.
20 *ορ̣ι̣ο̣ι̣ϲ̣ ερμου*: the editors read *ὁρίοις ἡλίου*.
21 *β*: the editors read ι̣ but there are not ten decans in a sign. The reading *β* is not only required by the context but fits the traces very well.
30: The final line is actually drawn below line 31 which is the first line of the next horoscope (No. 376 b).

TRANSLATION

1. With good fortune.
2. Nativity of Hermesion
3. 55th (year) of Diocletian
4. Choiak 28th to 29th; 3rd (hour) of night
5. Horoscopos in Leo, 13 degrees
6. sun in Capricorn, 50(!) degrees
7. moon in Scorpio, 35(!) degrees [1]
8. Saturn in Sagittarius, 9 degrees, 1 minute

[1] If we interpret *λε* not as a numeral but as abbreviation for *λεπτα* (*cf.* lines 16 and 18) we could translate "moon in Scorpio, degrees, [..] minutes," assuming omission of the number of degrees.

9. Jupiter in Pisces, 22 degrees
10. Mars in Capricorn, 14 degrees, 6 minutes
11. Venus in Scorpio, 29 degrees
12. Mercury in Capricorn, 19 degrees
13. Lot of Fortune in Virgo, 25 degrees
14. Lot of Daimon in Gemini, 20 degrees
15. Lot of Eros in Sagittarius, 15 degrees
16. Lot of Necessity in [Taurus], 10 degrees, 9 minutes
17. The preceding ἀπόκρουσις
18. occurred in Gemini, 2 degrees, 49 minutes
19. House of the sun, terms of Saturn, decan 2
20. House of Saturn, terms of Mercury, decan 1
21. House of Mars, terms of Mercury, decan 2
22. House of Jupiter, terms of Jupiter, decan 1
23. House of Jupiter, terms of Mars, decan 2
24. House of Saturn, terms of Venus, decan 2
25. House of Mars, terms of Saturn, decan 3
26. House of Saturn, terms of Venus, decan 2
27. House of Mercury, terms of Saturn, decan 3
28. House of Jupiter, terms of Venus, decan 2
29. House of Mars, terms of Mercury, decan 2
30. House of Mercury, terms of Mars, decan 3

338	Hermesion
351	Hermeione
366	Nilammon
370	Cyrillous
373	Hermeion
376a, b	Joannes (or Joanna)
381a, b	Cyrus
385	Nestia (also called Apollonia)

suggest a family connection between all these persons.

The lines 19 to 30 continue, line by line, the lines 5 to 16. For example, one has to connect line 5 "Horoscopos in Leo, 13 degrees," and line 19: "house of the sun, terms of Saturn, decan 2," and similarly in all other cases. This error in arrangement confirms the character of this document as a later copy made from individual texts.

The decans are denoted by δι μ/ which the editors suggested may stand for διμοιρῶν; because this makes no sense as a synonym of decans we suggest διαμερισμῶν "division." For the omission of an interior a see Mayser, *Gramm.* I p. 149.

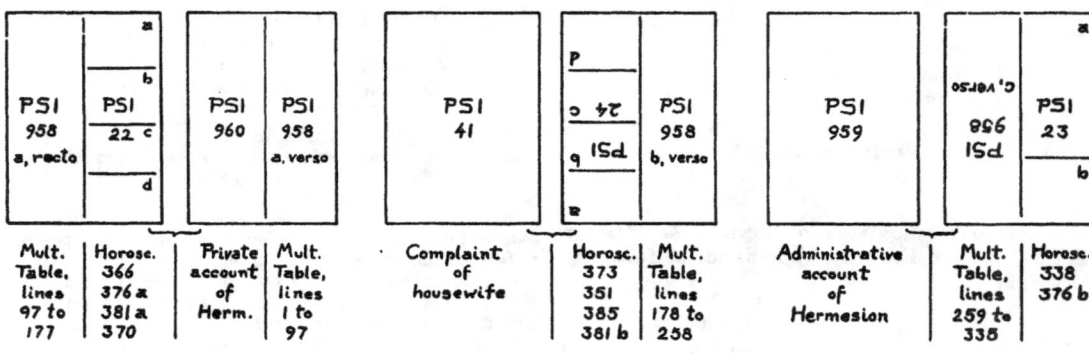

FIGURE 13.

COMMENTARY

Diagram: pl. 7

The horoscope of Hermesion (born 338) is the first, and most elaborate, of a group of eight horoscopes which were published as PSI 22, 23, and 24. All these horoscopes were copied down, in all probability by Hermesion himself, on three leaves of papyrus which contain also two accounts, written by Hermesion, and a large multiplication table in the same hand. One page was previously inscribed with a complaint of a housewife against her husband, but shows no connection with Hermesion or the other persons whose horoscopes were collected here. The arrangement of these texts is schematically represented in fig. 13. The dates of these collected horoscopes

Two obvious errors occur in the lines 6 and 7. In the first place we find ♑ 50, in the second ♏ 35, thus numbers greater than the maximum of 30° for positions within a zodiacal sign. One may assume a copyist's error either in the number of degrees or an omission of degrees and inclusion of minutes.

In lines 17 and 18 it is said that a "preceding ..." occurred in Gemini 2;49°. The editors read ἀπόκρασις but no such term seems to be attested elsewhere. We suggest ἀπόκρουσις and assume a connection with a doctrine which appears in Vettius Valens V.8 (p. 216 ff. Kroll). There the moon is said to ἀποκρούειν in the sign opposite to the sign of conjunction.[2] Now the preceding

---

[2] The preceding full moon happened December 14 about ♊ 24,

conjunction happened November 28 in ♐ 7 which is very nearly opposite to ♊ 2;49. Though the details escape us by which this specific position was computed we feel that we might have here an application of the above mentioned doctrine of Vettius Valens.

The positions given agree fairly well with computation for the date Diocletian 55 IV(a) 28 = 338 December 24 about 8 P.M. One finds

	Text	Computed	Text—comp.
H	♌ 13	8 P.M.	
☉	♑ 50(!)	♑ 4	
☾	♏ 35(!)	♏ 27	
♄	♐ 9;1	♐ 10	−1
♃	♓ 22	♓ 23	−1
♂	♑ 14;6	♑ 15	−1
♀	♏ 29	♏ 29	0
☿	♑ 19	♑ 12	+7

It is possible to determine the positions of sun and moon within narrow limits in spite of the errors in lines 6 and 7. For the sun we know that its longitude belongs to the first decan of ♑ and to the terms of Mercury (line 20) thus we have

$$\lambda_\odot = 270 + x \qquad 0 \leq x \leq 7$$

and can expect on the basis of our calculation[3] that x is not far from 4. The moon is in the second decan of ♏ and in the terms of Mercury (line 21), thus

$$\lambda_☾ = 221 + y \qquad 0 \leq y \leq 8$$

and the elongation

$$\lambda_☾ - \lambda_\odot = -49 + y - x = -49 - z$$
$$-8 \leq z = x - y \leq 7.$$

We know, furthermore, that the longitude of the Horoscopos is $\lambda_H =$ ♌ 13 = 133, and the longitude of the Lot of Fortune $\Lambda_F =$ ♏ 29 = 179. For a birth at night we have the relation

$$\Lambda_F = \lambda_H - (\lambda_☾ - \lambda_\odot) = 179 = 133 + 49 + z$$

or $z = x - y = -3$. Consequently,

$$x = y - 3 \quad \text{and} \quad 0 \leq x \leq 5; \qquad 3 \leq y \leq 8.$$

In other words the longitude of the sun can only vary between ♑ 0 and ♑ 5 and the longitude of the moon correspondingly between ♏ 14 and ♏ 19. This agrees well with the result of our computation and we may therefore assume that $\lambda_\odot \approx$ ♑ 4 and hence $\lambda_☾ \approx$ ♏ 18.

For the remaining lots one meets with several errors of the text as shown in the following table:

Daimon ♊ 20	House of ♃ instead of ☿	Terms of ♀ instead of ♂	Decan 2 (or 3?)
Eros ♐ 15	House of ♂ instead of ♃	Terms of ☿ instead of ♀	Decan 2
Necessity[..]10;9	House of ☿ hence ♊ or ♍	Terms of ♂ excluding 10°	Decan 3 instead of 1 or 2

For the Lot of Daimon one obtains from $\lambda_H =$ ♌ 13 and from $\lambda_☾ - \lambda_\odot = -46$

$$\Lambda_D = \lambda_H + \lambda_☾ - \lambda_\odot = ♊\ 27$$

House of ☿ Terms of ♄ Decan 3

which also agrees very badly with the text. If we, however, compute with this value the place of the Lot of Eros, following the procedure of No. 138/161, we obtain

$$\Lambda_E = \lambda_H + (\Lambda_F - \Lambda_D) = ♏\ 15$$

House of ♂ Terms of ☿ Decan 2

thus complete agreement with the text except for the sign ♐ which should be ♏.

Disagreement arises again for the Lot of Necessity for which one would obtain

$$\Lambda_N = \lambda_H - (\Lambda_F - \Lambda_D) = ♉\ 11$$

House of ♀ Terms of ☿ Decan 2.

We see no way to explain the contradictory statements of the text.

No. 345

PSI 312

TRANSCRIPTION

Traces of writing near the upper edge of the fragment, read ζυγω by the editors.

2. αγαθη τυχη γενεcιc ε̣....
3. ∠ ξα // επειφ γ ⸋ η ημεραc
4. ⸋ ζευc ζυγω /
5. cεληνη cκ̣ορπιω /
6. κρονοc υδρηχοω /
7. ερμηc αφροδιτη διδυμοιc /
8. ηλιοc αρηc καλκινω̣ /
9. η προ ζ′ του̣ αρεωc ερχομενοc
10. δ̣ιπλουc εφ εαυτον ποηcι
11. ανωμαλιαν μηδενοc
12. ειcοντοc ταπινωμα γαρ ου
13. [...] εcτιν ο̣[υ]δ̣εν caπρον
14. [π]ο̣ιηcι εαν δε αφροδιτη αυτη
15. [κα]τ̣α̣λαβη και τω πατρι πραξι
16. και cυcταcιν και επιτυχιαν

thus the moment of αποκρουcειc precedes by about 1½ days the moment of opposition.

[3] We also know that the time was the 3rd (seasonal) hour of night but this restricts the solar longitude to an interval only from about λ = 269 to 285.

TRANSLATION

1.   In Libra(?)
2. With good fortune. Nativity of E[.....]
3. Year 61, Epiphi 3rd, 8th hour of day
4. Horoscopos (and) Jupiter in Libra
5. moon in Scorpio
6. Saturn in Aquarius
7. Mercury (and) Venus in Gemini
8. sun (and) Mars in Cancer
9. (The star) of Mars indeed approaching it (the sun) (at) 7 (degrees ?)
10. with double prerogative, will cause
11. an anomaly. If no (other planet)
12. is there, since a (planet in) depression is not (effective)
13. it will effect nothing very bad;
14. but if Venus herself
15. intercepts it, it will even bring to the father
16. both patronage and success.

COMMENTARY

Diagram: pl. 7

The date of this horoscope is Diocletian 61 XI(a) 3 — 345 June 27. The agreement with the computed positions is good.

Text		Computed	
H	♎	noon	
♃		♎	6
☾	♏	♏	21
♄	♒	♒	25
☿	♊	♋	4
♀		♊	10
☉	♋	♋	5
♂		♋	21

The translation of the second half of the text (lines 9 ff.) is extremely doubtful and is no more than a rendering of what seems to us the general meaning of the text.[1] For the significance of Venus for the fortune of the parents cf. *Tetrabiblos* III,4 (Robbins; III,5 Boll-Boer). For the "anomaly" cf. Vettius Valens IV,16, interpreting this term as astrological.[2] An astronomical interpretation is hardly reconcilable with the text though it is a peculiar accident that both anomalies of Mars are zero: since Mars is in conjunction with the sun and therefore is in the apogee of the epicycle; consequently, also the center of the epicycle is in ♋ where the apogee of the eccenter is located.[3]

---

[1] Particularly difficult is the beginning of line 9; a zone of influence of 7° for Mars is mentioned by al-Bīrūnī, *Astrol.* (p. 255 Wright). Seven days for the moon seem to play a role in No. L 40.

[2] For the concept of "double prerogative" cf. *Tetrabiblos* I,21 p. 105 Robbins.

[3] According to Almagest X,7 (p. 345,19 Heiberg) the apogee of the eccenter is ♋ 25 ;30.

No. 351

PSI 24,b

TRANSCRIPTION

11. ξ̄ζ̄ δ[ιοκλητιανου γενεϲιϲ] ερμειω
12. τυ[βι .... φ ..] ν[υ]κτοϲ φ [ζυγ]ω νη[ϲ]
13. κρ[ονοϲ] κριω
14. ζ[ευϲ ιχθ]υϲι
15. αρ[ηϲ] ταυρω
16. αφ[ροτ]ιδη ερμηϲ υδρηχοω
17. ηλι[οϲ] αιγογερω
18. ϲελ[ην]η λεοντι
19. κλ[η]ρ[οϲ] τυχηϲ υδρηχοω

TRANSLATION

11. 67th (year) of D[iocletian. Nativity] of Hermeione
12. Tybi [.......... hour of night;] Horoscopos in [Libra]
13. Saturn in Aries
14. Jupiter in Pisces
15. Mars in Taurus
16. Venus (and) Mercury in Aquarius
17. sun in Capricorn
18. moon in Leo
19. Lot of Fortune in Aquarius

COMMENTARY

Diagram: pl. 7

The earliest possible date is 351 January 1 — Diocletian 67 V(a) 6. The moon is then at the beginning of Leo. For two more days she remains in the same sign; hence January 3 is the latest possible date. For this whole period agreement between text and computation is maintained for all remaining positions. For January 1 one finds

Text		Computed	
♄	♈	♈	26
♃	♓	♓	28
♂	♉	♉	28
♀	♒	♒	21
☿		♒	1
☉	♑	♑	12
☾	♌	♌	6

We first assume that the longitude Λ_F of the Lot of Fortune is given by

$$\Lambda_F = \lambda_H + (\lambda_{☾} - \lambda_{☉}).$$

According to our computation $\lambda_{☾} - \lambda_{☉} = -156°$ and Λ_F is said to be in ♒. Thus we obtain for the Horoscopos $\lambda_H = 330 + 156 = ♌\ 6$ as upper limit and $\lambda_H = 300 + 156 = ♋\ 6$ as lower limit. Because the longitude of the moon might be about 24° greater than assumed here we might say that the Horoscopos fell somewhere between ♊ 10 and ♌ 10. The hour of the birth would therefore be near sunset.

But in case Λ_F is computed by

$$\Lambda_F = \lambda_H - (\lambda_{☾} - \lambda_{\odot})$$

we obtain for the Horoscopos a longitude between ♌ 24 and ♎ 18. The time would then be an hour of the night, in agreement with the type of formula used. The traces in the text seem to give preference to this solution. For related texts *cf.* No. 338.

No. 353

Abydos. Graffito in Temple of Sethos I, First Court, East Wall

Described in Perdrizet et Lefebvre, *Graffites Grecs du Memnoneion d'Abydos*, p. 114 f. No. 641 and pl. 16,A. Also reproduced in Daremberg-Saglio, *Dict.* vol. V p. 1053 fig. 7593 but without distinguishing between preserved and restored text. The graffito shows a circle, divided into 12 equal segments, which contain the following inscriptions in counter-clockwise arrangement, ♈ and ♉ flanking the vertical radius of the upper semicircle:

κρε̣ιος
[ταυρ]ος
[διδυμοι κρο]νος
[καρκινος ζευς]
λεω̣ν
παρθενω ερμ̣ης αφροδιτη
ζυγο[ς] ηλιος
[ς]κορπιο̣[ς] αρης
τοξοτης ςεληνη
αιγοκε̣[ρως] οροςκ̣[οπος]
υδρηχοο̣ς
ιχθυες

Below this drawing is written

1. αρτεμιδωρος
2. υγειαν
3. νη τον βηςαν ου μη εξαλειψω

COMMENTARY

Diagram: pl. 7

The date which satisfies the given conditions so far as preserved is 353 September 21, or perhaps 22. The editors restored without reason Jupiter in ♊, together with Saturn, but computation shows that this is excluded and Jupiter must have been located in ♋.

For September 21 one finds:

Text		Computed
♄	♊	♊ 16
[♃	♋]	♋ 16
♀	♍	♍ 27
☿	♍	♍ 26
☉	♎	♍ 29
♂	♏	♎ 28
☾	♐	♐ 19

The Horoscopos being in ♑, the sun at the boundary between ♎ and ♍, the hour is about noontime. One day later the sun is in ♎ 0, the moon has entered ♑, Mars remains just outside ♏.[1]

At the end we find the name of the subect of the horoscope, Artemidoros, followed by the words "Health![2] By Bes! May I not be wiped out." The Horoscopos being in a house of Saturn, may be the cause of his alarm.[3]

No. 366

PSI 22,a

(Photo: Roberts GLH pl. 24)

TRANSCRIPTION

1. [γε]νεςις νιλαμμων
2. [π]β̣ (ετους) διακλητιανου θ' ϊντικτιον[ος]
3. π̣ροεδρευοντος ϊςιδωρου παμε[νους]
4. τυβι οψε εις ιβ
5. ɸ λεοντι
6. κρονος ςκορπιω
7. αφροτιδη : τοξοτη
8. ηλιος ; ερμης : αιγοκερω
9. ςεληνη : κριω :
10. αρης : ταυρω

TRANSLATION

1. Nativity of Nilammon
2. [82]nd (year) of Diocletian, 9th indiction
3. presidency of Isidorus son of Pamenes,
4. Tybi (11th), late, to 12th.
5. Horoscopos in Leo
6. Saturn in Scorpio
7. Venus in Sagittarius
8. sun (and) Mercury in Capricorn
9. moon in Aries
10. Mars in Taurus

[1] It is only because of the anomaly that Mars is not in ♏. The mean position is about ♏ 6.

[2] For the discussion of υγεια *cf.* P. Perdrizet, *Revue des études grecques* 18 (1914) pp. 266-280. Also Edelstein, *Asclepius II*, Index p. 274.

[3] Cumont considered the horoscope as favorable; *cf.* Perdrizet-Lefebvre, *GG* p. 115.

COMMENTARY

Diagram: pl. 7

For the position of this horoscope within a larger group *cf.* the commentary to No. 338. The date of the present horoscope is not preserved but can easily be restored. We know from No. 370 that Saturn was in Sagittarius in the year 86 of Diocletian = Indiction 13. The present horoscope shows Saturn in Scorpio for Indiction 9, thus in a position which corresponds very well to a difference of 4 years. This obvious guess is fully confirmed by computation.[1] One finds for Diocletian 82 V(a) 11 = 366 January 6:

Text		Computed
H	♌	about 8 P.M.
♄	♏	♏ 15
♀	♐	♐ 1
☉ }	♑	♑ 17
☿ }	♑	♐ 23(!)
☾	♈	♈ 28
♂	♉	♉ 15

The small discrepancy for Mercury is not significant. Jupiter is omitted; his place would be at the end of the list in Cancer. The Horoscopos in ♌ shows that the sun in ♑ 17 is about one zodiacal sign below the horizon. Thus we are dealing with the second or third hour of night, expressed in the text by "late." A similar terminology occurs in No. 373.

[1] The same result was reached by Goodrich [1] (1935).

No. 370

PSI 22,d

TRANSCRIPTION

29. γενεcιc κυριλλουτοc̣
30. $\overline{\pi\varsigma}$ διοκλητιανου $\overline{\iota\gamma}$ ιντικτιονοc
31. τυβι $\overline{\iota\gamma}$ ειc $\overline{\iota\delta}$ Φ $\bar{\delta}$ νυκτοc
32. Φ παρθενω: ζευc: cεληνη:
33. cκορπιω: κρονοc: τοξοτη:
34. ηλιοc: αιγογερω:
35. ερμηc: τοξοτη:
36. αφροτιδη: υδρηχοω:
37. αρηc: λεοντι
38. κληροc τυχηc cκορπιω

TRANSLATION

29. Nativity of Cyrillous
30. 86th (year) of Diocletian, 13th indiction
31. Tybi 13th to 14th, 4th hour of night
32. Horoscopos in Virgo, Jupiter (and) moon
33. in Scorpio. Saturn in Sagittarius
34. sun in Capricorn
35. Mercury in Sagittarius
36. Venus in Aquarius
37. Mars in Leo
38. Lot of Fortune in Scorpio

COMMENTARY

Diagram: pl. 7

Except for Mercury the positions given in the text are correct for Diocletian 86 V(a) 13 = 370 January 8. This year has indiction 13 given in the text (not 14 as the editors suggest emending the text).

Text		Computed
H	♍	about 10 P.M.
♃ }	♏	♏ 18
☾ }	♏	♏ 12
♄	♐	♐ 26
☉	♑	♑ 19
☿	♐	♒ 7(!)
♀	♒	♒ 15
♂	♌	♌ 10

It should be noted that the planets are arranged according to their position in the zodiac and consequently that Mercury would be out of place if ♐ were its position. Hence ♐ is a copyist's error either for ♑ or for ♒ as expected from computation.

The Lot of Fortune is computed from

$$\lambda_{☾} = 222 \qquad \lambda_{☉} = 289 \qquad \lambda_{H} \approx 165$$

by means of the formula

$$\Lambda_F = \lambda_H - (\lambda_{☾} - \lambda_{☉}).$$

Subtraction is chosen because we are concerned with a night birth; *cf.* also No. 373.

No. 373

PSI 24,a

TRANSCRIPTION

1. γενεcιc ερμειον
2. $\overline{\pi\theta}$ διοκλητιανου
3. τυβ̣ι̣ η̣ Φ οψε ειc $\bar{\theta}$
4. Φ καρκινω
5. cεληνη ζυγω
6. ερ[μ]ηc̣ δοξοτη
7. ηλιοc α̣φ̣ροδιτη αιγογε ρω
8. κρο̣ν̣ο̣c̣ ζε̣υ̣c̣ α̣ρ̣ηc
9. υ̣δρ̣η̣χοω
10. κλ(ηροc) τ̣[υχηc ζ]υ̣[γ]ω

TRANSLATION

1. Nativity of Hermeion
2. 89th (year) of Diocletian
3. Tybi 8th, late hour, to 9th

4. Horoscopos in Cancer
5. moon in Libra
6. Mercury in Sagittarius
7. sun (and) Venus in Capricorn
8. Saturn, Jupiter (and) Mars
9. in Aquarius
10. Lot of Fortune in Libra

COMMENTARY

Diagram: pl. 8

For the date Diocletian 89 V(a) 8 = 373 January 3 there is good agreement between text and computation:

Text		Computed
☾	♎	♎ 25
☿	♐	♐ 20
☉	} ♑	♑ 15
♀		♑ 22
♄		♑ 28
♃	} ♒	♒ 9
♂		♒ 13

The Horoscopos is said to be in ♋; the sun being in ♑ would therefore indicate that the hour corresponds to sunset. The question arises whether this moment is considered as belonging to daytime or to the night. The phrase "Tybi 8th, late hour, to the 9th" favors night, because the night is usually denoted by the date of the current Egyptian day and the date of the next day which begins at the following morning.

This result is supported by the following consideration. The Lot of Fortune is said to be in ♎, in the same sign as the moon. Because

$$\lambda_{☾} = 205 \qquad \lambda_{☉} = 285 \qquad \lambda_{H} = 105 \qquad \lambda_{☾} - \lambda_{☉} = -80$$

we find

$$\Lambda_F = 105 \begin{cases} +80 = 185 = ♎\ 5 \\ -80 = 25 = ♈\ 25 \end{cases}$$

Thus the Lot of Fortune was computed according to the formula

(1) $$\Lambda_F = \lambda_H - (\lambda_{☾} - \lambda_{☉})$$

and not acccording to

(2) $$\Lambda_F = \lambda_H + (\lambda_{☾} - \lambda_{☉}).$$

The first formula is used for a birth at night, the second for daytime. *Cf.* Nos. 351, 370, 376, 381, and 385.

For related texts *cf.* the commentary to No. 338; furthermore No. 81 and No. 366.

No. 376,a

PSI 22,b

(Photo of lines 11 to 18: Roberts, *GLH* pl. 24)

TRANSCRIPTION

11. γενεcιc ϊωαννηc φαωφι ιε̄
12. ϛ̄' ιντικτιονοc ⏀ ᾱ ημεραc
13. ϛ̄γ̄ διοκλητιανου ερμογενηc
14. ⏀ : cκορπιω / θεοτιμου προεδροc
15. αφροτιδη: τοξοτη
16. αρηc : αιγογερω
17. κρονοc : cεληνη : ϊχθυcι :
18. ζευc : διδυμοιc
19. ηλιοc : ερμηc : ζυγω :
20. κληροc τυχηc : κριω :

TRANSLATION

11. Nativity of Joannes (or Joanna?), Phaophi 15
12. 6th indiction, 1st hour of day.
13. 93rd (year) of Diocletian. Hermogenes
14. Horoscopos in Scorpio / son of Theotimos president
15. Venus in Sagittarius
16. Mars in Capricorn
17. Saturn (and) moon in Pisces
18. Jupiter in Gemini
19. Sun (and) Mercury in Libra
20. Lot of Fortune in Aries

COMMENTARY

Diagram: pl. 8

This horoscope is the second one of a group of four, PSI 22, which belong to the larger group of texts, discussed in No. 338. It is duplicated by No. 376,b.

The agreement between text and computation is perfect for Diocletian 93 II(a) 15 = 376 October 12. The indiction, however, should be 5 instead of 6. One finds the following positions

Text		Computed
H	♏	about 8 A.M.
♀	♐	♐ 7
♂	♑	♑ 20
♄	} ♓	♓ 9
☾		♓ 28
♃	♊	♊ 25
☉	} ♎	♎ 20
☿		♎ 6

Also the Horoscopos and the Lot of Fortune are correctly placed, following the formula

$$\Lambda_F = \lambda_H + (\lambda_{☾} - \lambda_{☉}).$$

In No. 376,a the "president" or mayor Hermogenes is mentioned where we find in No. 376,b the "night-watch commander" Macarius. *Cf.* for these offices Oertel, *Liturgie*, p. 281 ff. and p. 343 ff.

No. 376,b

PSI 23,b

TRANSCRIPTION

31. αγαθυ τυχη ϊωαννης

---

32. ϛ // ιντικτιονος επι νυκτοστρατηγο[υ]
33. .. μακαριος ερμειου ϭεcϫoc
34. ϛ̅γ̅ διο̣κλητιανου ϕ α ημερας φαω̣φι ι̅ε̅
35. ϕ c̣κ̣ορπιω
36. αφροτιδη: τοξοτη
37. αρης αιγογερω
38. κρονος̣ c̣ε̣λ̣ηνη ϊχθυcι:
39. ζ̣[ευ]c δ[ιδυμοις]
40. ηλιος ερμ̣η̣c ζυγω:
41. κληρος τυχης κριω

31: The line drawn between lines 31 and 32 belongs actually to the end of the preceding horoscope (No. 338).
32 ϭεcϫoc: ϭ = *g* and ϫ = *ḍ*.

TRANSLATION

31. With good fortune. Joannes (or Joanna?)
32. 6th indiction in the time of Night-watch commander
33. Macarius, son of Hermias Gesḍos (?)
34. 93rd (year) of Diocletian, 1st hour of day, Phaophi 15
35. Horoscopos in Scorpio
36. Venus in Sagittarius
37. Mars in Capricorn
38. Saturn (and) moon in Pisces
39. Jupiter in Gemini
40. sun (and) Mercury in Libra
41. Lot of Fortune in Aries

COMMENTARY

A duplicate of this horoscope is No. 376,a. This should be consulted for the commentary to both texts.

No. 381,a

PSI 22,c

TRANSCRIPTION

21. γενεcιc κυρος [ϛ]ζ διοκλητιανου
22. μεχειρ κ̅ε̅ ϕ : θ̅ ημερας
23. ϕ : καρκινω:
24. ζευc : ζυγω:
25. η̅λιος : αφροτιδη: ερμης : ϊχθυcι:
26. αρης : κρονος : ταυρω:
27. cεληνη: διδυμοις
28. κληρος τυχης ζυγω:

TRANSLATION

21. Nativity of Cyrus, [9]7th (year) of Diocletian
22. Mekheir 25, 9th hour of day
23. Horoscopos in Cancer
24. Jupiter in Libra
25. sun, Venus, (and) Mercury in Pisces
26. Mars (and) Saturn in Taurus
27. moon in Gemini
28. Lot of Fortune in Libra

COMMENTARY

Diagram: pl. 8

This is the third horoscope of PSI 22; *cf.* the commentary to No. 338. A duplicate is No. 381,b.

Text and computation agree perfectly for Diocletian 97, VII(a) 25 = 381 February 19.

Text		Computed
H	♋	about 2 P.M.
♃	[♎]	♎ 28
☉	♓	♓ 2
♀		♓ 20
☿		♓ 20
♂	♉	♉ 24
♄		♉ 6
☾		♊ 18

Also the Lot of Fortune is correctly computed, following the relation

$$\lambda_{F} = \lambda_{H} + (\lambda_{☾} - \lambda_{\odot}).$$

No. 381,b

PSI 24,d

TRANSCRIPTION

30. γ̣ε̣ν̣εcιc κυρος

---

31. ϛ̅ζ̅ δ̣ι̣ο̣κ̣λ̣ητιανου μεχειρ κ̅ε̅
32. θ̅ ϕ ημ[ερας ϕ καρκ]ινω
33. ζευ̣ς̣ [ζυγω ηλι]ο̣ς̣ : αφροτιδη
34. ερ̣μης ι̣χ̣θ̣υ̣[cι] αρης : κρονου:
35. ταυρω : cελ[ην]η : διδυμοις :
36. κλ(ηρος) τ̣υ̣χ̣[ης] ζ̣υγω

30: The line drawn below line 30 belongs to the end of the preceding horoscope (No. 385).

TRANSLATION

30. Nativity of Cyrus
31. 97th (year) of Diocletian, Mekheir 25
32. 9th hour of day. Horoscopos in Cancer
33. Jupiter [in Libra.] Sun, Venus,
34. (and) Mercury in Pisces. Mars (and) Saturn
35. in Taurus. Moon in Gemini
36. Lot of Fortune in Libra

COMMENTARY

A duplicate of this horoscope is No. 381,a.

No. 385

PSI 24,c

TRANSCRIPTION

20. γεν[εcι]ç νηcτια η και απολλωνια
21. $\overline{\text{ρα}}$ α[πο] διοκλητιανου επι cερηνου
22. αρτ[ε]μιδωρου προεδρευοντος
23. φ[αρμο]υθι $\overline{\text{ιδ}}$ ⏀ ημερας ∫ φανουcιρε
24. [⏀ διδ]υμοιc
25. κρο[νοc :] αρηc : καρκινω :
26. cε[λ]ηνη : παρθενω :
27. ζευ[c :] αφροτιδη : ϊχθυcι :
28. ηλι[ο]c : κριω : ταυρω :
29. κληρος τυχηc ζυγω :

23: The sign before φανουcιρε indicates that φανουcιρε belongs to the end of line 22.

23 $\overline{\text{ιδ}}$: Clearly $\overline{\text{ιδ}}$ is not ῑ δ̄; *cf.* commentary.

TRANSLATION

20. Nativity of Nestia, also called Apollonia
21. 101st (year) from Diocletian in the presidency
22. of Serenus (son of (?)) Artemidoros Phanousire
23. Pharmouthi 14, (?) hour of day.
24. Horoscopos in Gemini
25. Saturn (and) Mars in Cancer
26. moon in Virgo
27. Jupiter (and) Venus in Pisces
28. sun in Aries, (Mercury) in Taurus
29. Lot of Fortune in Libra

COMMENTARY

Diagram: pl. 8

The text agrees closely with the computation for Diocletian 101 VIII(a) 14 = 385 April 9.

	Text	Computed
H	♊	about 10 A.M.
♄	} ♋	♋ 3
♂		♋ 23
☾	♍	♍ 18
♃	} ♓	♓ 4
♀		♓ 10
☉	♈	♈ 20
☿	♉	♉ 8
Lot of Fort.	♎	From ♎ 28 to ♏ 28

The position of the moon shows that one must read "Pharmouthi 14 (?)th hour of the day" and not "Pharmouthi 10, 4th hour of the day" because on the latter date the moon would be in ♋ 6. Computing the longitude Λ_F of the Lot of Fortune using the formula

$$\Lambda_F = \lambda_H + (\lambda_{☾} - \lambda_{☉})$$

one obtains for Λ_F a point in ♎ if λ_H falls in the last two degrees of ♊. Hence the hour must be somewhat before noon.

No. 394

P. Oslo 165

TRANSCRIPTION	TRANSLATION
1. (ετουc) ρια	1. (Year) 111
2. μηνοc χοιακ λ	2. month Choiak 30
3. ⏀ γ νυκτοc	3. 3rd hour of night.
4. ⏀ λεοντι	4. Horoscopos in Leo,
5. κρονοc cκορ	5. Saturn in Scorpio,
6. ζευ̣ç [....	6. Jupiter [(and) Venus]
remainder destroyed	[7. in Sagittarius]
	....

COMMENTARY

Diagram: pl. 8

For the date given in the text, Diocletian III IV(a) 30 = 394 December 26, one obtains agreement with the preserved date:

	Text	Computed
H	♌	7 P.M.
♄	♏	♏ 10
♃	} ♐]	♐ 20
[♀		♐ 5

This result was established by Heegaard [1] p. 100 (1936).

No. 431

P. Mich. 153

TRANSCRIPTION

1. ρμη διοκλητια̣νου
2. θωθ κζ ⏀ ε νυκτοc
3. ηλιοc ζυγω
4. ερμηc ζυγω
5. αφροδιτη ζυγω
6. κρωνοc αιγοκερω
7. ζευc υδρυχωου
8. αρηc λεωντι
9. cεληνηc cκορπιω
10. ⏀ διδημηc

TRANSLATION

1. (Year) 148 of Diocletian
2. Thoth 27, 5th hour of the night
3. sun in Libra
4. Mercury in Libra
5. Venus in Libra
6. Saturn in Capricorn
7. Jupiter in Aquarius

8. Mars in Leo
9. moon in Scorpio
10. Horoscopos Gemini

COMMENTARY

Diagram: pl. 8

This text is unusually well preserved and there cannot be the slightest doubt of the correctness of the readings. Computation, however, shows that several elements given in the text are at fault. For 431 September 25 one obtains:[1]

	Text	Computed
☉	♎	♎ 3
☿	♎	♍ 20(!)
♀	♎	♍ 2(!)
♄	♑	♑ 21
♃	♒	♑ 14(!)
♂	♌	♋ 8(!)
☾	♏	♏ 9
H	♊	about 10 P.M.

The cause of these errors is difficult to detect. No improvement results from calculations based on the assumption that the scribe still wrote Diocletian 148 when he should have written 149. Another source of error might have been the use of the Egyptian calendar for the computation of the planetary longitudes, but again no improvement is gained. In fact no date reasonably close to the year Diocletian 148 can be made to correspond to the data of the text.

No. 465

PSI 25

TRANSCRIPTION

1. απο διοκλητιανου ρπ̣[α]
2. μεϲορη κθ ⳼ α η[μεραϲ]
3. ⳼ παρθενω
4. κρονοϲ ϊχθηϲι
5. ζ̣ε̣υ̣ϲ̣ ϲκορπιω
6. αρ[η]ϲ [δι]δ̣υ̣μ[οιϲ]
7. α[φρο]δ̣ιδη λεοντι
8. ερ̣[μ]ηϲ λ̣ε̣οντ̣ι̣
9. η̣λιοϲ παρθεν̣ω̣
10. [ϲελ]η̣ν̣η υδροχοω

TRANSLATION

1. (Year) 1[81] from Diocletian
2. Mesore 29th, 1st hour of day.
3. Horoscopos in Virgo
4. Saturn in Pisces

[1] The longitudes of the planets, given in vol. III p. 122 of the P. Mich., are incorrect.

5. Jupiter in Scorpio
6. Mars in Gemini
7. Venus in Leo
8. Mercury in Leo
9. sun in Virgo
10. moon in Aquarius

COMMENTARY

Diagram: pl. 9

The only possible date within the century in question is Diocletian 181 XII(a) 29 = 465 August 22. One finds for this date the following positions:

	Text	Computed
H	♍	6 A.M.
♄	♓	♓ 22
♃	♏	♏ 23
♂	♊	♋ 8(!)
♀	♌	♌ 29
☿	♌	♌ 27
☉	♍	♍ 0
☾	♒	♒ 26

No. 478

P. Oxy. 2060

TRANSLATION[1]

No. 478

1. Christos(?)
2. [........] of the lord Anubis(?).
3. Year of the blessed reign of Diocletian
4. 194, month Epiphi 5, day [......]
5. Sun in Cancer 5 degrees 50 minutes
6. Moon in Sagittarius 17 degrees 10 minutes
7. Saturn in Virgo 6 degrees [..] minutes
8. Jupiter in Capricorn 0 degrees [..] minutes
9. Mars in Taurus 18 [degrees ..] minutes
10. Venus [in Cancer ....]
11. [..... ....

4: the editors restore η[μεραϲ ωρα ..]. The almost contemporary horoscopes, Nos. L 479, L 486, L 487, however, suggest the mention of the day of the week in the form η[μερα διοϲ ωρα ..].
In line 9 the editors read λ̣η (38) but no zodiacal sign can contain more than 30 degrees. The emendation ιη (18) is confirmed by computation.

[1] The original of this text is unfortunately lost.

COMMENTARY

Diagram: pl. 9

According to the printed text line 1 contains the monogram ☧ which is customary for χριϲτοϲ. The second line, written by another hand, ends in "the lord Anup." This might be an invocation of "the lord Anubis"; to call Anubis κύριος would be in line with common Egyptian usage to call a god "the lord" of a locality.

The date of the horoscope is Diocletian 194 XI(a) 5 which corresponds to 478 June 29.[a] For this date one finds:

	Text	Computed	Text—comp.
☉	♋ 5;50	♋ 8	−2°
☾	♐ 17;10	♐ 17	0
♄	♍ 6;[..]	♍ 7	−1
♃	♑ 0;[..]	♑ 4	−4
♂	♉ 18;[..]	♉ 14	+4
♀	[♋ .....]	♋ 12	

---

[a] Given erroneously as 498 in the edition.

II. THE HOROSCOPES FROM LITERARY SOURCES

No. L — 71

CCAG 8,4 p. 236,24 to 237,10

INTRODUCTION

Through an excerpt from Balbillus on the "Method concerning the length of life from starter and destroyer" two very early horoscopes are preserved, one for — 71 December 27, the other for — 42 January 16. Balbillus is well known as the astrologer of Nero and Vespasian. His father Thrasyllus served Tiberius and Claudius in the same capacity.[1] Thus it is not surprising to find Balbillus in possession of case histories from the first century B.C.

The text of the chapter which contains the two horoscopes, as published in *CCAG* 8,4 p. 235,1 to p. 238,2, is based on two manuscripts: A = Paris. 2506 and P = Paris. 2425.[2] In A the elements for the positions of the celestial bodies are given for both horoscopes as a rather chaotic sequence of numbers and planetary symbols. The text of P proceeds in sentences, but is very defective for the first horoscope (which is the chronologically later one), thus making it necessary to rely for the first horoscope almost entirely on the data in A. Since the method of investigation of the text A is the same for both horoscopes, we combine here the discussion of both cases, following the order of the text, without anticipating the establishment of the actual chronological order, which would give the second horoscope precedence over the first.

It is on the basis of two remarks that it becomes possible to bring sense into the data of the version in A: (a) except for some obvious errors and omissions, all zodiacal signs are enumerated in their proper order, starting with the sign of the Horoscopos—and (b) these signs are counted consecutively by numbers from 1 to 12, or, to use astrological terminology, the numbers of the twelve loci are cited one by one.

That this is the case becomes immediately evident if one writes down the text of A in twelve single lines, one for each sign and locus, but without otherwise disturbing the order given in the text;[3] two omissions are restored in [ ]. For the sake of clarity the numbers of the loci are printed in bold face; only the remaining numbers can present degrees of longitude. We ignore, of course, the punctuation of the printed text.

			H	♑	☉	9	1	♄	5	♀	11	♂	12	9	τόπος
	[2]	♒													
	3	♓	♃	20											
	4	♈													
♏ ♃	5	♏	♌	5	4	(emend: **5** ♉ ♌ 4)									
	6	♊													
	7	♃	♋	9	1	(emend: **7** ♋ 9)									
	8	νύξ				(emend: **8** ♌)									
	9	♍													
	[10]	M	♎	3											
♏	11	♏	♐	5	4	(emend: **11** ♏ ♐ 4)									
	12	♐	♉	17	☾	19									

A sequence, like the above, can be easily explained as the result of converting into a running list, the numbers and symbols read off from a diagram, sometimes confusing, omitting, or repeating the entries found in the consecutive fields of the diagram. For a typical example of a diagram, as they appear in manuscripts, *cf.* figure 20, p. 156.

In the first line we have to ignore the repetition of 9, and we must combine τόπος with **1** and probably place it at the beginning:

τόπος **1** H ♑ ☉ 9 ♄ 5 ♀ 11 ♂ 12.

The same scheme is then followed throughout. In locus **5** ♏ is an easily explicable misreading of a common form of the symbol for Taurus, written with an upwards-bent tail, exactly as the symbol for Scorpio.[4] In locus **7** ♃ must be an error, since the position of Jupiter was given before in locus **3** as ♓ 20, which is confirmed by a subsequent reference (p. 236,21 f.). The explanation of the error comes from the next horoscope. There we find, in the same locus, ♃ ♐ instead of **7** ♐. Obviously ζ' = 7 was misinterpreted as the symbol ζ(εύ)ς. Thus we have in the first case simply a repetition of the number 7 of the locus. In locus **8** the symbol for νύξ, which is the inverted symbol for ἡμέρα, is easily explained as a misreading of the symbol for Leo.[5] In locus **11** for ♐ the copyist was obviously influenced by the entries in locus **5** for ♌, and similarly the number 1 in locus **7** for △ probably refers to locus **1** for H.

In the same fashion we can also analyze the data of the second horoscope, though here the disturbances in A are more serious. Fortunately P is in this case complete and can be used in combination with A. The loci,

---

[1] *Cf.* e.g., Cramer, *ARLP*, *passim*.
[2] *CCAG* 8,4 p. 235, apparatus.
[3] *Cf.* the critical apparatus in *CCAG* 8,4 p. 236 concerning the lines 9 to 14 and concerning p. 236,24 to p. 237,4.

[4] *Cf.*, e.g., cod. Laur. Plut. 28,14 fol. 309ʳ, 4 f., reproduced *CCAG* 4, plate 1. The inverse error, ♉ for ♏, is found *CCAG* 8,1 p. 253,14. *Cf.* also fig. 1 p. 1.
[5] *CCAG* 4 plate 1, line 7 and 5; νύξ instead of ♀ is found *CCAG* 8,1 p. 253,14.

however, are not mentioned in P,[6] a fact which supports our assumption that A does not reproduce a running text but a diagram with consecutively numbered fields.

In the following list we represent, as before, the text of A in the order of the manuscript. The order of P is the planetary order (except for interchanging ♀ and ☿) and is therefore not shown in the following table.

			A	P
H♓	1		(emend: H ♊ 1)	H ♊
	2	☿ 12 ♂ . . .[7] 2 τόπος ♋ (cf. locus 9)		
	3	♌		
	4	♍ · ♌ · ζ′α′		☾ 4, ♃ 14
	5	♎		
	5	♏ ☾ 4	(read 6)	
	♃	♐ α′β′	(read ζ′ = 7, not ζ(εύ)ς)	
	ἡ	♑ ♄ 4 ☉ 22	(read η′ = 8)	♄ 4, ☉ 22
	9	♒	(cf. locus 2)	☿ 12, ♂ 14
	[10]	M ♓ ζ′α′ ♀ 25		♀ 25
	11	♈		
	11	♉	(read 12)	

The entry in locus 2 is obviously misplaced, since Mercury cannot be in locus 2, when the sun is in locus 8. Undoubtedly, the version of P is here the correct one. We have no explanation for the entries ♌ ζ′ α′ in locus 4, and for ζ′ α′ in the diametrically opposite locus 10, nor for α′ β′ in locus 7. Since these letters certainly do not represent degree numbers they have no importance for the establishment of the astronomical conditions. The only element that remains doubtful is the position of the moon: ♍ 4 according to P, ♏ 4 according to A. The version of A is supported by the statement that neither sun nor moon was in a center, whereas ♍ would be in Lower Midheaven.

TRANSLATION OF L — 71

And again he says that for another theme the moon was in Scorpio[8] 4 degrees, the sun in Capricorn 22,[9] and Saturn in Capricorn 4 degrees, and Jupiter in Virgo 14 degrees, and Mars in Aquarius 14 degrees, and Mercury in Aquarius 12 degrees, and Venus in Pisces 25 degrees, and the Horoscopos in Gemini.[10] And, since the luminaries (sun and moon) did not fall in a center, he went to the epanaphorai, and he did not take the Horoscopos as starter, nor the sun, which was in epanaphora of the setting point, but he took Saturn in Capricorn as starter. And this, I think, because it (Saturn) had the greater claim in the theme and was in his own house (♑). And he says that Mars in Aquarius is the destroyer and he computed the distance from Aries to Mars, and so long, he said, would be the length of life.

COMMENTARY

For — 71 January 21 one finds on the basis of the positions given in A (*cf.* figure 13*a*):

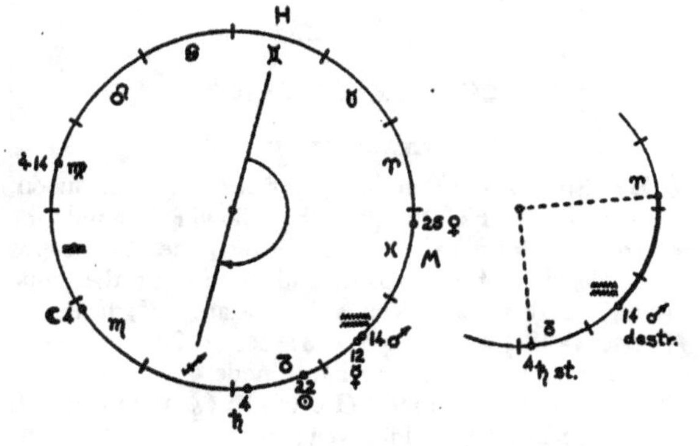

FIGURE 13*a*.

Text		Computed	Text — comp.
H	♊	about 4 P.M.	
♃	♍ 14	♍ 14	0
☾	♏ 4	♎ 29	+ 5
♄	♑ 4	♑ 7	— 3
☉	♑ 22	♑ 30	— 8
☿	♒ 12	♒ 17	— 5
♂	♒ 14	♒ 6	+ 8
M	♓		
♀	♓ 25	♓ 15	+ 10

In the version of P the moon is placed in ♍ 4. This would lead to a date 5 days earlier, a change without influence on the positions of the outer planets. Otherwise one would have for January 16:

Text		Computed	Text — comp.
☾	♍ 4	♌ 26	+ 8
☉	♑ 22	♑ 25	— 3
☿	♒ 12	♒ 14	— 2
♀	♓ 25	♓ 10	+ 15

The date Jan. 21 seems preferable for Venus and it avoids having the moon in a center (♍ — IMC), contrary to the text.

In the astrological section it is explained why Saturn in ♑ 4 is considered to be the starter, Mars in ♒ 14 the destroyer. The length of life is then determined by the number of degrees "from Aries to Mars." In view of the treatment of the next horoscope, No. L — 42, we are probably entitled to interpret this procedure more

---

[6] In the printed text of P some of the numbers of the loci were erroneously joined to the degree numbers of the positions: this is the case for Horoscopos, sun, and Venus.

[7] *CCAG* 8,4 p. 236 ad locum: ♂ deinde signum dubium.

[8] Following A; P has Virgo.

[9] The Editor added here, incorrectly, from the manuscript A the number 9.

[10] Again the numbers 1 2, added by the editor from A, do not belong here.

accurately as measuring the arc between Mars (♒ 14) and the quartile of the starter, i.e. ♈ 4 (*cf.* figure 13*a*). The result would be an arc of 50°, indicating a lifetime of 50 years. Judging from many similar discussions of the duration of life in the work of Vettius Valens, we may assume that our present horoscope is taken from an actual case history and would not have been recorded earlier than 22 B.C.

No. L — 42

CCAG 8,4 p. 236,8 to 23

TRANSLATION

He also says that there are four destroyers: sun, moon, Saturn, Mars. He also gives the following example for starter and destroyer: {The sun (and) the Horoscopos in 9 degrees of Capricorn, and Saturn in the same (sign), 5 degrees},[1] Venus 11°, Mars 12°, locus 1.[2] [(Locus) 2:] ♒. (Locus) 3: ♓, ♃ 20°. (Locus) 4: ♈. (Locus) 5: ♉, ascending node 4°. (Locus) 6: ♊. (Locus) 7: ♋ (Δ) 9°. (Locus) 8: (♌). (Locus) 9: ♍. [(Locus) 10,] Midheaven: ♎ 3°. (Locus) 11: ♏, descending node 4°. (Locus) 12: ♐, Mercury 17°, moon 19°. And he says that the sun is the starter of the theme, whereas Mars could not become destroyer, because Venus is rising after it (the sun) within 8°,[3] although he (Mars) is in his own exaltation (♑). And so one comes to the destructive sign, which is the one rising before the starter, i.e., Sagittarius. And finding there the moon (in ♐ 19), he says that it becomes the destroyer. Taking the orbit of the sun up to the quartile of the degree of the moon (♐ 19), i.e. Pisces 19, then, he says, comes the destruction. Jupiter, being in Pisces, 20 degrees,[4] was not able to assist, because the moon is located in the destructive sign (♐).

COMMENTARY

For the restoration of the text *cf.* the introduction to No. L — 71.

The first element to be discussed is the statement that Midheaven (M) is in ♎ 3 while the Horoscopos (H) is in ♑ 9 (*cf.* figure 13*b*). But if ♑ 9 rises at the latitude of Alexandria, about ♎ 26;30 is culminating and not ♎ 3; and the discrepancy could only be worse for places farther north. If we ask, however, which point is rising at Alexandria if ♎ 3 culminates, then we obtain ♐ 19, i.e. the position of the moon. Since the sun is also located in H, the place of the moon coincides with the Lot of Fortune. Thus M is reckoned from the Lot of Fortune,[5] not from H.

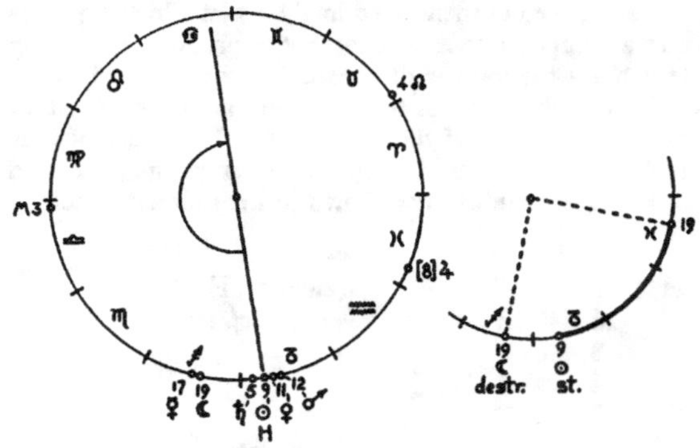

FIGURE 13*b*.

For the remaining positions we get sufficiently good agreement for — 42 December 27 to secure this date:

Text		Computed	Text — Comp.
H	♑ 9	sunrise	
☉	♑ 9	♑ 4	+ 5
♄	♑ 5	♑ 4	+ 1
♀	♑ 11	♑ 14	— 3
♂	♑ 12	♑ 4	+ 8
♃	♓ 20 [or 8]	♓ 5	+ 15 [or + 3]
☊	♉ 4	♈ (!) 1	[+ 3]
☿	♐ 17	♐ 10	+ 7
☾	♐ 19	♐ 11	+ 8

The entry for the node was probably made in the wrong sector of the diagram or right on the boundary, such that it could be interpreted as belonging to ♉ instead of to ♈. In the case of Jupiter, 20 is probably a misreading for 8, since κ and η are often indistinguishable in manuscripts of the Byzantine period.

Since the present horoscope as well as No. L — 71 belongs to a chapter on the length of life, "starter" and "destroyer" are determined: the sun in ♑ 9 is the starter, the moon in ♐ 19 the destroyer. Then the quartile ♓ 19, to the destroyer, is considered (*cf.* figure 13*b*) and analogy to the procedure in No. L — 71 leads us to restore the conclusion that the span of life allotted was 70 years, corresponding to the arc from ♑ 9 to ♐ 19. This would mean a recording of our horoscope not before A.D. 28.

Nos. L 37 and 68

Vettius Valens V,11 p. 233,25 to 234,30

TRANSLATION AND SUMMARY

The preceding is the table of the common return of the planets to the (same) intervals and (the same)

[1] Following P.

[2] Following, from now on, A. We ignore the repetition of 9 (the longitude of ☉ and H) and combine τόπος with α'.

[3] A synodic period of Venus is 8 years; *cf.* *CCAG* 8,4 p. 235,7 f. and p. 237,18 ff.

[4] Probably error for 8 degrees; *cf.* the commentary.

[5] *Cf.* Introduction p. 8.

combinations.[1] As for example: sun, Mars, Mercury (and) Horoscopos in Sagittarius. Moon in Leo, Saturn in Virgo, Jupiter in Scorpio, Venus in Capricorn. The moon is allotted the second (interval) since she is second (counting zodiacal signs) from Saturn; likewise also the sun and Mars, Mercury (and) Horoscopos (with reference) to Venus.

[In the same fashion all "intervals" are enumerated, up to 12 signs. The "trias" is assigned to Saturn, Jupiter, and Venus, because Jupiter is in the third sign counting from Saturn, Venus in the third sign counting from Jupiter. Thus to these three planets "is allotted the triad of the nativity."

The enumeration of all possible intervals leads somehow to the result that the configurations 31 years later[2] must be considered. Then the following situation is encountered:]

So in case of the preceding nativity, the stars at the time were sun, Jupiter, (and) Mercury in Gemini, Saturn in Virgo, Mars (and) Venus in Taurus, moon in Pisces, . . .

[There now follows again a discussion of intervals combined with "returns" which probably means that the combined intervals should make a total of 12. Because in the initial horoscope ☉ ♂ ☿ and Horoscopos are associated with the interval 10 (from ♄), whereas in the new configuration this interval is found for ☉ ☿ ♃ ♄, the first quadruple turns over its influence to the second.]

COMMENTARY

Diagrams: pls. 10 and 11

The given positions lead to the dates 37 December 15 and 68 June 11 respectively. The interval,[3] however, is not exactly 31 Egyptian years but only 30 Egyptian years 6 months and 6 days.

For the earlier horoscope one finds excellent agreement for 37 December 15:

	Text	Computed
☉	♐	♐ 23
♂	♐	♐ 27
☿	♐	♐ 1
H	♐	about sunrise
☾	♌	♌ 9
♄	♍	♍ 27
♃	♏	♏ 14
♀	♑	♑ 24

[1] This table is missing in the manuscripts, though space has been left free for it. Perhaps it is taken from Critodemus (*cf.* Kroll p. 233,22).

[2] In IV,2 (Kroll p. 159,4) a period of 32 years 3 months is mentioned after which a new cycle begins.

[3] It amounts to 11136 days.

The later configuration leads to 68 June 11:

	Text	Computed
☉	♊	♊ 18
♃	♊	♊ 13
☿	♊	♉ 27(!)
♄	♍	♍ 27
♂	♉	♈ 29(!)
♀	♉	♉ 18
☾	♓	♓ 18

Notice the fact of the accurate return of Saturn to the same longitude.

No. L 40

CCAG 8,2 p. 84,1 to p. 85,10

TRANSLATION

This is what Antigonus says in his abridgment of the writings of the ancients about the said nativity; a second one[1] he quotes as follows: Assume that somebody has the sun in Aries about[2] 19° in terms of Mercury, the moon in Gemini 15°, terms of Venus; Saturn in Libra about 20°, terms of Jupiter, in opposition,[3] Jupiter in Aquarius 6°, terms of Mercury at[4] morning rising, Mars in Aries about 15°, terms of Mercury, Venus similarly in Aries about 5°, terms of Jupiter, Mercury in Aries about 6°, and the three (planets) Mercury, Venus, Jupiter are (located) towards setting. And the Horoscopos is in Cancer 24°, the setting point similarly in Capricorn 24°. The upper culmination in Aries 10°, the lower (culmination)[5] in Libra 10 and, as he says, Mars will be the ruler of the nativity. After 3 days from the birth the moon will be in Cancer, after 7 in Virgo, after 40 in Scorpio.

The (person) who has the stars in this way (as his nativity) will be very distinguished, of very distinguished (ancestors), a person of authority and punisher of many, and very wealthy, because both the moon and Jupiter were in the triangle of Saturn (♊ ♎ ♒), the moon in Gemini and Jupiter in Aquarius, but unjust and not accused because of the position of Mars and Mercury near Midheaven, both being invisible,[5a] very distinguished because of the four stars (☉ ♂ ♀ ☿) being in Midheaven (♈). The preceding conjunction also was in the same sign; for the preceding conjunction has no little power when it happens to be at a center and especially when the Horoscopos or Midheaven is concerned. And he was indifferent to female(?) inter-

[1] The first one is the nativity of Hadrian; *cf.* No. L 76.

[2] περὶ here and in the following.

[3] ἀκρόνυχος; *cf.*, e.g., P. Mich. 149, XI, 9-11.

[4] ἀπὸ.

[5] The terminology is of interest: ὑπέργειον μεσουράνημα and ὑπόγειον respectively.

[5a] ὕπαυγος i.e., under the rays of the sun.

course⁶ and sordid toward males, especially because Mars⁷ had its position with Venus and Mercury in Aries, which is a vicious (sign), and (because) Venus (in ♈) was looked upon by Saturn (in ♎) in quartile,⁸ and (because) all these stars were in male signs. The sun being in its own exaltation (♈) and at Midheaven and attended in the same sign by the stars (♀ ☿ and ♃) which precede it, and Jupiter being in Aquarius, the stars which precede the sun⁹ up to quartile position "attend" the sun¹⁰ and make the nativity magnificent and illustrious. The moon (in ♊) waxing in the trigonal configuration with Saturn (in ♎) and Jupiter (in ♒) also effected a happy and very wealthy (theme) and a person who provided many dedications and gifts for his fatherland.

[At the lower margin of fol. 133ᵛ is given a diagram (in the same hand as the text) which shows the same positions as given in the text, except for the moon ♊ 25 (text: ♊ 15) and for Saturn ♑ 5 (text: ♎ 20). In the edition no mention is made of this diagram and the variants it contains]

COMMENTARY

Diagram: pl. 10

The three horoscopes Nos. L 40, L 76, and L 113,IV are preserved through Hephaistion (fourth century) who took them from a work of Antigonus. As Kroll realized,¹¹ No. L 76 is the horoscope of Hadrian, who died in July 138. Cumont suggested for No. L 113,IV Pedanius Fuscus, Hadrian's grandnephew as the subject,¹² who was executed in 137/138. We shall see presently that the relationship between the rising (H) and the culminating degree (M) in No. L 40 as well as in No. L 76 agrees exactly with the geographical latitude of southern Spain,¹³ the place of origin of Hadrian and his family. One might therefore conjecture that the subject of L 40 was also a member of Hadrian's family, e.g., his father. Cramer (*ARLP* p. 177) considers also Servianus, the grandfather of Pedanius Fuscus.

The data of the horoscope No. L 40 lead to A.D. 40, April 5 as date of birth

	Text	Computed	Text—comp.
☉	♈ 19	♈ 14	+5°
☽	♊ 15¹⁴	♊ 11	+4
♄	♎ 20¹⁵	♎ 20	0
♃	♒ 6	♑ 29	+7
♂	♈ 15	♈ 9	+6
♀	♈ 5	♓ 29	+6
☿	♈ 6	♓ 26	+10
H and Δ	♋ and ♑ 24	about noon	
M and IMC	♈ and ♎ 10	♈ and ♎ 10	
prec. conj.	♈	March 30: ♈ 8	

For the culminating sign, agreement is obtained if one uses the clima of Rhodes, to which southern Spain belongs. It is of course possible to question the value of such an agreement since we do not know the tables which were used by Antigonus.¹⁶ It is only because of the repetition of the same type of agreement in No. L 76 that a mere accident appears less likely.

Except for Saturn¹⁷ the deviations between ancient and modern longitude are of the order of magnitude to be expected for the first century and also agree with the figures in No. L 76.

It is obviously correct to say that ♄ (in ♎ 20) is in opposition to the sun (in ♈ 19). Jupiter is beyond its morning rising (Γ) but still proceeding with direct motion. Because ♈ 10 is culminating, ☿ (in ♈ 6), ♀ (in ♈ 5), and ♃ (in ♒ 6) are beyond culmination and moving toward setting.

The preceding conjunction of the moon (on March 31) is correctly located in ♈. The future positions also are in the proper signs:

after 3 days:	Apr. 8	♋ 17
after 7 days:	Apr. 12	♍ 7
after 40 days:	May 15	♏ 25.

The lunar positions after 7 days and after 40 days are also investigated in No. L 113,IV. The moon after 7 days is considered in No. L 479, after 3 days in No. L 487. The rising of Jupiter after 7 days is noted in No. L 76, Hadrian's horoscope, and the positions of the moon and of Mars and Saturn after 40 days are computed. An interval of danger of 40 days because of Mars occurs in No. — 3. Mercury and the 7th day are mentioned in No. 81 (col. VII); *cf.* also No. 345.

---

⁶ πρὸς δὲ τὰς θελείας (i.e. θηλείας ?) μίξεις.
⁷ Incorrectly given in the edition (p. 84, 25) as Jupiter.
⁸ According to the text, which is confirmed by computation, in ♎ 20 at diameter. The diagram, however, shows Saturn in ♑ 5 thus in quartile to Venus.
⁹ In the direction of the daily rotation.
¹⁰ *Cf.* for this concept Bouché-Leclercq, *AG* p. 252 ff.
¹¹ *CCAG* 6 p. 67 and *CCAG* 8,2 p. 83 note 1.
¹² *CCAG* 8,2 p. 85 note 1.
¹³ *Cf.* below p. 91 and p. 136 respectively.
¹⁴ In diagram: ♊ 25. The value of the text is confirmed by our computation.
¹⁵ In diagram: ♑ 5(!). This would be in the terms of Venus and not of Jupiter as required by the text, nor in trine with the moon in ♊ and Jupiter in ♒, though in quartile to Venus in ♈.
¹⁶ It must be admitted that the procedure of Paulus Alexandrinus (*cf.* p. 10) leads, for Alexandria and System A, to M = ♈ 10;20. No such agreement is found, however, for No. L 76.
¹⁷ No emendation of the text is possible since Saturn (in ♎) is in the terms of Jupiter which extend only to ♎ 21.

No. L 50

Vettius Valens II,21 p. 83,10-26

INTRODUCTION

Chapter II,21 consists of 14 examples, the first and earliest of which is No. L 50. Three of them are used once more in later chapters and for different purposes. The following is the complete list of their dates:

(*a*)	50 Oct. 25	p. 83,10-26
(*b*)	95 May 18	p. 83,27-p. 84,2
(*c*)	85 Febr. 5	p. 84,3-11
(*d*)	83 Apr. 28	12-22
(*e*)	74 Nov. 26	23-33
(*f*)	72 Jan. 6	p. 85,1-8
(*g*)	82 July 9	9-18
(*h*)	97 Nov. 6	19-24
(*i*)	95 May 14	p. 85,25-p. 86,3
(*j*)	65 Oct. 31	p. 86,4-9
(*k*)	105 Jan. 1	10-16
(*l*)	61 May 1	17-23
(*m*)	109 June 3	24-31
(*n*)	62 Jan. 22	p. 86,32-p. 87,4

For the details see the commentaries to the individual horoscopes.

TRANSLATION

To make clear the preceding (chapters) we shall use examples, taking first a distinguished nativity. Let the sun be in Scorpio, the moon in Cancer, Saturn in Aquarius, Jupiter in Sagittarius, Mars in Scorpio, Venus in Libra, Mercury in Scorpio, Horoscopos in Libra. Since, then, the nativity is nocturnal, I look for the moon; this happens to be in Cancer, in trine with Mars. And we find the (star) of Mars in epanaphora in his own house (♏), triangle (♏, ♓, ♋), and sect;[1] then (we find) Venus sharing (rulership) with Mars, since she is in the Horoscopos and in her own house (♎); and third (we find) the moon in Midheaven in its own house (♋). It is evident, then, that the nativity is a notable one, since the rulers are in their own domains. And looking for the Lot (of Fortune) I found it in Aquarius;[2] Saturn is there, is the ruler, is in the (5th) locus of Good Fortune (♒), and is in his own house and triangle. Similarly Jupiter (in ♐) has the 11th locus with respect to the Lot of Fortune, that is the locus of accomplishment. And I took also the exaltation of the nativity[3] which, from the moon to Taurus, is 11 (signs) and the same number (counted) from the Horoscopos in Libra fell in Leo in the Locus of the Agathos Daimon; the sun is ruler of this,[4] and since it was found in Midheaven with respect to the Lot of Fortune,[5] it made the nativity more illustrious and distinguished.

COMMENTARY

Diagram: pl. 10

Computation shows agreement for A.D. 50 October 25:

Text[6]		Computed
⊙	♏	♏ 1
☾	♋	♋ 16
♄	♒	♒ 10
♃	♐	♐ 3
♂	♏	♎ 27(!)
♀	♎	♎ 21
☿	♏	♏ 13
H	♎	about 4 A.M.

No. L 61,V

Vettius Valens II,21 p. 86,17-23

TRANSLATION

Sun, Venus, (and) Horoscopos in Taurus, moon in Aquarius, Saturn in Cancer, Jupiter in Libra, Mars (and) Mercury in Gemini. In the first periods (or times) he had great political prestige and affairs and positions of trust; for the rulers (♀ and ☾) of the triangle (♉ ♍ ♑) happened to be in centers (H and M respectively). But later he was brought down in life and became a vagabond; for Mars and Mercury were in opposition to (the Locus of) accomplishment[1] and the rulers (♄ and ♃) of the Lot (of Fortune[2]) and of the (Locus of) accomplishment[3] were in apoklima.

COMMENTARY

Diagram: pl. 10

For 61 May 1 one finds

Text		Computed
⊙ }	♉	♉ 9
♀ }	♉	♉ 22
H }	♉	about sunrise
☾	♒	♒ 6
♄	♋	♋ 2
♃	♎	♎ 27
♂ }	♊	♊ 2
☿ }	♊	♊ 2

For related horoscopes *cf.* No. L 50.

---

[1] *Cf.* Introduction p. 12.

[2] From $\lambda_F = \lambda_H - (\lambda_☾ - \lambda_⊙) =$ ♎ $+ 4° =$ ♒.

[3] *Cf.* Introduction p. 7.

[4] Leo is the house of the sun.

[5] If ♒ (the Lot of Fortune) is rising then ♏ (the place of the sun) is in Midheaven.

[6] Note that the arrangement follows strictly the order from ⊙ to H instead of combining those which fall in the same sign.

No. L 61,V

[1] The 11th locus from the Lot of Fortune (♒), thus ♐.

[2] Computing for daytime one finds $\lambda_F = \lambda_H + (\lambda_☾ - \lambda_⊙) =$ ♉ $+ 9° =$ ♒; and ♒ is a house of Saturn.

[3] A house of Jupiter is ♐.

No. L 61,X

Vettius Valens III,8 (p. 143,5 ff.) and VIII,8 (p. 318,15 ff.)

INTRODUCTION

Chapter III,8 and the first half of VIII,8 are parallel versions of the same text. Kroll seems not to have recognized this fact, or, if he did, made no attempt to restore an intelligible text. Calling the two versions A and B respectively (A = p. 143,5 to p. 144,12; B = p. 318,15 to p. 319,17) we shall give in the following translation a text which is relatively consistent, quoting in the notes the version which we reject, though only in cases of significant differences.

It is an immediate and obvious consequence of a comparison of the two versions III,8 and VIII,8 that their titles refer in fact to two different chapters: one "On the hostile places and stars" the text of which is given by A and B, the second "On the critical places according to the first table of Critodemus" the text of which is only preserved in the second half of VIII,8 beginning with the words "We must first speak about the construction of the table . . ." (p. 319,18 Kroll). Both versions have combined these two titles into one, though B is closer to the original meaning: "On the hostile places and stars; on the critical places according to the first table." In A we read "On the hostile stars and critical places; on the first table of Critodemus."

We shall see that the example quoted in this chapter can in all probability be dated October 61 A.D. In it a system of "terms" is used which assigns "terms" not only to the planets but also to sun and moon. This system is so far only known from Vettius Valens III,9 (p. 144 f. Kroll; cf. Bouché-Leclercq AG p. 213 f.). Perhaps the association with a chapter which seems to come from Critodemus points to the same author as inventor of this particular procedure.

TRANSLATION

On the hostile places and stars.

It is necessary to investigate the hostile places and stars not only with respect to the other (planets) but also for the Horoscopos and sun and moon. Because these, also, when [1] they come into diameter during their travel [2] indicate crisis and death. Thus one must investigate, in the case of Saturn, the degrees at diameter, to which god the terms [3] belong, as given in the table. And the person will die when Saturn is there or in the square of the Horoscopos or (in the places) with the same rising time, according to the combination of time in the squares of the Horoscopos or (in the places) with the same rising time.[3] The same must be done also for the other stars because the rulers of the terms of the diametrically opposite degrees are hostile. If these planets come to the places (where the rulers of the diameter-terms are) or in (the places) with the same [4] rising time, then they indicate destruction.

Thus let Saturn be in Cancer 21 degrees, terms of Venus. At diameter is Capricorn (21°), terms of Mars, who was [5] in Taurus 27 degrees. He will die when Saturn is there. He died (when Saturn was) in Virgo because it is in its square, reckoned by degrees.[6]

Jupiter [7] in Scorpio 14 degrees, terms of Saturn. The 14th (degree) of Taurus is in the terms of Saturn. But it does not become hostile to itself. Now Leo has the same rising time as Scorpio and the 14th degree of Leo is in the terms of the sun. Thus Jupiter destroys when it comes to the places of the sun.[8]

Mars in Taurus 27, terms of the sun. The same degrees of Scorpio are in the terms of the sun.[9] But it does not become hostile to itself. Thus one must investigate the 27 degrees in Leo or in (the sign of) equal rising time (with ♉) which is Aquarius [10] according to hourly distinctions; [11] but the 27 (degrees) of Aquarius are in the terms of Venus. Thus he will die when Mars is in Scorpio or in Pisces,[12] which are of equal rising time, or in their squares. If anyone reckons the 27 of Leo he will find the terms of Saturn. Saturn was in Cancer; thus he will die when Mars is in Cancer or in Sagittarius or their squares.

Venus in Scorpio 27 degrees, terms of the sun. The 27 degrees of Taurus at diameter are in the terms of the sun. But it does not become hostile to itself. Thus I investigate in (the sign of) equal rising time, Scorpio, the 27 (degrees); they are in the terms of Mercury. He will die when Venus is in Virgo where Mercury was; or in their squares. The same should be done with Mercury.

Also for illnesses it is necessary to investigate the diametrically opposite places and what planet is located in the hostile places [13] and (the stars) which cause the crises of months, days, and hours according to the degree [14] of the moon from which the opposing star is found.

---

[1] B καί, A ἐάν.
[2] om. A.
[3] Last part of sentence omitted in A.
[4] om. A.
[5] A has here ἥλιος which is excluded by the positions of Venus and Mercury given later on.
[6] The data are too incomplete to check what is meant by "reckoned by degree."
[7] om. B.
[8] B: . . . becomes hateful in (the place) of the same rising time and there it kills.
[9] B: of the moon.
[10] Gemini in B. The rest of the sentence is omitted in B until "terms."
[11] Perhaps referring to time units instead of degrees.
[12] Error for Aquarius.
[13] A: ἐπὶ τοὺς ἐχθροὺς τροπὰς (!) ὄντας instead of ἐπὶ τοῖς ἐχθροῖς τόποις ὄντα in B.
[14] Lit.: thirtieth.

Then[15] from sun, moon, and Horoscopos the starter will be determined or else from the star found following the Horoscopos and so on in order to the other (stars) as they happen to be by sign and degree at time of birth; making the decision for (every period) of 10 years and 9 months.[16]

COMMENTARY

Diagram: pl. 10

We have here a doctrine (with example) which requires not only the consideration of the longitude λ of a planet but also its diameter $\bar{\lambda} = \lambda + 180°$, the signs of equal rising time α, the "squares," and the terms according to the scheme given in Vettius Valens III,9. Whenever the latter are found to be correct they are the terms at night. Thus we must assume that the example was a case of night birth. It turns out that many of the terms do not agree with the table. In several cases the reason for the disagreement is obvious: in order to discuss also the need for equal rising times and for squares, the terms of λ and $\bar{\lambda}$ should be the same (in order to apply the principle "nobody is hostile to himself"). Thus one term is found correctly, the other one is then simply assumed to be the same.

The following data can be extracted from the example.

♄	$\lambda =$ ♋ 21	terms of ♀	(correct)
	$\bar{\lambda} =$ ♑ 21	terms of ♂	(correct)
	♂ in ♉ 21	(*cf.* below)	
♃	$\lambda =$ ♏ 14	terms of ♄	(no: ☾)
	$\bar{\lambda} =$ ♉ 14	terms of ♄	(correct)
	α(♌) = α(♏)		(correct)
	♌ 14	terms of ⊙	(no: ♀)
♂	$\lambda =$ ♉ 27	terms of ⊙	(correct, on boundary)
	$\bar{\lambda} =$ ♏ 27	terms of ⊙	(no: ♄)
	α(♌) = α(♏)		(correct)
	♌ 27	terms of ♄	(no: ♂)
	♄ in ♋		(*cf.* above)
	α(♒) = α(♉)		(correct)
	♒ 27	terms of ♀	(no: ♄)
♀	$\lambda =$ ♏ 27	terms of ⊙	(correct, on boundary)
	$\bar{\lambda} =$ ♉ 27	terms of ⊙	(no: ♂)
	α(♌) = α(♏)		(correct)
☿	in ♍		

For an attempt to date this horoscope we have to rely on the following elements:

♄	♋ 21	♀	♏ 27
♃	♏ 14	☿	♍

Saturn and Jupiter for the period of Vettius Valens lead to only three possibilities: A.D. 61, 120, and perhaps 179 or 180. The elongation of at least 57 degrees between Venus and Mercury is only possible when the sun is between these planets. Because of Mercury the sun cannot be beyond about ♎ 25. Because of Venus the sun must be at least in ♎ 10. This leaves us with the narrow interval from about October 4 to October 19. Computing, e.g., for October 7, rules out the two later dates. For 61 October 7 we find excellent agreement.[17]

	Computed	Text
♀	♏ 30	♏ 27
☿	♍ 24	♍

Similarly for Saturn and Jupiter:

	Text	Computed
♄	♋ 21	♋ 17
♃	♏ 14	♏ 9

For Mars, however, there is total disagreement: ♍ 27 instead of ♉ 21 and similar differences for the other years.[18] Thus we must assume that the agreement for four planets for 61 October 7 is a mere accident and all five positions were selected arbitrarily. Or, and this seems more plausible, the data except for Mars were taken from a horoscope very close to 61 October 7 and only Mars was differently placed for some reason connected with the doctrine to be exemplified.

The fact that a whole group of horoscopes between 65 and 123 are supposedly taken from Critodemus (*cf.* No. L 65,V) might be taken as a supporting factor of the date 61 of the present horoscope and its association with Critodemus.

No. L 62

Vettius Valens II,21 p. 86,32 to p. 87,4

TRANSLATION

Sun in Aquarius, moon (and) Jupiter in Scorpio, Saturn in Cancer, Mars, Venus (and) Mercury in Capricorn, Horoscopos in Pisces. This man was a eunuch priest assigned to the goddess; for the ruler of the Lot (of Fortune[1]) happened to be in Scorpio, the (9th) locus of the God; and the rulers of the Sect,[2]

---

[15] This whole final paragraph is missing in B.

[16] This refers to a doctrine described in Vettius Valens VI,5 (p. 251 ff. Kroll).

[17] A closer investigation shows that Venus restricts the dates to the narrow interval from October 4 to October 8 in which it moves from ♏ 27 to ♐ 2. Mercury remains during this interval at about ♍ 24 or 25.

[18] It is perhaps not an accident that the position of Mars is also entirely wrong in the horoscopes No. L 62 (January 23) and No. L 65 X, both in Vettius Valens II,21.

No. L 62

[1] That is ♐ from $\Lambda_{\mathfrak{F}} = \lambda_{H} + (\lambda_{☾} - \lambda_{⊙}) =$ ♓ $- 3° =$ ♐. Sagittarius is a house of ♃ which is located in ♏.

[2] The birth is diurnal and Saturn belongs to the diurnal sect (*cf.* Introduction p. 11) and the same holds for Mercury which is morning star. One expects, however, the rulers of the triangle ♒ ♊ ♎ (♄ and ☿) because the sun is in ♒.

Saturn and Mercury, were found in the (11th) Locus Agathos Daimon (♑), but in opposition.[3] Wherefore he fell into a great many troubles and losses and offenses against governor and king.

COMMENTARY

Diagram: pl. 10

A date for this horoscope can be found within the historically plausible limits only if we assume that Mars is incorrectly placed in ♑ with ♀ and ☿ instead of in ♏ with the moon and Jupiter. For 62 January 22 one finds

	Text	Computed
☉	♒	♒ 1
☾	♏	♏ 13
♃	♏	♏ 29
♄	♋	♋ 14
♂	♑	♏ 24(!!)
♀	♑	♐ 29(!)
☿	♑	♑ 6
H	♓	about 10 A.M.

The earliest horoscope in this chapter is No. L 50.

No. L 65,V

Vettius Valens II,41, p. 130,1 to 5 = CCAG 5,2 p. 121,13 to 16

INTRODUCTION

The chapter[1] "On violent deaths with examples" contains a series of eleven horoscopes, eight of which are also published in *CCAG* 5,2, this time in a chapter headed "On the yielding of chronocratorship of the planets; (Chapter) 2: The assignments. From the (books) of Critodemus." As all these horoscopes can be dated within the interval from A.D. 65 to 123 one might be tempted to assign Critodemus to this period, were it not known that he lived before Pliny.[2] Consequently one has to prefer the version of the title which is given in Kroll's text.

The collection of horoscopes is not identical in the two versions as is shown by the following list:

Kroll	*CCAG* 5,2	Date
(*a*) p. 128,31-37		123 July 2
(*b*) p. 129,1-7		97 Feb. 23
(*c*) 8-14		87 July 9
	(A) p. 120,30-32	112 Aug. 17
(*d*) p. 129,15-17	(B) 33 f.	86 Dec. 27
(*e*) 18-23	(C) p. 120,35-p. 121,2	101 Jan. 28
(*f*) 24-28	(D) 3-7	103 Jan. 10
(*g*) 29-33	(E) 8-12	115 Dec. 26
(*h*) p. 130,1-4	(F) 13-16	65 May 24
(*i*) 6-10	(G) 17-21	88 May 5
(*k*) 11-16	(H) 22-26	89 July 29
(*l*) 17-22	(I) p. 120,27-31 and p. 52,24-28	91 April 4

The first three examples of the first version are missing in the second. The first horoscope in *CCAG* 5,2 is missing in II,41 of the first version but appears in II,36 (Kroll p. 114,16 to 18 [3]). Again Kroll's text appears as the better one because (A) does not concern "violent death" as do all the other examples, its subject being only hunchbacked.

The closeness of the dates clearly shows that we are dealing here with a uniform collection. This is confirmed by certain details, e.g., by the reckoning of the "Locus of Death" from the Lot of Fortune, not from the Horoscopos.[4]

TRANSLATION

Sun, moon, (and) Mercury in Gemini, Saturn in Leo, Jupiter in Pisces, Mars in Cancer, Venus in Aries,[5] Horoscopos in Capricorn, where also the Lots (of Fortune and Daimon) were located.[6] The ruler (of the Lots), Saturn,[7] was in the (8th) Locus of Death (♌) and was in aspect with Venus (in ♈). Mars (in ♋) was in opposition to the Horoscopos (in ♑). This person died by poison.

COMMENTARY

Diagram: pl. 11

For 65 May 24 one finds excellent agreement.

	Text	Computed
☉	♊	♊ 1
☾	♊	♊ 10
☿	♊	♊ 13
♄	♌	♌ 22
♃	♓	♓ 16
♂	♋	♋ 11
♀	♈ [5]	♈ 15
H	♑	about 8 P.M.

[3] Mercury is in ♑, Saturn in ♋ in opposition to ♑.

No. L 65,V

[1] It is counted as Chapter 40 on the page titles of Kroll's edition, but as Chapter 41 in the Greek text (p. 123,25) and in the index (p. XIV).

[2] *Cf.* the discussion below p. 185.

[3] *Cf.* below p. 108.

[4] *Cf.* Introduction p. 8.

[5] Following *CCAG* 5,2 p. 121,14. Kroll p. 130,2 adopted the incorrect version (Taurus), relegating the correct text to the apparatus.

[6] Because ☉ and ☾ are in the same sign.

[7] A house of Saturn is ♑.

No. L 65,X

Vettius Valens II,21 p. 86,4-9

TRANSLATION

Sun (and) Mercury in Scorpio, moon in Aries, Saturn in Virgo, Jupiter in Pisces, Mars in Leo, Venus (and) Horoscopos in Sagittarius. Even in childhood the nativity was destined to very great property; for the (11th) locus [1] of accomplishment was in Pisces with Jupiter in his own domain (house). And Venus co(ruler) of the triangle (of the sun, ♏ ♓ ♋) and ruler of the Lot (of Fortune [2] in ♉, a house of ♀) and of the exaltation (of the nativity, again in ♉) was . . .[3]

COMMENTARY

Diagram: pl. 11

The date of this horoscope is very questionable. From the dates of the other examples in this chapter [4] and from the general form, a date in the first century seems to be plausible. And indeed, for 65 October 31 one can satisfy the first five data:

	Text	Computed
☉	♏	♏ 7
☿	♏	♏ 17
☾	♈	♈ 18
♄	♍	♍ 9
♃	♓	♓ 10

Mars and Venus, however, should at this date both be in ♎ (27 and 23 respectively) whereas the text locates Mars in ♌, Venus in ♐. And no other date, down to the middle of the third century will bring all the three outer planets in the right position with the sun in Scorpio. Thus some of the elements must be wrong but there seems to be no obvious emendation which could restore a plausible date, leaving 65 as a tentative solution.

The time of birth was about 8 A.M.—For related horoscopes *cf.* No. L 50.

No. L 68

Vettius Valens V,11 p. 234,7-11

Cf. No. L 37

[1] Counted from ♉, the Lot of Fortune.
[2] Found from $A_F = \lambda_H + (\lambda_☾ - \lambda_☉)$ = ♐ − 7° = ♉.
[3] Lacuna in the text followed by "Horoscopos by degree(?)."
[4] *Cf.* the Introduction to No. L 50.

No. L 72

Vettius Valens II,21, p. 85,1-8

TRANSLATION

Sun (and) Mercury [1] in Capricorn, moon (and) Venus [1] in Sagittarius, Saturn in Scorpio, Jupiter in Libra, Mars in Aquarius, Fortune in Aries,[2] Horoscopos in Taurus. This nativity was also at the beginning mixed and mediocre but later was raised and had a share of chaplets and became archpriest. For the triangle (♑ ♉ ♍) rulers (☾ and ♀) were found to be in epanaphora [3] and the third ruler (♂) of the triangle and the ruler (♂) of (♈)[4] the Lot (of Fortune) was in Midheaven (♒). Likewise the ruler (☉) of (♌), the exaltation (of the nativity), was in Midheaven with reference to the Lot (of Fortune,[5] and (also) the ruler (☿) of Daimon (♊)[6] [text broken].

COMMENTARY

Diagram: pl. 11

This is the sixth horoscope in a collection of examples, the first of which is No. L 50. For the present horoscope one finds the date *72 January 6*:

	Text	Computed
☉	♑	♑ 15
[☿]	♑	♑ 10
☾	♐	♐ 6
♀	♐	♐ 11
♄	♏	♏ 18
♃	♎	♎ 5
♂	♒	♒ 3
H	♉	about 2 P.M.

No. L 74,IV

Vettius Valens III,13 (p. 154,1 to 10)

TRANSLATION

As an example let the sun be in 1 degree of Taurus, the moon in 16 degrees of Gemini, Horoscopos in 14 degrees of Taurus, the (preceding) conjunction in about 27 degrees of Aries, the ascending node in 25 degrees of Capricorn. I counted upwards [1] from the conjunction

[1] Kroll interchanges in the text Venus and Mercury and removes to the apparatus the correct reading of S "moon (and) Venus." All manuscripts seem to have "sun (and) Venus" instead of "sun (and) Mercury."
[2] From $A_F = \lambda_H + (\lambda_☾ - \lambda_☉)$ = ♉ − 1° = ♈.
[3] A better translation would be "rising," but both moon and Venus approach invisibility.
[4] A house of ♂ is ♈.
[5] If ♈ is rising ♑ is culminating.
[6] A house of ☿ is ♊.

No. L 74,IV

[1] That means in the direction of the daily rotation.

(♈ 27) to the ascending node (♑ 25); it makes 92. I subtracted it upwards [1] from the degree of the Horoscopos (♉ 14); it ends in 12 degrees of Aquarius. Thus the 92 (degrees) total for the clima of Alexandria 70 years.[2] He lived seventy years and one month. [There follows a mutilated passage concerning the situation at conception. The corresponding full moon is said to have been in Capricorn 21.]

COMMENTARY

Diagram: pl. 11

For 74 April 19 one finds excellent agreement with the data of the text:

	Text	Computed
☉	♉ 1	♈ 27
☾	♊ 16	♊ 12
H	♉ 14	about 7 A.M.
prec. conj.	♈ 27	♈ 23 (April 15)
☊	♑ 25	♑ 21
full moon of conception	♑ 21	♑ 14 (73 July 8)

A deviation of 4° between text and computation agrees with the correction to be expected for the first century.[3]

A similar fragmentary horoscope is No. L 115,II and No. L 75,1.

No. L 74,XI

Vettius Valens II,21, p. 84,23-33 and VII,2, p. 270,6-17

TRANSLATION

First Version

Sun, Mercury, Saturn (and) Jupiter in Sagittarius, moon in Cancer, Mars in Virgo, Venus (and) Horoscopos in Libra. As it is a night birth we find the moon, in the triangle of Mars (♋ ♏ ♓), and Mars itself in apoklima, as are both the Lot (of Fortune, i.e., ♓)[1] and (its) ruler (♃). Wherefore he lived his first (years) humbly and in poverty and (experienced) captivity and servitude and was involved in many perils. But since the (planets) of the same sect [2] happened to be in operative positions he came into friendships and associations and received positions of royal trust. Indeed, he became highly esteemed because the exaltation (of the nativity) was found in Leo and the ruler (of Leo), the sun, was in Midheaven (with respect) to the Lot (of Fortune [3]) and he was found worthy of leadership and a position of power.

Second Version

Sun, Saturn, Jupiter (and) Mercury in Sagittarius,[4] moon in Cancer, Mars in Virgo, Venus (and) Horoscopos in Libra. First clima. In the 69th year he was deemed worthy of the governorship [5] and became feared and most highly regarded and blessed by many but was (also) envied and became involved in mob uproar and riots and did not complete his term of office and was overtaken by painful illness and death. For operative were the rising time of Libra (the Horoscopos) 38;20° and that of Cancer (Midheaven) 31;40, which total 69; and the same (total) is found for the maleficent (planets): (namely) 30 for Saturn and 19 for the sun and 20 for Mercury [6] make 69. And again 38;20 of Virgo, 12 of Jupiter and 19 of the sun make 69. Thus all the stars were operative and each of them, according to its own nature and to its place in the scheme, had its own effect.

COMMENTARY

Diagram: pl. 11

The first version is the fifth horoscope in a series of 14 examples, the first and earliest of which is No. L 50. For the present horoscope one finds the date 74 November 26:

	Text	Computed
☉	} ♐	♐ 3
☿		♐ 6
♄		♐ 14
♃ [7]		♐ 17
☾	♋	♋ 10
♂	♍	♍ 29
♀	} ♎	♎ 18
H		about 2 A.M.

The second version is the earliest horoscope in chapter VII,2 which contains the following six horoscopes:

74 Nov. 26	p. 270,6 to 17
114 Sept. 24	p. 268,7 to 15
117 June 30	p. 269,11 to 28
118 Nov. 26	p. 268,29 to p. 269,2
120 Dec. 8	p. 267,28 to p. 268,4
122 Dec. 4	p. 268,16 to 28.

---

[2] This is probably taken from tables similar to the tables in book VIII (Kroll p. 321 ff.).

[3] The larger deviation in the case of the full moon is not significant since it can be explained as the result of the computation of the lunar longitude for the expected *time* of the opposition, where a small error has a great effect on the longitude of the moon.

No. L 74,XI

[1] From $\Lambda_F = \lambda_H - (\lambda_{☾} - \lambda_{☉}) = ♎ + 5° = ♓$. And ♓ is a house of Jupiter.

[2] Probably moon and Venus, co-rulers of the triangle ♋, ♏, ♓.

[3] If ♓ is rising then ♐ is in Midheaven.

[4] Note the different order of enumeration as compared with the first version.

[5] Perhaps Valerius Eudaemon who became ἔπαρχος Αἰγύπτου in 142 but was no longer prefect in 144; *cf.* Stein *Präf.* p. 74 ff. See also W. L. Westermann. The prefect Valerius Eudaemon and the indigent liturgist, *JEA* 40 (1954) pp. 107-111.

[6] *Cf.* Bouché-Leclercq, *AG* p. 493.

[7] In the second version the arrangement is ☉ ♄ ♃ ☿.

From the second version we obtain the information that the birthplace belongs to the first clima (Alexandria). The rising times quoted

$\alpha_6 = \alpha_7 = 38;20$ and $\alpha_4 = 31;40$

belong to System A.

No. L 75

Vettius Valens III,6 (p. 141,28 to 142,7) and III,14 (p. 156,7 to 13) and IV,8 (p. 167,11 to p. 169,3 — CCAG 5,2 p. 82,30 to p. 84,11)

TRANSLATION

First Version (III,6)

For example let the sun be in Cancer degree 29;30, the moon in Pisces degree 12, Saturn in Sagittarius degree 27;8, Jupiter in {Virgo degree 22;13},[1] Mars in Scorpio degree 7;23, Venus in Cancer degree 28;13, Mercury in Leo degree 11;25, Horoscopos in Pisces degree 17, Midheaven in Sagittarius degree 25. The nativity was without house-ruler because Venus, ruler of the terms of the moon,[2] had fallen below Dysis. The Horoscopos was the starter, the ruler of the terms of Venus was itself found in apoklima below Dysis.[3] Thus the start[4] (extends) from the Horoscopos to the (end of the) side of the quartile (♓ 17) and the diametrical radiation (from ♓) toward Saturn (which is) in the terms of a maleficent (star).[5] But Mars altered its diametrical radiation because Jupiter was found in an equal degree[6] and hindered the destructive (influence). He died then at 69 years but if Jupiter in quartile[7] had not hindered, he would have lived only 64 years.

Second Version (III,14)

For example: let sun (and) Venus be in Cancer, moon and Horoscopos in Pisces, Saturn in Sagittarius, Jupiter in Capricorn, Mars in Scorpio, Mercury in Leo, the Lot of Fortune in Cancer[8] (and in the 5th Locus) Agathos Daimon.[9] The ruler of the Lot (of Fortune in ♋, the moon[10]) was found at center (H). I set down the minimum (period) of the moon, 25 years, and the rising time of Cancer in the second clima, 32 years, and 12 years of the house-ruler of the moon, Jupiter.[11] They add to the same [12] 69 years; he died at that age.

Third Version (IV,8)

For example let the nativity be this: sun (and) Venus in Cancer, moon (and) Horoscopos in Pisces, (Saturn)[13] in Sagittarius, (the nearest) full moon [14] (and) Jupiter (in Capricorn),[15] Mars in Scorpio, Mercury in Leo, the Lot of Fortune in Leo,[15] the Daimon in Scorpio.[15] I investigate year 70.

I took the physical (life) periods from Leo,[16] giving first to Leo itself 19 years,[17] then to Virgo 20 years,[18] then to Libra 8 years,[19] then to Scorpio 15 years;[20] it makes 62 years. In these (years) he had many crises, both falls from a height and broken limbs. And the remaining 8 years (until 70) from Sagittarius, Saturn being located there by sect (?). In these years he had to expect shipwrecks and bodily disorders. We get the cause of the injury from the sign where the ruler of the Lot of Fortune was found passing through; viz. the Lot was found in Leo and the ruler of Leo, the sun, was found in Cancer;[21] and Cancer indicates breast and throat; we say then that the injury was from Cancer.

For take the apportionments of the years and count them as 360 days; and computing the 5¼ (days) apart, add them to the years.[22] Then [23] it gave to Sagittarius one year,[24] to Capricorn 2 years 3 months,[25] to Aquarius 2 years 6 months.[26] To Pisces one year;[24] then to Aries

---

[1] Should be Capricorn, about 12°; cf. Commentary.
[2] The terms of Venus end at ♓ 12.
[3] The only planet "in apoklima below Dysis" is Mercury. Since ♓ 17, the Horoscopos, belongs to the terms of Mercury, Kroll emends the text to "(Mercury) the ruler of the terms (of the Horoscopos)." Consequently "of Venus" must be considered an error. Another possibility would consist in removing "below Dysis" and to read "(Mars) the ruler of the terms of Venus was in apoklima" because ♏ 7;23 where Mars is, belongs to the terms of Venus, and is in apoklima. Neither one of these emendations is really convincing.
[4] That is the first chronocratorship.
[5] Saturn in ♐ 27;8 is in the terms of Mars.
[6] Meaning doubtful; cf. Commentary.
[7] Cf. Commentary.
[8] From $\lambda_F = \lambda_H - (\lambda_{\mathrm{C}} - \lambda_{\odot}) = \text{♓} - 8° = \text{♋}$.
[9] Actually not Agathos Daimon but Agathe Tyche (same error in No. L 135,I).

[10] Cancer is the house of the moon.
[11] The moon is in ♓, a house of Jupiter.
[12] Cf. the first version.
[13] Cf. below p. 88 f.
[14] Kroll, for no reason, changes the text and places the full moon in ♐. Computation shows that the text is right.
[15] Cf. note 21.
[16] The Lot of Fortune; cf. note 21.
[17] The period of the sun; and Leo is its house.
[18] The period of Mercury; ♍ is a house of ☿.
[19] The period of Venus; ♎ is a house of ♀.
[20] The period of Mars; ♏ is a house of ♂.
[21] We see here how the data of a nativity were arbitrarily modified to suit the purpose. In the second version the Lot of Fortune was located in ♋, and the moon its ruler (cf. notes 8 and 10). Now we must use the accurate positions of the first version in order to find the Lot of Fortune in Leo. Indeed we then have $\lambda_F = \lambda_H - (\lambda_{\mathrm{C}} - \lambda_{\odot}) = \text{♓} 18 - \text{♓} 12 + \text{♋} 29;30 = \text{♌} 5;30$. For the Daimon we find similarly ♏ 0;30 thus luckily the same sign for accurate and for crude counting.
[22] This rule is not followed in the subsequent procedure.
[23] Continuing the zodiacal signs from Scorpio where we left off.
[24] Sagittarius and Pisces are the houses of Jupiter whose period is 12 (months = 1 year).
[25] Cf. note 26.
[26] Aquarius is a house of Saturn whose ordinary period is 30 (months = 2 years + 6 months). The maximal period of Saturn is 57 years; thus the remaining 27 (months = 2 years + 3 months) are given to the second house of Saturn, Capricorn.

the remainder of the 9 years.[27] Mars[28] then, being ruler of physical (life) periods, taking over from Saturn, which is located in Sagittarius, accomplished the end; he died diseased in the throat and afflicted with coughing; for the (8th) Locus of Death (from ♌, the Lot of Fortune) was in Pisces where the moon was located and Saturn (in ♐) was in dominant aspect (Quartile); hence the power (of ♄ to bring death). And besides, the ruler of the full moon,[29] Saturn, after turning away[30] caused the type violent death. But the disease of the throat and the cough (happened), because the ruler of the Lot of Fortune (♌), the sun, was found in Cancer; and Cancer indicates breast and throat.

I reckoned the periods of activities from Scorpio,[31] giving to Mars 15 years,[32] then to Sagittarius 12,[33] Saturn being there. Up to 27 years (the total so far), he was an intermittent vagrant. His adequate property was squandered by guardians; also the Locus of Accomplishment[34] was in Gemini and none of the beneficent (planets) were in aspect, but only Saturn (in ♐) was in opposition. Next then Capricorn took over 27 years,[35] Jupiter being located there and in (the 11th locus) Agathos Daimon and in aspect of opposition to sun and Venus. And he lived this whole period in success and was entrusted with public and royal affairs, and was a friend of governors and kings and became accordingly rich, but destined to reversals and ups and downs in the course of time, according to the successive taking over or aspects of the maleficent (planets), and to transient wealth, because Jupiter was found to be retrograde[36] and in his own depression (♑). After Capricorn, Aquarius took over the period, Mars (in ♏) and Mercury (in ♌) in aspect (quartile and opposition respectively) and the beneficent (planets) not in aspect; and he came to the end of his career and lost much through misplaced trust, and he undertook pledges for relatives and slaves, through whose negligence and need he fell into debt and was found abject, because the whole foundation (of the nativity) bore in this direction.

Aquarius itself took 2 years 6 months,[37] then Jupiter 1 year, then Mars 1 year and 3 months, then Venus 8 months, then Mercury 1 year 8 months.[38] Then his affairs gave way to great decline. Then next the moon took over 2 years 1 month.[39] In these it was destined that he recover some of the pledges and get the help of friends. Then similarly the sun took 1 year 7 months in Leo and Mercury in Virgo 1 year 8 months.[40] Then, since maleficent (planets) were in aspect to the loci[41] and to Mercury,[42] he was brought down in this period. For the Lot (of Fortune, ♌) was found in apoklima and the ruler of the triangle of the moon (♓ ♋ ♏) was Mars.[43] After the (period) of Mercury, Venus took 8 months, next Mars 1 year 3 months,[44] then Sagittarius 1 year;[45] and it was the end.[46]

COMMENTARY

Diagram: pl. 12

The dating of the first version caused great difficulties. Saturn and Jupiter are nowhere near the assigned positions in ♐ 27 and ♍ 22 respectively except as late as 546 July 21 and then again in 605:

	546	605	Text
☉	♋ 29;30	♋ 30	♋ 29;30
♄	♐ 22	♐ 23	♐ 27;8
♃	♍ 25	♍ 17	♍ 22;13

In the first case, however, all three remaining planets and the moon occupy positions far away from those given in the text; for 605 the same holds except for Venus which is then in conjunction with the sun as required.

Consequently it is necessary to assume an error in the text. Since the lifetime of the person in question was 69 years it seemed plausible to investigate whether disregarding Saturn's position in ♐ or Jupiter's in ♍ would lead to a date near A.D. 70 such that the remaining positions would be represented correctly (of course with a deviation of about $+5°$ for a date in the first century). The investigation of all dates which placed Jupiter in ♍ (disregarding Saturn) led to negative results. For Saturn in ♐, however, the year 75 not only showed all planets in their proper place when the sun had the expected longitude of about ♋ 25 but at the same time

---

[27] Which complete 70 years from the 62nd year on. The total so far amounts to 68 years 9 months.

[28] As ruler of its house Aries.

[29] The preceding full moon occurred July 16 in ♑ 22. Since ♑ is a house of Saturn, this planet is the ruler of the full moon.

[30] ἀπόστροφος γενόμενος, here perhaps "turning away" from the full moon (in ♑). Cf., however, Introduction p. 13.

[31] The Lot of Daimon (cf. note 21) following Vettius Valens IV,7 (Kroll p. 166, 10 ff.).

[32] Cf. note 20; Mars was in ♏.

[33] The period of ♃ because ♐ is a house of ♃.

[34] The 11th locus from the Lot of Fortune (♌) is ♊.

[35] Cf. note 26.

[36] Obviously correct since Jupiter is in opposition to the sun.

[37] Cf. note 26. Now we are operating again with months (24 + 6 = 30) since 30 years would be too much after the accumulation of 27 + 27 = 54 years so far.

[38] The periods of the planets (counted as months) in the order of their houses from ♓ (♃) to ♊ (☿).

[39] The period of the moon is 25, the house is ♋.

[40] The periods of sun and Mercury are 19 and 20 respectively. The signs, however, are not the locations but again the houses.

[41] Saturn in ♐ in opposition to the locus of accomplishment in ♊; Mars in ♏ in trine to H and in quartile to the Lot of Fortune in ♌.

[42] Mars in ♏ is trine to Mercury in ♌.

[43] Mars is the ruler of the triangle ♋ ♏ ♓.

[44] The periods of the planets in the order of the signs, which are their houses, from ♍ (☿) to ♏ (♂); cf. note 38.

[45] Cf. note 24.

[46] The total is 69 years 4 months.

the moon was correctly located. Thus, within more than six centuries only 75 July 19 satisfied all requirements, except for Jupiter which had to be in ♑ 7(+ 5 — 12) instead of ♍ 22;13. Now for 75 July 19 two other versions are attested, one in III,4 the other in IV,8 quoting all zodiacal signs correctly, including ♑ for Jupiter. And in both cases the duration of life was again 68 years.[47] Thus all three versions concern undoubtedly the same individual, who was born A.D. 75 July 19 and died early in A.D. 144.

This result is confirmed by a direct reference [48] to the same day, Vespasian 7 Epiphi (XI) 25/26 = 75 July 19, for an individual who died in the middle of his 69th year. As longitude of the sun we find ♋ 27;43 (as compared with ♋ 29;30 in the first version) and for the moon ♓ 12;42 (♓ 12 in the first version). The discrepancy of almost 2 degrees in the solar position is difficult to explain, if not as computing error in at least one of the two cases.[49]

The relationship between the first version and the two others is of importance because we have here a case where we can see that Vettius Valens computed his horoscopes with an accuracy of degrees and minutes even if he normally quotes zodiacal signs only.[50] It is of interest, furthermore, that Jupiter is entirely misplaced (both sign and degree) and that this position in ♍ instead of in ♑ is used in the astrological commentary in which Jupiter appears to be in quartile to Saturn in ♐. This shows that the error is old and even suggests an intentional modification of the correct data, though such a suspicion is not at all supported by the experience with all the other horoscopes quoted by Vettius Valens. Equally inexplicable is the number of degrees, 22;13°, unless the term ἰσόμοιρος (p. 142,5) should indicate that both Mars and Jupiter have the same number of degrees in their respective signs,[51] and thus allow us to emend 22 to 7—which would be the longitude obtained by modern computation for Jupiter.

Otherwise the agreement is very good. For the first version one finds for 75 July 19:

	Text	Computed
☉	♋ 29;30 [52]	♋ 24
☾	♓ 12	♓ 10
♄	♐ 27;8	♐ 19
♃	♍ 22;13(!!)	♑ 7(!)
♂	♏ 7;23	♏ 9
♀	♋ 28;13	♋ 21
☿	♌ 11;25	♌ 9
H	♓ 17	about 10 P.M.
M	♐ 25	♐ 22 [53]

In the second version we have the zodiacal signs only (in complete agreement with computation) while the third version adds the nearest full moon in ♑, together with Jupiter. This is indeed correct since the full moon of July 16/17 happened at ♑ 22.

The second version precedes No. L 135,I. A second horoscope in the same chapter, III,6, is No. L 110,III.

No. L 75,1

Vettius Valens III,13 (p. 153,15 to 21)

TRANSLATION

As an example, let the ascending node be in Gemini 23 degrees, the (preceding) conjunction in Pisces about 8 degrees, the Horoscopos in Pisces 4 degrees, the Midheaven in Sagittarius 13 degrees. I took (the distance) from the conjunction to the ascending node; it makes 105 degrees. These (I count) from the Horoscopos since it was a day-birth. It ends in 19 degrees of Gemini. For clima 2 it totals 76 years . . . months; so long he lived.

COMMENTARY

Diagram: pl. 12

The two following horoscopes in this chapter, No. L 74 IV and No. L 115 II, record as length of life 70 and 32 years respectively, thus leading to A.D. 144 and 147 as the period of writing. Since our horoscope belongs to the same group and gives 76 years as length of life one may expect a date in the decade from A.D. 65 to 75.

The only data given in the text are

	☊ ♊ 23	
Preced. conj.	♓ 8	distance 105°
	H ♓ 4	
	M ♐ 13	
	clima 2	

---

[47] According to the first and second versions he died in his 69th year. The third version shows that he should have died at least 68 years 9 months old and at the most 69 years 4 months.

[48] In VIII,7 (p. 317,15-31); cf. our list p. 180 No. 5.

[49] In VIII,7 p. 317 one must read "3rd hour of night" instead of 6th hour of night as follows from the position of H. This error is, of course, palaeographically explicable.

[50] The only other instances of explicit references to degrees and minutes by Vettius Valens are the horoscopes for 110 March 15, for 114 May 14, and for 127 November 22. Degrees (no minutes) are given in Nos. L 61 X, 74 IV, 75,1 and 115 II.

[51] This interpretation is supported by Vettius Valens II,16 (p. 70,31) where κατ' ἰσόμοιρον στάσεις τῶν τετραγώνων καὶ διαμέτρων are mentioned.

---

[52] Or 27;43 in VIII,7 (p. 317,16).

[53] Computation with the Handy Tables gives for Alexandria M = ♐ 21;53, for Rhodes ♐ 22;35, thus for Babylon a value of about ♐ 22. The approximate procedure of Paulus Alexandrinus (cf. p. 10) leads, for clima 2 (and System A), to ♐ 21;20. Thus neither method explains satisfactorily the result of the text.

Since the last conjunction fell in ♓ 8, the sun can only be between ♓ 8 and ♈ 8, or, with a correction of about 4° for longitudes in the first century, between ♓ 4 and ♈ 4. This means about February 23 for the date of conjunction and the following 30 days as range for the date of the horoscope.

For the node at conjunction we should expect to find by modern computation about ♊ 19 or 20 since the motion of the node during one month amounts to about —1;30°. The only years in which new moon falls close to February 23 are

A.D. 57	☊ about	♐ 22;15
65				♋ 17;30
76				♐ 14;45
84				♋ 10

Since the node recedes 7;30° (plus one complete rotation) in 19 years, it is evident that nowhere near the expected period do node and new moon have the required longitudes.

This result can be confirmed also from the determination of the longitudes of the nodes, computed, e.g., for March 1:

A.D. 47	☊ =	♋ 5;40
48				♊ 16;20
66				♊ 28;10
85				♊ 20;40
104				♊ 13;10

The node is near the expected position ♊ 19 or 20 only for A.D. 48 and 85. The preceding conjunctions, however, are February 3 and February 13 respectively, thus 20 or 10 days too early.

Probably the text contains an error. Indeed H = ♓ 4 and ☉ between ♓ 8 and ♈ 8 would indicate the hours before sunrise whereas the text speaks about day birth. On the other hand, the 105° from the conjunction to the ascending node are counted from H towards IMC thus in the direction associated with daytime according to this very chapter (p. 153,3-5 Kroll). This suggests that at least H is incorrectly given, though M agrees with H.[1] Thus no simple emendation seems possible which would solve these contradictions.

No. L 76

CCAG 6 p. 67 to p. 71 and CCAG 8,2 p. 82 to p. 84

TRANSLATION

There was, he (Antigonus of Nicaea) says, a person having the sun in Aquarius 8[1] degrees, the moon and Jupiter and the Horoscopos, the three together at the first degree of the same sign, namely Aquarius; Saturn in Capricorn 5[2] degrees and Mercury with it (at) 12 degrees; Venus in Pisces 12 degrees and Mars with it (at) 22 degrees; Midheaven in Scorpio 22.[3] {In this configuration the house ruler of the moon (in ♒), Saturn,[4] being in its own house (♑), allots its maximum number of years of life (namely) 56 (sic); and since Venus (in ♓) is in aspect (sextile) to it (Saturn in ♑), it allots additional 8 years, so that the total of years is 64. After 61 years 10 months the degrees of the Horoscopos and the moon (in ♒ 1) come into quartile to Saturn,[5] which, however, is not destructive because Venus is in aspect to it (Saturn) (also) the second time.[6]}[7]

Such a person {whose nativity he recorded,}[7] was adopted by a certain emperor (Trajan), akin to him, and having lived with him two years, became emperor about his 42nd year and was wise and educated, so that he was honored by shrines and temples; and he was married to one wife from maidenhood and was childless; and he had one sister.[7a] And he was at discord and conflict with his own relatives.[8] When he had reached about his 63rd year he died,[9] a victim to dropsy and asthma.

And why it happened in this way is explained as follows. He became emperor because the two luminaries (☉ and ☾) were with the Horoscopos and especially because the moon was of the (same) sect[10] and in conjunction to the degree with the Horoscopos (both in ♒ 1) and with Jupiter which was also due to make its morning phase after 7 days. And its (the moon's) attending stars[11] themselves were found in favorable

---

[1] The Handy Tables give for Alexandria M = ♐ 13;30, for Rhodes ♐ 14;50. The procedure described by Paulus Alexandrinus (cf. p. 10) gives for System A and Babylon M = ♐ 12;40 (text: ♐ 13).

[1] Variant from CCAG 8,2 (Paris. 2501): 20.
[2] Variants: 10 in Paris. 2417 (P) and 16 in the text and in the diagram of the Vindob. (V). This latter text, including the diagram, is published in part in Cramer, ARLP p. 165 and p. 166 left half, corresponding to CCAG 6 p. 67 to p. 69,25 (what follows is F. 302 and 302ʳ of V and has nothing to do with this horoscope).
[3] Following the text of P. The diagram in V has 24.
[4] Aquarius is a house of Saturn.
[5] This is obviously wrong since 61 years is only slightly more than two periods of Saturn. Consequently the longitude of Saturn changes only from the beginning of ♑ to the beginning of ♒. Thus Saturn is not in quartile with H and moon in ♒ 1 but in conjunction.
[6] Venus is then at the beginning of ♐, thus again in sextile to ♄ in ♒.
[7] Omitted in V and in CCAG 8,2.
[7a] Domitia Paulina.
[8] This probably refers to the case of Pedanius Fuscus whose horoscope (No. L 113,IV) belongs to the same group of three horoscopes of which No. L 40 is the earliest one.
[9] Hadrian died 138 July 10, thus at the age of 62½ years.
[10] Meaning not clear.
[11] That is the planets which follow the rising of the moon; cf. Bouché-Leclercq, AG p. 253. Here sun, Venus and Mars.

positions, Venus in her own exaltation (♓), Mars in its own triangle (♓, ♋, ♏) and located in its own degrees,[12] both in their own domains and in epanaphora with respect to the moon. And besides the cosmos-ruling sun was its (the moon's) attendant in the subsequent degrees (♒ 8°) and had as attending stars Saturn in its own house (♑), and Mercury (in ♑), both at their morning rising. It is also significant that the moon was about to come in conjunction with a certain bright fixed star which is at the 20th degree (of ♒). For it is necessary to look at the conjunctions of the moon not only with the planets but also with the fixed stars.

[In a similar fashion characteristics of Hadrian and events in his life are discussed; [13] the following astronomical data occur:] Mercury and Saturn were in the morning phase and in the 12th locus.[14] Mars, the ruler of M,[15] is in Pisces, in its own triangle and degrees. Jupiter with sun and moon in the Horoscopos (♒). Saturn at morning rising, Mars at evening rising.[16] After 40 days the moon in Cancer, Mars in Aries,[17] Saturn in Capricorn at diameter to the moon in Cancer, Mars in trine.

COMMENTARY

Diagram: pl. 12

Kroll realized that this horoscope concerns the emperor Hadrian. Indeed for A.D. 76 January 24, the birthday of Hadrian, one obtains very good agreement between text and computation:

	Text	Computed	Text—comp.
☉	♒ 8	♒ 3	+5
☾ }		♑ 28	+3
♃ }	♒ 1	♑ 27	+4
H }		about 6 A.M.	
♄	♑ 5	♑ 0	+5
☿	♑ 12	♑ 7	+5
♀	♓ 12	♓ 15	—3
♂	♓ 22	♓ 19	+3
M	♏ 22(?)[18]	♏ 21;23[19]	

Jupiter is said to be 7 days before heliacal rising. This statement is probably based on the then elongation of 14 or 15 degrees; corresponding to a mean interval of invisibility of 30°. Later on it is stated several times that Saturn and Mercury are in their morning rising, an essentially correct statement, and Mars in evening rising, which is true only in a looser sense, the elongation being 45°. The moon is said to approach a bright fixed star in ♒ 20, but there are only stars of 4th magnitude in this region.[20]

Forty days later the moon is in Cancer, Mars in Aries as stated in the text. *Cf.* the commentary to No. L 40 for similar cases.

[12] Mars in ♓ 22 is in his own terms which extend from ♓ 19 to ♓ 28.

[13] *Cf.* also Stegemann [1] cols. 370-372.

[14] *CCAG* 6, p. 69,6.

[15] Called κατὰ κορυφὴν τόποs (p. 69,15); here ♏, a house of Mars.

[16] P. 69,23 f.

[17] P. 71,17 ff.

[18] Variant: 24; *cf.* note 3.

[19] Since Hadrian was born in Halica (southern Spain, near Seville, latitude about 37;30) the above value was computed for the clima Rhodes with the Theon tables (cod. Vat. gr. 1291). The procedure of Paulus Alexandrinus (*cf.* p. 10) leads to M = ♏ 14;10 for Alexandria and M = ♏ 17;40 for the latitude of Spain.

[20] The only bright star in this whole region is α Pisc. austr. (1st magn.) the longitude of which is, however, ♒ 7 according to Ptolemy (Almagest VIII,1; β = —20;20) and ♒ 10;26 in the Liber Herm. (Gundel, *HT* p. 159).

No. L 78

Vettius Valens II,26, p. 93,25 to p. 94,2

INTRODUCTION

Chapter II,26 is headed by the title "Examples for the preceding topics" on "famous and distinguished" and on "obscure and debased" nativities. Four horoscopes are collected in II,26 whose dates are

188 Aug. 10
86 Aug. 2
78 April 1
101 March 5.

TRANSLATION

Sun, moon, Jupiter, (and) Horoscopos in Aries, Saturn (and) Venus in Aquarius, Mars in Gemini, Mercury in Pisces. This person was commanding and despotic, because the rulers (☉ and ♃) of the triangle (of the sun, ♈ ♌ ♐) were found to be at a center (H) and in the Horoscopos, and in it also fell the Lot (of Fortune) and the Daimon and the Basis and the exaltation (of the nativity). But Mars, the ruler of these,[1] being unfavorably situated and not in aspect to the place [2] had opposite effects, both exile and violent death; for it was ruler of the conjunction (of sun and moon in ♈).

COMMENTARY

Diagram: pl. 12

Except for Mars (which is between its second station and setting) good agreement is reached for 78 April 1:

No. L 78

[1] A house of ♂ is ♈.

[2] As is seen from Kroll's apparatus, the passage is not read with certainty and we do not know how to restore it. Kroll assumes a lacuna and supplies [γ'], obviously assuming the meaning "not in aspect (being) in the [3rd] locus." Though the 3rd locus (♊) is in sextile to ♈, Kroll may be right because sextile often seems not to be considered as "aspect" by Vettius Valens; *cf.*, e.g., Vettius Valens I,24.

Text		Computed
☉	♈	♈ 10
☾		♈ 10
♃		♈ 8
H		sunrise
♄	♒	♑ 28(!)[3]
♀		♒ 25
♂	♊	♉ 21(!!)[4]
☿	♓	♓ 17

The conjunction of sun and moon happened on the same day, in Aries. Consequently the lots of Fortune and Daimon coincide and therefore also the "basis" (*cf.* p. 9). The same is said about the exaltation of the nativity. This, however, is only correct when we are dealing with daytime,[5] or if the Horoscopos fell before ♈ 10. Similarly, daytime is assumed in the choice of the triangle of the sun and not that of the moon.[6]

No. L 80

Manetho, Apotelesmaticorum III, 738-750

TRANSLATION

But I will proceed by a new turn in verse and recall the stars of my own nativity when and in what sign it was that oft-sought Eilithyia delivered me from the womb so that for all time they may teach and prove what Fate granted me to teach, the wisdom and beautiful poetry of the stars. The sun was in Gemini and there too was beautiful Kypris (Venus) and also beloved Phaeton (Jupiter) and golden Hermes (Mercury), and in Aquarius at the time (were) the moon and Phainon (Saturn), and Mars was in many-footed Cancer and Centaur was turning about Midheaven, trailing his weapon. Thus the Fates determined my nativity.

COMMENTARY

Diagram: pl. 12

One obtains complete agreement with the data of the text for 80 May 28:

Text		Computed
☉	♊	♊ 5
♀		♊ 8
♃		♊ 15
☿		♊ 26
☾	♒	♒ 16
♄		♒ 23
♂	♋	♋ 10

[3] The Stobart tables agree with our text: Saturn is in ♒ from February 25 to July 27 (A, rev. V, 16/17).

[4] According to the Stobart tables (A, rev. V,28) Mars enters ♊ two days later.

[5] In daytime we have to consider the distance from ☉ to ♈ which is here zero; at night the distance from ☾ to ♉ which would here bring the exaltation in ♉ (*cf.* p. 7).

[6] *Cf.* Vettius Valens III,2 p. 26,20 ff.

Obviously the subtraction of about 13° does not remove the moon from Aquarius; consequently also May 27 is equally possible.

We know furthermore that Centaur[1] was near culmination. Because this constellation covers in longitude the sign of Libra, we know that ♑ must be rising, ♋ setting. Thus the time was about two hours after sunset.

No. L 82

Vettius Valens II,21 p. 85,9-18 and III,16 p. 157,15-19

TRANSLATION

First Version

Sun (and) Mercury in Cancer, moon in Taurus, Saturn in Pisces, Jupiter (and) Mars in Leo, Venus in Virgo, Horoscopos in Libra. This nativity was also brilliant and distinguished. He was entrusted with royal office and deemed worthy of the high priesthood. For the ruler of the triangle (of the sun, ♋ ♏ ♓, i.e., Mars) was found with the ruler (♃) of the Daimon (♐)[1] in the (11th) Locus Agathos Daimon (♌) and with the Lot of Fortune (♌),[2] and the sun, which was (located in) the Lot of Fortune in Midheaven; and the ruler of the exaltation (of the nativity, i.e., ♋), the moon[3] (in ♉), was in Midheaven with respect to the Lot (of Fortune)[4] and the (11th Locus of) accomplishment was irregular and unstable, sometimes overfull and sometimes deficient, for Saturn and Venus were in aspect to it.[5]

Second Version

Let sun, Venus(!) (and) Mercury be in Cancer, moon in Taurus, Saturn in Pisces, Jupiter (and) Mars in Leo, Horoscopos in Virgo(!). In the first clima the rising time of Virgo is 38⅓. And because Mercury in Cancer is in Agathos Daimon (the 11th locus) it gave the rising time (of ♋) 31⅔. The total is 70; so long he lived.

COMMENTARY

Diagram: pl. 12

For the date of this horoscope one finds 82 July 9:

[1] *Cf.* for this constellation Boll-Gundel in Roscher VI col. 1012 ff.

No. L 82

[1] A house of ♃ is ♐ and $\Lambda_{D} = \lambda_{H} - (\lambda_{☾} - \lambda_{☉}) =$ ♎ $+ 2^{s} =$ ♐.

[2] From $\Lambda_{F} =$ ♎ $- 2^{s} =$ ♌; *cf.* note 1.

[3] The house of the moon is ♋.

[4] If ♌ is rising then ♉, the place of the moon, is culminating.

[5] Counting from ♌, the Lot of Fortune, one finds the 11th locus in ♊ which is in quartile to Venus in ♍ and to Saturn in ♓.

	Text	Computed
☉	♋	♋ 14
☿	♋	♋ 28
☾	♉	♉ 13
♄	♓	♓ 19
♃	♌	♌ 13
♂	♌	♌ 17
♀	♍	♌ 29(!)
H	♎	about noon

The first version is the seventh example in a group of 14 examples, the earliest of which is No. L 50.

The second version places Venus by mistake with sun and Mercury in Cancer and then, instead of Venus, the Horoscopos in Virgo. The rising times for Virgo and Cancer are given as 38⅓ and 31⅔ which is correct for Alexandria and System A.[6]

The second version (in III,16) is associated with Nos. L 102 IV (a) and (b).

No. L 83

Vettius Valens II,21 p. 84,12-22 and II,36 p. 114,1-5

TRANSLATION

First Version

Sun (and) Mercury in Taurus, moon in Aquarius, Saturn (and) Venus in Aries, Jupiter in Virgo, Mars in Pisces, Horoscopos in Leo. We find the sun in the triangle of Venus and moon (♉ ♍ ♑) and Venus in apoklima. And so at first (his life) was troubled and lowly; but the moon being found in a center (Δ) he came into military and advantageous circumstances. Likewise the Lot of Fortune was found in Taurus[1] and the exaltation (of the nativity) in Cancer. Its ruler, the moon, was found in Midheaven with respect to the Lot (of Fortune),[2] hence he came to the greatest fortune and governorship. But Mars was found in (♓) the (11th) Locus of accomplishment[3] . . . (lacuna) . . . property by plunder and stealing and violence, (and) they were plundered wickedly after his death.

Second Version

[(Positions unchanged)], the Lot of Fortune in Taurus. Its ruler, Venus,[4] in Aries with Saturn. This person had mange on the head and leprosy and lichens on the body for (♂) the ruler of the Daimon (♏)[5] was in Pisces.

COMMENTARY

Diagram: pl. 13

For 83 April 28 one obtains

	Text	Computed
☉	♉	♉ 5
☿	♉	♈ 15(!!)[6]
☾	♒	♒ 15
♄	♈	♓ 27(!)
♀	♈	♈ 11
♃	♍	♍ 0
♂	♓	♓ 21
H	♌	about noon

About one month later (May 24) one would obtain the correct longitudes of all planets, including sun and moon, except for Mars which in the meantime had entered ♈, and for ♀ in ♉. Were it not for the moon May 14 would be the best date, because Mercury, Venus, and Mars would then be very close to the boundary of their respective signs; but the moon in ♍ excludes this possibility.

The first version (II,21) belongs to a group of horoscopes the list of which is given in No. L 50. The second version (II,36) is the earliest horoscope of eleven:

(*a*)	87	Jan. 9	p. 111,33-112,4
(*b*)	118	Nov. 26	p. 112,5-10
(*c*)	106	Jan. 16	p. 113,2-10
(*d*)	85	Nov. 24	11-18
(*e*)	116	Jan. 21	19-24
(*f*)	117	Nov. 30	25-31
(*g*)	92	Nov. 17	32-36
(*h*)	83	Apr. 29	p. 114,1-5
(*i*)	104	Apr. 23	6-9
(*j*)	108	Mar. 28	10-15
(*k*)	112	Aug. 17	16-19

No. L 85,II

Vettius Valens II,21 p. 84,3-11

TRANSLATION

Sun, Mars, Venus, (and) Mercury in Aquarius, moon (and) Jupiter in Scorpio, Saturn in Aries, Horoscopos in Leo. And this nativity came from lowly and ordinary fortune to a fortune of governorship and wealth. For, since it was a daytime birth, we find (first) the sun in the triangle (♒ ♊ ♎) of Saturn, but Saturn in apoklima, wherefore the first circumstances were ordinary. But the co(ruler) of this (triangle), Mercury, was in a center (Δ). And we find also the Lot of Fortune in Taurus[1] and the exaltation (of the nativity) in Libra[2]

---

[6] *Cf.* p. 4, table 2.

No. L 83

[1] From $\Lambda_F = \lambda_H + (\lambda_{☾} - \lambda_{☉})$ = ♌ + 9° = ♉.

[2] Cancer is the house of the moon which is in ♉. When ♉ is rising, ♒ is in Midheaven.

[3] Counted from the Lot of Fortune (♉).

[4] A house of ♀ is ♉.

[5] From Λ_D = ♌ − 9° = ♏ (*cf.* note 1) and ♏ is a house of ♂.

[6] Mercury is retrograde, after inferior conjunction.

No. L 85,II

[1] From $\Lambda_F = \lambda_H + (\lambda_{☾} - \lambda_{☉})$ = ♌ − 3° = ♉, assuming daytime.

[2] Indeed ☉ − ♈ = H − ♎.

and the ruler (Venus[3]) in Midheaven with respect to the Lot (of Fortune)[4] and in a center (Δ) besides.

COMMENTARY

Diagram: pl. 13

For 85 February 5 one finds excellent agreement:

	Text	Computed
☉	♒	♒ 16
♂	♒	♒ 5
♀	♒	♒ 8
☿	♒	♒ 29
☾	♏	♏ 8
♃	♏	♏ 10
♄	♈	♈ 12
H	♌	shortly before sunset[5]

For related horoscopes *cf.* No. L 50.

No. L 85,XI

Vettius Valens II,36 p. 113,11 to 18

TRANSLATION

Sun in Sagittarius, moon in Cancer, Saturn in Taurus, Jupiter (and) Mercury in Scorpio, Mars in Leo, Venus in Capricorn, Horoscopos in Aquarius, Lot of Fortune in Leo.[1] Mars is there (and) Saturn in dominant aspect (to it). And the sun found in the places (i.e., the house) of Jupiter, which indicate things concerning the groin and the thighs, caused infirmity in the (parts) and gout; for it (the sun) is the ruler of sinews; and since Saturn was found in Lower Midheaven he had visions of gods and of the spirits of the dead.[2]

COMMENTARY

Diagram: pl. 13

For 85 November 24 one finds

	Text	Computed
☉	♐	♐ 2
☾	♋	♋ 20
♄	♉	♈ 25(!)
♃	♏	♏ 24
☿	♏	♏ 11
♂	♌	♌ 10
♀	♑	♑ 8
H	♒	about 10 A.M.
Λ_F	♌	♍ (!)

Related horoscopes are listed with No. L 83.

No. L 86,VIII

Vettius Valens II,26 p. 93,21-24

TRANSLATION

Sun, Mercury, Venus, (and) Horoscopos in Leo, Saturn in Taurus, Jupiter in Sagittarius, Mars in Libra, moon in Capricorn. Such a person (was) commanding and master over life and death,[1] because the stars were found in their own domains.[2]

COMMENTARY

Diagram: pl. 13

This is the second horoscope of the group discussed under No. L 78.

The text adopted by Kroll places the moon in Virgo whereas the manuscript M gives Capricorn. That the latter version is the correct one is shown by the following comparison:

	Text	+86 August 2	August 11
☉	♌	♌ 6	♌ 16
☿	♌	♌ 0	♌ 19
♀	♌	♋ 20(!!)	♌ 2
H	♌	sunrise	sunrise
♄	♉	♉ 14	♉ 14
♃	♐	♐ 4	♐ 4
♂	♎	♍ 29(!)	♎ 6
☾	S: ♍	♍ 20	—
	M: ♑	—	♑ 15

No. L 86,XII

Vettius Valens II,41, p. 129,15 to 17 (= CCAG 5,2 p. 120,33 f.)

TRANSLATION

Sun, Mercury, Mars, Jupiter, (and) Venus in Capricorn, moon in Aquarius, Saturn in Taurus, Horoscopos in Aries. This person was beheaded.

COMMENTARY

Diagram: pl. 13

For 86 December 27 one finds

	Text	Computed
☉	♑	♑ 5
☿	♑	♑ 14
♂	♑	♑ 19
♃	♑	♐ 27(!)
♀	♑	♑ 25
☾	♒	♒ 7
♄	♉	♉ 9
H	♈	about noon

[3] The houses of Venus are ♎ and ♉.

[4] If ♉, the Lot of Fortune, is rising then ♀ in ♒ is culminating.

[5] *Cf.* note 1.

No. L 85,XI

[1] From $\Lambda_F = \lambda_H + (\lambda_☾ - \lambda_☉)$ = ♒ − 5° = ♍ it follows that ♌ is an error. The result remains the same if one computes the elongation to the degree. The Horoscopos cannot be the cause of the error since ♄ in ♉ is said to be in Lower Midheaven.

[2] *Cf.* Cumont, *EA* p. 166.

No. L 86,VIII

[1] Perhaps a strategos; *cf.* Cumont, *EA* p. 51.

[2] *Cf.* also for οἰκείως Bouché-Leclercq, *AG* p. 391. We do not see how the criterion, given in Vettius Valens p. 93,5 f. applies here.

This horoscope belongs to the same group as No. L 65 V, which text should be consulted for further details.

No. L 87,I
Vettius Valens II,36 p. 111,33-112,4

TRANSLATION

For example, so that we may not be found to have spoken cryptically, let sun, Jupiter (and) Mars be in Capricorn, moon (and) Horoscopos in Leo, Saturn in Taurus, Venus (and) Mercury in Aquarius, the Lot of Fortune [in Capricorn]. The ruler of the (Lot of) Fortune, Saturn,[1] in Taurus. He was disabled on account of the Pleiades[2] and because of maleficent Saturn and he had unmentionable vices on account of both signs (♑ and ♉). In addition, Jupiter, the ruler of the Daimon in Pisces, was found in Capricorn. From these (configurations) it was evident that he had gout. It was sufficient[3] to have learned from the Lot (of Fortune) and its ruler both the infirmity and the disease.

COMMENTARY

Diagram: pl. 13

For 87 January 9 one finds

Text		Computed	
☉ ⎫		♑	18
♃ ⎬	♑	♑	0
♂ ⎭		♒	1(!)
☾ ⎫	♌	♌	13
H ⎭		about 8 P.M.	
♄	♉	♉	8
♀ ⎫	♒	♒	11
☿ ⎭		♒	4

For related horoscopes cf. No. L 83.

No. L 87,VII
Vettius Valens II,41 p. 129,8 to 14

TRANSLATION

Sun in Cancer, moon in Pisces, Saturn, Mars (and) Mercury in Gemini, Jupiter in Capricorn, Venus in Leo, Horoscopos in Libra, Lot of Fortune in Gemini.[1] In this (sign) Saturn, Mercury (and) Mars were accompanied by one another (all being) destructive and in aspect to the moon (in ♓). Likewise the ruler (♄) of the full moon (in ♑) was turning away[2] and Jupiter, being in the (8th) Locus of Death (♑) and being in opposition,[3] had no power to help. This person was beheaded.

COMMENTARY

Diagram: pl. 14

For 87 July 9 one finds

Text		Computed	
☉	♋	♋	14
☾	♓	♓	15
♄ ⎫		♉	26(!)
♂ ⎬	♊	♊	12
☿ ⎭		♊	0
♃	♑	♑	13
♀	♌	♌	8
H	♎	about noon	

Counting from the Lot of Fortune (♊) eight signs, one obtains ♑ for the Locus of Death where Jupiter is located in opposition to the sun (in ♋). The nearest full moon, about 5 days earlier was in ♑, a house of Saturn. This planet being located in ♊, 5 signs away, is therefore not in aspect with ♑.

For related horoscopes cf. No. L 65,V.

No. L 88
Vettius Valens II,41, p. 130,6 to 10
(= CCAG 5,2 p. 121,17-21)

TRANSLATION

Sun, Mercury, (and) Horoscopos in Taurus, moon in Pisces, Saturn in Gemini, Jupiter in Aquarius, Mars in Virgo, Venus in Aries, the Lot of Fortune in Pisces.[1] There the moon was in aspect with Saturn (in ♊) and Mars (in ♍). The rulers of the Daimon (♋) and of the (nearest) full moon were in opposition. This person was drowned in bilge water.

COMMENTARY

Diagram: pl. 14

This is a horoscope from a group which is discussed in the introduction to No. L 65,V. For 88 May 5 one finds

---

[1] A house of Saturn is ♑ and $\Lambda_F = \lambda_H - (\lambda_{\mathrm{C}} - \lambda_{\odot}) = \mathrm{♌} + 5° = \mathrm{♑}$.
[2] Cf. Vettius Valens I,2 p. 6,23 ff. where ♉ 1 to 5 is associated with the Pleiades. Similarly in II,36 p. 109,34 to 110,2. It must be admitted, however, that the mention of the Pleiades is strange in this group of horoscopes.
[3] I.e., without checking on Daimon.

No. L 87,VII
[1] From $\Lambda_F = \lambda_H + (\lambda_{\mathrm{C}} - \lambda_{\odot}) = \mathrm{♎} + 8° = \mathrm{♊}$.

[2] Cf. Introduction p. 13.
[3] Literally "in evening (rising)" which is the technical term for opposition to the sun; cf. e.g., Bouché-Leclercq, AG p. 112 or P. Mich. 149, XI,9-11.

No. L 88
[1] From $\Lambda_F = \lambda_H + (\lambda_{\mathrm{C}} - \lambda_{\odot}) = \mathrm{♉} + 10° = \mathrm{♓}$ and thus $\Lambda_D = \mathrm{♉} - 10° = \mathrm{♋}$.

	Text		Computed
☉	⎱		♉ 13
☿	⎰ ♉		♉ 27
H	⎱		about sunrise
☾		♓	♓ 14
♄		♊	♊ 0
♃		♒	♒ 20
♂		♍	♍ 2
♀		♈	♉ 9(!)

Thus there is agreement with the exception of Venus.

The Lot of Fortune and the moon fall in the same sign, Pisces, which is in quartile to Saturn in Gemini and in opposition to Mars in Virgo. The nearest full moon was the preceding one in Scorpio[2] which is a house of Mars. The Daimon is in ♋, the house of the moon. Consequently Mars is the ruler of the full moon, the moon the ruler of the Daimon, and Mars (in ♍) and moon (in ♓) are in opposition to each other.

No. L 89

Vettius Valens II,41, p. 130,11 to 16
(= CCAG 5,2 p. 121,22-26)

TRANSLATION

Sun in Leo, moon (and) Mercury in Virgo, Saturn in Gemini, Jupiter in Aries, Mars, Horoscopos (and) Venus in Cancer, the Lot of Fortune in Gemini;[1] Saturn the ruler of (the 8th Locus of) Death (♑) was there (in ♊) and he was in dominant aspect to Mercury (in ♍) the ruler of the Lot (of Fortune) and to the moon (in ♍); besides Mars (in ♋) was opposite to the (Locus of) Death (♑). This person hanged himself.

COMMENTARY

Diagram: pl. 14

For 89 July 29 one finds

	Text		Computed
☉		♌	♌ 4
☾	⎱		♍ 15
☿	⎰ ♍		♍ 1
♄		♊	♊ 24
♃		♈	♈ 1
♂			♋ 7
H	⎱ ♋		about 4 A.M.
♀	⎰		♊ 29(!)

Gemini is a house of Mercury, so the planet is called ruler of the Lot of Fortune. Saturn in ♊ prevails over moon and Mercury (in ♍) because it is in quartile to them. Mars in ♋ is opposite to the eighth Locus of

---

[2] April 24, the moon being in about ♏ 3.

No. L 89

[1] From $\Lambda_F = \lambda_H - (\lambda_☾ - \lambda_☉) = ♋ - 1^\circ = ♊$.

Death (♑), counting from the Lot of Fortune (♊).

The victim of such a bad nativity properly hanged himself. His horoscope was incorporated in the same collection with No. L 65, V.

No. L 91

Vettius Valens II,41 p. 130,17-22
(= CCAG 5,2 p. 121,27-31 and CCAG 5,2 p. 52,24-28)

TRANSLATION

Sun and Mercury in Aries, moon (and) Venus in Pisces, Saturn in Cancer, Jupiter (and) Mars in Taurus, the Horoscopos in Scorpio, the Lot of Fortune in Sagittarius,[1] the ruler (♃ [2]) with Mars in Dysis,[3] the (8th) Locus of Death in Cancer,[4] Saturn, the ruler of the full moon,[5] turning away;[6] also Mars (in ♉) was in opposition to his own house (♏). This person was killed by wild beasts.

COMMENTARY

Diagram: pl. 14

This is the last in order of a group of horoscopes the earliest of which is No. L 65,V. For 91 April 4 one finds

	Text		Computed
☉	⎱		♈ 12
☿	⎰ ♈		♓ 19(!!)
☾	⎱		♓ 9
♀	⎰ ♓		♓ 14
♄		♋	♋ 5
♃	⎱		♉ 10
♂	⎰ ♉		♉ 3
H		♏	about 8 P.M.

thus very good agreement with exception of Mercury, which is retrograde.

No. L 92,II

Vettius Valens V,12 p. 234,33 f.

TRANSLATION

As: sun in Aquarius, moon in Leo, Saturn in Cancer, Jupiter in Gemini, Mars in Scorpio, Venus in Aries,

---

[1] From $\Lambda_F = \Lambda_H - (\lambda_☾ - \lambda_☉) = ♏ - 11^\circ = ♐$.

[2] A house of Jupiter is ♐.

[3] CCAG 5,2 p. 52,26 has here: "with the (Agathos) Daimon in Virgo," CCAG 5,2 p. 121,29: "with Mars in Taurus."

[4] Counted from the Lot of Fortune.

[5] Following Kroll's text (p. 130,20); CCAG 5,2 p. 52,27 and p. 121,30 has "of the moon" only. The nearest, preceding, full moon falls at the boundary between ♍ and ♎, the following one near ♎/♏. Perhaps Saturn is the ruler because he is ruler of the triangle ♎, ♒, ♊.

[6] Cf. Introduction p. 13.

Mercury in Pisces. [This is followed by a long discussion of "critical years," based on the number of signs which the celestial bodies are distant from each other; e.g., Mars from moon is the fourth sign and "4 distant is the simple critical point . . . ; for the squares are simple, the product of unequal factors composite," etc.]

COMMENTARY

Diagram: pl. 14

This is the first of two examples in a chapter, entitled "Differently, on critical (years), as Critodemus takes the start from the moon." The second example is No. L 104,VII. For the first horoscope one finds 92 February 11 as is shown by the following comparison.

Text		Computed
☉	♒	♒ 20
☾	♌	about ♌ 18
♄	♋	♋ 16
♃	♊	♊ 1
♂	♏	♏ 22
♀	♈	♈ 5
☿	♓	♓ 13

No position is given for the Horoscopos. This is unusual in Vettius Valens, as well as in original documents.

No. L 92,XI

Vettius Valens II,36 p. 113,32-36

TRANSLATION

Sun, moon, Mercury (and) Horoscopos in Scorpio, Saturn in Leo, Jupiter in Cancer, Mars in Capricorn, Venus in Libra, the Lots (of Fortune and Daimon) in Scorpio. He was disabled in the penis (because) in particular Saturn was in dominant aspect to the conjunction [1] (in ♏) and the luminaries (sun and moon), and the ruler Mars [2] was unfavorably placed.

COMMENTARY

Diagram: pl. 14

Except for Mars the conditions of this horoscope are well satisfied for 92 November 17:

Text		Computed
☉	} ♏	♏ 25
☾		♏ 13
☿		♏ 6
H		sunrise
♄	♌	♌ 11
♃	♋	♋ 13
♂	♑	♒ 13(!!)
♀	♎	♎ 28

Mars is between second station and conjunction.

For related horoscopes *cf.* No. L 83.

[1] The conjunction occurs November 18 about 1 A.M. The data of the horoscope are equally satisfied for November 17 and November 18 about 6 A.M. Consequently no decision is possible as to whether the conjunction precedes or follows the moment of birth.

[2] Whose house is ♏.

No. L 95,V,14

Vettius Valens II,21 p. 85,25-86,3

TRANSLATION

Sun (and) Mercury in Scorpio,[1] moon in Aquarius, Saturn in Leo, Mars (and) Venus in Cancer, Jupiter in Virgo, Horoscopos in Sagittarius. It was a night birth, and the rulers of the triangle (of the moon, ♒ ♊ ♎, thus ♄ and ☿) were in apoklima. He had many ups and downs in his first period of life and lived in debt although the property of his parents was good. Then later, getting an inheritance and improving his means by shrewd enterprises, he became ambitious, dominant, and munificent and popular and a friend of kings and governors, and he provided temples and (public) works and gained perpetual remembrance. For the Lot of Fortune [2] and the exaltation [2] (of the nativity) were in Pisces and the ruler of this (♃ [3]) was in Midheaven (♍).

COMMENTARY

Diagram: pl. 15

The text obviously contains an error because Venus in Cancer and sun and Mercury in Scorpio is astronomically impossible. The question must then be answered whether the position of Venus or that of the sun is incorrectly given. Now we are told that it was a night birth with H = ♐. But sun in Scorpio with H = ♐ would mean daytime. Thus the position of sun and Mercury must be wrong whereas Venus can be right. The exaltation of the nativity falls in ♓. This is correct for nighttime because ♉ — ☾ = ♉ — ♒ = 3^s = ♓ — H = ♓ — ♐. For daytime and ☉ in ♏, the exaltation would be ♉. Thus again the position of the sun must be altered. In order to determine it correctly we use the information that the Lot of Fortune falls in ♓. Thus we have for nighttime

$$\lambda_F = \lambda_H - (\lambda_{☾} - \lambda_{☉})$$

or

$$♓ = ♐ - ♒ + \lambda_{☉} = \lambda_{☉} - 2^s$$

thus

$$\lambda_{☉} = ♉.$$

[1] Error for Taurus; *cf.* Commentary.

[2] *Cf.* Commentary.

[3] A house of ♃ is ♓.

This shows that we have to correct for sun and Mercury ♏ to ♉, the diametrically opposite sign. This correction restores not only a proper elongation to Venus but also is consistent with the requirement of nighttime.

With the corrected elements the following comparison shows agreement for 95 May 14:

	Text	Computed
☉	[♉]	♉ 21
☿	[♉]	♉ 11
☾	♒	♒ 13
♄	♌	♌ 29
♂	♋	♊ 26(!)
♀	♋	♋ 7
♃	♍	♍ 5
H	♐	about 8 P.M.

This horoscope is only four days earlier than No. L 95,V,18 and both belong to the group of No. L 50.

No. L 95,V,18

Vettius Valens II,21 p. 83,27 to 84,2

TRANSLATION

Sun (and) Mercury in Taurus, moon in Aries, Saturn [in Leo], Mars, Venus, (and) Horoscopos in Cancer, Jupiter in Virgo, Lot of Fortune and exaltation of the nativity in Gemini. It (the nativity) rose from mediocrity and became that of a governor and strategus. For since it was a daytime birth I found (first) the sun in the triangle of the moon [1] (♉ ♍ ♑) and the moon and its partners, Venus and Mars, at centers [2] and the Lot of Fortune [3] and the exaltation (of the nativity) in Gemini, in apoklima—hence the beginning was mediocre—and the ruler [4] (of ♊, ☿) in the (11th locus) Agathosdaimon (♉).

COMMENTARY

Diagram: pl. 15

A systematic investigation shows that no date in the first three centuries of our era corresponds to the data of the text according to which Saturn, Mars, and Venus should be in Cancer. Assuming a simple omission of Leo [5] after Saturn solves the problem. The text then becomes almost a parallel to No. L 95,V,14 of the same group in which No. L 50 is the earliest horoscope. With this emendation we obtain for 95 May 18

	Text	Computed
☉	♉	♉ 24
☿	♉	♉ 9
☾	♈	♈ 6
♄	[♌] [5]	♌ 28
♂	♋	♊ 27(!)
♀	♋	♋ 10
H	♋	about 10 A.M.
♃	♍	♍ 4

[1] One would expect ♀ which is ruler of the triangle at daytime, not the moon, which is the ruler at night.

[2] The moon in M, Mars and Venus in H. The latter are partners of the moon because they are rulers of the same triangle (*cf.* Vettius Valens II,1 p. 55,22-25).

[3] From $A_F = \lambda_H + (\lambda_☾ - \lambda_☉)$ = ♋ − 1° = ♊.

[4] A house of ☿ is ♊.

[5] Also "Virgo" would be possible since a modern longitude ♌ 28 might well be counted as belonging to ♍.

No. L 97,II

Vettius Valens II,41 p. 129,1-7

TRANSLATION

Sun, Mercury, Venus in Pisces, Saturn in Virgo, Jupiter in Aries, Mars in Taurus, moon in Sagittarius, Horoscopos in Leo. The Lot of Fortune in Taurus.[1] Mars was located so that he ruled the Daimon (in ♏) to which he was opposite (in ♉). The (8th) Locus of Death [2] was in Sagittarius and the moon was there, in an inferior aspect to Saturn (in ♍), which was in the sign of the full moon (in ♍). Similarly Mercury (in ♓), the ruler [3] of the full moon (in ♍), was in opposition (to ♄ and full moon in ♍). This person was beheaded.

COMMENTARY

Diagram: pl. 15

This is the second example in a group of horoscopes which is discussed in the introduction to No. L 65,V.

The positions as given in the text cannot be satisfied for any date within the first three centuries. On the other hand the astrological data confirm the text with respect to sun, moon, and all planets except Venus and Jupiter. The fact that Venus is mentioned before Saturn would make it difficult to emend its position. Thus Jupiter remains as the sole planet for which an emendation of its place does not affect the remaining text. Under this condition a satisfactory solution can be found as 97 February 23:

	Text	Computed
☉	♓	♓ 4
☿	♓	♓ 16
♀	♓	♓ 15
♄	♍	♍ 28
♃	♈ (!)	♏ 17(!!)
♂	♉	♉ 23
☾	♐	♐ 10
H	♌	about 4 P.M.

No. L 97,II

[1] From $A_F = \lambda_H + (\lambda_☾ - \lambda_☉)$ = ♌ + 9° = ♉.

[2] Counted from the Lot of Fortune.

[3] Because ♍ is a house of Mercury.

The preceding full moon occurs eight days earlier, the moon being in ♌ 26, the next full moon on March 16 in ♍ 25. Considering the difference of about 4° or 5° between ancient and modern longitudes both positions could be called " in ♍ " by the text.

No. L 97,XI

Vettius Valens II,21 p. 85,19-24

TRANSLATION

Sun, Jupiter, Mars, (and) Venus in Scorpio, Saturn in Libra, moon in Aries, Mercury in Sagittarius, Horoscopos in Leo. Also the ruler of the exaltation (of the nativity, ♍), Mercury,[1] was found in Midheaven with respect to the Lot (of Fortune, ♓ [2]) and improved the nativity as regards livelihood. The rulers (☉ and ♃) of the triangle (of the moon, ♈ ♌ ♐) and of the Lot (of Fortune, ♃ [3]) being found in Lower Midheaven (♏) made him miserly, unambitious and niggardly.

COMMENTARY

Diagram: pl. 15

Perfect agreement is found for 97 November 6:

	Text	Computed
☉	♏	♏ 13
♃	♏	♏ 24
♂	♏	♏ 4
♀	♏	♏ 3
♄	♎	♎ 7
☾	♈	♈ 24
☿	♐	♐ 3
H	♌	about midnight

For related horoscopes *cf.* No. L 50.

No. L 101,I

Vettius Valens II,41, p. 129,18 to 23 (= CCAG 5,2 p. 120,35 to p. 121,2)

TRANSLATION

Sun (and) Venus in Aquarius, moon in Gemini, Saturn in Scorpio, Jupiter in Pisces, Mars in Cancer, Mercury (and) Horoscopos in Capricorn, the Lot of Fortune in Virgo,[1] the Locus [2] of Death in Aries. The lords of these (places) were in opposition (to each other) in watery signs [3] and furthermore Mars (in ♋) happened to be in the Dysis. This person was burned to death in a bath.

COMMENTARY

Diagram: pl. 15

The given positions lead to 101 January 28 as date of this horoscope as is shown by the following comparison:

	Text	Computed
☉	♒	♒ 8
♀	♒	♒ 29
☾	♊	♊ 10
♄	♏	♏ 15
♃	♓	♓ 4
♂	♋	♋ 11
☿	♑	♑ 17
H	♑	about 4 A.M.

Virgo, where the Lot of Fortune is located, is ruled by Mercury whose house lies there. The 8th locus, counted from the Lot of Fortune, is Aries, a house of Mars. And Mercury (in ♑) and Mars (in ♋) are in opposition to each other.

For related horoscopes *cf.* No. L 65,V.

No. L 101,III

Vettius Valens II,26 p. 94,3-11

TRANSLATION

Sun, Jupiter, (and) Venus in Pisces, moon in Libra, Mars in Cancer, Mercury in Aquarius, Saturn in Scorpio, Horoscopos in Leo. This person was famous and wealthy. For the sun was accompanied by the beneficent (planets) and was found situated in the Lot of Fortune (♓),[1] together with its ruler (i.e., Jupiter). But since the co-rulers (♂ and ☾) of the triangle (♓ ♋ ♏) were unfavorably located and the ruler (♄) of Daimon (in ♑)[2] was turning away,[3] this person was banished and committed suicide.[4] In addition, Mars was in opposition to (♑) the (Locus of) accomplishment [5] and the ruler (☿) of the exaltation (of the nativity, ♏ [6]) did not have a suitable locus (△) but was in inferior aspect and vitiated by Saturn.

---

[1] A house of Mercury is ♍.
[2] From $A_F = \lambda_H - (\lambda_☾ - \lambda_☉)$ = ♌ − 7° = ♓; and ♐ is culminating when ♓ is rising.
[3] A house of ♃ is ♓.

No. L 101,I

[1] From $A_F = \lambda_H - (\lambda_☾ - \lambda_☉)$ = ♑ − 4° = ♍.
[2] Omitted in *CCAG*.

[a] *Cf.* Vettius Valens I,2 (p. 8,32; p. 11,12/13 Kroll).

No. L 101,III

[1] From $A_F = A_H + (\lambda_☾ - \lambda_☉)$ = ♌ − 5° = ♓.
[2] A house of ♄ is ♑.
[3] *Cf.* Introduction p. 13.
[4] Following Kroll, Index p. 404b. Literally "he removed of his own accord."
[5] This is the 11th locus, counting from the Lot of Fortune (♓).
[6] A house of ☿ is ♍.

COMMENTARY

Diagram: pl. 15

For 101 March 5 one finds excellent agreement:

Text		Computed
☉ ⎱	♓	♓ 14
♃ ⎬	♓	♓ 12
♀ ⎰		♓ 14
☽	♎	♎ 12
♂	♋	♋ 10
☿	♒	♒ 25
♄	♏	♏ 15
H	♌	about 4 P.M.

For three other horoscopes which are contained in the same chapter cf. No. L 78.

No. L 102,IV,a

Vettius Valens III,16 p. 157,20-23

TRANSLATION

Sun in Taurus, Mercury in Taurus, moon in Pisces, Saturn in Scorpio, Jupiter, Mars (and) Venus in Aries, Horoscopos in Gemini. The rising time (of ♊) in the second clima is 28°. Mercury in Taurus (which has) the rising time 24 and {Venus} Mars in Aries 15. He died in the 67th year.

COMMENTARY

Diagram: pl. 16

The date of this horoscope is 102 April 30:

Text		Computed
☉	♉	♉ 8
☿	♉	♉ 5
☽	♓	♓ 12
♄	♏	♏ 25
♃ ⎱		♈ 23
♂ ⎬ ♈		♈ 18
♀ ⎰		♓ 29(!)
H		about 8 A.M.

The rising times for Babylon, System A, are

$$a_2 = 24 \qquad a_3 = 28.$$

If we add to the sum of these two numbers the numerical value of the 15 year period of Mars one obtains $52 + 15 = 67$.[1]

The horoscope immediately following concerns again 102 April 30 but about 10 P.M. For the subjects of both

horoscopes Vettius Valens records death at the age of 67 years,[2] a coincidence which he certainly considered as an important confirmation of the theory.

Another horoscope in this chapter is No. L 82.

No. L 102,IV,b

Vettius Valens III,16 p. 157,24-27

TRANSLATION

Another. The same configuration of the stars [as in No. L 102,IV,a] of another nativity, only the Horoscopos in Capricorn, the Lot of Fortune in Pisces.[1] The rising time (of ♓) in the second clima is 20° and the period of Jupiter 12 (years). And since (Jupiter) is in Aries whose rising time is 20° and (the period) of Mars is 15 (years) the total is 67. So long he lived.

COMMENTARY

Diagram: pl. 16

This second horoscope for 102 April 30 is about 14 hours later than No. L 102,IV,a. Consequently only the position of the moon is appreciably different (♓ 21). As rising times are used

$$a_1 = a_{12} = 20$$

and the periods of Jupiter (ruler of ♓) and Mars (ruler of ♈) are added, leading to a total of $40 + 12 + 15 = 67$.

The third and earliest horoscope in this chapter is No. L 82.

No. L 102,XII,4

Vettius Valens V,10 p. 230,14 to 17

TRANSLATION

Sun, Mercury (and) Saturn in Sagittarius, moon in Pisces, Mars in Leo, Venus in Capricorn, Horoscopos (and) Jupiter in Taurus. In the 45th year birth of twin children non-effective;[1] in the same (year) also high priesthood; in the 51st (year) a distinguished public office; in the 52nd year death of a child.

COMMENTARY

Diagram: pl. 16

For 102 December 4 one finds

Text		Computed
☉ ⎱		♐ 12
☿ ⎬ ♐		♐ 30
♄ ⎰		♐ 1

---

[1] Gemini is taken because it is the Horoscopos and the Horoscopos is the starter; the ruler of ♊ is ☿ (whose house is ♊). Thus ♉, the location of ☿, is considered. Finally we find for the Lot of Fortune $\Lambda_F = \lambda_H + (\lambda_☽ - \lambda_☉) = ♊ + 10° = ♈$ and ♂ is its ruler.

[2] One died in his 67th year, the other 67 years old.
No. L 102,IV,b
[1] From $\Lambda_F = \lambda_H - (\lambda_☽ - \lambda_☉) = ♑ - 10° = ♓$.
No. L 102,XII,4
[1] ἄχρηστος; perhaps stillborn.

Text		Computed
☾	♓	♓ 9
♂	♌	♌ 27
♀	♑	♐ 28(!)
H	♉	about 4 P.M.
♃	♉	♉ 3

This example is the earliest one in a group of thirteen, collected in V,10. The following is their list in chronological order:

102 Dec. 4	p. 230,14-17
107 May 8	p. 228,31-229,5
110 Sept. 27	p. 230,7-10
111 Apr. 24	p. 228,15-30
112 July 27	p. 229,13-36
113 July 1	p. 231,32-232,3
114 Nov. 10	p. 230,32-231,4
120 May 12	p. 230,18-21
122 Jan. 22	p. 230,22-31
123 Jan. 3	p. 231,5-31
134 Nov. 4	p. 227,27-228,14
135 Oct. 27	p. 229,6-12
153 May 8	p. 230,11-13

No. L 102,XII,14

Vettius Valens VII,5 p. 283,14 to p. 284,2

Chapter VII,5 "Method concerning successful and unsuccessful times (determined by means) of half, third, and two-thirds of the rising times and of the planetary periods" contains as examples 18 horoscopes the earliest of which is the present one. Their list follows:

p. 280,34 to p. 281,23	124 July 29
p. 281,24 to p. 282,2	134 June 23
p. 282, 3 to 15	108 Nov. 6
16 to 27	110 Dec. 15
p. 282,28 to p. 283,2	113 July 1
p. 283, 3 to 13	129 Jan. 16
p. 283,14 to p. 284,2	102 Dec. 14
p. 284, 3 to 11	105 Jan. 1
p. 284,12 to p. 285,3	158 Aug. 14
p. 285, 4 to 15	111 Sept. 30
p. 286,29 to 37	114 July 26
p. 287, 1 to 10	120 Feb. 8
11 to 16	118 Nov. 26
17 to 23	127 July 18
24 to 29	122 Jan. 30
30 to 34	133 Apr. 24
p. 288,20 to p. 289,14	142 Mar. 25
p. 290, 6 to 14	120 Sept. 28

TRANSLATION

Sun, Saturn (and) Mercury in Sagittarius, moon in Cancer, Jupiter in Taurus, Mars in Leo, Venus in Capricorn, Horoscopos in Virgo, clima 2. This person began as deputy governor but falling into the governor's disfavor in his 34th year was condemned to the quarry. For Mars with the sun ruled the period—of sun 19 and Mars 15 are 34. In the 36th year by the aid of greater (persons) he was released from confinement as disabled and at that time the rising time of Leo [1] (location of Mars) was operative in the 36th year, and 12 of Jupiter in dominant aspect and 24 [2] of Taurus are 36. But also of Capricorn (location of Venus) 28 [3] and of Venus 8 are 36. The beneficent (stars) then were strong. In the 39th year his circumstances were upset through the former enmity and he was condemned to an oasis. For the stars in Sagittarius ruled the period, of the sun 19, of Mercury 20 make 39; and take two-thirds of 58 (years) which is 38 (years) 8 months.[4] Again of sun 19, Mercury 20, and Leo, because (♌ is the location) of Mars in trine, 19,[5] make 58. (Again:) for Mars (in ♌) 19,[5] and of the sun 19 and of Mercury 20 are 58. In the 40th year he lived precariously and fell ill. However, his wife accompanying him affectionately comforted him and shared her possessions with him. The 40 years were indicated by the approaching conjunction of the moon (in ♋) with Mars (in ♌)—for 25 of the moon and 15 of Mars are 40—and the triangular position of Jupiter and Venus indicated the same, for 12 of Jupiter and 28 of Venus [3] are 40. And these are the prognostics I myself found.[6]

COMMENTARY

Diagram: pl. 16

We have here a horoscope for 102 December 14:

Text		Computed
☉	♐	♐ 22
♄	♐	♐ 2
☿	♐	♐ 22
☾	♋	♋ 10
♃	♉	♉ 2
♂	♌	♌ 30
♀	♑	♑ 10
H	♍	about midnight

No. L 103

Vettius Valens II,41, p. 129,24 to 28 (= CCAG 5,2 p. 121,3 to 7)

TRANSLATION

Sun (and) Venus in Capricorn, moon in Cancer,

[1] In clima 2, System A and B, $a_5 = 36$.
[2] In clima 2, for both systems, $a_2 = 24$.
[3] In clima 2, System A, $a_{10} = 28$.
[4] Thus again about 39 years; the text (or Kroll) has 58 (years) 8 months.
[5] Leo is the house of the sun, whose period is 19.
[6] Similar references to evidence from Vettius Valens' own experience are found in Nos. L 114,IX; L 122,XII; L 158.

Saturn (and) Mercury in Sagittarius, Jupiter in Taurus, Mars in Leo, Horoscopos in Aquarius, Lot of Fortune in Leo. Mars was located here in a fiery and solar sign [1] opposite to the Horoscopos, Saturn and Mercury were in a dominant aspect to the Locus of Death. This person was burned alive.

COMMENTARY

Diagram: pl. 16

For 103 January 10 one reaches good agreement except for Venus and a small deviation for Mars:

Text		Computed
☉	♑	♑ 19
♀	♑	♒ 14(!!)
☾	♋	♋ 9
♄	♐	♐ 4
☿	♐	♐ 24
♃	♉	♉ 1
♂	♌	♍ 1(!)
H	♒	about 8 A.M.

This horoscope is one of a group in which No. L 65,V is the earliest.

Because the elongation of the moon is about 180° the Lot of Fortune coincides with the setting sign. The 8th locus from the Lot of Fortune (♌), the Locus of Death, is ♓, which is in quartile to Saturn and Mercury in Sagittarius.

No. L 104,IV

Vettius Valens II,36 p. 114,6-9

TRANSLATION

Sun in Taurus, Mars in Taurus, moon in Virgo,[1] Saturn in Sagittarius, Jupiter in Gemini, Mercury, Venus (and) Horoscopos in Aries, the Lot of Fortune in Sagittarius,[2] its ruler (♃ [3]) in Gemini. Further, Daimon in Leo, its ruler (☉ [4]) in Taurus. This person was short-armed.

COMMENTARY

Diagram: pl. 16

For 104 April 23 one finds

Text		Computed
☉	♉	♉ 1
♂	♉	♈ 29(!)
☾	♍	♍ 6
♄	♐	♐ 19
♃	♊	♊ 14
☿	♈	♈ 7
♀	♈	♈ 26
H	♈	about 4 A.M.

For related horoscopes *cf.* No. L 83.

[1] *Cf.* for these qualities of Leo: Vettius Valens I,2 p. 9,14 Kroll. See also Bouché-Leclercq, *AG* p. 138.

No. L 104,IV

[1] The variant, moon in Capricorn, is incorrect.

[2] From $\lambda_F = \lambda_H - (\lambda_{☾} - \lambda_{☉}) = ♈ - 4° = ♐$.

[3] Whose house is ♐.

[4] Whose house is ♌.

No. L 104,VII

Vettius Valens V,12 p. 237,24 to 31

TRANSLATION

For example let sun, Jupiter (and) Mars be in Cancer, moon in Libra, Saturn in Sagittarius, Venus (and) Mercury in Leo, Horoscopos in Gemini. He died in the 54th year. For the cycle of Saturn was (in its) 18(th critical period), that of Jupiter (in its) sixth, that of the sun (in its) third; [1] a common return of these (stars, to the same distances). And sun and Jupiter were found (to be) in the death-bringing month, that is Sagittarius.[2] And besides, sun and Jupiter and Mars yielded the 54th year from the (8th) Locus of Death [3] (in ♏) to Saturn in Sagittarius. The yielding was grievous.

COMMENTARY

Diagram: pl. 17

One finds for 104 July 17

Text		Computed
☉	♋	♋ 23
♃	♋	♋ 3
♂	♋	♊ 25(!)
☾	♎	♎ 13
♄	♐	♐ 14
♀	♌	♌ 11
☿	♌	♌ 8
H	♊	about 4 A.M.

This is the second of two horoscopes in a chapter for which Critodemus is quoted as author. The first horoscope is No. L 92,II.

[1] Previously in this chapter Vettius Valens assigned to each planet a critical number. e.g. ♄ = 3, ♃ = 9, and ☉ = 18. Because 54 = 18·3 = 6·9 = 3·18 we have the above statements.

[2] This, probably, is related to a doctrine stated in V,6 (p. 214, 19 ff.) according to which the point opposite to the sun is significant, thus in our case ♐ where the harmful ♄ is located.

[3] Counted from the Lot of Fortune.

No. L 105

Vettius Valens II,21, p. 86,10-16 and VII,5 p. 284,3-11

TRANSLATION

First Version

Sun (and) Mercury in Capricorn, moon (and) Saturn in Sagittarius, Jupiter in Cancer, Mars in Virgo, Venus in Aquarius, Horoscopos in Libra. The rulers (♃ and ☉) of the Triangle (of the moon, ♐ ♈ ♌) were found in center (M and IMC) but in opposition. Wherefore the nativity, though well provided and prosperous at first was later found to be reversed and needy, with forecast of burning and plunder. For the ruler of the Lot (of Fortune, ♏ [1]) Mars,[2] was found in the (11th) locus [3] of accomplishment, in apoklima and in aspect with Saturn.

Second Version

(Positions as before, adding only) second clima. In the 48th year he witnessed a very grave sorrow, the death of a beloved child, and in the same (year) also (the death) of the mother. Virgo and Libra indicated it since they are of equal rising time. For Libra 8 years,[4] for Virgo 40, make 48. Since Mars was found within the Horoscopos [5] he caused grief in spirit. And besides, 40 of Virgo and 32 of Sagittarius make 72, two-thirds of which is 48.

COMMENTARY

Diagram: pl. 17

The date is found to be 105 January 1:

Text		Computed
☉	♑	♑ 11
☿	♑	♑ 0
☾	♐	♐ 13
♄	♐	♐ 24
♃	♋	♋ 14
♂	♍	♍ 30
♀	♒	♒ 27
H	♎	about midnight

Additional information comes from the second version (VII,5). The second clima indicates a geographical latitude of Babylon-Syria. The following rising times are quoted:

$$a_6 = a_7 = 40 \quad \text{and} \quad a_9 = 32$$

correct for Babylon, System A.

For related horoscopes *cf.* No. L 50 and No. L 102 XII 14 respectively.

[1] From $\lambda_F = \lambda_H - (\lambda_☾ - \lambda_☉) = ♎ + 1° = ♏$.

[2] A house of ♂ is ♏.

[3] Counted from ♏, the Lot of Fortune.

[4] Because ♎ is a house of Venus which has a period of 8 years.

[5] Mars is found, for the date in question, to be very close to entering Libra whereas the rising point was about ♎6 (*cf.* Commentary). Hence Mars was certainly within the 30° section which makes the "Horoscopos."

No. L 106

Vettius Valens II,36 p. 113,3-10

TRANSLATION

Sun (and) Saturn in Capricorn, moon in Scorpio, Jupiter in Leo, Mars in Pisces, Venus (and) Mercury in Aquarius, Horoscopos in Virgo, the Lot of Fortune in Scorpio,[1] the Daimon in Cancer. Then in opposition to the Daimon, which forecasts the intellectual and the spiritual, was Saturn and he was in dominant aspect to the (preceding) full moon (in ♋) and to the phase at that time, and the ruler of the Lot of Fortune (♂ [2]) was in opposition to the Horoscopos. Thus this person had in the fated places injury and tender feet and most of all he was lunatic.

COMMENTARY

Diagram: pl. 17

One finds for 106 January 16:

Text		Computed
☉	♑	♑ 26
♄	♑	♑ 5
☾	♏	♏ 16
♃	♌	♌ 16
♂	♓	♓ 2
♀	♒	♒ 4
☿	♒	♒ 9
H	♍	about 10 P.M.

For related horoscopes *cf.* No. 83.

No. L 107

Vettius Valens V,10 p. 228,31 to p. 229,5

TRANSLATION

Sun in Taurus, moon, Venus (and) Horoscopos in Aries, Saturn in Capricorn, Jupiter in Virgo, Mars in Scorpio, Mercury in Gemini. In the 51st year (of his life) he went abroad and going to the (court of the) king he won a case for the high priesthood of a friend. For the yielding was from moon and Venus and Horoscopos [to] Mercury [1] in the (3rd locus) of god and king. In this same year also the death of a child occurred, for Mars in the (8th) locus of death transmitted to Saturn in the (10th) locus concerned with children.

COMMENTARY

Diagram: pl. 17

The date of this horoscope is 107 May 8 as is seen from the following comparison

No. L 106

[1] From $\lambda_F = \lambda_H - (\lambda_☾ - \lambda_☉) = ♍ + 2° = ♏$.

[2] Whose house is ♏.

No. L 107

[1] Because 51 ≡ 3 mod. 12 and ☿ in ♊ is in the 3rd sign from ☾ ♀ H in ♈.

Text		Computed
☉	♉	♉ 15
☾ ⎫		♈ 14
♀ ⎬	♈	♈ 25 [2]
H ⎭		about 4 A.M.
♄	♑	♑ 24
♃	♍	♍ 9
♂	♏	♏ 23
☿	♊	♊ 8

The assignment of the loci is strange. The locus of "god" is ordinarily locus 9, locus 3 being "goddess"; the locus of "children" is ordinarily locus 5 but we know from Vettius Valens IV,12 that also the 10th locus (M) is associated with children.[3]

For related horoscopes cf. No. L 102,XII,4. The horoscope of the son who died in the 51st year of the father is No. L 135,X.

No. L 108,III

Vettius Valens II,36 p. 114,10-15

TRANSLATION

Sun (and) Mercury in Aries, moon in Pisces, Saturn (and) Horoscopos in Aquarius, Mars (and) Venus in Taurus, Jupiter in Libra, the Lot of Fortune in Pisces,[1] the Lot of Daimon in Capricorn. This person was possessed of a god like a madman (because) Jupiter, the ruler[2] of the Lot (of Fortune) (was found) in Libra, the (9th) locus concerning god, and Saturn, the ruler[3] of the Daimon, in the Horoscopos, and Venus in Lower Midheaven.

COMMENTARY

Diagram: pl. 17

Perfect agreement is found for 108 March 28:

Text		Computed
☉ ⎫	♈	♈ 7
☿ ⎭		♈ 18
☾	♓	♓ 10
♄ ⎫	♒	♒ 4
H ⎭		about 2 A.M.
♀ ⎫	♉	♉ 8
♂ ⎭		♉ 20
♃	♎	♎ 16

For related horoscopes cf. L 83.

---

[2] According to the Stobart tables, C₁ obv. III,24, Venus is entering ♉ the next day: Trajan 10 month 9 day 14 = 107 May 9. Cf. Neugebauer [4] p. 223.
[3] Cf. Introduction p. 8.

No. L 108,III
[1] From $\Lambda_F = \lambda_H - (\lambda_\mathbb{C} - \lambda_\odot) = \text{♒} - 11° = \text{♓}$.
[2] A house of ♃ is ♓.
[3] A house of ♄ is ♑.

No. L 108,XI

Vettius Valens VII,5 p. 282,3 to 15

TRANSLATION

Sun, Mercury (and) Jupiter in Scorpio, moon in Taurus, Saturn in Aquarius, Mars in Virgo, Venus in Libra, Horoscopos in Leo, second clima. In the 52nd year he had a very great quarrel and lawsuit with his sister[1] about property and inheritance, and at the royal (court) he won. The operative position was that diametrical to the moon (in ♉, with period) 25, of Jupiter (in ♏) 12, of Scorpio 15,[2] making 52. Besides, for the second clima Scorpio 36,[3] of Jupiter 12, and of Saturn, located (in ♒) in the Lot of Fortune,[4] 30 making 78, two-thirds of which (is) 52. And again, of Taurus 24,[5] of Aquarius 24, and of Saturn 30, make 78, two-thirds of which (is) 52. Thus all the stars were operative except Mars. Accordingly he was ill at this time and had a narrow escape at sea and had great expenditures, but the beneficent (stars) were destined to be in dominant aspect to Saturn and were the stronger.

COMMENTARY

Diagram: pl. 17

For 108 November 6 we find:

Text		Computed
☉ ⎫		♏ 14
☿ ⎬	♏	♏ 19
♃ ⎭		♏ 3
☾	♉	♉ 17
♄	♒	♒ 0
♂	♍	♎ 2(!)[6]
♀	♎	♎ 16
H	♌	about midnight

Related horoscopes are listed with No. L 102,XII,14 and especially No. L 110,XII which gives the horoscope of the sister who was unsuccessful in the law suit.

No. L 109

Vettius Valens II,21 p. 86,24-31

TRANSLATION

Sun (and) Mercury in Gemini, moon in Capricorn, Saturn (and) Mars in Aquarius, Venus (and) Horoscopos in Cancer, Jupiter in Scorpio. This person was born a slave and having entered a gens, attained to

---

[1] Her horoscope is No. L 110,XII.
[2] The period of Mars whose house is ♏.
[3] For both systems $a_5 = 36$ in clima 2.
[4] From $\Lambda_F = \lambda_H - (\lambda_\mathbb{C} - \lambda_\odot) = \text{♌} + 6° = \text{♒}$.
[5] For both systems $a_2 = a_{11} = 24$ in clima 2.
[6] According to the Stobart tables, C₁ rev. I,6 (Neugebauer [4] p. 224) Mars entered ♎ on November 1 (= Trajan 12 month 3 day 5).

political offices and enjoyed honors. For the rulers (♄ and ☿) of the triangle of the sun (♊ ♎ ♒) and of the Lot (of Fortune, ♒) (♄)[1] and of the exaltation (of the nativity,[2]) were found in their own domains[3] and in aspect with Jupiter,[4] but Mars and Saturn and Mercury were unfavorably located[5] and therefore they reduced his livelihood and made it embarrassed.

COMMENTARY

Diagram: pl. 18

Good agreement is found for 109 June 2:

Text		Computed
☉ }	♊	♊ 10
☿ }		♉ 25(!)
☾	♑	♑ 0
♄ }	♒	♒ 18
♂ }		♒ 4
♀ }	♋	♋ 5
H }		about 8 A.M.
♃	♏	♏ 14

For related horoscopes *cf.* No. L 50. The same date is quoted as example[6] in VIII,6 (p. 310,11 ff.). For Trajan 12 Payni (X) 8 = 109 June 2, second hour of day, the sun is assumed to be in ♊ 13, the moon in ♑ 4, with H in ♋. This person died in his 45th year.[7]

According to the Stobart tables Mercury would be in ♉, entering ♊ two days later.[8]

No. L 110,III

Vettius Valens III,6 p. 142,8-20

TRANSLATION

Sun in Pisces degree 25;8, moon in Gemini degree 16;53, Saturn in Pisces degree 1;25, Jupiter in Sagittarius 24;18, Mars in Taurus, 21;8, Venus in Aquarius degree 9, Mercury {Saturn} in Taurus 32(!), Horoscopos in Libra 15, Midheaven in Cancer 16. The luminaries were in apoklima, the Horoscopos was the starter, the terms of Jupiter and Jupiter (itself) were unsuitably located, the nativity was without ruler, the start was up to the point diametrically opposite to Mars in Scorpio 21 degrees. For, located in the starting terms[1] and sending rays toward the same terms, (Mars) cancelled it. He died in the 51st year.

COMMENTARY

Diagram: pl. 18

This and the preceding horoscope (No. L 75) are among the few cases in Vettius Valens where the planetary positions are given with degrees and minutes. The date is undoubtedly 110 March 15 in spite of the deviations which amount to several degrees and which are in part caused only by the difference in norm of longitudes (*cf.* p. 180).[2] In the case of Mercury the text is in confusion as is shown by the mention of Saturn and of an impossible number of degrees (32).

Text		Computed	Text—comp.
☉	♓ 25;8	♓ 23	+2
☾	♊ 16;53	♊ 15	+2
♄	♓ 1;25	♒ 23(!)	+8;30
♃	♐ 25;18	♐ 21	+4
♂	♉ 21	♉ 13	+8
♀	♒ 8	♒ 7	+1
☿	♉ 32(!)	♈ (!) 12	
H	♎ 15	about 7 P.M.	
M	♋ 16	♋ 15;52[3]	

An explicit confirmation of our computation can be found in Vettius Valens I,20 (p. 35,15 to p. 36,2; *cf.* also *CCAG* 2 p. 104,25 to p. 105,6). There a simple method for the cyclic computation of planetary longitudes is explained and illustrated by means of examples. As a first example is chosen the date Trajan 13 Phamenoth 18 which corresponds to 110 March 14/15. This precedes our horoscope by only a few hours, a difference which would only be recognizable for the moon. For this date Vettius Valens finds

♄ in ♓ ♂ in ♉
♃ in ♐ ♀ in ♒[4]

that is the same signs which he found by more accurate computation for the horoscope.

The Demotic planetary tables,[5] Stobart C_1, give the

[1] From $\Lambda_F = \lambda_H + (\lambda_☾ - \lambda_☉)$ = ♋ + 7° = ♒. And ♒ is a house of ♄.

[2] Counting from the Lot of Fortune (♒) one finds ♐ (whose ruler is ♃); with respect to the Horoscopos one finds ♉ (whose ruler is ♀).

[3] Saturn is in his house (♒) and so is Mercury (♊) but not ♃ or ♀.

[4] Saturn in quartile, Venus in trine, but not Mercury.

[5] *Cf.* Vettius Valens p. 89,12: *μὴ ἰδιοτοπῇ ἀλλὰ παραπέσῃ.*

[6] *Cf.* p. 180 No. 12.

[7] Vettius Valens p. 310,31.

[8] C_1 rev. II,5; *cf.* Neugebauer [4] p. 224.

[1] Both Horoscopos and Mars are in terms of Jupiter.

[2] The same signs, but no degrees, are given for the planets from Saturn to Venus in I,20 (p. 35,15 Kroll) for the day before: Trajan 13 Phamenoth (VII) 18 = 110 March 14 (*cf.* p. 180 No. 13). The position of the moon excludes the identification of the two examples.

[3] Computed by use of the Handy Tables (cod. Vat. gr. 1291). Because the vernal point is almost in the horizon it is obviously correct that the culminating point is practically 90° distant in longitude. The method of Paulus Alexandrinus (*cf.* p. 10) would lead to M = ♋ 19;10.

[4] *CCAG* 2, p. 105,5 has here, incorrectly, ♓. The error goes back to the general rules where (p. 104,16) Phaophi (II) is associated with ♑ instead of with ♐.

[5] *Cf.* p. 173. Neugebauer [4] p. 224.

following positions for 110 March 12 = Trajan 13 month 7 day 16[6]

	Text
♄ in ♒, 8 days before entering ♓	♓ 1;25
♃ in ♐	♐ 25;18
♂ in ♉	♉ 21
♀ in ♒, 2 days after entering ♒	♒ 8
☿ in ♈, 5 days after entering ♈	♉ 32(!)

No. L 110,IX

Vettius Valens V,10 p. 230,7-10

TRANSLATION

Sun, Mercury (and) Venus in Libra, Saturn in Aquarius, Jupiter (and) Horoscopos in Sagittarius, Mars in Virgo, moon in Leo. In the 47th year he was heir to a friend and in the same (year) he was separated from his wife because of jealousy and abuse.

COMMENTARY

Diagram: pl. 18

The given data correspond to 110 September 27:

	Text	Computed
☉	♎	♎ 3
☿	♎	♎ 9
♀	♎	♎ 4
♄	♒	♒ 24
♃	♐	♐ 17
H	♐	about 10 A.M.
♂	♍	♍ 18
☾	♌	♌ 11

For related horoscopes *cf.* No. L 102,XII,4.

According to the Stobart tables Mercury entered ♎ the day before.[1]

No. L 110,XII

Vettius Valens VII,5 p. 282,16 to 27

TRANSLATION

The defeated woman[1] had the following theme. Sun (and) Mercury in Sagittarius, moon in Cancer, Saturn in Aquarius, Jupiter {in Virgo} [and][2] Venus in Capricorn, Mars in Scorpio, Horoscopos in Gemini, clima 4. The 54th year is indicated by Saturn and Venus, 30 and 24[3] (respectively), making 54. And, again, 36 (the rising time) of Scorpio, and (the period) of Mars, located in the Lot (of Fortune, ♏[4]), which[5] had effective influence, 15, and of Saturn 30, making 81, two-thirds of which is 54, for also the dominant aspect[6] of the maleficent (stars) itself was operative. In the 53rd year she had the prospect and expectation of victory through the help of influential persons, wherefore she was persuaded to go to court, since the moon counted 25 and (the rising time of) Capricorn (the place of Venus) 28, making 53. But in the 54th year she was abandoned by her helpers, for the beneficent (stars) did not combine their times (to this total).

COMMENTARY

Diagram: pl. 18

The position of the sun (♐) requires a date in December. For 106 December 1 one finds Jupiter in the middle of Virgo, but Saturn is about 20° short of Aquarius. The situation is equally bad near A.D. 47 or in 165 and 166. Thus it is clear that for centuries no solution can be found such that all data of the text are satisfied. At the same time it is easy to see that the text contains errors; for example, the rising times of Scorpio and Capricorn

$$\alpha_8 = 36 \qquad \alpha_{10} = 28$$

are only correct for clima 2, Babylon, (System A) but not for "clima 4" as stated in the text. Similarly the enumeration Venus-Mars instead of Mars-Venus is suspicious.

In order to emend the text, the experience can be utilized that the date of the latest event quoted varies only in very small limits within each individual chapter. In our present case the limits of the dates quoted in 16 horoscopes from this chapter[7] are 154 and 164 respectively. The latest year quoted is the 54th which leads us back to a birth between 100 and 110. Within this interval Jupiter is correctly in ♍ in December 106 but Venus, Mars, and Saturn are wrong, whereas the astrological commentary agrees with the initial positions of Mars and Venus. Saturn is correctly located in ♒ for the years 108 to 110 but Mars and Venus are wrong for 108 and 109. For 110 December 15, however, one finds perfect agreement, Jupiter alone excepted:

	Text	Computed
☉	♐	♐ 23
☿	♐	♐ 0
☾	♋	♋ 9

[6] Alexandrian calendar.

No. L 110,IX

[1] Stobart C_1 rev. IV,25 (Neugebauer [4] p. 224): Trajan 14 month 1 day 29 = 110 September 26.

No. L 110,XII

[1] Referring to the law suit in the preceding horoscope (No. L 108,XI).

[2] *Cf.* the Commentary.

[3] The origin of this figure for Venus is doubtful; it is the rising time of ♐, the place of Mercury, not of Venus.

[4] From $A_F = A_H - (\lambda_{☾} - \lambda_{☉})$ = ♊ + 5° = ♏.

[5] That is Mars.

[6] *Cf.* Vettius Valens p. 102,16.

[7] *Cf.* No. L 102,XII,14.

Text		Computed
♄	♒	♒ 26[a]
♃	♍(!)	♑ 3
♀	♑	♑ 12
♂	♏	♏ 12
H	♊	sunset

This also explains the apparent error of enumerating Venus before Mars; one has simply to read "Jupiter (and) Venus in Capricorn."

No. L 111,IV

Vettius Valens V,10 p. 228,15 to 30

TRANSLATION

Sun in Taurus, moon (and) Mercury in Aries, Saturn in Pisces, Jupiter (and) Mars in Aquarius, Venus in Gemini, Horoscopos in Virgo. In the 42nd year (of his life) he was heir to a female person; for the paradosis at (interval) 6[1] was from moon and Mercury in the (8th) locus of Death in Aries, to Virgo, house of Mercury, and from a sign of exaltation (♈, exaltation of the sun) to an exaltation (of ☿, i.e., ♍). In the 45th year he held a distinguished office for public affairs; for Venus at Midheaven gave over to Mars, which indicated trouble, and to Jupiter,[2] which indicated esteem. And the start was from the Horoscopos to the sun (in M) wherefore he gained royal recognition at that time. And in the same year he freed (his) concubines because Jupiter in the (6th) locus of slaves took over from Venus. In the 46th year he had troubles and reverses in affairs of property[3] and because of female persons, quarrels, and the death of two concubines; for the transfer (of power) was from Venus to Saturn in the (7th) locus of marriage and from sun to Mars and Jupiter. From these quarrels then he ceased.

COMMENTARY

Diagram: pl. 18

The strenuous life of this person began in 111 April 24 as is seen from the following comparison:

Text		Computed
☉	♉	♉ 2
☾ }	♈	♈ 18
☿ }		♈ 10
♄	♓	♓ 10

[a] The Stobart tables as well as P. Tebt. 274 would give ♓. According to Stobart C_1 rev. III,24 Saturn enters ♓ in Trajan 14 month 3 day 24 (= 110 November 20) whereas T gives day 23 (*cf.* Neugebauer [4] p. 224 and p. 241 respectively).

No. L 111,IV

[1] Because 42 ≡ 6 mod. 12.

[2] Because 45 ≡ 9 mod. 12 and ♂ and ♃ in ♒ are in the 9th sign from ♀ in ♊.

[3] Reading κτητων for τηκτων in line 27.

Text		Computed
♃ }	♒	♑ 27(!)
♂ }		♒ 15
♀	♊	♊ 18
H	♍	about 2 P.M.

Both the Demotic Stobart tables and the Greek planetary tables P. Tebt. 274 lead to ♒ as position of Jupiter. According to Stobart C_1 rev. III,28 Jupiter entered ♒ at Trajan 14 month 8 day 15 = 111 April 10 while T gives a date 3 days earlier.[4]

For related horoscopes *cf.* No. L 102,XII,4.

No. L 111,IX

Vettius Valens VII,5 p. 285,4 to 15

TRANSLATION

Sun (and) Mercury in Libra, moon in Aquarius, Saturn in Pisces, Jupiter in Capricorn, Mars in Aries, Venus in Leo, Horoscopos in Cancer, clima 6. This person was rich, but about the 47th year he lost (his riches). For Saturn was operative, being in Midheaven (in the) 9(th locus),[1] and the rising of Pisces (the place of Saturn): of Saturn 30, of Pisces 17,[2] making 47. Further, the diameter of sun (in ♎) and Mars (in ♈) and Mercury (in ♎): of Mars 15 and of the rising time of Aries 17[2] and of the sun 19 and of Mercury 20 {and of Leo 38}[3] making altogether 71, two-thirds of which is 47 (years) and 4 months. And the side of the square of Libra and Capricorn: of Capricorn 27,[4] of Mercury (in ♎) 20, making 47. In appearance then he seemed to be guarded all round but he had to expect reverses and loss because of a woman.

COMMENTARY

Diagram: pl. 18

The date 111 September 30 satisfies the conditions of the horoscope:

[4] *Cf.* Neugebauer [4] p. 224 and p. 241 respectively.

No. L 111,IX

[1] Counting by zodiacal signs, beginning with H = ♋ as locus 1, Saturn is found in the 9th locus (♓). Counting in the same fashion, Midheaven would be locus 10 = ♈. In order to justify the statement that Saturn was in M we would have to assume accurate computation ("by degrees"), which indeed may place Saturn in the culminating segment (a real check is made impossible by the omission of degrees for H and the sun). But then also the counting of the loci should have been modified accordingly.

[2] The rising time of Pisces and Aries is $a_{12} = 17 = a_1$ in clima 6, System B.

[3] This is an interpolation, which, however, could have indicated that also Venus (period 8) in ♌ plus the rising time of ♌, namely 39 for clima 6 (System B) give the total 47. Thus 38 is probably an error for 39 + 8.

[4] The rising time of ♑ is $a_{10} = 27$ for clima 7(!) in System B or for clima 2(!) in System A.

Text		Computed
☉ }	♎	♎ 6
☿ }		♎ 24
☾	♒	♒ 14
♄	♓	♓ 9
♃	♑	♑ 19
♂	♈	♈ 14
♀	♌	♌ 19
H	♋	about midnight

For related horoscopes *cf.* No. L 102,XII,14.

No. L 112,VII

Vettius Valens V,10 p. 229,13-36

TRANSLATION

Sun, Mars (and) Venus in Leo, moon in Aquarius, Saturn in Aries, Jupiter in Pisces, Mercury in Cancer, Horoscopos in Virgo. In the 24th year he profited from the dead and (from) friends; in the 26th year marriage and woman's help, in the 29th year by the death of another person's slave and allegation of poison he had (legal) affairs and quarrels. For Saturn (in ♈) made paradosis to sun, Mars (and) Venus in (♌) the (12th) locus concerned with slaves, but he got help by the friendship of greater (persons), male and female. [In similar fashion this discussion continues down to his 45th year when] he was released by the influence of greater persons on ground of illness; for the paradosis of each the first period and the second period was mixed, beneficent and maleficent. And besides the maleficent (stars) were under the sun's ray retrograde or dimmer [1] and fell in weaker places.

COMMENTARY

Diagram: pl. 19

One finds as date of this horoscope 112 July 27:

Text		Computed
☉ }		♌ 2
♂ }	♌	♌ 19
♀ }		♌ 24
☾	♒	♒ 9
♄	♈	♓ 26(!)
♃	♓	♓ 2
☿	♋	♋ 17
H	♍	about 8 A.M.

Saturn is about 13 days past its first station. Both Stobart tables and P. Tebt. 274 agree with the present text in the position of Saturn since they place the entry of the planet into ♈ at Trajan 15 month 9 day 4 = 112 April 29; [2] ♓ is reentered on September 28. The case of ♀ also is of interest since the Stobart tables record the entry of this planet into ♍ for a date only two days later than our horoscope.[3]

For related horoscopes *cf.* No. L 102,XII,4.

[1] Saturn is retrograde and Mars close to the sun.

[2] *Cf.* Neugebauer [4] p. 222 (C_1 rev. V,10) and p. 241.

[3] C_2 obv. I,6: Trajan 15 month 12 day 5 = 112 July 29.

No. L 112,VIII

Vettius Valens II,36 p. 114,16 to 18

(= p. XI note 1 = CCAG 5,2 p. 120,30 to 32)

TRANSLATION

Sun (and) Mercury in Leo, moon in Scorpio, Saturn (and) Horoscopos in Aries, Jupiter in Pisces, Mars (and) Venus in Virgo, the Lot of Fortune in Capricorn,[1] the (Lot of) Daimon in Cancer.[2] This person was hunchbacked.[3]

COMMENTARY

Diagram: pl. 19

For 112 August 17 one finds:

Text		Computed
☉ }	♌	♌ 23
☿ }		♌ 27
☾	♏	♏ 18
♄ }	♈	♓ 25(!)
H }		about 10 P.M.
♃	♓	♓ 0
♂ }	♍	♍ 4
♀ }		♍ 20

Saturn is retrograde; *cf.* the commentary to the preceding horoscope.

For related horoscopes *cf.* No. L 83.

No. L 112,VIII

[1] From $\lambda_F = \lambda_H - (\lambda_☾ - \lambda_☉) = ♈ - 3° = ♑$.

[2] From $♈ + 3° = ♋$.

[3] Or any disease with humps.

No. L 113,IV

CCAG 8,2 p. 85,18 to p. 86,12

TRANSLATION

And a third[1] nativity he[2] records as follows and he says about this person that he was born to destruction of himself and of his parents at about the age of 25; and the theme of this nativity was as follows: Sun and Horoscopos were in Aries, moon in Scorpio, 3 days after its evening rising, Mercury and Saturn in Aries at morning setting, Jupiter in Pisces at morning rising. The Lot of Fortune fell in Scorpio. The ruler, he says, in this case will be Mars. The moon after 7 days will be in [Leo],[3] after 40 days in Libra.

No. L 113,IV

[1] Referring to No. L 40 and to No. L 76 as the two other ones.

[2] Antigonus of Nicaea.

[3] Left blank in manuscript, correctly supplied by the editors.

[There follows a short biographical sketch. The causes of this fate are found in the astrological conditions.]

Furthermore, all of his family suffered in the same reversals because the sun was in the Horoscopos, attended by Mercury and Jupiter. He says about it: He was ill advised because Mercury and Saturn were found in a male sign (♈) and destroyed someone of those of the family because of this configuration, and he himself was cut down because the moon was in Scorpio, a misfortune-causing sign, and projecting its rays[4] against Mars which was in Epanaphora with respect to Midheaven. The damage would come to a person because Mars was in a human-shaped sign[5] (Aquarius). He was erotic and fond of gladiators, erotic through Venus and Jupiter, fond of gladiators because Mars was in Epanaphora with respect to Midheaven and Mercury was in the house of Mars (♈). He died about his 25th year, in a bad way, because this (the number 25) and the rising time of Aquarius were the same.[6]

COMMENTARY

Diagram: pl. 19

We must first correct an obvious error in the text. Twice it is stated that the moon was in ♏. This is excluded, however, by all other data concerning the moon: (*a*) the moon was 3 days after evening rising and the sun in ♈; this puts the moon in ♉ or ♊; (*b*) after *7* days the moon would be in ♌; but *7* days correspond to a progress of about 3 signs hence to an original position in ♉; (*c*) after 40 days the moon would be in ♎ or one complete rotation and 5 signs from its starting point, which thus must have been ♉. Hence ♉ is the correct place for the moon. The cause of the error is probably a simple misreading of the symbol for Taurus, as it also occurred in No. L — 71.

The remaining data amply suffice to determine the date of the horoscope as 113 April 6, that is 3 days after conjunction on April 3. One finds:

	Text	Computed
☉	♈	♈ 15
H	♈	sunrise
☾	[♉]	♉ 19
☿	♈	♈ 6
♄	♈	♈ 2
♃	♓	♓ 23
♂	♒	♒ 22

The moon is at the end of ♌ after *7* days and at the end of ♎ after 40 days.[7] It is furthermore correct that Mars is in ♒ before culmination in ♑ when the Horoscopos is with the sun in ♈. Mercury is retrograde as morning star and thus in "morning setting" and the same terminology is crudely applied to Saturn which is very close to Mercury. Jupiter is correctly at morning rising. Thus all astronomical data are correct.[8] The Lot of Fortune, however, is said to be in ♏ and is therefore based on the wrong lunar position in ♏, computed according to

$$\Lambda_{F} = \lambda_{H} + (\lambda_{☾} - \lambda_{☉}) = ♈ - 7^{s} = ♏.$$

Cumont assumed (*CCAG* 8,2 p. 85 note) that this horoscope refers to Pedanius Fuscus who was executed with his parents, supposedly, in 136 by Hadrian. According to Cassius Dio his age was then 18, not "about 25" as in the above text. Our result shows that the age given in the horoscope requires the year 137/138 as the date of execution.

No. L 113,VIII

Vettius Valens V,10 p. 231,32 to p. 232,3 and VII,5 p. 282,28 to p. 283,2

TRANSLATION

First Version

Sun (and) moon in Cancer, Saturn, Jupiter (and) Mars in Aries, Venus, Mercury (and) Horoscopos in Gemini. In the 20th year the parents of this person were both killed by a guardsman in an assembly of robbers; the paradosis was then from the Horoscopos to the (8th) Locus of Death (♑) and even more operative was the (paradosis) at (the interval) 4 from Saturn and Mars (in ♈) to sun and moon (in ♋) which happened to be in the (2nd) Locus of Death, indicating both father and mother. He himself, however, (although) in the melee, escaped the danger, so that the paradosis of Jupiter (in ♈) was powerful in the same (time).

Second Version

[Positions as before] clima 1. In the 47th and 48th and 49th (year) he was ill and was weakened, suffering a great flow of blood and at the same time he was banished. The 47 years were indicated by Cancer 31;40 and Mars 15 making 47 (years) and 8 months; and the 48 (years), again of Cancer 31;40 and of the moon 25 and of Mars 15 making 72, two-thirds of which is 48; operative then in the 48th year were both the rising time of Gemini 28;20 and the period of Mercury 20, making 48 (years) 4 months wherefrom he gained a little lenience and support in the 49th year. Sun and Saturn allotted 19 and 30 making 49 (years); then the maleficent (planets in ♈) being in dominant aspect to

[4] That is the quartile aspect ahead of the moon, counted in the zodiacal order. *Cf.* Bouché-Leclercq. *AG* p. 249.

[5] *Cf. CCAG* 11, p. 139,6 f.

[6] In both systems the rising time of Aquarius is 25°, but only for Alexandria.

[7] *Cf.* also No. L 40.

[8] The Stobart tables (Neugebauer [4] p. 225) indicate that Jupiter is near entering ♈ (C_4 obv. I,30: Trajan 16 month 8 day 15 = 113 April 10) and Mars near ♓ (C_4 obv. II,7: day 18 = April 13).

the luminaries (in ♋) were the cause of great danger and if Jupiter had not been there (in ♈) too they would have brought him violent death absolutely.

COMMENTARY
Diagram: pl. 19

For 113 July 1 one finds:

Text		Computed
☉ }	♋	♋ 7
☾ }		♋ 11
♄ }		♈ 7
♃ }	♈	♈ 9
♂ }		♈ 23
♀ }		♉ 27(!)
☿ }	♊	♊ 17
H }		about 4 A.M.

According to the Stobart tables, C_2 obv. II,20, Venus entered ♊ the day before (Trajan 16 month 11 day 6 = 113 June 30[1]). Mars, however, is just one day ahead of moving from ♈ into ♉ (C_2 obv. II,9 day 8 = July 2). The rising times

$$a_3 = 28;20 \qquad a_4 = 31;40$$

are correct for Alexandria (clima 1), System A.

For related horoscopes cf. No. L 102,XII,4 and No. L 102,XII,14 respectively.

No. L 114,V
Vettius Valens III,10 p. 147,11 to 23

TRANSLATION

For example let the sun be in degree 25;18 of Taurus, the moon 7;10 of Aquarius, Saturn degree 24 of Aries, Jupiter degree 4;25 of Taurus, Mars degree 22;53 of Cancer, Venus 28;16 of Gemini, Mercury degree 6 of Gemini, Horoscopos degree 27 of Capricorn. I took (the distance) from the degree of the moon to the next conjunction[1] which happened to be in Gemini degree 2;25; it makes 115. These (degrees) I subtract from the Horoscopos (and) it falls in the second degree of Libra. And the start is from here to the radiation of the maleficent (stars). For Saturn radiating at diameter and Mars at quartile about the same degrees caused the death; and Jupiter was turning away[2] and Venus was not well located and had no power to aid. The (subject of the) nativity lived 28 years 9 months.

COMMENTARY
Diagram: pl. 19

Fair agreement is reached for 114 May 13:

---

[1] Neugebauer [4] p. 225.

No. L 114,V
[1] Cf. for this concept Vettius Valens p. 145,32 ff.
[2] Cf. Introduction p. 13.

Text		Computed	Text—comp.
☉	♉ 25;18	♉ 20	+ 5
☾	♒ 7;10	♒ 2	+ 5
♄	♈ 24	♈ 19	+ 5
♃	♉ 4;25	♉ 0	+ 4;30
♂	♋ 22;53	♋ 15	+ 8
♀	♊ 28;16	♊ 27	+ 1
☿	♊ 6	♊ 1	+ 5
H	♑ 27	about 10 P.M.	

The next conjunction is placed at ♊ 2;25 as compared with ♉ 29 on May 22 according to computation.

It is of interest to note that the Stobart tables list the entry of Mars into the next sign for 13 days later,[3] of Venus into ♋ for the next day.[4]

A second horoscope from the same chapter is No. L 127,XI.

No. L 114,VII
Vettius Valens VII,5 p. 286,29 to 37

TRANSLATION

Sun (and) Mercury in Leo, moon in Libra, Saturn in Aries, Jupiter in Taurus, Mars (and) Venus in Virgo, Horoscopos in Capricorn, clima 2. In the 40th year he had a crisis: of moon 25, of Aries 15,[1] making 40. Or of its diameter, Aries 20,[2] of Libra 40,[3] two-thirds of which is 40. Thus the crisis was double. And at the same time I found Capricorn, the Horoscopos, 28[4] and of Jupiter (in ♉) in trine 12, making 40. And again of Capricorn 30[5] and of Taurus 22[6] and of period 8,[7] making 60, two-thirds of which is 40.

COMMENTARY
Diagram: pl. 19

For 114 July 26 one finds

Text		Computed
☉ }	♌	♌ 1
☿ }		♌ 15
☾	♎	♎ 22
♄	♈	♈ 24

---

[3] Neugebauer [4] p. 225: C_2 obv. III,16 Trajan 17 month 10 day 1 = 114 May 26 entry into ♌.
[4] C_2 obv. III,29: month 9 day 19 = May 14.

No. L 114,VII
[1] Aries is a house of Mars, which has the period 15 years.
[2] The rising time of Aries is $a_1 = 20$ for clima 2, System A.
[3] The rising time of Libra is $a_7 = 40$ for clima 2, System A.
[4] The rising time of Capricorn is $a_{10} = 28$ for clima 2, System A.
[5] Capricorn is a house of Saturn, whose period is 30 years.
[6] The rising time of Taurus has the value $a_2 = 22$ only for clima 5(!) (in both Systems). Everything would have been correct if the rising times of both ♑ and ♉ had been taken, since for clima 2 and System A holds

$$a_{10} + a_{12} = 28 + 24 = 52.$$

[7] Because of Venus whose house is Taurus.

Text		Computed
♃	♉	♉ 13
♂	} ♍	♍ 1
♀		♍ 17
H	♑	about 4 P.M.

This date is confirmed by an example [8] quoted in I,23 (p. 52, 16-29) where for Trajan 17 Mesore (XII) 2 = 114 July 26 the sun is said to be in ♌ 5, the moon in ♎ 18 and H in ♑ 24.

The Stobart tables seem to assign to Saturn a position in ♉ since the planet entered this sign in Trajan 17 month 11 day 4 = 114 June 28 [9] being back in ♈ the next New Year day [10] (= August 29). In all probability, however, the date of the reentry in ♈ should have been listed at the end of year 17.—According to the same tables Mercury would have entered ♌ only 4 days earlier.[11]

For related horoscopes *cf.* No. L 102 XII 14 and in particular No. L 133.

No. L 114,IX

Vettius Valens VII,2 p. 268,7 to 15

TRANSLATION

Sun (and) Venus in Libra, Saturn in Aries, Jupiter in Taurus, Mars [in Libra], Mercury in Virgo, moon in Sagittarius, Horoscopos in Libra; clima 3. In the 39th year he was banished; rising time of Aries 20, Saturn there, and the sun diametrically opposite 19, for both (♄ and ☉) were in opposition to their own exaltations. He had, then, crises in the earlier times; but for the comparison with the preceding nativity,[1] for it is a brother, we watched the 39th year. One must marvel then at the natural law [2] that it brought the periods to the same (result)[3] even though they were born in different climata.[4]

COMMENTARY

Diagram: pl. 20

We find the date 114 September 24 from

Text		Computed
☉	} ♎	♎ 0
♀		♎ 8
♄	♈	♈ 22
♃	♉	♉ 12
♂	[♎]	♎ 10
☿	♍	♍ 18
☾	♐	♐ 21
H	♎	about sunrise

One would expect both Horoscopos and Mars to be mentioned with sun and Venus; [5] and also the moon should come before Saturn.

The rising time of Aries, $a_1 = 20$ belongs to clima 3 only in System B. For System A one would need clima 2; *cf.* also No. L 120,XII.

For related horoscopes *cf.* No. L 74,XI.

No. L 114,XI

Vettius Valens V,10 p. 230,32 to p. 231,4

TRANSLATION

Sun, Mars (and) Mercury in Scorpio, Saturn in Aries, moon in Virgo, Jupiter in Taurus, Venus in Libra, Horoscopos in Sagittarius. In the 42nd year, because of a woman, quarrels and confusion and notoriety and in the 44th (year) violent death of a slave and crisis of (his) father and accusation of ignoble descent and of rape; but he got help and gifts from friends. He got in trouble with the law about the writings(?)[1] and he went through punishment and fraud [2] and false accusation and was harrassed by treachery and was physically ill. Thus each paradosis had its own influence and likewise each locus.

COMMENTARY

Diagram: pl. 20

The date is determined as 114 November 10 from

Text		Computed
☉	} ♏	♏ 17
♂		♏ 11
☿		♏ 18
♄	♈	♈ 19
☾	♍	♍ 21
♃	♉	♉ 11
♀	♎	♎ 4
H	♐	about 8 A.M.

For related horoscopes *cf.* No. L 102,XII,4.

---

[8] No. 14 in the list on p. 180.

[9] Neugebauer [4] p. 225: C_3 obv. III,7.

[10] C_3 obv. IV,20.

[11] C_3 obv. IV,16: Trajan 17 month 11 day 28 = 114 July 22.

No. L 114.IX

[1] *Cf.* No. L 120,XII.

[2] φύσις.

[3] The brother was banished in his 33rd year; as Kroll correctly noticed he is 6 years younger and this is confirmed by our dating of the two horoscopes.

[4] The younger brother was born in clima 2.

[5] According to the Stobart tables (Neugebauer [4] p. 225) Venus is moving retrograde, entering ♍ two days later: C_3 obv. V,6 Trajan 18 month 1 day 29 = 114 Sept. 26.

No. L 114,XI

[1] Perhaps religious writings (*cf.* p. 199,34) or legal documents; *cf.* *CCAG* 5,2 p. 143, 144.

[2] For ἐπίθεσις *cf.* the remarks in *CCAG* 8,1 p. 161.

No. L 115,II

Vettius Valens III,13 p. 154,10 to 15

TRANSLATION

Sun in 29 degrees of Aquarius, moon in the beginning of Aries, Horoscopos in 18 degrees of Capricorn, (preceding) conjunction in 26 degrees of Aquarius, ascending node in 16 degrees of Scorpio. I counted upwards [1] from the degree of the conjunction (♒ 26) to the degree of the Horoscopos (♑ 18); it makes 38. These downward from the degrees of the Horoscopos.[2] It totals for the first clima approximately 33 years. He lived 32 years five months. The place of the conception did not have the start; for the full moon of the conception and the ascending node happened to be at the same degrees. Therefore, he had a dangerous birth and a violent end.

COMMENTARY

Diagram: pl. 20

The date of this horoscope is 115 February 15:

Text		Computed	Text—comp.
☉	♒ 29	♒ 25;30	+ 3;30
☾	beg. of ♈	♓ 26	+ 4
H	♑ 18	about 4 A.M.	
prec. conj.	♒ 26	Feb. 12: ♒ 22	+ 4
☊	♏ 16	♏ 11	+ 5

The "full moon of the conception" is the full moon which occurs about 114 June 6, the sun being in ♊ 13 and the moon in ♐ 13. The ascending node is then at about ♐ 5;45 hence at a longitude which is reached by the moon just before or during the day of full moon. Thus the statement of the text that full moon and node coincide is closely justified as far as time is concerned but about 7° wrong with respect to the longitudes.

Two related, and incomplete, horoscopes are No. L 74,IV and No. L 75,1.

[1] That is in the direction of the daily rotation.

[2] The text is here confused; *cf.* the procedure of No. L 74,IV.

No. L 115,XII

Vettius Valens II,41, p. 129,29 to 33

(= CCAG 5,2 p. 121,8 to 12)

TRANSLATION

Sun in Capricorn, moon in Libra, Saturn in Taurus, Jupiter in Gemini, Mars (and) Horoscopos in Cancer, Venus in Aquarius, Mercury in Sagittarius, the Lot of Fortune in Libra.[1] In this sign the moon was in inferior aspect to Mars (in ♋) which is [2] in opposition to the sun (in ♑). The (8th) Locus of Death [3] was in Taurus, and Saturn is there. This person was killed by wild beasts.

COMMENTARY

Diagram: pl. 20

All conditions of this horoscope are satisfied for 115 December 26:

Text		Computed
☉	♑	♑ 4
☾	♎	♎ 12
♄	♉	♉ 2
♃	♊	♊ 13
♂	♋	♋ 6
H	♋	sunset
♀	♒	♒ 13
☿	♐	♐ 16

For related horoscopes see No. L 65,V.

No. L 115,XII

[1] From $\lambda_F = \lambda_H - (\lambda_☾ - \lambda_☉)$ = ♋ + 3° = ♎.

[2] Kroll p. 129,32 incorrectly ἐναντιουμένη instead of -μένου (CCAG).

[3] Counted from the Lot of Fortune (in ♎).

No. L 116

Vettius Valens II,36 p. 113,19-24

TRANSLATION

Sun in Aquarius, moon in Virgo, Saturn in Taurus, Jupiter (and) Horoscopos in Gemini, Mars in Cancer, Venus in Pisces,[1] Mercury in Capricorn, the Lot of Fortune in Capricorn,[2] the Daimon in Scorpio. To these (lots) the maleficent (stars ♂ and ♄) were in opposition. This person was effeminate and he had unmentionable vices, for Capricorn is lascivious and its ruler (Saturn [3]) was in Taurus, the sign (which indicates the kind of) weakness, and Scorpio indicates the kind of lasciviousness.

COMMENTARY

Diagram: pl. 20

For 116 January 21 one finds

Text		Computed
☉	♒	♒ 0
☾	♍	♍ 23
♄	♉	♉ 1
♃	♊	♊ 10
H	♊	about 2 P.M.
♂	♋	♊ 27(!)
♀	♓	♓ 13
☿	♑	♑ 6

Mars is close to the second station. According to the Stobart tables it had entered ♋ four months earlier (C_2 rev. I,9: Trajan 19 month 1 day 22 = 115 Sep-

No. L 116

[1] Variant (incorrectly): Aquarius.

[2] From $\lambda_F = \lambda_H + (\lambda_☾ - \lambda_☉)$ = ♊ − 5° = ♑.

[3] Capricorn is a house of Saturn.

tember 20⁴). Mercury, according to the same tables, reached ♑ only 5 days earlier.⁵

For related horoscopes *cf.* No. L 83.

No. L 117,VI

Vettius Valens VII,2 p. 269,11 to 28

TRANSLATION

Sun, Jupiter (and) Horoscopos in Cancer, moon in Sagittarius, Saturn in Gemini, Mars in Taurus, Venus (and) Mercury in Leo; clima 3. After serving in a distinguished army post he was involved in an accusation in the 38th year and lost his standing. For there was operative the rising time of Taurus 23 and the period of Mars 15, making 38, and also the side of the hexagon of Saturn (30) and Venus (8), 38. Then from the 37th year he had to expect enmities and oppositions; for the moon indicated the 25 and Sagittarius the 12,¹ making 37, Saturn being in opposition. Further Cancer 25² and Jupiter 12 from which he got some small assistance. In the 39th year he was accused and bound and had no power to set things straight, for Mercury being in Leo allotted the 20 and for the sign 19,³ [and] being in inferior aspect to the (star) effecting the accusation (♄), was the weaker. And besides, Saturn was about to begin another reverse; rising time of Gemini 28 and for Sagittarius 12,¹ making 40. Then at this time he was abroad and was betrayed by a woman by writings⁴ and became poor and was troubled about slaves, some of them by alienation, and some of them by loss and penalty, and he himself was physically ill.

COMMENTARY

Diagram: pl. 20

The date is found to be 117 June 30:

Text		Computed
☉		♋ 6
♃ } ♋		♋ 27
H		sunrise
☾	♐	♐ 19
♄	♊	♊ 1
♂	♉	♉ 25
♀ } ♌		♌ 8
☿		♌ 3

⁴ *Cf.* Neugebauer [4] p. 226.
⁵ C₁ rev. I,31: Trajan 19 month 5 day 20 = 116 January 16.

No. L 117,VI
¹ The moon is as usual (*cf.* p. 10) given the period of 25 years, Sagittarius, a house of Jupiter, the period of this planet, i.e., 12.
² Cancer is the house of the moon, whose period is 25.
³ Mercury has the period 20 and Leo, the house of the sun, its period of 19.
⁴ Perhaps indicating a lawsuit.

The locality is given as clima 3 and the following rising times are quoted

$$a_2 = 23 \qquad a_3 = 28.$$

The first value is only found, however, in clima 4, the second for clima 2 (Syst. A).

According to the Stobart tables Jupiter is to enter ♌ seven days later,⁵ and Mars is only two days from entering ♊,⁶ while Mercury had moved into ♌ two days before the date of the horoscope.⁷

For related horoscopes *cf.* No. L 74,XI.

No. L 117,X

Aelius Aristides, Sacred Discourses IV,26¹

TRANSLATION

[Description of a dream in which Plato appears to Aristides.] This dream came to me in Smyrna, and the other a little while before this in Pergamum. And he who explained it credited to Jupiter the dream and the manifest solicitude of the gods; for (he said) the star of Jupiter was cleaving the middle degree of Midheaven when I was born; and moreover the astronomers say that Leo was then at Midheaven and the star of Jupiter in Leo, in quartile to the right to Mercury, and both (Leo and Jupiter) in morning rising.

COMMENTARY

All that this very fragmentary horoscope contains is the following data: Jupiter was exactly culminating; the sign in which this planet was located is Leo; Mercury was in quartile to Jupiter three signs ahead, thus in Scorpio. Both Jupiter and Leo were in morning rising, thus not far to the east of the sun.

The last detailed discussion of this horoscope is to be found in Boulanger's large work on Aristides (1923). The statement² that Aristides "nous donne en bonne et due forme son thème astrologique complet" is a somewhat surprising description of our text. Boulanger is also wrong in interpreting "ἑῷους" as "l'observation est faite à l'aurore." Since the concept "morning" has nothing to do with individual planets, it is clear that we are dealing with the technical terminology of "morning rising" which indicates a position to the west of the sun. Consequently, the sun must be in ♍ or, at the most, in ♎ (*cf.* fig. 14). On the other hand, Mercury in ♏ requires, because of its limited elongation, a solar longitude of at least ♎ 10 or a distance from

⁵ Neugebauer [4] p. 226: C₁ rev. II,14 ♃ entering ♌ in Hadrian 1 month 11 day 13 = 117 July 7.
⁶ C₁ rev. II,23: month 11 day 8 = July 2.
⁷ C₁ rev. III,21: month 11 day 4 = June 28.

No. L 117,X
¹ *Aristides*, ed. Keil, vol. II p. 440, 21-28.
² Boulanger, *Aristides*, p. 465.

Leo of at least 40°. It requires some poetical license to speak still of morning rising at so great a distance from the sun.

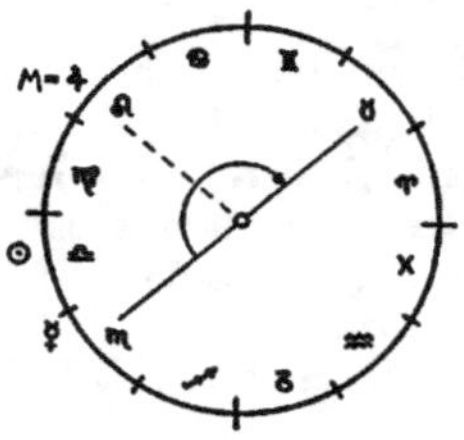

FIGURE 14.

The data

☉ in ♎ ♃ in ♌ = M ☿ in ♏ = H

alone would not suffice to determine the date of a horoscope. But we know from Suidas that Aristides died under Commodus (180 to 192) and from Philostratus that he reached an age of 60 or 70 years.[3] Thus he was born between about 110 and 132. Since Jupiter is in Leo during this interval only in 117 and in 129, the only question remaining is to decide between these two possibilities. The position of the sun requires a day in October and it is obvious that we must assume ☿ at the very beginning of ♏ in order to keep the distance of the sun from ♌ as small as possible. These conditions are satisfied in both years

	117 Oct. 8	129 Oct. 16
☉	♎ 14	♎ 22
♃ = M	♌ 17	♌ 22
☿	♏ 0	♏ 1
H[4]	♏ 9	♏ 13

but since in the first case Jupiter is 57° distant from the sun, in the second case 60° (and the end of Leo 44° and 52° respectively), the year 117 seems to be the preferable date by a very narrow margin. It is clear that so small a difference, based on so slender and inaccurate elements, cannot be considered as a decisive astronomical argument in favor of either one of these two dates. According to Boulanger the year 117 agrees better than 129 with the other biographical data of Aristides.

BIBLIOGRAPHICAL NOTE

The horoscope in Aristides' text was first utilized for the biography of the author by J. Masson in the introduction " De Aristidis vita collectanea historica " which was printed in vol. I[5] of *Aelii Aristidis . . . opera omnia*, edited by Samuel Jebb, Oxford, 1722. Masson consulted Edm. Halley about the possible date and Halley gave him on the basis of his own tables the years 117 and 129, roughly October since Mercury in Scorpio required a position of the sun in Scorpio or Sagittarius.[6] Masson then preferred 129 on biographical grounds while later investigators, e.g. Lenormant,[7] suggested 117. W. Schmid in *Pauly-Wissowa, RE* 2,1, col. 886 (No. 24) adopted 129 again but changed the month to " March or April " which is, of course, astronomically excluded.

[3] Boulanger, p. 465.
[4] Computed for clima 5.
[5] Signature g 2, § IV.

No. L 117,XI

Vettius Valens II,36 p. 113,25-31

TRANSLATION

Sun (and) Venus in Sagittarius,[1] moon in Cancer, Saturn in Gemini, Jupiter (and) Mars in Leo, Mercury in Scorpio, Horoscopos in Capricorn. The Lot of Fortune in Leo,[2] the Daimon in Gemini. Saturn located in this (sign) caused him to be castrated. For the ruler (of Gemini [3]), Mercury, was in Scorpio, which indicates the part, and the sun in Sagittarius indicates the region about the groin.

COMMENTARY

Diagram: pl. 21

The date is 117 November 30:

	Text	Computed
☉	♐	♐ 8
♀	♐	♐ 28
☾	♋	♋ 15
♄	♊	♊ 1
♃	♌	♌ 23
♂	♌	♌ 16
☿	♏	♏ 16
H	♑	about 8 A.M.

For related horoscopes *cf.* No. L 83.

No. L 118

Vettius Valens II,36 p. 112,5 to 10; VII,2 p. 268,29 to 269,2; VII,5 p. 287,11 to 16; CCAG 2 p. 168,24 to p. 169,8

TRANSLATION

First Version (Vettius Valens II,36)

Sun, Venus (and) Mars in Sagittarius, moon in Libra, Saturn in Cancer (error for Gemini [1]), Jupiter

[6] No reason is given for excluding Libra.
[7] *Recherches pour servir à l'histoire de l'Égypte* (1823) pp. 254-259.

No. L 117,XI

[1] Erroneously omitted in the manuscript V[a].
[2] From $\Lambda_F = \lambda_H + (\lambda_{☾} - \lambda_{☉})$ = ♑ − 5° = ♌.
[3] A house of Mercury.

No. L 118

[1] Correct in the second and the third version, again wrong (Aquarius), in the fourth version.

in Virgo, Mercury in Scorpio, Horoscopos in Capricorn, the Lot (of Fortune) in Scorpio.[2] The genitals were injured (because) the ruler of Scorpio (♂ [3]) was in Sagittarius. This person was baldheaded and disabled in the penis. But Jupiter, the ruler of the Daimon (♓ [4]) was found in the (9th) locus of the god and effected his recovery through the god; and he became a seer.[5]

Second Version (Vettius Valens VII,2)

Sun, Mars (and) Venus in Sagittarius, moon in Libra, Saturn in Gemini, Jupiter in Virgo, Mercury in Scorpio, Horoscopos in Capricorn. In the 19th year his father's violent death, and he himself was afflicted by eye trouble. In the same year he went abroad and was in danger at sea. For operative was the period of the sun (19 years) with Mars with it (in ♐) and Saturn opposite (in ♊). And in the 20th year he recovered, (following an) oracle of the god by treatment and ointment. For at that time Saturn (in ♊) was operative with Gemini giving 20,[6] wherefore he suffered many ills. And Virgo indicated 20 with Jupiter in it; or of Jupiter 12 and of Venus (in ♐) in quartile 8, making 20. Thus the period belonged to many stars, but the powerful ones were those operative. [Other general considerations follow.]

Third Version (Vettius Valens VII,5)

[Same positions as before except error at the end: "Horoscopos (and) Mercury in Capricorn" instead of "Mercury [in Scorpio], Horoscopos in Capricorn"]; clima 6. The 36th year midway was critical. Of Gemini 27 (years) 6 months and of Venus 8 (years) makes 35 (years) 6 months. Of the sun 19 (years) [and of Sagittarius 35 (years) 6 months] makes 54 (years) 6 months, two-thirds of which makes 36 (years) 4 months. There the beneficent (stars) shared (in power).

Fourth Version[7] *(CCAG 2 p. 168,24 to p. 169,8)*

Sun, Mars (and) Venus in Sagittarius, moon in Libra, Saturn in [Gemini [8]], Jupiter in Virgo, Mercury in Scorpio, Horoscopos in Capricorn. In the 19th year the father died a violent death. For in this prognostication the period of the sun (19 years) was effective through its being with Mars. For Mars with the sun indicates a bad death for the father. And they also cause injury to the eye and indicate danger. All this happened: his father died by violence in the 19th year and he himself was injured in the eyes and going abroad had danger at sea. For the Ship Argo is in Sagittarius.[9] [After some further discussion:] But in the 20th year by divine providence he recovered his eyesight. For Venus was present with the sun (in ♐) and Jupiter (in ♍) was in quartile aspect; and the 20 years were indicated by Virgo, because of its ruler Mercury, in which (sign, ♍) Jupiter was in aspect to the sun. And besides Jupiter (indicated) 12 and Venus, in quartile, 8, and both portended eye injury. [Followed by additional arguments.]

COMMENTARY

Diagram: pl. 21

The date of this horoscope is 118 November 26

Text		Computed
☉	♐	♐ 4
♂	♐	♐ 29
♀	♐	♐ 19
☾	♎	♎ 4
♄	♊	♊ 20
♃	♍	♍ 20
☿	♏	♏ 14
H	♑	about 8 A.M.

For related horoscopes *cf.* for the first version No. L 83, for the second version No. L 74 XI and for the third version No. L 102 XII 14 respectively. The subject is one of six survivors of a shipwreck in A.D. 155; *cf.* No. L 133. The same date, Hadrian 3 Athyr (III) 30 = 118 November 26, is quoted as an example[10] in IX,19 (p. 363, 5 ff.) giving for the sun ♏(!)7, for the moon ♍ 30 as longitude, and as time the 4th hour of day, which is possible if the Horoscopos was near the end of ♑.

The rising times quoted for clima 6

$$\alpha_3 = 27;30 \qquad [\alpha_9 = 35;30]$$

disagree with both systems since

System A	System B
$\alpha_3 = 27;6,40$	$\alpha_3 = 25;40$
$\alpha_9 = 32;53,20$	$\alpha_9 = 34;20.$

The fact that the fourth version with the three preceding ones belongs obviously to Vettius Valens,[11] though incorporated into the work of an anonymous, disproves Kroll's hypothesis[12] of an origin from the collection of horoscopes by Antigonus of Nicaea.

[2] From $\Lambda_F = \lambda_H + (\lambda_{☾} - \lambda_{☉}) = ♑ - 2° = ♏$.
[3] Whose house is ♏.
[4] A house of Jupiter.
[5] *Cf.* Cumont, *EA* p. 159, note 3.
[6] Gemini is a house of Mercury whose period is 20.
[7] From *De planetis* of an anonymous author who frequently quotes Vettius Valens. *Cf.* Commentary.
[8] Kroll here writes "Aquarius(?)." *Cf.* the preceding versions.

[9] *Cf.* Gundel, *HT* p. 202 and p. 217 f.
[10] Example No. 20 on p. 180.
[11] The same holds for the preceding horoscope (No. L 122,XII, second version).
[12] *CCAG* 2, p. 168 note 1.

No. L 120,II

Vettius Valens II,30 (p. 101,22 to 33), IV,11 (p. 174,8 to 15), V,9 (p. 223,1 to 33), and VII,5 (p. 287,1 to 10)

TRANSLATION

First Version (II,30, p. 101,22-33)

The title of this section is "on the predecease of the parents according to an example." The first paragraph describes a method for the prediction of an earlier death of father or mother on the basis of the relation between sun, moon, Saturn, and Venus. A short "alternate method" considers the "Lots" of father and mother. Thereafter we read again:

"Alternate (method). Take the days from the rising of Sirius to the day of birth; divide the total by 12 [1] and count the remainder one by one beginning from the (longitude of the) moon. If it falls in a male sign the father will die earlier, if in a female, the mother. In the present [2] nativity on Mekheir (VI) 13 there are 203 days [3] from Epiphi (XI) 25 [4] to the 13th (of Mekheir). Subtract 16 times 12 and the remainder is 11. Counting this from the moon in ♏ it fell in ♍, a female sign. But Mars was there too. Therefore, the earlier death referred to the mother.

{Alternate (method)}.[5] For example let the nativity be: Sun (and) Mercury in Aquarius, moon in Scorpio, Saturn in Cancer, Jupiter in Libra, Venus in Capricorn, Mars and Horoscopos in Virgo. ..." [6]

Second Version (IV,11, p. 174,8-15 = CCAG 5,2 p. 88,27-34)

[Following the enumeration of all positions as in the first version]: We investigate the year 35.[7] I subtracted twice twelve, which is 24, remainder 11. Then we count the 11 (signs) from whatever star to whatever star it overtakes.[8] We then find the 11 signs from the Horoscopos and Mars (in ♍) to Saturn in Cancer, and again from the moon (in ♏) to Mars (in ♍), and from Venus (in ♑) to the moon (in ♏). All these yieldings are effective in the 35th year. . . .

Third Version (V,9 p. 223,1-33)

[Same positions as before.[9] Then follows a count which begins:] "there will be from Mars and Horoscopos (in ♍) two (signs) to Jupiter (in ♎), and from Jupiter (in ♎) to the moon (in ♏) two, and from Venus (in ♑) to Mercury and sun (in ♒) two. (Add) 2 to 4, to 6, to 8, to 10 together" and this process is similarly continued for all distances up to 12 ending with the total 12 + 24 + 36 + 48 + 60. "And one will find the number by addition as far as one wishes."

Fourth Version (VII,5, p. 287,1-10)

[Same arrangement as in first and second version; followed by] "Clima 7. In the 35th year he had a critical period, because the period of Mars (in ♍, the Horoscopos) was operative: 15 years and 20 for Virgo,[10] makes 35; but also the 8 of Venus (in ♑) and the rising time 27 of Capricorn makes 35. And again 30 for Saturn (in ♋) opposite (to ♀) and 32;30 for Cancer and 8 of Venus makes 70 (years) and 6 months, the half of which makes 35 (years) 3 months. Besides also Jupiter (in ♎) and Saturn (in ♋) shared the period, for Libra 42 (years) 6 months, for Cancer 27 (years) 6 months makes 70, of which the half is 35."

COMMENTARY

Diagram: pl. 21

The positions given in these four versions lead to the date of 120 February 8:

Text		Computed
☉ ⎱	♒	♒ 18
☿ ⎰		♒ 18
☾	♏	♏ 3
♄	♋	♋ 0 [11]
♃	♎	♎ 23
♀	♑	♑ 25
♂ ⎱	♍	♍ 22
H ⎰		about 7 P.M.

This result is confirmed by an example [12] using the above configuration but now giving explicitly its date as Hadrian 4 Mekheir (VI) 13 which is the exact equivalent of 120 February 8.[13] In V,7 (p. 215,30 to p. 216,6) we read:

"For example let it be year 4 of Hadrian, Mekheir 13, first hour of night, to investigate (the situation) for

---

[1] Literally: subtract multiples of 12.

[2] Kroll remarks there "the scribe was asleep when he copied this from his source." Actually, however, Kroll was asleep when he did not recognize that the example referred to follows right after the explanation.

[3] Kroll complains that the computation mixes Alexandrian with Egyptian dates; actually both dates use the Alexandrian calendar.

[4] Kroll wrote 20, relegating the correct number 25, which is given by S, to the apparatus. Epiphi 25 (= July 19) as date of the rising of Sirius is also attested by Hephaestion XXIII (p. 91,31 Engelbrecht) and by the Demotic P. Cairo 31222 (cf. Hughes [1] p. 257b). July 20 is assumed in CCAG 6, p. 154,30 (Antiochus) and in the Geoponica (cf. Bidez-Cumont, Mages hellénisés, II p. 179,7).

[5] This heading should have been omitted.

[6] The final sentences give additional reasons for the earlier death of the mother.

[7] For the reason cf. the 4th version.

[8] The procedure is simply this: reduce the given number of years modulo 12. The resulting number (here 11) is then considered as a number of signs and we look for all pairs of planets which are so many signs distant from each other.

[9] But mentioning Mars before Venus.

[10] Virgo is a house of Mercury, whose period is 20.

[11] Saturn is retrograde, entering ♊ from ♋.

[12] This is No. 26 in the table on p. 180.

[13] This can be seen, e.g., by using Schram's tables.

the 20th year of Antoninus, Phaophi 10.[14] They (the days from Mekheir 13 to Phaophi 10) are 243 [15] and make altogether 432.[16] I subtract 360, remainder 72. These I count from the Horoscopos (in) Virgo; it falls in Leo.[17] (Thus) the day (under investigation is) in apoklima (in ♌). The ruler (of ♌, the sun) was according to the nativity in opposition to the day (under investigation) for the sun at the time of birth was in Aquarius, and the one entering into (chronocratorship)[18] was Mars, and the moon in Capricorn was not in aspect."

All data agree with the previous versions, except the last, according to which the moon was in ♑. But for the given date the moon was in ♏ and Venus was in ♑.

The same day, Mekheir (VI) 13 = 120 February 8 is quoted in three other examples.[19] In I,5 (p. 20, 9 ff.) we are given the positions of sun and moon as ♒ 22 and ♏ 7 respectively; in I,10 (p. 26,21 f.) the hour is given as the first hour of night; in I,20 (p 36,15 ff.) the positions of Venus and Mercury are computed according to an approximate method, leading to ♑ 16 and ♒ 25 respectively, whereas one should expect about ♑ 29 and ♒ 22.

In the fourth version one can observe some confusion in the text. According to it we should have for the 7th(!) clima the rising times:

$$a_4 = 32;30 \text{ or } 27;30(!)$$
$$a_7 = 42;30 \quad a_{10} = 27.$$

The only correct figure is $a_{10} = 27$ for clima 7, System A. All the other numbers are taken from different places of the schemes of rising times:

$$32;30 = a_4 \quad \text{clima 1, System B}$$
$$42;30 = a_7 \quad \text{clima 6, System B}$$
$$27;30 = a_{10} \quad \text{clima 1, System B.}$$

It is difficult to see how these figures could get into the text.

For related horoscopes of the fourth version cf. No. L 102,XII,14 and especially No. L 133.

---

[14] That is 156 October 7 or in the 36th year of the person in question. In the second and fourth version the 35th year is considered critical.

[15] Actually 242 only (at least for an ordinary year). Thus one has to count both end points (or to assume a leap year).

[16] According to the rule which is stated at the beginning of this chapter (V,7) one has to multiply the number of years (in our case 36) by 5¼ and add the result (here 189) to the number of days (here 243 + 189 = 432). This is obviously equivalent to saying: determine the number of days between the two extremal dates and reduce the result modulo 360. In this way the remainder for each year is 5¼ days.

[17] Counting in the order of the signs, beginning with ♍ = 1, to 72, each unit representing one zodiacal sign, one ends in ♌ because 72 ≡ 0 mod. 12.

[18] For the concept of ἐπέμβασις "enter into (chronocratorship)" see Vettius Valens IV,2 (p. 159,2), VI,4 (p. 251,8 ff.), and p. 182,15.

[19] Nos. 23 to 25 in the list on p. 180.

No. L 120,V

Vettius Valens V,10 p. 230,18 to 21

TRANSLATION

Sun (and) Venus in Taurus, moon in Aries, Saturn in Cancer, Jupiter [in Libra], Mars in Virgo, Mercury in Gemini, Horoscopos in Sagittarius. In the 36th year he had disputes and (legal) affairs on account of his wife and hostility of friends.

COMMENTARY

Diagram: pl. 21

For 120 May 12 one finds complete agreement if one assumes omission of the position for Jupiter. Otherwise no plausible date can be assigned to this horoscope.

Text		Computed
☉ }	♉	♉ 20
♀ }		♉ 21
☾	♈	♈ 23
♄	♋	♋ 4
♃	[♎]	♎ 16
♂	♍	♍ 9
☿	♊	♊ 13
H	♐	about 8 P.M.

For related horoscopes cf. No. L 102,XII,4.

No. L 120,IX

Vettius Valens VII,5 p. 290,6 to 14

TRANSLATION

For example let sun, Jupiter (and) Mercury be in Libra, moon in Taurus, Saturn in Cancer, Mars in Sagittarius, Venus (and) Horoscopos in Scorpio, clima 2. This person in the course of the 40th year was condemned to exile. The 40 years were indicated by the dominant aspect (quartile) of Saturn (in ♋) to Libra: 32 degrees of Cancer [1] and 8 of Libra,[2] making 40. And the diameter from moon (in ♉) to Venus (in ♏): of Taurus 25,[3] of Scorpio 15,[4] making 40. Thus the forecast was effective by a female (person) and prospect of profit. And the approaching contact of the moon with Mars:[5] of moon 25 and of Mars 15, making 40.

---

[1] The rising time of Cancer is $a_4 = 32°$ for clima 2, System A.

[2] Libra is a house of Venus, whose period is 8.

[3] The moon (period 25) is in ♉.

[4] Scorpio is a house of Mars, whose period is 15.

[5] As is explained by Vettius Valens (p. 290,20) this means that the moon (♉) is in contact with the diameter of Mars (♊ — ♐).

COMMENTARY

Diagram: pl. 21

For 120 September 28 we have

	Text	Computed
☉	♎	♎ 4
♃	♎	♎ 29
☿	♎	♎ 2
☾	♉	♉ 18
♄	♋	♋ 19
♂	♐	♐ 0
♀	♏	♏ 11
H	♏	about 8 A.M.

For related horoscopes *cf.* No. L 102,XII,14.

No. L 120,XII

Vettius Valens VII,2 p. 267,28 to 268,4

TRANSLATION

As for example: sun, moon (and) Mercury in Sagittarius, Saturn in Cancer, Jupiter (and) Horoscopos in Scorpio, Mars in Capricorn, Venus in Aquarius, Lot of Fortune in Scorpio;[1] second clima. In the 33rd year he was exiled, for the rising time of the sign in which Saturn was (♋), being in opposition to Mars, and the sextile position of the moon (in ♐) to Venus (in ♒) indicated [treachery] by a female person at 33 years.[2] And in the 27th year he was in danger because of Capricorn and in the 30th year and in the 40th year bodily illness of eyes and feet; for the period of Cancer (is) 25[3] and of Mars, in opposition (to ♋), 15. And until the 42nd year the period of Mars (15) and the rising time of Capricorn (27) were operative, in which there were many causes (of trouble).

COMMENTARY

Diagram: pl. 21

The date of this horoscope is 120 December 8, six years later than the horoscope No. L 114,IX of an older brother.

	Text	Computed
☉	♐	♐ 16
☾	♐	♐ 10
☿	♐	♐ 26
♄	♋	♋ 26
♃	♏	♏ 14
H	♏	about 4 A.M.
♂	♑	♑ 25
♀	♒	♒ 4

[1] Sun and moon being in conjunction bring the Lot of Fortune in the same sign as the Horoscopos.

[2] The rising time of ♋ and ♐ is in both cases 33, for clima 2 and System B.

[3] The period of the moon because ♋ is the house of the moon.

As rising times occur

$$a_4 = 33 \qquad a_{10} = 27$$

both correct for clima 2 and System B. This is in agreement with the fact that also No. 114,IX quotes a value from System B.

Related horoscopes are listed with No. L 74,XI.

No. L 121

Vettius Valens V,1, p. 210,8 to 17

TRANSLATION

For example let sun (and) Jupiter be in Scorpio, moon, Mercury, Venus (and) Horoscopos in Libra, Saturn in Leo, Mars in Capricorn. Both the maleficent (stars) were in sextile to sun and moon (respectively). If then the luminaries had happened to be without the company of beneficent (stars) it would have indicated confinement; but now the scheme was most promising and honorable. This man at 35 years happened to be a soldier, and he began among prisoners and in prison, and was loved by a woman in the prison through whom he got out of a troublesome accusation, and he made his way out and escaped the danger; and at the same time he captured and bound a runaway slave.

COMMENTARY

Diagram: pl. 22

A.D. 121 is the earliest date, within historical possibilities, such that Saturn is in ♌, Jupiter in ♏. These conditions are also satisfied in A.D. 180, 239, and 298 but in all three cases Mercury and Mars do not occupy the positions given in the text. For 121 October 27 one obtains

	Text	Computed
☉	♏	♏ 3
♃	♏	♏ 29
☾	♎	♎ 8
☿	♎	♎ 30
♀	♎	♎ 13
H	♎	about 4 A.M.
♄	♌	♌ 3
♂	♑	♍(!) 6

Thus we find good agreement except for Mars. Because for a horoscope from Vettius Valens the later dates are excluded (except, perhaps 180) one must assume that the position of Mars is mistakenly given as Capricorn, instead of Virgo, probably on the basis of a misreading of the two symbols, which look often very similar (*cf.* fig. 1 p. 1).

No. L 122,I,22

Vettius Valens V,10 p. 230,22 to 31

TRANSLATION

Sun (and) Venus in Aquarius, moon (and) Jupiter in Sagittarius, Saturn in Leo, Mercury in Capricorn, Mars (and) Horoscopos in Libra. In the 35th year he was in danger of being imprisoned for sedition and violence; for the moon had sextile scheme to Mars and Mars itself having taken over from it (the moon) and handed over to Saturn.[1] These successions caused hardship and disturbances. But Jupiter was strong, happening to be in company with the moon (in ♐) and taking over from sun and Venus (in ♒) the same year, which were located in the (5th) locus concerned with friendship. It (Jupiter) then happening to be in the (3rd) locus concerned with travel, portended travel, which was voluntary but risky, and help and cooperation of friends.

COMMENTARY

Diagram: pl. 22

One obtains perfect agreement for 122 January 22:

Text		Computed
☉	♒	♒ 2
♀	♒	♒ 12
☾	♐	♐ 21
♃	♐	♐ 18
♄	♌	♌ 1
☿	♑	♑ 6
♂	♎	♎ 22
H	♎	about 10 P.M.

For related horoscopes cf. No. L 102,XII,4.

No. L 122,I,30

Vettius Valens VII,5 p. 287,24 to 29

TRANSLATION

Sun in Aquarius, moon in Aries, Saturn in Leo, Jupiter in Sagittarius, Mars in Libra, Venus (and) Mercury in Capricorn, Horoscopos in Pisces, clima 6. In the 33rd year he had a crisis. Of moon (in ♈) 25 and of Libra 8[1] makes 33. And of Saturn (in ♌) 30 and of sun (in ♒) 19 makes 49, two-thirds of which is 32 (years) 8 months. And also the rising time of Sagittarius was operative, Jupiter being located there, that is 33.[2]

---

[1] Because 35 ≡ 11 mod. 12 all such pairs of zodiacal signs are investigated where the second sign is No. 11, counting the first one as No. 1. Such pairs form the configuration of Sextile.

No. L 122,I,30

[1] Libra is a house of Venus whose period is 8.
[2] The rising time of Sagittarius is α₀ = 33 only in clima 7(!), System A, or in clima 2(!), System B. Exactly the same error occurs in No. L 162.

COMMENTARY

Diagram: pl. 22

For 122 January 30 one finds

Text		Computed
☉	♒	♒ 10
☾	♈	♈ 7
♄	♌	♌ 0
♃	♐	♐ 20
♂	♎	♎ 26
♀	♑	♒ 22(!!)
☿	♑	♑ 15
H	♓	about 8 A.M.

Related horoscopes are listed with No. L 102 XII 14 and particularly in No. L 133.

No. L 122,VI,12

Vettius Valens VII,4 p. 275,12 to 27

TRANSLATION

For example let sun (and) Mercury be in Gemini, moon in Aquarius, Saturn (and) Venus in Leo, Jupiter in Sagittarius, [Mars (and)] Horoscopos in Libra, clima 1. Suppose we investigate the 42nd year. I took the rising time of Libra 38;20,[1] which is 38 years 4 months and again the equal number of months (i.e.) 38, which are 3 years 2 months, making altogether 41 years 6 months. In this year fleeing from battle and falling from his horse as the enemy approached, and many killed and he himself wounded, he was mixed up with the rest of the fallen and, thought to be dead, escaped the danger and remained in the enemy's country until the 44th year, leading the campaign. Then I added to the previous years the 8 months of Venus on account of Libra [2] and 15 months of Mars, located there. Besides, the diameter of Saturn (in ♌) and moon (in ♒) was operative at that time: rising time of Leo 35 (years)[3] and of Saturn 57[4] months, and of Venus 8 months, and of moon 25 months and of Leo 19 months,[5] makes altogether 44 years 1 month.

COMMENTARY

Diagram: pl. 22

The data of the text are satisfied for 122 June 12:

Text		Computed
☉	♊	♊ 19
☿	♊	♊ 17
☾	♒	♒ 16

---

[1] Correct for clima 1, System A.
[2] Libra is the house of Venus.
[3] Correct for clima 1, System A.
[4] Saturn's "maximum period."
[5] Leo, as the house of the sun, obtains the latter's period.

Text		Computed
♄	♌	♌ 2
♀	♌	♌ 2
♃	♐	♐ 24
♂	♎	♎ 21
H	♎	about 2 P.M.

The battle in question took place 41 years 6 months later, thus at the end of A.D. 163, or in the beginning of 164. This could be the campaign conducted by Avidius Cassius against the Parthians, which ended in 166, again in agreement with our horoscope.[6]

No. L 122,VI,30

Vettius Valens VII,3 p. 274,4 to 13 = CCAG 5,2 p. 45,16 to 24

TRANSLATION

The stars share with each other when favorably located and of (the same, diurnal) sect. As: sun (and) Mercury in Cancer, moon (and) Mars in Libra, Saturn (and) Venus in Leo, Jupiter in Sagittarius, Horoscopos in Capricorn, clima 6, the Lot of Fortune in Sagittarius.[1] Jupiter, ruler of this,[2] was located in the Lot (of Fortune). It allotted the rising time of Sagittarius, 33 years,[3] and its own period, 12 years, making 45 years. Saturn, being in trine to this (locus) allotted 57 months of its own period,[4] making (a total of) nearly 50 years; in this (year) he died. And if one reckons also the months of the rising of Leo, 38,[5] and of the sun 19 months, it will total the same number of months.

COMMENTARY

Diagram: pl. 22

For 122 June 30 one finds

Text		Computed
☉	♋	♋ 6
☿	♋	♋ 18
☾	♎	♎ 17
♂	♎	♎ 26
♄	♌	♌ 3
♀	♌	♌ 21
♃		♐ 22
H	♑	about sunset

The Lot of Fortune is located in ♐ whereas we obtain from

$$\lambda_H \pm (\lambda_{☾} - \lambda_{\odot}) = ♑ \pm 3^s$$

♈ or ♎ and not ♐. The planetary positions exclude a change in the solar position by two signs; thus the positions of the moon and the Horoscopos (or both) are suspect. The moon's association with Mars makes an emendation at this place very implausible. Thus the only correction possible is H in ♓ or ♍ which makes the hour about 10 P.M. or 10 A.M. without altering the planetary positions. The lunar positions would remain essentially the same (about ♎ 19 or ♎ 13 respectively). A confusion between the symbols of ♑ and ♍ can easily happen (*cf.*, e.g., No. L 121 and fig. 1 p. 1).

The other examples in this chapter, VII,3, are:

p. 273,10-22	173 Febr. 3
23-32	159 July 18
p. 273,33-p. 274,3	162 Febr. 9

All were also published in *CCAG* 5,2 p. 44,26 to p. 45,24.

No. L 122,XII

Vettius Valens VII,2 p. 268,16 to 28 and CCAG 2, p. 168,4 to 23

TRANSLATION

First Version (Vettius Valens VII,2)

Sun, Mercury (and) Horoscopos in Sagittarius, moon in Cancer, Saturn in Leo, Jupiter in Capricorn, Mars in Aquarius, Venus in Scorpio. Clima 2. In the 34th year, death of wife (because of) the 19 of Leo (place of ♄) and the 15 of Scorpio[1] (place of Venus) or of Mars itself (with period 15). Both the maleficent stars hemmed in Venus. And in the 36th year by reason of the death of his wife being supposedly plotted, he was to go on trial, accused at the royal (court), but he fled. For the 36 years are the rising time of Leo and likewise of Scorpio, where Venus happened to be in inferior aspect[2] to Saturn. But the (next) period of time was to become more kindly in the 37th year, (because) of Jupiter (in ♑) 12 and of moon (in ♋), diametrically opposite, 25. Many other (facts) were significant for past and future but I have thought it necessary to set forth (only) those which I myself knew more exactly and which I happened upon.

Second Version (CCAG 2 p. 168,4 ff.)

Let the nativity be in clima 2. Sun, [Mercury,][3] (and) Horoscopos in Sagittarius, moon in Cancer, Saturn in

[6] *Cf. RE* II,2 col. 2379 f.

No. L 122,VI,30

[1] *Cf.* the Commentary.

[2] A house of Jupiter is ♐.

[3] For Sagittarius $a_s = 33$ holds only for clima 2(!) in System B, or clima 7(!) in System A.

[4] The large period of Saturn.

[5] For Leo $a_s = 38$ only for clima 5(!) (both Systems).

[1] Leo, the house of the sun, takes the period 19 of the sun and similarly Scorpio, a house of Mars, its period 15.

[2] Quartile.

[3] *CCAG* 2, p. 168,5 gives Taurus (confusion of symbols in the 14th century codex?) which is obviously wrong, if Venus is in Scorpio.

Leo, Jupiter in Capricorn, Mars [in Aquarius],[4] Venus in Scorpio. In the 34th year there occured the death of his wife; for the quartile of Saturn (in ♌) and Venus (in ♏), Saturn in dominant aspect as in this case, brings troubles and griefs because of women. But the time was propitious; for of Leo 19,[1] in which (sign) Saturn was, and of Scorpio, where Venus (was), 15;[1] altogether 34. For the periods were taken from the stars ruling Leo (☉) and Scorpio (♂). But when (the periods were taken from) the rising times: in the 36th year he was about to be tried on accusation of having plotted his wife's death, but he escaped; for Saturn in dominant aspect to Venus causes reverses and sorrows because of female persons. [After similar arguments]: and in the 37th year he enjoyed even greater kindness because of the diametrical aspect of moon (in ♋) and Jupiter (in ♑): moon 25 and Jupiter 12; altogether 37.

COMMENTARY

Diagram: pl. 22

The date of this horoscope is 122 December 4 with the following positions:

	Text	Computed
☉	♐	♐ 12
☿	♐	♐ 24
H	♐	sunrise
☾	♋	♋ 19
♄	♌	♌ 17
♃	♑	♑ 4
♂	♒	♒ 9
♀	♏	♎ 25(!)

Venus is morning star in greatest elongation. The rising times $\alpha_5 = \alpha_6 = 36$ are correct for clima 2 in both systems.

The first version belongs to a group of horoscopes of which No. L 74,XI is the oldest one. The second version is preserved in an anonymous *De Planetis*. The horoscope immediately following, No. L 118, is preserved in several versions: in a second version in Vettius Valens VII,2, in the fourth version in *De Planetis*.

No. L 123,I

Vettius Valens V,10 p. 231,5 to 31

TRANSLATION

Sun (and) Jupiter in Capricorn, moon (and) Saturn in Leo, Mars in Pisces, Venus (and) Horoscopos in Scorpio, Mercury in Sagittarius. The responsible places[1] were Pisces and Scorpio, in Scorpio Venus, in Pisces Mars. He was a dancer and in his 25th year, he was put in confinement in the course of a public riot but he was defended before the governor and released through the help of friends and the entreaty of the crowd and became more esteemed. For the paradosis of the year was from Saturn and moon (in ♌) to Mars and the responsible place (♓), and from Jupiter and sun in ♑) in the (3rd) locus, concerned with property, to Saturn and moon (in ♌) in Midheaven and the (10th) locus of activity. And besides it indicates according to the division by 4 or that by 8 from Saturn and moon (in ♌) to Venus and Horoscopos (in ♏) that there would be the riot and quarrelsomeness and rivalry through the affair, and (so does the paradosis) from Mercury to Mars and the responsible place. Therefore, all the stars were operative in the 2[5]th year. The nativity was precarious as regards loss of reputation and condemnation and danger of life. But Venus being found in the Horoscopos (♏) and in the responsible sign and Jupiter with the sun (in ♑), it (the nativity) had the best imaginable outcome and was successful in action. For the Lot of Fortune was in Aries[2] and the ruler of the exaltation of the nativity (♌),[3] the sun, was found in Midheaven (♑) with respect to the Lot (of Fortune in ♈) and Mars (in ♓) with respect to the Daimon (♊). Later, in the 32nd year he was deprived of honor and reputation and livelihood, and lived unhonored, because the Lot (of Fortune) happened to be in apoklima and Saturn being in Midheaven (in ♌) by sect[4] was in opposition to the (11th) locus of accomplishment[5] in Aquarius, his own house. Wherefore he was to blame for his own fall, having become a braggart and pretender. For the ruler of the Daimon (♊) and the place of intellect,[6] Mercury (in ♐), was in opposition to itself, that is Gemini.[7]

COMMENTARY

Diagram: pl. 23

We find as date of this horoscope 123 January 3:

	Text	Computed
☉	♑	♑ 12
♃	♑	♑ 10
☾	♌	♌ 19
♄	♌	♌ 16
♂	♓	♓ 1
♀	♏	♏ 26
H	♏	about 2 A.M.
☿	♐	♐ 17

For related horoscopes *cf.* No. L 102,XII,4.

---

[4] Omitted, *cf.* first version.

No. L 123,I

[1] *Cf.* for this concept Vettius Valens V,1.

[2] From $\Lambda_F = \lambda_H - (\lambda_{☾} - \lambda_{☉})$ = ♏ + 5° = ♈.

[3] For a night birth: ♉ from the moon has the same distance as ♌ from H = ♏.

[4] That is the actual M, not with respect to the Lot of Fortune(?)

[5] With reference to the Lot of Fortune (♈).

[6] *Cf.* Vettius Valens II,15 (p. 69, 15).

[7] We observe here an interesting type of astrological argu-

No. L 123,VII

Vettius Valens II,41 p. 128,31 to 37

TRANSLATION

For example let sun, Mars, (and) Venus be in Cancer, Saturn (and) Mercury in Leo, Jupiter in Aquarius, the moon in Pisces, the Horoscopos in Scorpio. The Lot of Fortune in Leo, the Locus of Death in Pisces; the moon was (also) there, and Saturn was at the Lot (of Fortune, in Leo, whose) ruler, the sun, was with Mars in Cancer, a wet sign.[1] This person died in the bath, choked (to death) in water. Mars (in ♋) was in opposition to the full moon (in Capricorn) and Saturn the ruler (of the full moon in Capricorn[2]), was turning away.[3] Therefore the violent death.

COMMENTARY

Diagram: pl. 23

This horoscope belongs to the same group as No. L 65,V. Its date is 123 July 2:

Text		Computed
☉		♋ 8
♂ } ♋		♋ 5
♀		♋ 4
♄ } ♌		♌ 16
☿		♌ 3
♃	♒	♒ 0
☾	♓	♓ 19
H	♏	about 3 P.M.

The Lot of Fortune fell in ♌ only if the Horoscopos was located in the last third of ♏ (hence the hour was about 3 P.M.):

$$\Lambda_F = \lambda_H + (\lambda_{\mathbb{C}} - \lambda_{\odot}) = \text{♏ } 25 + 8^s 11^\circ = \text{♌ } 6.$$

The ordinary counting would lead to ♋. The 8th sign from Leo is ♓, the Locus of Death. The nearest full moon, about 6 days earlier occurred in ♑, whose ruler, Saturn in ♌, was 5 signs away, thus not in aspect.

No. L 124

Vettius Valens VII,5 p. 280,34 to p. 281,20

TRANSLATION

For example let sun (and) moon be in Leo, Saturn in Virgo, Jupiter in Pisces, Mars in Sagittarius, Venus in Gemini, Mercury in Cancer, Horoscopos in Libra, clima 7. In the 37th year, on account of his wife, through whom he expected great profit, he had to stand a suit about inheritance and was worsted at the king's (court); however the time was not altogether injurious to him though (at the end) disappointing to his hope. The 37 years then were indicated by the triangular aspect of moon to Mars (in ♐), of moon 25, of Sagittarius 12,[1] making 37. And (the triangular aspect) of Jupiter to Mercury: of Jupiter 12, of Cancer 25.[2] Investigating as we have explained, we find the opposition of Mars and Venus thus operative for the 7th clima, as is also the side of the square of Saturn (in ♍): for Gemini 27[3] and of Venus (in ♊) 8, and of Virgo 20,[4] making 55, two-thirds of which is 36 (years) 8 months. And Pisces indicated the same (number).[5] But the maleficent (stars) were powerful. Since Venus was ruler of the Lot (of Fortune),[6] he had won the same suit in the 35th year but the (case) being appealed he was worsted (later on, in the 37th year). For the 35 indicated good (luck): of the rising time of Pisces (♃) 15[7] and for Gemini (♀) 20;[8] and again of Gemini 27[3] and of Venus 8, making 35; and again of Gemini 20[8] and of Mars 15, making 35; and again of Virgo (♄) 20[4] and of Mars 15, making 35. Thus both the beneficent and the maleficent (stars) were operative, and according as they were in bicorporeal signs he will often be in need.

COMMENTARY

Diagram: pl. 23

For 124 July 29 one finds

Text		Computed
☉ } ♌		♌ 4
☾		♌ 14
♄	♍	♌ 29(!)
♃	♓	♓ 7
♂	♐	♐ 26
♀	♊	♊ 18
☿	♋	♋ 21
H	♎	about 10 A.M.

For generally related horoscopes cf. No. L 102,XII,14. The other part of the story, derived from the horoscope of the wife, is found in No. L 134,VI: she was 10 years younger than her husband, but both were born in the 7th clima. The year of the first law suit was A.D. 159, that of the appeal 161.

---

mentation: in spite of a strictly deterministic attitude the victim is given responsibility for his own downfall. This is reminiscent of "je me laisse surprendre par ce que j'ai prévu."

No. L 123,VII
[1] Cf. Vettius Valens p. 8, 25/26 and 32.
[2] Capricorn is a house of Saturn.
[3] Cf. Introduction p. 13.

[1] A house of Jupiter is ♐ which thus allots its period, 12.
[2] Cancer is the house of the moon whose period is 25.
[3] For clima 7 we have $a_3 = 27$ in System A.
[4] Virgo is a house of Mercury whose period is 20.
[5] Probably 20 for Mercury + 8 for Venus + 12 for Jupiter (in ♓) + 15 for the rising time of ♓ = 55.
[6] Because sun and moon are in the same sign $\Lambda_F = H = ♎$ and ♎ is a house of Venus.
[7] Correct for clima 7, System A.
[8] Gemini is a house of Mercury whose period is 20.

No. L 127,VII

Vettius Valens VII,5 p. 287,17 to 23

TRANSLATION

Sun, Mercury (and) Venus in Cancer, moon in Aries, Jupiter (and) Horoscopos in Gemini, Saturn in Libra, Mars in Leo, clima 1. In the 27th (year) he had a crisis. Of the sun 19 and of Libra[1] 8, making 27; further also of Gemini 28 (years) 4 months,[2] and of Jupiter (in ♊) 12, making 40 (years) 4 months, two-thirds of which is approximately 27.[3] And again of Leo 19[4] and of Cancer 31 (years) 8 months,[5] of Saturn 30 making 80 (years) 8 months, the third of which is approximately 27.

COMMENTARY

Diagram: pl. 23

For the date one obtains 127 July 18 from

Text		Computed
☉	} ♋	♋ 23
☿		♋ 27[6]
♀		♋ 16
☾	♈	♈ 12
♃	} ♊	♊ 12
H		about 4 A.M.
♄	♎	♎ 2
♂	♌	♌ 9

For related horoscopes *cf.* No. L 133 and No. L 102, XII, 14.

[1] Libra is a house of Venus whose period is 8.

[2] The rising time of ♊ is $\alpha_3 = 28;20$ in clima 1, System A.

[3] Text (or misprint?) 47.

[4] The sun's period, and ♌ is the solar house.

[5] The rising time of ♋ is $\alpha_4 = 31;40$ in clima 1, System A.

[6] According to the Stobart tables (Neugebauer [4] p. 227) Mercury would leave ♋ on July 23: Stobart table E obv. III,5 Hadrian 11 month 11 day 29 = 127 July 23, ☿ entering ♋.

No. L 127,XI

Vettius Valens III,10 p. 147,23 to 30

TRANSLATION

Sun in Sagittarius 12;16, moon in Sagittarius 17;24, Saturn in Libra 11;33, Jupiter in Gemini 19;11, Mars in Scorpio 4;20, Venus in Libra 26, Mercury in Scorpio 27, Horoscopos in Libra 20. I took (the distance) from the degree of conjunction to that of the moon; it makes 5.[1] These (counted) from the Horoscopos, it fell in Libra 25. The start (was) from there to the 5 degrees of Mars in Scorpio. He died in the 12th year. If this encounter (with Mars) had not occurred he would have lived the years of Venus, 84.

COMMENTARY

Diagram: pl. 23

For 127 November 23 one obtains

	Text	Computed	Text—comp.
☉	♐ 12;16	♐ 0	+ [2]
☾	♐ 17;24	♐ 19	— 1;30
♄	♎ 11;33	♎ 14	— 2;30
♃	♊ 19;11	♊ 22	— 3
♂	♏ 4;20	♏ 4	+ 0;30
♀	♎ 26	♎ 22	+ 4
☿	♏ 27	♐ 6(!)	
H	♎ 20	about 3 A.M.	

Probably one has to emend the number for the longitude of the sun by deleting the initial ι = 10. The deviations for the moon, Saturn, and Jupiter are numerically small but do not show the expected sign. This is still worse for Mercury. Yet there is no other date possible within historically admissible limits.

The immediately preceding example is No. L 114,V.

No. L 127,XI

[1] This is obviously a correct estimate since the moon in ♐ 17;24 is only 5;8° distant from the sun in ♐ 12;16 or less than one half of the moon's daily motion. Thus the conjunction happened close to ♐ 12 or 5° before ♐ 17.

No. L 129

Vettius Valens VII,5 p. 283,3 to 13

TRANSLATION

Sun (and) Mercury in Capricorn, moon, Mars (and) Horoscopos in Taurus, Saturn in Scorpio, Jupiter in Cancer, Venus in Pisces, clima 6. In the 35th year he escaped from slavery and committed many thefts, and though undetected for a short time was caught in the same year. For both diameters (♑ — ♋ and ♏ — ♉) were operative; they added up to 60[1] the half of which is 30; of Capricorn 28[2] and of Mercury (in ♑) 20 and of Jupiter (in ♋) 12 making 60, the half of which is 30. And of Saturn (in ♏) 30 and of Mars (in ♉) 15, two-thirds of which is 30. And of Cancer 25[3] and of Jupiter (in ♋) 12 and of Venus (in ♓) in trine 8 making 45, two-thirds of which is 30. It was fated then that, because of the beneficent (stars), he escaped danger for a short time and made himself comfortable with the thefts but fell because of the maleficent (stars).

COMMENTARY

Diagram: pl. 23

The date of this horoscope is 129 January 16:

Text		Computed
☉	} ♑	♑ 26
☿		♑ 1

[1] For all climata and for all systems the rising times of ♋ and ♑ and of ♉ and ♏ total 60°.

[2] Only for clima 2(!), System A, $\alpha_{10} = 28$.

[3] Cancer is the house of the moon whose period is 25.

Text		Computed
☾ }		♉ 11
♂ }	♉	♉ 10[4]
H }		about 2 P.M.
♄	♏	♎ 29(!)[5]
♃	♋	♋ 22
♀	♓	♓ 7

For related horoscopes *cf.* No. L 102,XII,14.

According to the Stobart tables[6] Venus seems to be in ♒ from Hadrian 13 month 4 day 8 (= 128 December 4) to month 8 day 10[7] (= 129 April 5). Actually, however, two lines were omitted which one may restore approximately as follows:

5 16 ♓ (= 129 January 11)
7 2 ♒ (= 129 February 26).

No. L 132

Vettius Valens VI,5[1] p. 253,6 to p. 254,8 (= CCAG 5,2 p. 119,16 to p. 120,27)

TRANSLATION

Let us take as an example a nativity, so that we may explain the procedure more briefly. The moon in Pisces 18 degrees, Venus in Aries, Jupiter in Libra, Saturn (and) Horoscopos in Sagittarius, Mars in the beginning of Aquarius, sun in Aquarius,[2] Mercury in Pisces; for in this way they are arranged one after the other.[3] We investigate 52 years and of the 53rd (up to) Payni (X) 15. Since in this division days are treated [like degrees and hence] the (complete) circles as 360 days, but the years from birth as 365¼, I took for the full years 52 times 5¼, it makes 273 (days). And from Mekheir (VI) 12[4] to Payni (X) 15 makes 124 (days);[5] altogether it makes 397 days. I subtracted one circle (of) 360, remainder 37 days. Then, as the nativity was nocturnal, the moon was chosen to be the starter of the periods (of life), and it was in Lower Midheaven, in a female sign, in a triangle of (a star of) the same sect[6] and in its own (domain). It (the moon) first took 10 years 9 months[7] of the stated years, then, in direct order, Venus in Aries 10 years 9 months [and similarly Jupiter and Saturn, totalling so far 4 cycles or 43 years. The addition of one more cycle of 10 years 9 months for Mars would exceed the given total of 52 years 4 months 2 days. Consequently, beginning with Mars, the single periods are now added up in the order of the horoscope, counted as months: Mars 15, sun 19, Mercury 20, moon 25, Venus 8, Jupiter 12. This brings us to a total of 51 years 3 months. The addition of the next period, 30 months for Saturn, would again exceed the given total. The remaining 667 days[8] are now divided in cycles of $10 \cdot 12 + 9 = 129$ days, beginning with Saturn. After four such cycles we reach a remainder of 22 days for Venus. Then again the single periods, but counted as days: Venus 8, Jupiter 12, to be followed by a remainder of 2 days for Saturn. This brings the procedure to a close. The rest is devoted to explanations of events in the life of this person, corresponding to the passing of rulership according to the above-mentioned cycles.]

COMMENTARY

Diagram: pl. 24

The data of the horoscope are excellently satisfied by the positions found for 132 February 7:

Text		Computed	Text—comp.
☾	♓ 18	♓ 16	+2
♀	♈	♈ 3	
♃	♎	♎ 27	
♄ }	♐	♐ 2	
H }		about 2 A.M.	
♂	[beg. of] ♒	♒ 3[9]	−3
☉	♒	♒ 17	
☿	♓	♓ 3[10]	

The text discusses an interval which begins with Mekheir (VI) 12. Though not explicitly stated it seems natural to consider this as the day of birth. Indeed (A.D. 132) February 7 = Mekheir 12. The date of investigation, Payni (X) 15, 52 years later, is 184 June 9.

[4] Neugebauer [4] p. 227 column V line 10 restores incorrectly month 5 instead of month 6. The correct date for the entry of Mars into ♊ is therefore Hadrian 13 month 6 day 11 = 129 February 5.

[5] Stobart tablet E obv. V,3 gives for the entry of Saturn into ♏ the date Hadrian 13 month [5] day 11 = 129 January 6.

[6] Neugebauer [4] p. 227.

[7] Tablet E obv. V,19/20.

No. L 132

[1] Called VI,6 in *CCAG* 5,2 p. 118,8; *cf.* also *CCAG* 5,2 p. 8.

[2] The text has the opposite arrangement: "Mars in Aquarius, the sun in the beginning of Aquarius." Our emendation is required by the order of enumeration by increasing longitude, and confirmed by computation.

[3] Following increasing longitude, beginning with the moon as "starter" because of night birth; *cf.* p. 252,2 f.

[4] This is the day of the horoscope, February 7; *cf.* the commentary.

[5] Both ends included.

[6] The triangle ♓ ♋ ♏ is ruled by Venus and moon, both of which are of nocturnal sect.

[7] This is the total 129 of all planetary periods, counted as months. *Cf.* p. 11.

[8] To complete the 52nd "year" (of 360 days each) one needs 270 days to which the previously computed correction of 397 days is to be added. Result: 667 days.

[9] The Stobart tables give for the entry of Mars into ♒ the date Hadrian 16 month 6 day 4 = 132 January 30. (Neugebauer [4] p. 228, E rev. IV,20).

[10] Stobart E rev. V,14 gives month 6 day 8 = February 3 as date of entry of Mercury into ♓.

No. L 133

Vettius Valens VII,5 p. 287,30 to p. 288,3

TRANSLATION

Sun, Mercury, Venus (and) moon in Taurus, Saturn in Sagittarius, Jupiter in Scorpio, Mars in Leo, Horoscopos in Pisces, clima 2. In the 22nd year he had a crisis: of Leo 19[1] and of moon 25, making 44, half of which is 22. In addition, 36 of Scorpio[2] and 8 of Taurus[3] are 44, the half of which is 2[2].

These six men on a voyage, with many others, encountered a violent storm and, the rudder being lost, were in danger of death by drowning as the ship took in water. But by the draught of the blowing wind and the steersman's management of the sails they escaped; and they encountered other dangers at the same time from a roving pirate (ship).

COMMENTARY

Diagram: pl. 24

This is the last of a series of six horoscopes of six men whose common fate in the year 154 is under investigation. This date results from adding the latest quoted year to the date of the horoscope

114 (July 26) + 40th year — 153/154
120 (Feb. 8) + 35th year — 154/155
118 (Nov. 26) + middle of 36th year — 154
127 (July 18) + 27th year — 153/154
122 (Jan. 30) + 32 years 8 months — 154
133 (Apr. 24) + 22nd year — 154/155.

The data of the last horoscope are

	Text	Computed
☉	♉	♉ 2
☿		♉ 18
♀		♉ 19
☾		♉ 17
♄	♐	♐ 13
♃	♏	♏ 28
♂	♌	♌ 0
H	♓	about 2 A.M.

For other related horoscopes *cf.* No. L 102,XII,14.

No. L 134,VI

Vettius Valens VII,5 p .281,24 to p. 282,2

TRANSLATION

And the nativity of the wife (of the owner of the preceding horoscope[1]) defeated (at court) was this: {Another:}[2] sun in Cancer; moon, Saturn (and) Jupiter in Sagittarius; Mars, Venus (and) Mercury in Gemini, Horoscopos in Capricorn, clima 7. In the 25th year she was destined to have the best of the affair. For the moon, being with Jupiter and Saturn (in ♐) indicated the 25 and likewise the sun in Cancer indicated the 25.[3] But the other stars were operative by association and opposition. Namely of Saturn 30 years and of Mercury at diameter (in ♊) 20 make 50, the half of which is 25. And again of Saturn 30 and of Mercury 20 and of the moon 25 make 75, the third of which is 25. Afterwards the affair came to reversal and then being appealed in the 27th year the verdict was adverse. For operative was the rising time of Gemini (— 27[4]) where the Lot (of Fortune was[5]) with Mars and Mercury in it. And besides, of the moon (in ♐) 25 and of Mars (in ♊) at diameter 15 make 40, of which two-thirds is 26 (years) 8 months. Here too[6] the stars happened to be in bicorporeal signs (♊ and ♐).

COMMENTARY

Diagram: pl. 24

This horoscope is 10 years later than the horoscope of the husband (No. L 124) who was involved in the same law suit. For 134 June 23 we have

	Text	Computed
☉	♋	♋ 0
☾	♐	♐ 25
♄		♐ 22
♃		♐ 28
♂	♊	♊ 0
♀		♊ 6
☿		♊ 16
H	♑	sunset

For other related horoscopes *cf.* No. L 102,XII,14.

No. L 134,XI

Vettius Valens V,10, p. 227,27 to p. 228,14

TRANSLATION

And we will explain nativities by way of example for the quick understanding of those approaching the theory. As: sun, moon, Venus, Mercury (and) Horoscopos in Scorpio, Saturn in Sagittarius, Jupiter in Capricorn, Mars in Leo. In the 20th year the paradosis (takes place) from Jupiter in Capricorn to Mars in Leo via

[1] The period of the sun, whose house is Leo.
[2] The rising time a_8 for clima 2 in both systems.
[3] A house of Venus, whose period is 8.

[1] No. L 124.
[2] The word "another" is customary in these series of examples and was added also here, out of context.
[3] Cancer is the house of the moon.
[4] Correct for clima 7, System A.
[5] From $\Lambda_F = \lambda_H + (\lambda_{☾} - \lambda_{☉}) = ♑ + 5° = ♊$.
[6] *Cf.* No. L 124 at the end.

8 (signs).[1] Hence Jupiter gave over to Mars from the 3rd to the 10th sign (counted from H), that is to Midheaven. A petition for honor to the king's (court) was made but did not succeed for Jupiter giving over to Mars was grievous. Strong is also the distribution by 4, that is from Mars (in ♌) to sun and moon and Horoscopos and Mercury and Venus (in ♏). He was ill then in the 20th year and fell from an animal and was dragged so as almost to lose his eyesight. And fault and deception and penalty had to do with a female person, so that each of the stars had its own effect taking over from maleficent stars.

And in the 23rd year Jupiter (in ♑), in the locus of the king—for the third and the ninth locus from the Horoscopos indicate god and king—giving over to the luminaries and Venus and Horoscopos and Mercury (in ♏ [2]) provided by way of gifts a powerful colleague. Thus nothing is able to make a man possessed of the friendship of kings and great men if the periods (of life) work against it.

COMMENTARY

Diagram: pl. 24

The date of this horoscope is 134 November 4:

	Text	Computed
☉	♏	♏ 11
☾	♏	♏ 16
♀	♏	♏ 22
☿	♏	♏ 11
H	♏	sunrise
♄	♐	♐ 22
♃	♑	♑ 1
♂	♌	♌ 20

For related horoscopes *cf.* No. L 102,XII,4.

No. L 135,I

Vettius Valens II,14 p. 156,14 to 19

TRANSLATION

Sun in Aquarius, moon (and) Horoscopos in Virgo, likewise also Mars; Saturn, Jupiter (and) Mercury in Capricorn, Venus in Pisces, the Lot of Fortune in Aquarius [1] in apoklima. The ruler (♄ [2]) in (♑, the 5th locus) Agathos Daimon [3] according to division (of periods) allotted its own 30 years and the same number of months, being in its own sign.[4] And Jupiter being with it in this (sign) had one year; [5] he died in the 34th year.[6]

COMMENTARY

Diagram: pl. 24

The date, 135 January 20, is found from

	Text	Computed
☉	♒	♒ 0
☾	♍	♍ 12
H	♍	about 8 P.M.
♂	♍	♍ 10
♄	♑	♑ 2
♃	♑	♑ 18
☿	♑	♑ 3
♀	♓	♒ 29(!)

The preceding example is No. L 75.

No. L 135,X

Vettius Valens V,10 p. 229,6 to 12

TRANSLATION

Sun (and) Mercury in Scorpio, moon (and) Mars in Sagittarius, Saturn in Capricorn, Jupiter in Aquarius, Venus in Virgo, Horoscopos in Taurus. As in the preceding nativity [1] the son of the person was indicated in the interpretation, so from the present division (of periods) he reached the 22nd year (in agreement) with the father's (horoscope).[2] For Jupiter handed over to the sun, which indicates the father, and the handing over from Mars in (♐ the 8th) Locus of Death to Venus was the cause of death.

COMMENTARY

Diagram: pl. 24

For 135 October 27 one finds

	Text	Computed
☉	♏	♏ 3
☿	♏	♏ 11
☾	♐	♐ 15
♂	♐	♐ 15
♄	♑	♑ 3
♃	♒	♒ 1
♀	♍	♍ 20
H	♉	about sunset

[1] Because 20 ≡ 8 mod. 12 one has to look for a pair of planets which are in the 1st and 8th signs respectively, counting in the direction of the signs.

[2] Because 23 ≡ 11 mod. 12, and ♏ is 11 signs ahead from ♑, the place of Jupiter.

No. L 135,I

[1] From $A_F = \lambda_H - (\lambda_{☾} - \lambda_{☉}) =$ ♍ $+ 5^s =$ ♒.

[2] A house of ♄ is ♒.

[3] Error for Agathe Tyche (*cf.* also the preceding horoscope, No. L 75, which committed the same error).

[4] Its house ♑.

[5] The period of Jupiter is 12 (years), counted here for 12 months.

[6] 30 years + 30 months + 12 months = 33½ years.

No. L 135,X

[1] The horoscope No. L 107 (*cf.* p. 103).

[2] *Cf.* Commentary.

The father's horoscope has the date A.D. 107 May 8, a date which falls in the year Augustus 137. The son's horoscope, A.D. 135 Oct. 27, falls in the year Augustus 166; the 22nd year of the son is therefore Augustus 188 which equals the 51st year of the father. This is indeed the year in which the "death of a child" is reported in the father's horoscope (No. L 107).

For other horoscopes in the same chapter cf. No. L 102,XII,4.

No. L 142

Vettius Valens VII,5 p. 288,20 to p. 289,33

TRANSLATION

For example let sun, Saturn (and) Mercury be in Aries, moon (and) Jupiter in Leo, Mars in Taurus, Venus in Aquarius, Horoscopos in Virgo, clima 1. In the 18th year he went abroad with a distinguished woman because of friendship and esteem and for further intimacy and erotic passion. For 25 years [1] were indicated by the rising time of Aquarius [2] where Venus was. And if it (Venus) alone had been operative, it would have effected success quickly. But now Mars (in ♉) also ruled the 25 because of the rising time of Taurus,[2] and the period of the moon (in ♌ [3]) also carried the 25, two-thirds of which is 16 (years) 8 months. If then the other aspects of the beneficent (stars) after the time mentioned had agreed, the hope would have been realized, but now the same (stars) were in power, for in the 18th year the female person died and he himself failing of his hope returned home with little profit. For Jupiter (in ♌) had allotted 12 and Mars (in ♉) in dominant aspect (quartile) to it 15, making 27 two-thirds of which is 18 years. Besides, of Leo 19 [4] and Taurus 8 [5] make 27, two-thirds of which is 18.

In the 19th year he got help and gain but he had disagreements and mental troubles and enmities with relatives. For the 19 was indicated by Leo [4] with Jupiter and moon there, and besides, the sun itself being with Saturn in Aries indicated the 19.[6] And again of Aquarius 30 [7] and of Venus (in ♒) 8 the half of which is 19.

And in the 20th year because of a woman's friendship he went abroad and looked forward to greater hopes and profits but then was disappointed because she died. For Mars indicated 15 and moon 25, making 40, half of which is 20. And besides, the rising time of Taurus 25 [2] and of Leo 35 [2] make 60, the third of which is 20. And also Mercury in Aries with Saturn indicated the 20.

Likewise in the 21st year the same (stars) prevailed. Of Leo 19 [4] and of Taurus 8,[5] of Mars (in ♉) 15 make 42, the half of which is 21. And the rising time of Aries 21 (years) 8 months [8] and of Mercury (in ♈) 20, make 41 (years) 8 months the half of which is 20 years 10 months.

And in the 22nd year likewise the same (stars) prevailed. Of Leo 19 [4] and of Taurus 25 [2] make 44 the half of which is 22. The 23rd year Venus and Mars: Venus 8 Mars 15, make 23.

Thus many times by broad and complete observations the stars seeming to be in favorable configurations are indications of opposite things, being counteracted by the power of the period.[9] For instance the 21st year was indicated by Leo 19 [4] and of Jupiter, located there, 12, (making 31) two-thirds of which is 20 (years) 8 months. Besides, the triangle (aspect) of Jupiter (in ♌, period 12) and sun (in ♈, period 19) indicated the same (31 — 12 + 19). Thus he had friendship with a prominent and royal person from whom he expected the right to wear the wreath and archpriesthood. And this would undoubtedly have happened if Mars had not been in dominant aspect and operative. And also Saturn being (in ♈) with the stars (☉ and ♉) that provide a gift, inquiries and delays and expenditures and jealousies attended (him); and the obstacle to his expectation was not so much the triangle, which was that of Jupiter (♌ ⚹ ♈), as Mars (in ♈) being in dominant aspect by sect.[10] Thus climacteric was illness and bloodshed and frustration of expectation, treachery of slaves, attacks, injuries, wants. Then later this person fared well and with an upward trend for the following period he was transferred to the influence simultaneously of the beneficent and maleficent (stars). There will be troubles, great expenses, or there will be independence resulting from preceding affairs or from some other income and the assistance of friends.[11]

COMMENTARY

Diagram: pl. 25

The date of birth of this enterprising young man was 142 March 25:

---

[1] This is a preliminary prediction; the last actual information in this horoscope concerns the 21st year. The upper limit for data in this chapter is A.D. 164/5; in our case we have 142 + 21 = 163. Cf. also note 9.

[2] Correct for clima 1 in both systems.

[3] Thus in opposition to ♀ in ♒.

[4] Leo is the house of the sun whose period is 19.

[5] Taurus is a house of Venus whose period is 8.

[6] The period of the sun.

[7] Aquarius is a house of Saturn whose period is 30.

[8] For clima 1 and System A we have correctly $a_1 = 21;40°$.

[9] This seems to be the introduction to a renewed investigation of this horoscope whose first discussion included events of the 18th and the 19th year with a general outlook until the year 23. Now we are told about events in the year 21 and probably shortly after, bringing us right up to the limit 142 + 22/23 = 164/5 of this chapter.

[10] The nocturnal Mars is in the nocturnal sign ♉, the diurnal Jupiter in the diurnal sign ♌.

[11] This is one of the rare cases of a prediction for the future.

Text		Computed
☉	♈	♈ 3
♄	♈	♓ 25(!)
☿	♈	♈ 23
☾	♌	♌ 19
♃	♌	♌ 22
♂	♉	♉ 22
♀	♒	♒ 20
H	♍	about 4 P.M.

For related horoscopes *cf.* No. L 102,XII,14.

No. L 152

Vettius Valens IV,10, p. 170,15 to p. 171,3 (= CCAG 5,2 p. 85,17 to 40)

TRANSLATION

As for example, let sun (and) Mercury be in Capricorn, Saturn (and) Jupiter in Leo, Mars (and) Venus in Aquarius, moon in Gemini, Horoscopos in Leo, the Lot of Fortune in Pisces,[1] that of Daimon in Capricorn.

Let the start be the (Lot of) Fortune in Pisces. To investigate Mesore 16 of the fourth year[2] including the additional 5 (days) of each year. Since for Pisces (the 12th sign) the division by 12 leaves no remainder,[3] I gave to it 1 year,[4] then to Aries 1 year 3 months,[5] to Taurus 8 months,[6] making (a total of) 2 years 11 months. Then next it will give to Mercury 1 year 8 months[7] up to the completion of four years and 7 months,[8] but the nativity has not completed this period.[9] So let Mercury be chronocrator, having already 8 months and 15 days, making 255 days.[10] [This total is now distributed among the planets.]

[1] From A♈ = $\lambda_H - (\lambda_{☾} - \lambda_{☉})$ = ♌ + 7° = ♓.

[2] Since the date of the horoscope is 152 December 27, the fourth year, Mesore (XII) 16 corresponds to 156 August 11.

[3] This must be the meaning of the text which allows no literal translation.

[4] Because 1 year = 12 months and ♓ being a house of Jupiter, is represented by 12, the period of Jupiter.

[5] Or 15 months, because ♈ is a house of Mars whose period is 15.

[6] The period of Venus since ♉ is a house of Venus.

[7] Or 20 months, 20 being the period of Mercury whose house is ♊.

[8] The date of the horoscope is the night of December 27 = Choiak 30 (IV)/Tybi (V) 1. From Tybi to Mesore (XII) are 7 months. On the other hand 2y. 11m. + 1y. 8m. = 4y. 7m.

[9] Mesore (XII) of the 4th year is only 3 years 7 months later than Tybi (V) of the 1st year.

[10] From Tybi (V) 1 to Mesore (XII) 16 elapsed 7 months 15 days = 225 days. The 8 months 15 days = 255 days of the text are probably an error caused by having quoted the date of the horoscope in the form "Choiak (IV) 30 to Tybi (V) 1" and then erroneously counting the 8 months from Choiak to Mesore.

COMMENTARY

Diagram: pl. 25

Complete agreement is reached for 152 December 27:

Text		Computed
☉	♑	♑ 6
☿	♑	♑ 17
♄	♌	♌ 25
♃	♌	♌ 4
♂	♒	♒ 14
♀	♒	♒ 21
☾	♊	♊ 14
H	♌	about 8 P.M.

No. L 153

Vettius Valens V,10 p. 230,11-13

TRANSLATION

Sun and Mars in Taurus, moon and Horoscopos in Aries, Saturn in Leo, Jupiter in Cancer, Venus in Pisces, Mercury in Gemini. In the 4th year death of the father occurred.

COMMENTARY

Diagram: pl. 25

For 153 May 8 one obtains

Text		Computed
☉	♉	♉ 16
♂	♉	♉ 20
☾	♈	♈ 9
H	♈	about 4 A.M.
♄	♌	♌ 18
♃	♋	♋ 29
♀	♓	♓ 30
☿	♊	♊ 19

For related horoscopes *cf.* No. L 102,XII,4.

No. L 158

Vettius Valens VII,5 p. 284,12 to p. 285,3

TRANSLATION

Sun, Mercury (and) Venus in Leo, moon in Virgo, Saturn in Libra, Jupiter in Capricorn, Mars in Aries, Horoscopos in Cancer, clima 6. I interpreted from investigation also these periods and since it was an infant I counted months instead of years. For at the end of the 8th month and during part of the 9th he was subject to convulsions lacking little of being dangerous. Libra indicated the 8th[1] (month), Saturn being located there; and in the middle of the 9th (month) the rising time of Aries 17[2] (indicated) up to the completion of 8 months

[1] Libra is a house of Venus whose period is 8.

[2] Correct for clima 6 and System B.

and 15 days. The rising time of Capricorn was also operative thus: of the 27³ the third is 9, and of the sun 19 and of Venus 8 make 27, the third of which is 9. And there were indications for the other months being controlled by the maleficent (stars) and he was afflicted by eruptions and eczema, namely in the 15th and 17th and 23rd (month) and in the rest, but especially about the 27th month. And supposedly(?) he fell into an animal snare(?) and was struck in parts of the body(?). The 27th was indicated by Mars (in ♈) and moon (in ♍), of Mars 15 and of moon 25 make 40, two-thirds of which is 26 (months) 20 days; I reckon with Aries and Virgo at diameter, because (the moon) was moving toward it. And besides, of Mars 15 and of Horoscopos, Cancer, 25⁴ are 40, two-thirds of which is 26 (months) 20 days. And also then the rising time of Capricorn was operative, the 27,³ and of Mars 15 and of Jupiter 12 are 27. And again, of the sun 19⁵ and of Venus 8 make 27.

Then from the 28th month he lived precariously, for the rising time of Libra and the period of Saturn (in ♎) coincided.⁶ And in the 32nd month he was dangerously ill and was in convulsions; for of Mars 15 and of the rising time of Aries 17² are 32. And in the 33rd month he died, for the rising time of Cancer⁷ (H) effected it; and besides of Cancer (H) 25⁴ and of Libra (♄) 8¹ are 33, or of moon (in ♍) 25 and of Libra⁸ 8 are 33. Thus I reckoned as if it (the moon) were also in Libra (which is permissible) because of the equality of rising times (of ♍ and ♎).

COMMENTARY

Diagram: pl. 25

The victim of this horoscope was born 158 August 14:

Text		Computed
☉		♌ 20
☿ }	♌	♌ 20
♀		♌ 12
☾	♍	♍ 17
♄	♎	♎ 21
♃	♑	♑ 4
♂	♈	♈ 10
H	♋	about 4 A.M.

The death of the child occurred in May 161. The upper limit of records utilized in this chapter is A.D. 164/5.

For related horoscopes cf. No. L 102,XII,14.

---

³ Capricorn has the rising time $a_{10} = 27$ only for clima 7(!) in System B or for clima 2(!) and System A.
⁴ Cancer is the house of the moon whose period is 25.
⁵ Text (misprint?) 12.
⁶ Libra has in clima 7(!) and for System B a rising time of $a_7 = 43;30$ which is numerically equal to the "mean period" of Saturn (cf. p. 11).
⁷ Cancer has the rising time $a_4 = 33$ in clima 7(!) System A or clima 2(!) System B.
⁸ Because the moon was approaching ♎.

No. L 159

Vettius Valens VII,3 p. 273,23 to 29
(= CCAG 5,2 p. 44,38 to p. 45,7)

TRANSLATION

Sun, Venus (and) Mercury in Cancer, moon in Capricorn, Saturn in Scorpio, Jupiter in Aquarius, Mars in Leo, Horoscopos in Taurus, the Lot of Fortune in Scorpio,¹ the ruler (of it,² ♂) in Leo. The ruler (☉) of this (♌)³ in Cancer exceeded⁴ the hours and days of the stars and signs; for this (reason) then, in the third place (from the Horoscopos), I counted the months of Mars 66 months and 15 (months)⁵ and of the sun in Cancer 25 months,⁶ making 106 months which is 8 years 10 months.

COMMENTARY

Diagram: pl. 25

For 159 July 18 one finds

Text		Computed
☉		♋ 23
♀ }	♋	♋ 2
☿		♋ 12
☾	♑	♑ 20
♄	♏	♎ 29(!)
♃	♒	♒ 15
♂	♌	♌ 12
H	♉	about 2 A.M.

Related horoscopes are listed in No. L 122,VI,30.

No. L 162

Vettius Valens VII,3 p. 273,33 to p. 274,3
(= CCAG 5,2 p. 45,8 to 15)

TRANSLATION

Sun (and) Mercury in Aquarius, moon in Taurus, Saturn in Sagittarius, Jupiter in Aries, Mars in Capricorn, Venus (and) Horoscopos in Pisces, clima 6. Lot of Fortune in Gemini.¹ I took 76 months of Mercury² and, since Mercury is in Aquarius, I took the rising time of Aquarius, 23³ months and, Saturn being in

---

¹ From $\Lambda_F = \lambda_H - (\lambda_{\mathbb{C}} - \lambda_{\odot}) = \text{♉} - 6^s = \text{♏}$.
² Scorpio is a house of Mars, which is therefore its ruler.
³ Leo is the house of the sun.
⁴ The sun (in ♋) is beyond the side of the square (from ♏ to ♌).
⁵ These are the extremal periods assigned to Mars.
⁶ Cancer is the house of the moon whose period is 25.

No. L 162
¹ From $\Lambda_F = \lambda_H + (\lambda_{\mathbb{C}} - \lambda_{\odot}) = \text{♓} - 9^s = \text{♊}$.
² The maximum period of Mercury is 76 years.
³ Aquarius has the rising time $a_{11} = 23$ for clima 4(!) in System A and B.

Sagittarius, of Sagittarius 33[4] months. They make altogether 132 months which is 11 years. He died at that age.

COMMENTARY

Diagram: pl. 25

The date of this horoscope, 162 February 9, is found from

Text		Computed
☉	♒	♒ 20
☿	♒	♒ 7
☾	♉	♉ 12
♄	♐	♐ 9
♃	♈	♈ 26
♂	♑	♑ 23
♀	♓	♓ 8
H	♓	about 8 A.M.

For the three other horoscopes in this chapter see No. L 122,VI,30.

No. L 173

Vettius Valens VII,3 p. 273,10 to 23 (= CCAG 5,2 p. 44,26 to 37)

TRANSLATION

For example let sun (and) Venus be in Aquarius, moon (and) Jupiter at the beginning of Aries, Saturn in Aries, Mars in Sagittarius, Mercury in Capricorn, Horoscopos in Scorpio, Lot of Fortune in Libra,[1] clima 6. The apportioning (stars) were: Venus because of Libra,[2] Saturn because Venus was in Aquarius,[3] Mars because Saturn was in Aries.[4] Then I reckoned with respect to periods and to rising times: first hours, then days, then months. Thus: I took for Libra 8 days and 8 hours[5] and again for the rising time of Libra, clima 6, [4]3 hours,[6] and, since Venus is in Aquarius,[3] I took for Saturn 57 hours[7] and again since Saturn is in Aries,[4] I took for Mars 15 hours, summing up to 8 days and 123 hours, which are 5 days 3 hours, making altogether 13 days 3 hours. He died 13 days 3 hours (old).

[4] Sagittarius has the rising time $a_9 = 33$ for clima 7(!) in System A or for clima 2(!) in System B. The same error occurs also in No. L 122,I,30. *Cf.* also No. L 173 note 7.

No. L 173

[1] From $A_F = \lambda_H - (\lambda_☾ - \lambda_☉) =$ ♏ — slightly less than 11° = ♎. Note that the moon's position is at the beginning of Aries.

[2] A house of ♀ where the Lot of Fortune is located.

[3] A house of Saturn.

[4] A house of Mars.

[5] Libra as a house of Venus is represented by its period 8.

[6] Following System B.

[7] That is 30 for the period and 27 for Capricorn which is a house of Saturn. The rising time of Capricorn is $a_{10} = 27$ in clima 7(!) System A and in clima 2(!) System B. *Cf.* for a similar error, also for clima 6, No. L 162 and No. L 122,I,30.

COMMENTARY

Diagram: pl. 26

For 173 February 3 we find:

Text		Computed
☉	♒	♒ 15
♀	♒	♒ 13
☾	beginning of ♈	♈ 3
♃	beginning of ♈	♈ 1
♄	♈	♈ 10
♂	♐	♐ 9
☿	♑	♑ 24
H	♏	about midnight

The child died 173 February 17. In the same year died, 11 years old, the owner of another horoscope recorded in this chapter (No. L 162) which thus is one of the latest sections in the whole work. *Cf.* also No. 149 and No. L 122,VI,30.

No. L 188

Vettius Valens II,26 (p. 93,13 to 20)[1]

TRANSLATION

As an example let the sun, the moon, Jupiter, (and) Mercury be in Leo, Saturn (and) Horoscopos[2] in Libra, Mars in Aquarius, Venus in Cancer;[3] such a person (will be) fortunate, a leader, despotic, possessed of royal fortune, and established in great property. For the Lot of Fortune and the Daimon and the Basis fell in the same (sign ♎) and the ruler of these, Venus, was in Midheaven in Cancer, and the rulers (♃ and ☉) of the triangle (of the sun, ♌ ♐ ♈) and of the exaltation (♊, ruler ☿) were found to be in Agathos Daimon (♌)[4] in the (11th) Locus[5] of accomplishment (♌).

COMMENTARY

Diagram: pl. 26

This is the first in a group of four examples (*cf.* No. L 78). For this text one finds the date 188 August 10, which is the latest date of a horoscope in Vettius Valens.

Text		Computed
☉	♌	♌ 17
☾	♌	♌ 15
♃	♌	♌ 3
☿	♌	♍ 6(!)
♄	♎	♎ 27
H	♎	about 10 A.M.
♂	♒	♒ 4
♀	♋[3]	♋ 6

[1] The three first lines also in *CCAG* 2, p. 120,41-121,1.

[2] Kroll, p. 93,14 gives ὡροσκόπος, *CCAG* 2 p. 120,41 ὡροσκοπῶν.

[3] Kroll p. 93,14 (supported by line 18) gives καρκίνῳ. *CCAG* 2, p. 120,42 wrongly διδύμοις.

[4] The 11th locus from the Horoscopos (♎).

[5] The 11th locus from the Lot of Fortune (♎).

Mercury which is approaching inferior conjunction would be in Leo if one disregards the anomaly. Because sun and moon are practically in conjunction we have for the Lots of Fortune and of Daimon the same longitude as for the Horoscopos, that is Libra, the house of Venus. Because these two Lots coincide, the "Basis" also falls in the same sign (cf. p. 9). The "exaltation" (of the nativity, cf. p. 7) falls in ♊, a house of Mercury.

No. L 380
CCAG 8,1 p. 146,3 to 20 and
CCAG 8,2 p. 58,3 to 14, p. 67,34 to p. 68,31

TRANSLATION

First Version (CCAG 8,1 and Engelbrecht, Heph., p. 22 f.)

From *Hephaestion* II,1

As an example suppose somebody to have been born in the 97th year from the reign of Diocletian, Athyr (III) 30, 6th hour (of the day), in the third clima. For this hour we found the sun in approximately the fourth degree of Sagittarius, the moon in Taurus [1] 47 minutes, the Horoscopos in the 25th degree of Aquarius. Since the moon (in ♉) was found in the signs which follow [2] the Horoscopos (in ♒), one must count from the Horoscopos (♒ 25) to the degree of the moon (♉ 0;47) and (thus) we found 65 degrees 47 minutes. These we divided by the (mean) daily motion of the moon, that is 13°, and found 5 days; and the remaining 47 minutes we divided similarly by the hourly mean motion of the moon and found 1 hour and $\frac{1}{3}$ approximately.[3] The 5 days and 1$\frac{1}{3}$ hours we added to the 273$\frac{1}{3}$ days of pregnancy,[4] which made 278 days and 9$\frac{1}{3}$ hours.[5] Subtracting this from the day and hour of birth, Athyr (III) 30 6th hour, it extended to Mekheir (VI) 26 to 27 8th hour of night.[6] Subtracting also the fourth part of the intercalary day,[7] that is 6 hours, one finds the

---

[1] The printed text gives Capricorn. According to a photo of the manuscript only very faint traces of writing are preserved. The next sentence requires ♉ and computation confirms it.
[2] Counted in the direction of increasing longitude.
[3] This result is inaccurate. Since the mean motion of the moon is 13;10,35°/d and 0;32,56°/h we should find 65;47° : 13;10,35°/d = 4d plus 13;4,40° : 0;32,56°/h = 23;50h thus 10 minutes less than 5 days. The text computes 65° : 13°/d = 5d plus 0;47° : 0;33°/h (= 1;25,..)h = 1$\frac{1}{3}$ (!)h.
[4] Cf. the scheme in Vettius Valens I,25 (p. 50 Kroll) which follows a different method, but leads, for the present case, to the same result.
[5] Sic, instead of $\frac{2}{3}$. One manuscript has $\frac{1}{3}$!
[6] The figure of the text, 9$\frac{2}{3}$, would lead to 8$\frac{2}{3}$; the correct figure, 9$\frac{1}{3}$, to 8$\frac{1}{3}$. The date 26 to 27 means the night of the Egyptian 26th which precedes the 27th and is for once correctly computed.
[7] To account for the difference between Egyptian years (upon which the tables were based) and Alexandrian years. Leap years are Diocletian 95 and 99.

year 96, Mekheir (VI) 26 to 27 2nd hour (of night)[8] at which hour we found the sun in [........] degree, the moon in [........] degree.[9]

Second Version (CCAG 8,2 p. 58,3 to 14)

From *Hephaestion* II,2

For the sake of clarity let us have an example. Some (person) was born in the day(time) of the 30th of Athyr (III), the sun being in the 4th degree of Sagittarius. To these 4 degrees I added 2° [10] and it made 6°; these 6° multiplied by twelve made 78.[11] Counting these from the degree of the sun (♐ 4) I arrive at about the end of Aquarius (♒ 22). Consequently either Aquarius or Gemini or Libra [12] is the Horoscopos at conception. But since the sun was near the beginning of Sagittarius (♐ 4), the hemisphere of daylight above the earth extends from it to the beginning of Gemini. Thus the Horoscopos (of conception) will not be in Gemini, nor can it be in Libra because also Libra is below the earth. Thus we must believe that the Horoscopos can only fall in Aquarius.

Third Version (CCAG 8,2 p. 67,34 to p. 68,31)

From *Hephaestion* II,11

Let us also set forth an example for the sake of clarity. Suppose that a person has 25 degrees of Aquarius as Horoscopos, Midheaven 5 degrees [of Sagittarius]. Counting from the fifth degree of Sagittarius (M) to the 25th degree of Aquarius (H) I have 80 degrees: I have namely 25 of Sagittarius and 30 of Capricorn and 25 of Aquarius. Thus there remain 10 degrees or $\frac{1}{9}$ of a quadrant.[13] We now take from the 5 degrees of Sagittarius 25 degrees toward the east and of the resulting 30 degrees the 9th part, that is 3;20, and subtract this from 30 degrees leaving 26;40 degrees of Sagittarius, which is the end of the twelfth [14] of Midheaven, and the beginning of the 11th Locus, that is Good Daimon. Taking again $\frac{1}{9}$ of one-twelfth (= 30°) that is 3;20 degrees, I subtract these from 30 and add the remaining 26;40 to the 26;40 degrees of Sagittarius which were found before, and I subtracted from the resulting 53;20 the 30 degrees of Sagittarius. The remaining 23;20 belong to Capricorn, making the end of Good Daimon and the beginning of Bad Daimon. And again adding to the 23;20 of Capricorn the aforesaid 26;40 of one-twelfth and from the resulting 50 I sub-

---

[8] This is now, supposedly, the hour of conception.
[9] Cf. the Commentary for the expected positions.
[10] The symbol which the editors print as "s" must mean "degree."
[11] Added to the original 6°; cf. p. 6 s.v. "dodekatemoria."
[12] In trine to Aquarius.
[13] Here and in the following our translation is somewhat freer than usual.
[14] "Twelfth" is here used in the sense of "locus," i.e., one of the 12 parts of the whole circle.

tract the 30 degrees of Capricorn. The resulting 20 of Aquarius is the end of the Bad Daimon and the beginning of the twelfth of the Horoscopos. [In the same fashion the limits of other loci are determined, not without some minor arithmetical errors.]

COMMENTARY
Diagram: pl. 26

For Diocletian 97 Athyr (III) 30 = A.D. 380 [15] November 26 one finds:

Text		Computed	Text—comp.
☉	≈ ♐ 4	♐ 6	−2°
☾	♉ 0;47	♉ 3	−2
H	♒ 25	noon	
M	♐ 5	♐ 7	

For the date of conception, 278^d 9^h earlier, one finds Diocletian 97 Mekheir (VI) 26 9th hour of night = A.D. 380 February 22, 3 A.M., the sun being in ♓ 4, the moon about ♒ 23.

As locality we may assume Alexandria since it belongs to the "third clima" in Ptolemy's terminology.[16] Using Theon's tables for this latitude, from H = ♒ 25 one would obtain as culminating point M = ♐ 7;4 and not ♐ 5.

The hour given in the text does not agree too well with the position of H (♒ 25) and sun (♐ 4). From these data it would follow that the moment in question was 6;16 seasonal hours (Alexandria) or 5;21 equinoctial hours after sunrise. The text gives for it "6(th) hour" which would imply the use of equinoctial hours, thus constituting the only known exception from our general experience which leads us to interpret "hours" as meaning seasonal hours. But even without computation the text itself is contradictory since ♐ 5 is given as Midheaven, ♐ 4 as solar position. This implies a moment after noon or within the 7th hour and not the 6th hour.

The procedure carried out in the third version can be explained much more simply than it is expressed by the text. Suppose the arc from H to M were exactly 90°. Then each of the three "loci" from H to M would be exactly 30° long. It is furthermore the accepted rule that H as well as M should lie 5° inside of its respective locus, counted in the direction of the zodiacal signs.[17] In our case, however, the arc from H to M is 10° or $\frac{1}{9}$ of a quadrant shorter than a quadrant. Thus we take $\frac{1}{9}$ away from each 30° section leaving us with three loci each of which is 30 − 3;20 = 26;40° long. Since M = ♐ 5 its locus, the 10th, begins exactly at ♐ 0°

---

[15] Engelbrecht, *Heph.* p. 22, and *CCAG* 8,1 p. 146 give incorrectly A.D. 381.
[16] *Cf.* p. 5. This is explicitly confirmed by *CCAG* 8,2 p. 79,10 (the length of a seasonal hour for ♓ 1 is 17;8) and p. 78,17/18 ("clima of Lower Egypt").
[17] *Cf.* No. L 428.

and ends at ♐ 26;40. Counting from this point again 26;40° leads to ♑ 23;20 as the end of the 11th locus; plus 26;40 gives ♒ 20 for the end of the 12th locus, thus 5° distant from H = ♒ 25, as it should be. Similarly, for the remaining centers (*cf.* fig. 15).

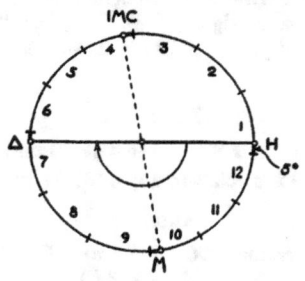

FIGURE 15.

No. L 401

CCAG 8,1 p. 232,1 to p. 234,26

TRANSLATION

15.[1] On the Progression [of the moon].[2]

Clima 4. Moon Taurus 23;30. Saturn Capricorn 16;40,[2a] Jupiter Cancer 2;50, Mars Capricorn 4;42, Venus Pisces[3] 3;14, Mercury Gemini[4] 0;18. Horoscopos Virgo 4, Midheaven Taurus 30.

It is necessary then, to find the "progressions" of the moon toward each planet.[5] Rising time of sphaera recta for the moon 141;1. Sphaera recta for Midheaven 147;44. Difference 6;43. Time degrees of daylight for the moon 17;24. Divide the 6;43 by this. It makes 0;23 seasonal hours.[6] Thus the moon is 0;23 seasonal hours distant from Midheaven.

Left Quartile of Venus: Gemini[7] 3;14. Sphaera recta 151;8. Excess over the time-degrees of Midheaven 3;24. Time degrees of daylight of the Quartile of Venus 17;23.[8]

---

[1] From Rhetorius.
[2] Restoration by the editors, obviously based on the first sentence of the second section. What is meant is the succession of chronocratorship beginning with the moon. — The two manuscripts on which our text is based (P = Paris.; V = Venet.) write the title between the first and the second section. At least in P the first section is written in direct continuation of the preceding chapter.
[2a] P: 16;8.
[3] P: Aquarius; omitted in V. Emended by the editors to Pisces because later on the Quartile of ♀ is located in ♊ 3;14.
[4] *Sic*, instead of Capricorn. The correct position is used later on.
[5] A similar procedure, called περίπατος, is described in Vettius Valens IV,29 (p. 205, 13-16 Kroll).
[6] Accurately: 0;23,9,...
[7] P: Aquarius.
[8] *Sic*, instead of 17;43.

Divide the 3;24 by this. It makes 0;8 seasonal hours.[9] Thus the Quartile of Venus is 0;8 seasonal hours distant from Midheaven. Add this to the 0;23 of the distance of the moon from Midheaven; it makes 0;31 seasonal hours [10] which the Quartile of Venus is distant from the moon. And the excess of the moon over the Quartile of Venus was 16 which is 10 years and 2 days.[11]

The Diameter of Mercury is Cancer 0;18. Sphaera recta 180;20. Sphaera recta of Midheaven was 147;44. Difference 32;36. Time degrees of daylight for the Diameter of Mercury 18;7. Divide the 32;36 of the excess by the 18;7 time degrees. It makes $1\frac{1}{2}$ [$\frac{1}{5}$ $\frac{1}{10}$] hours.[12] Consequently the Diameter of Mercury is $1\frac{1}{2}$ $\frac{1}{5}$ $\frac{1}{10}$ seasonal hours distant from Midheaven. And the Quartile of Venus was 0;35 seasonal hours distant from Midheaven.[13] This multiplied by 18;7 makes 3;18 time degrees.[14] Subtract this 3;18 from the 32;36 of Mercury. Remainder 29;18. Since it has been shown that the moon in its course will contact the Quartile of Venus at 10 years 2 days and that Venus, taking over the period by virtue of the Quartile, will be chronocrator from the 11th year for 29 years 3 months, Mercury will then take over by virtue of the Diameter the period after 39 years and will be chronocrator.[15]

Again, Jupiter in Cancer 2;50. Sphaera recta 183;7. Sphaera recta of Midheaven 147;44. Excess 35;23. Time degrees of daylight for Jupiter 18;6. Divide the 35;23 by this. It makes approximately 1;57 seasonal hours.[16] The distance of Jupiter from Midheaven is therefore 1;57 seasonal hours. [And the Diameter of Mercury was 1;48 seasonal hours distant from Midheaven.] Multiply this by the 18;6 seasonal hours of Jupiter. It makes 32;35 time degrees. Subtract it from the 35;23. Remainder 3;49.[17] Thus we say that from year 39 and 3 months the chronocratorship of the Diameter of Mercury lasts for 2 years 9 months approximately. And then Jupiter will take over the period, i.e., after [18] 42 years 0 months.

Again, the Trine of Venus in Cancer 3;14. Sphaera recta 183;33. Sphaera recta of Midheaven 147;44. Difference 35;49. Time degrees of daylight for the Trine of Venus 18;6. Divide the difference 35;49 by it. It makes approximately 1;58 seasonal hours. Thus the Trine of Venus is 1;58 seasonal hours distant from Midheaven. Jupiter is 1;57 [19] seasonal hours distant from it. Multiply this by the time degrees of the Trine of Venus, i.e., the 18;6 in the direction of direct motion. It makes 35;18 time degrees. Subtract this from the 35;49 of the difference. Remainder 0;31 time degrees. This is approximately 6 months 2 days. Hence Jupiter [20] will be chronocrator from year 41 month 10 to year 42 month $4\frac{1}{20}$ (?).[21]

Again the Diameter of Mars, Cancer 4;42. Sphaera recta 185;10. Sphaera recta of Midheaven 147;44. Difference 37;26. Time degrees of daylight of the Diameter of Mars 18;6. Divide the 37;26 by this. It makes $2\frac{1}{15}$ seasonal hours, which is the distance of the Diameter of Mars from Midheaven. But Venus is $1\frac{1}{2}$ $\frac{1}{3}$ $\frac{1}{12}$ $\frac{1}{15}$ seasonal hours [22] distant from Midheaven. Multiply this by the seasonal hours of Mars, i.e., by 18;6. It makes 35;36 time degrees.[23] Subtract it from 37;26. Remainder 1;51.[24] Thus, approximately after 1 year and 8 months, Venus, receiving by virtue of the trine the period from Jupiter in the year 42 month [4] day 26(?) will rule for about 1 year [8] [25] months until year 44 and 26 17(?) day.[26]

Diameter of Saturn Cancer 16;40. Sphaera recta {1[9]8;8. Sphaera recta of Midheaven 147;[44]}.[27] Difference 50;24. Time degrees of daylight of the Diameter of Saturn 17;57. Divide the difference 50;24 by it. It makes 2;41 seasonal hours.[28] Thus Saturn is

[9] This result is incorrect since 3;24 : 17;23 = 0;11,44,... The correct value would be 3;24 : 17;43 = 0;11,31,... *Cf.* the next note.

[10] Indeed 0;8 + 0;23 = 0;31. Since, however, 0;8 is an error for 0;11 or 0;12 (*cf.* the preceding note), the distance from the moon to the Quartile of Venus should be 0;34 or 0;35. Indeed, in the next section, 0;35 is quoted as distance of the Quartile of Venus from Midheaven (!). *Cf.* note 13.

[11] The whole procedure which leads to this result has fallen out. We can restore it as follows: "Multiply 0;23, the distance of the moon from M, by 17;43, the time degrees of the Quartile of Venus. Result: 6;47. Add this to the 3;24 of the Quartile of Venus. Result: 10;11 or approximately $10\frac{1}{6}$ which is 10 years and 2 months." Thus the 16 of the text is an error for $10\frac{1}{6}$ and the 2 days is an error for 2 months.

[12] Accurately: 1;47,58,0,... $\approx$ 1;48 = $1\frac{1}{2}$ $\frac{1}{4}$ $\frac{1}{20}$.

[13] Actually this distance was 0;11 or 0;12 (*cf.* note 9). The value 0;35 is the distance from the moon (*cf.* note 10). In the subsequent multiplication 0;11 was used. *Cf.* the next note.

[14] Actually 3;18 = 0;11 · 18 (instead of 0;11 · 18;7 = 3;19,17).

[15] The allotment of periods is thus as follows: the moon rules the first 10 years, to be followed by Venus from the 10th year to 39 years 3 months.

[16] Accurately: 1;57,24,...

[17] *Sic*, instead of 2;48.

[18] This is consistent with the preceding allotment to Mercury. *Cf.* note 15.

[19] The manuscripts have 1;58 (dittography from last figure).

[20] Consistent with the statement at the end of the preceding section; *cf.* note 18. The result, however, is incorrect, since 0;31 year = 6 months 6 days.

[21] *μῆνων δ' κ'*. Meaning of *κ'* not clear; *cf.* the similarly enigmatic ending of the next section.

[22] The Trine of Venus was found in the last section at 1;48 from Midheaven but $1\frac{1}{2}$ $\frac{1}{3}$ $\frac{1}{12}$ $\frac{1}{15}$ = 1;59. Therefore, A. Rome emended the text to $1\frac{1}{2}$ $\frac{1}{3}$ $\frac{1}{12}$ $\frac{1}{20}$ (= 1;58). *Cf.* the next note and note 29.

[23] For 1;59 · 18;6 one finds 35;53,54, for 1;58 · 18;6 the value 35;35,48 $\approx$ 35;36 as in the manuscripts; incorrectly emended in *CCAG* 8,1 p. 234,13 to 35;32.

[24] The manuscripts have 1;37,51 (V) and 1;37 with 51 over the 37 (P) respectively. 1;50 would be correct. The emendation 1;54 in *CCAG* is based on the previous incorrect emendation (*cf.* note 23).

[25] Following the emendation of A. Rome. The manuscripts have 2 months.

[26] *Cf.* note 21.

[27] The text is confused here; *cf.* the Apparatus in *CCAG*.

[28] Accurately 2;48,28 and not 2;41. *Cf.* the next note.

distant from Midheaven 2 seasonal hours $\frac{1}{2}\ \frac{1}{4}\ \frac{1}{15}$.[29] But Mars was distant from Midheaven 2;4 seasonal hours. Multiply this by the 17;57 time degrees of Saturn. It makes 37;6 time degrees. Subtract this from the 50;24 of the excess of Saturn. Remainder 13;18. Thus after the 44 years Saturn takes over and brings to an end (the period of Mars, i.e.,) approximately 13 years 2 months, at [5]7 years 2 months.[30]

ASTRONOMICAL COMMENTARY

Diagram: fig. 16

We encounter great difficulties in the dating of this horoscope. A systematic search leads to the result that the three outer planets cannot be located near the places given to them in the text except for December 400/January 401. The position of the sun is omitted but the moon restricts us within this interval either to December 15 or to January 11 and the first date gives the better results for Mars and Jupiter whereas Saturn and Mercury are better represented by the second:

	400 Dec. 15	401 Jan. 11	Text
☉	♐ 25	♑ 22	——
☾	♉ 25	♉ 23	♉ 23;30
♄	♑ 9	♑ 12	♑ 16;40
♃	♋ 0	♊ 26	♋ 2;50
♂	♑ 0	♑ 20	♑ 4;42
♀	♐ 4	♐ 9	♓ 3;14
☿	♐ 6	♑ 18	♑ 0;18

Venus is far from the position given in the text and also Mercury is only roughly correct.[31] The date 401 (or 400) is therefore not at all certain. No emendation of the figures of the text is possible since all numbers are extensively used in the astrological sections. It seems not impossible that we have here an artificially made up example of positions.

The astrological section contains a great number of rising times and related numbers for sphaera recta and for clima IV. M = ♉ 30;40 belongs to H = ♍ 4 for clima IV, rounded off in the text to ♉ 30. All other numbers, however, give degrees and minutes in exact agreement with the Handy Tables of Theon.

ASTROLOGICAL COMMENTARY

Diagram: fig. 16

The goal of the discussion, which follows the list of planetary positions, is the determination of the duration of chronocratorship of each planet. The total of these periods, ending with Saturn's rule, is then the length of life allotted to this nativity.

The procedure consists in finding the "distance" from Midheaven for each planet or for a point in aspect to the planet. In this way the following points are investigated (*cf.* fig. 16):

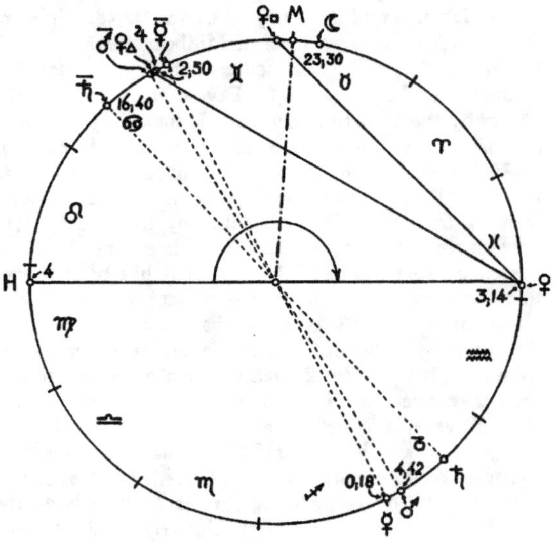

FIGURE 16.

n		λ_n
1.	☾	♉ 23;30
2.	Quartile of ♀	♊ 3;14
3.	Diameter of ☿	♋ 0;18
4.	♃	♋ 2;50
5.	Trine of ♀	♋ 3;14
6.	Diameter of ♂	♋ 4;42
7.	Diameter of ♄	♋ 16;40.

The "distance" of these points from M = ♉ 30 is to be measured in time degrees and found in the following way. Let n be the number of the point in question, following the above arrangement. For $n \geqq 2$ the "distance" Δ_n is given by

$$(1)\quad \Delta_n = [a(\lambda_n) - a(\mathrm{M})] - [a(\lambda_{n-1}) - a(\mathrm{M})]\,\frac{h(\lambda_n)}{h(\lambda_{n-1})}$$

where $a(\lambda)$ represents the right ascension of $\lambda + 90$ (as tabulated in the Handy Tables for sphaera recta), $h(\lambda)$ the length of one seasonal hour in time degrees for clima IV and a solar longitude λ.

The meaning of this procedure is roughly this: $a(\lambda_n) - a(\mathrm{M})$ is the distance in right ascension between the points n and M. Therefore,

$$[a(\lambda_n) - a(\mathrm{M})] - [a(\lambda_{n-1}) - a(\mathrm{M})] = a(\lambda_n) - a(\lambda_{n-1})$$

would be the ascensional difference between the points

[29] Which equals 2;49. *Cf.* note 22 for a similar approximation; perhaps $\frac{1}{15}$ is the smallest fraction of an hour to be reckoned.

[30] The role of the Diameter of Saturn is to destroy the life as is clear from the following example (No. L 488).

[31] Note that the elongation of 63° between Venus and Mercury is very near to the extremal possible amount.

n and $n-1$. Such a procedure would, however, not depend on M and on the geographical latitude. In order to bring these elements also into play, the second term is modified by the ratio of the seasonal hours associated with λ_n and λ_{n-1} for the given clima. The whole procedure is a drastic example of the ever increasing complication of astrological doctrines during late antiquity and the Middle Ages.

Formula (1) holds only from $n=2$ on. For $n=1$

$$(2) \quad \Delta_1 = [a(\lambda_1) - a(M)] \frac{1}{h(\lambda_1)}$$

is used as "distance" from M.

The numbers which result from (1) and (2) are counted in degrees and minutes. They are now supposed to signify years of rulership assigned to the planet in question.

In transforming the resulting numbers of degrees and minutes into years and months the text operates very carelessly indeed:

		Text		Total
$n=1$	10;11°	10^{y}2;12^m	10^{y}2^d(!)	10^y
2	29;18	29^{y}3;36^m	29^{y}3^m	39^{y}3^m
3	2;48	2^{y}9;36^m	2^{y}9^m	42^y
4	0;31	6;12^m	4^m(!)	42^{y}4^m
5	1;50	1^{y}10^m	1^{y}8^m(!)	44^y
6	13;18	13^{y}3;36^m	13^{y}2^m(!)	57^{y}2^m

This carelessness cost the nativity 9 months of lifetime since the correct total would be $57;57^y = 57^y11^m$.

No. L 412

Marinus, Life of Proclus

TRANSLATION

(35.)[1] In order that those who love the higher things may be able to conclude, from the configuration of the stars under which he was born, how the Lot (of Fortune) fell to him not in the lowest place but in the very best, we shall give their schemata as it occurred at birth:[2] ☉ in Aries 16;26[3] degrees, ☾ in Gemini 17;29 degrees, (♄)[4] in Taurus 24;23 degrees, ♃ in Taurus 24;41[5] degrees, ♂ in Sagittarius 29;50, ♀ in Pisces 23 degrees,[6] [☿] in Aquarius 4;42 degrees, Horoscopos in Aries 8;19 degrees, Midheaven in Capricorn 4;42 degrees, the ascending (node) in Scorpio 24;33 degrees, preceding conjunction in Aquarius 8;51 degrees.

(36.) He died in the 214th year from Julian's reign, under the Athenian archonship of Nicagoras, jr., in the Athenian month Munychion 17, according to the Romans April 17. . . .

(37.) There also occurred signs in the year before his death, e.g., a solar eclipse of such magnitude that it became night at daytime. Deep darkness fell and the stars became visible. This happened in Capricorn toward the rising center. The calendar makers recorded also another (eclipse) to happen at the completion of the first year (after his death). . . .

COMMENTARY

Diagram: pl. 26

An extensive literature has accumulated around this horoscope and the additional date of the death of Proclus. At first Fabricius in his edition of the Vita Procli by Marinus (London 1703) computed as date of birth 412 February 8, emending the solar longitude to ♒ 20;26 as is obviously necessary.[7] The same result was obtained by Delambre[8] and independently by Stein.[9] It can be immediately confirmed by computing with the approximate tables used throughout in this work.[10] Thus one finds for 412 February 8:

	Text	Computed	Text—comp.
☉	♈ 16;26(!)[11]	♒ 20;16	(?)
☾	♊ 17;29	♊ 16;48	+0;40°
♄	♉ 24;23	♉ 27;33	−3;10
♃	♉ 24;41[12]	♉ 24;45	0
♂	♐ 29;50	♑ 3;9	−3;20
♀	♓ 23[..][13]	♒ 23;45	[−0;50]
[☿][13]	♒ 4;42	♒ 6;47	−2
H	♈ 8;19	9 A.M.	
M	♑ 4;42	correct for Rhodes	
☊	♏ 24;33	♏ 26;58	−2;30
conj. 412 Jan. 29	♒ 8;51	♒ 9	−0;10

The date of his death being 485 April 17 he lived 73 years and 2 months.

The years of Julian must be counted in the Athenian calendar, resulting from the classical-pagan tradition of

---

[1] The division of the text into chapters and their numbering is modern.

[2] We have checked the following data with the two existing manuscripts, Vat. Graec. 1374 fol. 162 [V] and Pal. Graec. 404 fol. 100 [P].

[3] V: 16;26, P: 16. Stein [1] p. 72 states incorrectly " [V.] et [P.] portent tous les deux: le Soleil 16°26′ dans le Bélier. . . ."

[4] Both manuscripts have here καί instead of the symbol for Saturn.

[5] P: 24;44.

[6] Perhaps minutes lost in the lacuna.

[7] A diagram which shows the emended data was published by Fabricius and reproduced in Steck-Schönberger PD p. 10 and by Stein [1] p. 73. No diagram seems to be given in the manuscripts.

[8] Boissonade. Proclus p. 138 note; repeated Freudenthal [1] p. 487.

[9] Stein [1], with a slight error in the date of the preceding conjunction.

[10] Cf. Introduction p. 2.

[11] V: 16;26 P: 16; cf. note 3.

[12] 24;44 in P.

[13] Lacuna in text.

the circle of Proclus.[14] Now Hekatombaion corresponds to about July; Julian became Augustus in December 361, thus his first year will be counted from July 361, the 124th year from July 484. Thus the death of Proclus would fall in 485 April 17 in his 74th year of life.

The fact that Marinus assigns in these passages a life of 75 years to his hero, whereas the astronomical data require his death in the 74th year, has caused a long series of arguments. Up to 1888 they are listed by Freudenthal who himself proposes the absurd solution that " bei der Berechnung der Nativität waren Fehler schwer zu vermeiden " (p. 493), based on the judgment of an astronomical " Fachmann " who thinks that the " damaligen unpraktischen Zahlzeichen und schwerfälligen Rechnungsmethoden " favored an error.[15] In fact the tables of Ptolemy or Theon are almost identical in structure and accuracy with the modern tables used here and provide the data of a horoscope by means of a few additions exactly as in modern tables. And it is, of course, totally irrelevant whether one uses our numerals or the Greek ones.

The article on Proclus by R. Beutler in *RE* 23,1 (1957) col. 186 ff. again discards the precision of the astronomical data in favor of the loose statement of a 75-year lifetime. It furthermore gives incorrectly February 18 as the date of the horoscope.

The solar eclipse mentioned in Chapter 37 as having occurred is the eclipse of 484 January 14, the sun being in ♑ 26, shortly after sunrise in Athens.[16] The next, predicted, solar eclipse has the date 486 May 19.[17]

No. L 419

Vettius Valens, Additamenta, Kroll p. 365,29 to 366,5

Another example, (the) 3(rd).[1] Reign of Valentinian. Clima that of Spain. This person was killed at the age of 36. Year of Diocletian 135, Epiphi (XI) 8, beginning of 1st hour (of the day). Sun in Cancer 7;11, moon in Aries 22;30, Horoscopos in Cancer 7;20, (the sun at the time of)[2] full moon in Gemini 28;40, 7th hour of day, Payni (X) 29. The days from the full moon to the day of birth are 7½[3] which is the ½ of 15. Since the solar gnomon exceeds the horoscopic gnomon I add this ½ part to the beginning of the first hour and do thus: time-degrees of the sun 18;5 of which one half is 9; and the rising time of the sun 75;7[4] making together 88 time-degrees. With these I entered the table and found the Horoscopos approximately in Cancer 14.[5] With these 14 time (degrees) I entered the table and found [6] 34 periods (i.e., years) 10 months, corresponding to the (number) 10, which makes 6 sixtieths. Taking now the time-degrees of 14 degrees of Cancer I found [7] 18;0. Multiplying these by 12, I found 216;0 the 6th of which makes 36 periods (i.e., years); which he lived.

ASTRONOMICAL COMMENTARY

Diagram: pl. 26

Diocletian 135 Epiphi (XI) 8 = 419 July 2. For the time near sunrise one finds

	Text	Computed	Text—comp.
☉	♋ 7;11	♋ 9;40	—2;30°
☾	♈ 22;30	♈ 23	—0;30

The preceding full moon occurred at Payni (X) 29 7^h daytime = June 23 1 P.M. the sun being in ♋ 1;30, the moon in ♑ 1. The text gives for the sun at opposition ♊ 28;40 or again 2;50° less than with our norm.

Since the person in question was killed at the age of 36 our text cannot have been written before 455.

ASTROLOGICAL COMMENTARY

No. L 419 and No. L 431 perform a series of operations, which is attested in no other horoscope, to determine the length of life which belongs to the given nativity. In order to understand this procedure one has to realize that one is dealing with three essentially different steps:

(*a*): accurate fixation of the hour in question by means of purely astrological speculations which involve sun and moon;

(*b*): given the longitude of the sun and the seasonal hour (of daytime or night) as determined in the first step, find the rising point of the ecliptic, i.e., the Horoscopos. This is a straightforward astronomical problem.

(*c*): derive from the position of the Horoscopos, found in (*b*), the length of life. Here the tables from Vettius Valens VIII (pp. 321 to 324 Kroll) come into play. This step is, of course, again purely astrological in character.

Step a. Both No. 419 and No. 431 give the hour of the birth and both re-evaluate it on the basis of some only very incompletely understood procedure. Both texts operate with the otherwise unknown concepts of " solar gnomon " and " horoscopic gnomon." In No. 431 we are given:

(1)   solar longitude λ_s = ♑ 19;2
        lunar longitude λ_m = ♉ 23;30
        clima IV

---

[14] Hanell [1] p. 8 has remarked that the equation Munychion 17 = April 17 is generally correct as far as the months are concerned but merely formal for the day numbers.
[15] Freudenthal [1] p. 492 f.
[16] *Cf.* Ginzel, *Kanon*, p. 222. Also Stein [1] p. 74.
[17] Oppolzer, *Canon*, No. 4019.

No. L 419
[1] The "second" is No. L 431.
[2] The text gives erroneously "full moon" only.
[3] *Sic*, instead of 8½.
[4] Scribal error for 79;7. The next number is again correct.

[5] The Handy Tables give for ♋ 14 the value 88;23.
[6] *Cf.* the table Kroll p. 324 (line 16) for ♋ 14.
[7] From the Handy Tables, clima 4.

From this, it is said, follows a "solar gnomon" of 6° which is the difference of rising times from sun to moon. This is obviously impossible.[8] One arrives, however, at a reasonably close result if one takes the difference between the corresponding time-degrees from Theon's Tables which are used throughout in the following:

$$h_m - h_s = 17;24 - 12;5 = 5;19 \approx 6^q.$$

We now find $\bar{h}_s$, the time-degrees which correspond to the point ♋ 19;2 diametrically opposite the sun, and multiply the result by 12. The motivation for this procedure lies in the fact that we are dealing with a night-birth and that $\bar{h}_s$ gives the length of one seasonal hour of night. Consequently, $12 \bar{h}_s = 12 \cdot 17;55 = 215°$ is the length of the night.

It was furthermore given that the birth happened at the third hour of night. This gives the motivation to form $3 \cdot 215 = 645 \equiv 285 \pmod{360}$ though no astronomical sense can be connected with this operation. The result 285 is called the "horoscopic gnomon" and it is found to be greater than the solar gnomon. From this result, which can hardly be more than the statement $285 > 6$, we are told it may be concluded that somehow[9] the 3 hours must be reduced to $2\frac{1}{2}$ hours only and this is indeed the amount which is used in step (b).

In No. L 419 all these steps are suppressed. We are given

$\lambda_s = $ ♋ 7;11   Clima of Spain (= clima IV)
$\lambda_m = $ ♈ 22;30   beginning of 1st hour of daylight.

From this and the time elapsed since full moon, it is mysteriously concluded that the solar gnomon exceeds the horoscopic gnomon[10] and thus the "beginning of the first hour" extended to $\frac{1}{2}$ hour, to be used in the next step.

Step b. This step is fully intelligible. As the result of Step (a) we are left with the seasonal hours

No. L 419: $n = 0;30$ hours of daylight $\lambda_s \approx$ ♋ 7
No. L 431: $n = 2;20$ hours of night $\lambda_s \approx$ ♋ 19.

From the Handy Tables we find that the corresponding time-degrees are

$$h(\lambda_s) = 18;5 \quad \text{thus} \quad n \cdot h = 0;30 \cdot 18;5 \approx 9$$

and

$$\bar{h} = h(\lambda_n) = 17;55 \quad \text{thus} \quad n \cdot \bar{h} = 2;20 \cdot 17;55 \approx 41;48$$

respectively.

We now find from the same tables the corresponding rising times and add them to the previous results:

$$n \cdot h + \rho(\lambda_s) = 9 + 79;7 \approx 88°$$

---
[8] This difference would be 88;28°.
[9] The time interval of 10 days between birth and the preceding conjunction plays some role.
[10] Perhaps simply because the horoscopic gnomon is zero, ☉ and H being almost identical.

and

$$n \cdot \bar{h} + \rho(\lambda_s) = 41;48 + 93;7 = 134;55°.$$

It is easy to see that this total is the rising time of the point which is in the horizon (cf. the stereographic projection in fig. 17) at the given hour. This point is again

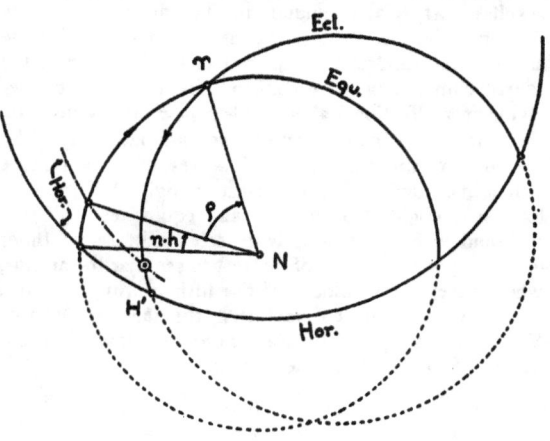

FIGURE 17.

called "Horoscopos" (H') though it is the Horoscopos for the hour n which is, as we have seen in Step (a), a modification of the hour originally given.

Having thus found the rising time $\rho(H')$, one uses the tables in the inverse direction and finds

in L 419:   H' = ♋ 14   [since $\rho(H') = 88;23$]

and

in L 431:   H' = ♌ 23   [since $\rho(H') = 134;55$].

These are the data to be used in Step (c).

Step c.
With these new horoscopic values we enter the first table which follows Book VIII of Vettius Valens. There we find[11]

for ♋ 14:   ἀριθμός 10   34ʸ10ᵐ15ᵈ
for ♌ 23:   ἀριθμός 22   73ʸ11ᵐ.

No. L 419 quotes the "number" 10 and the associated 34 years 10 months (without days) but makes no use of the latter, and the same holds for No. L 431. Both texts now proceed to compute the length of daylight which would correspond to a solar longitude of ♋ 14 and ♌ 23 respectively. This leads to

$$12 \cdot h(\text{♋} 14) = 12 \cdot 18;0 = 216$$
$$12 \cdot h(\text{♌} 23) = 12 \cdot 16;45 = 201.$$

Now the "numbers" come into play in the form of factors 0;10 and 0;22. The result is interpreted as years of lifetime. Thus one finds

---
[11] Kroll p. 324 lines 16 and 25.

in No. L 419:   216 · 0;10 = 36 years
in No. L 431:   201 · 0;22 = 73;42 = 73⅖ ¹⁄₃₀ years.

This is the final result of the astrological investigation.

One may ask why the lifetimes listed in Vettius Valens' table were disregarded and replaced by the result of a special computation. The answer is simply that the tables of Vettius Valens are computed for the clima of Alexandria whereas our examples concern the fourth clima. One may confirm this statement by deriving from Vettius Valens' table a new one by dividing the numbers in the column of "years-months-days" by the corresponding "ἀριθμοί." The result [12] is a scheme of length of daylight with shortest daylight 150° in ♑ 8°, longest daylight 210° in ♋ 8 and equinoxes (180) in ♈ 8 and ♎ 8. The underlying scheme of rising times follows System A. It is of interest to see that the author, who wrote in the middle of the fifth century or later, was still aware of the fact that the table in Vettius Valens was computed for the latitude of Alexandria and had to be modified in case of a different clima.

No. L 428

CCAG 8,1 p. 221,1 to p. 222,28

TRANSLATION

12.[1] How one has to find by degree the Kentra of the 12 Loci.

Example of a nativity: Clima 5, year of Diocletian 145, Thoth (I) 10 to 11, beginning of the 4th hour of night: sun 14;19 of Virgo, moon 3;4 of Pisces, Saturn 14;21 of Sagittarius, Jupiter 15;41 of Libra, Mars 21;6 of Virgo, Venus 25;40 of Leo, Mercury 3;37 of Libra, Horoscopos 25;16 of Taurus, Midheaven 3;37 of Aquarius, ascending node 3;41[2] of Capricorn, (preceding) conjunction 1;26 of Virgo, (Lot of) Fortune 6;31 of Sagittarius,[3] Daimon 14;1 of Scorpio.[4]

We shall put forward the explanation of the 12 loci by degree in order that we may show according to the natural system and its corollary the distances of the degrees which were in centers and in apoklima as follows:

Because, as Ptolemy says,[5] the 5 degrees preceding (the Horoscopos) [.....] we subtract these from the given degrees of the Horoscopos and thus find for the beginning of the center of the Horoscopos (♉) 20;16. To this correspond for sphaera recta 137;44 as rising time. The 20;16 degrees of Taurus make at the 5th Clima 17;44 time degrees, and doubling this we get 35;28 time degrees, and subtracting it from 137;44 we find 102;16 as remainder. Aries 13;21 (which is the beginning of the 12th locus) culminates at sphaera recta with 102;16. And again in the same way we subtract 35;28 from 102;16, finding 66;48 as remainder; and we find again that with them culminates at sphaera recta Pisces 4;8. Thus we shall have Pisces 4;8 for the beginning of the 11th locus.[6] {Consequently the moon (in ♉ 3;4) belongs rather to the 10th locus.} Again subtracting the 35;28 from 66;48 with 31;20 as remainder, we find that with them culminates in sphaera recta Capricorn 29;7. This then will be the beginning of the (locus) Midheaven. But at the beginning Midheaven was Aquarius 3;37 which precedes (♑ 29;7) by 4;30. We should have, however, 5°; thus we have a difference of 0;30. The reason for it is as follows. Ptolemy says neither that one must subtract from the center time degrees nor (that one must subtract) longitudinal degrees, but the (degrees) of the ideal [7] and fixed zodiac which (is the basis of the calculation) of the aspects, (i.e.) the variable longitudinal and time degrees. We, however, took the variable and perceptible 5 degrees. Thus, when Ptolemy says [8] "and the sextile which is to the right of these 30 degrees," [9] then he is speaking of the ideal (degrees), for these are (degrees) which always accurately make 30 degrees together; but the variable (degrees) give only an approximate (total).

But we must now determine the locus more accurately. While there are from ♑ 29;7 to ♋ 29;7 of culminating times 180 degrees of the semicircle, there are from ♑ 29;7 to ♉ 20;16 of simultaneously culminating times, six times 17;44, i.e., 106;24. Thus, the difference between the simultaneously culminating times from ♉ 20;16 to ♑ 29;7 and the semicircle is 73;36. Its third part is 24;32 times. If we now subtract the 24;32 from the given 31;20 we find the remainder 6;48. Simultaneously with these culminates ♑ 6;13°. Thus the beginning of the 9th (locus) will be ♑ 6;13.[10] If we again subtract from the 6;48 the 24;32 we obtain the remainder 342;16. With these culminates ♐ 16;46.[11] Thus the beginning of the 8th (locus) will be ♐ 13;48.[11]

---

[12] Example: for ♋ 8 the table gives 56 years 0 months (Kroll p. 324,10). The corresponding ἀριθμός is 16. Consequently, we have 56;0;16 = 210° = 14ʰ as length of daylight. This is the classical value for the length of daylight at the summer solstice at Alexandria. Cf. for details Ch. III,F (p. 174).

No. L 428

[1] From Rhetorius.
[2] CCAG 8,1 p. 221,7 gives incorrectly λ' μα' instead of γ' μα' (checked on photo).
[3] From Λᵧ = ♉ 25;16 − (♓ 3;4 − ♍ 14;19) = ♐ 6;31.
[4] From Λᴅ = ♉ 25;16 + (♓ 3;4 − ♍ 14;19) = ♏ 14;1.
[5] Tetrabiblos ed. Robbins III,10 (p. 272/273), ed. Boll-Boer III,11 (p. 129,18).

[6] The text has here the words "♑ 3;11 is the beginning of the 10th (locus)" which the editors replace by the equally meaningless statement "♒ 3;11 is the beginning of the 10th (locus)." Obviously this whole phrase does not belong to the text. The next sentence is probably only a scholion, though factually correct.
[7] νοητοῦ.
[8] Tetrabiblos III,11 Boll-Boer p. 129,20 = III,10 Robbins p. 272,6.
[9] The 5 + 25 degrees preceding and following the Horoscopos.
[10] The text has here 6;16.
[11] Should be 13;42.

Again subtract the 24;32 from the 342;16; the remainder is 317;44. With these culminates ♏ 20;16 which is diametric to the Horoscopos. The diametrically opposite points will be the beginnings of the loci below the horizon, i.e., of II, III, IV, V, VI.

COMMENTARY

Diagram: pl. 26

For Diocletian 145 Thoth (I) 11 = A.D. 428 September 8 one finds

	Text	Computed	Text—comp.
☉	♍ 14;19	♍ 16;45	−2;20
☾	♓ 3;4	♓ 6;36	−3;30
♄	♐ 14;21	♐ 17	−2;40
♃	♎ 15;41	♎ 18	−2;20
♂	♍ 21;6	♍ 24	−2;50
♀	♌ 25;40	♌ 29	−3;20
☿	♎ 3;37	♎ 8	−4;20
H	♉ 25;16 beg. 4^{h}n.	10 P.M.	
M	♒ 3;37	♒ 2;47	
☊	♑ 3;41	♑ 6;13	−2;30
conj.	♍ 1;26	Aug. 26: ♍ 4	−2;30

In the determination of the 12 loci Rhetorius follows a very refined procedure. First he ends the 12th locus 5° before the Horoscopos, that is, at ♉ 20;16, in accordance with a rule in the *Tetrabiblos*.[12] He then determines for the 5th clima ($\phi = 40;56$) the length of the seasonal hours corresponding to a solar longitude of ♉ 20;16. These seasonal hours are now arranged in 3 groups of two consecutive hours each, and the points of the ecliptic are found which culminate simultaneously with the end of the 2nd, 4th, and 6th hour. These points of the ecliptic are then the boundaries of the loci X, XI, and XII.

The meaning of these operations becomes clear if one uses stereographic projection of the celestial sphere from its North pole—a procedure well known in the theory of the Astrolabe since Ptolemy's *Planisphaerium*. In fig. 18 the arc AB of the parallel circle ABCD represents one half the length of daylight if the sun rises at A = ♉ 20;16. The projection from N of the points 2, 4, and 6 on the ecliptic gives the boundaries of the last three loci.

The corresponding numerical procedure is as follows. Using Theon's Handy Tables we find $h = 17;44°$ as the length of one seasonal hour for Clima V and $\lambda =$ ♉ 20;16. We determine furthermore the point of the equator whose right ascension is the same as for $\lambda =$ ♉ 20;16 by using the table for sphaera recta and find $a_0 = 137;44$ (actually: right ascension $+90°$).[13] We now form three arcs of equal length, totalling half the length of daylight for Clima V and $\lambda =$ ♉ 20;16:

$a_0 = 137;44$

$a_1 = a_0 - 2h = 137;44 - 35;28 = 102;16$

$a_2 = a_1 - 2h = 66;48$

$a_3 = a_2 - 2h = 31;20.$

Using again the table for sphaera recta we can find the corresponding points of equal right ascension on the ecliptic:

$\lambda_0 =$ ♉ 20;16

$\lambda_1 =$ ♈ 13;21[14]

$\lambda_2 =$ ♓ 4;8[15]

$\lambda_3 =$ ♑ 29;7.

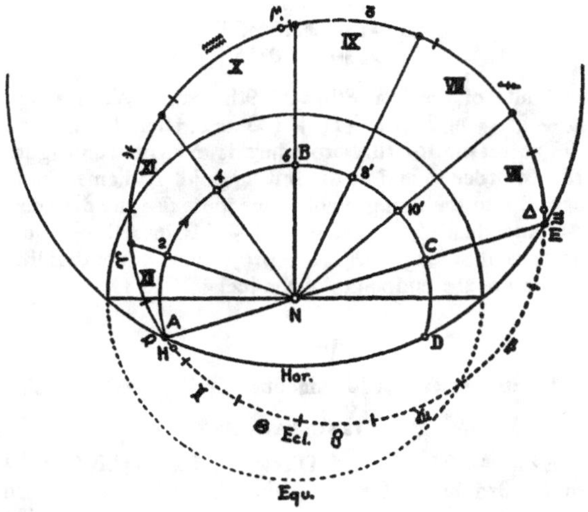

FIGURE 18.

These are the boundaries of the 10th, 11th, and 12th locus.

Rhetorius is surprised that λ_3 is not M − 5° = ♒ 3;37 − 5 = ♑ 28;37. The reason for this discrepancy escaped him as his irrelevant references to Ptolemy show. In fact, two combined effects explain the difference. First, Rhetorius' longitude for M is not correct and should be ♒ 2;47, thus M − 5 = ♑ 27;47. Secondly, if M is the culminating point to H, then M − 5° is in general not the culminating point to H − 5°. Indeed, using Theon's tables one finds that ♑ 29;4 culminates with ♉ 20;16 rising, thus a deviation of only 0;3°, caused by the rounding-off of the numbers in the tables.

Having found that ♑ 29;7 is the endpoint of the 10th

[12] *Cf.* note 5.

[13] For the reason *cf.* Almagest II,9 (p. 144 f. Heiberg).

[14] Should be 13;23. The text interpolates as if the tabular difference were 1 and not 0;55.

[15] Probably scribal error for 4;50 which would be the result of using 1 as tabular difference. The accurate value would be 4;53.

locus (which contains M) Rhetorius determines the excess of $180°$ [16] over $6h$ and finds

$$180 - 6h = 180 - 106;24 = 73;36$$

and from this

$$2h' = \tfrac{1}{3} \cdot 73;36 = 24;32.$$

Proceeding as before he then forms

$$a_4 = a_3 - 2h' = 31;20 - 24;32 = 6;48$$
$$a_5 = a_4 - 2h' = 366;48 - 24;32 = 342;16$$
$$a_6 = a_5 - 2h' = 317;44$$

and finds for the corresponding points of the ecliptic with equal right ascension from the table of sphaera recta

$$\lambda_4 = ♑\, 6;13$$
$$\lambda_5 = ♐\, 16;46\ ^{17}$$
$$\lambda_6 = ♏\, 20;16$$

as limits of the 7th, 8th, and 9th locus. We see that $\lambda_6 + 5° = ♏\, 25;16 = H + 180$ as it should be.

The reason for this procedure is evident from figure 18. In order to end at the setting point E diametrically opposite to the rising point A we must use the diameter AC and then divide $BC = 180 - AB$ into three equal parts each of length $2h'$. The projections of the dividing points are the endpoints of the loci VII to IX.

No. L 431

Vettius Valens, Additamenta, Kroll p. 365,3 to 28

TRANSLATION

Example 2.[1] Year of Diocletian 147, Tybi (V) 14 to 15, 3rd hour of night, clima 4. Sun in Capricorn 19;2, moon in Taurus 23;30, (preceding) conjunction in Capricorn 9;29. {From conjunction to the day and hour of birth [10] days}.[2] I inquire how many rising times there are from sun to moon in the 4th clima, and I found 6 time-degrees. These 6 time (-degrees) I wrote down as solar gnomon. Then again taking the time-degrees of the point diametrically opposite to the sun, because it was a nocturnal birth, I found for the 19th degree of Cancer 17;53 time-degrees.[3] Multiplying these by twelve I found 215 time-degrees.[4] These 215 time-degrees I multiplied by the 3 hours of the nativity, making 645. Subtract one circle, i.e., 360; remaining are 285 time-degrees. These I write down as horoscopic gnomon. Since the horoscopic gnomon exceeds the solar gnomon, the 3rd hour of the birth requires subtraction. Thus subtract two-thirds of one hour from the 10 days which the moon is distant from conjunction until the day and hour of birth and do not compute 3 hours of night but (only) $2\tfrac{1}{3}$. And so you do thus: sun in Capricorn 19;2. Time-degrees of Cancer, because it is a night birth, 17;55.[5] Multiply it by $2\tfrac{1}{3}$ hours; it makes 41;48 time-degrees.[6] Add the rising time of Cancer 93;7,[7] making together 134;55 time-degrees. With these I enter (the table) of the 4th clima and find as Horoscopos Leo 23;0 degrees.[8] With these 23 degrees I entered the table for Leo [9] and find correspondingly 22 sixtieth parts, that is $\tfrac{1}{3}\tfrac{1}{30}$, and 73 years 11 months. Looking for the proportional part I did thus. The Horoscopos Leo 23 (has) 16;45 time-degrees.[10] These I multiplied by 12 making 201. These I multiply by $\tfrac{1}{3}\tfrac{1}{30}$; the $\tfrac{1}{3}$ makes 67 years and the $\tfrac{1}{30}$ (is) 6 years two-thirds (and) $\tfrac{1}{30}$.[11] Together it makes 73 years two-thirds (and) $\tfrac{1}{30}$.

COMMENTARY

Diagram: pl. 27

For Diocletian 147 Tybi (V) 14 = 431 January 9 9 P.M. one obtains

	Text	Computed	Text—comp.
☉	♑ 19;2	♑ 21	$-2°$
☾	♉ 23;30	♉ 25;10	$-1;40.$

The preceding conjunction 9 days earlier (430 December 31 = Tybi 5) occurred in ♑ 11 (text: ♑ 9;29).

For the astrological commentary cf. the Commentary to No. L 419.

No. L 440

CCAG 8,4 p. 221,1 to p. 224,20

TRANSLATION

113.[1] Nativity of a grammarian.

This person [2] was a poor grammarian of Thebes until 32 years (of age). At 33 years he married and began to rise at Athens, and then fleeing to Byzantium was associated with a great man, and, pretending to be a wizard or one initiated, he became quaestor, then consul,

---

[16] Incorrectly interpreted as the arc from ♑ 29;7 to ♋ 29;7 rather than from ♉ 20;16 to ♏ 20;16.
[17] Error for 13;42.

No. L 431
[1] Example 1 is a simple numerical example (Kroll p. 364,24 ff.) concerning the determination of the Horoscopos. No specific date can be associated with it.
[2] The text here is confused.
[3] The Handy Tables of Theon give 17;55 and this is indeed the value used later on; cf. notes 4 and 5.
[4] The product would be 214;36 but $17;55 \cdot 12 = 215$.
[5] Correct for ♋ 19; cf. note 3.
[6] Accurately 41;48,20. Kroll's "correction" of 48 to 42 is wrong.
[7] Correct for ♋ 19 and clima IV.
[8] The rising time for ♌ 23 is indeed 134;55.
[9] It is the table given on p. 324 of Kroll's edition (line 25 which corresponds to 23°).
[10] Correct according to the Handy Tables.
[11] Indeed $201/30 = 6;42$.

No. L 440
[1] From Rhetorius.
[2] Pamprepios, as identified by A. Delatte; cf. the Commentary.

then patrician and later, at 44⅙ years was put to death in camp as traitor. He was a villain.

Sun in Libra 5;8, moon in Taurus 8;4, Saturn in Taurus 25, Jupiter in Libra 23;8, Mars in Capricorn 26;8, Venus in Scorpio 26, Mercury in Libra 23, Horoscopos in Aquarius 23;30, Midheaven in Sagittarius [3] 5;3, full moon in Aries 2;42, (Lot of) Fortune in Virgo 26;34,[4] Daimon in Cancer 20;34,[5] exaltation of the nativity in Virgo, Lot of destroyer in Aquarius.

Investigating the preceding nativity I found the moon and Saturn and Venus and Mars in apoklima by degree, and by sign moon and Saturn and Venus in center.[6] And the sun and Mercury and Jupiter (in ♎) were in apoklima by sign; [7] and by degree the sun happened to be at the epanaphora of the setting center.[8] I see then that Saturn is ruler of the Horoscope (♒ [9]) and the first triangle (♒, ♊, ♎) ruler [10] — the Horoscopos and the luminary (☉) being in trine, indicating the first period of life — in apoklima of lower center (♊), and is retrograde and injured by Mars because of the equilateral triangle, and further that [Venus] (in ♏) is also its (Saturn's) ruler and ruler of the moon [11] (in ♉) (both) being in opposition to it.[12] How could he not have had a hard first period of life and not been a fugitive in many places with the moon opposed by its own house ruler. [In this way the whole life of the grammarian is discussed; [13] of astronomical details we hear that Taurus has the rising time 25 and Scorpio 35,[14] Virgo and Libra 38;20,[15] that Jupiter is at evening rising [16] — actually the planet is near last visibility in the evening.]

COMMENTARY

Diagram: pl. 27

The date of this horoscope has been determined by Delatte and Stroobant [1] who also identified the "grammarian" as the poet Pamprepius who rose to a powerful position as follower of Illus and Leontius, but soon perished when disaster overtook their revolt. For the historical discussion *cf. RE* 18,3 col. 409 ff.

For 440 September 29, 4 P.M. one finds:

	Text	Computed	Text—comp.
☉	♎ 5;8	♎ 8	—3°
☾	♉ 8;4	♉ 10;30	—2;30
♄	♉ 25	♉ 27	—2
♃	♎ 23;8	♎ 26	—3
♂	♑ 26;8	♑ 28	—2
♀	♏ 26	♏ 24	+2
☿	♎ 23	♎ 27	—4
H	♒ 23;30	about 4 P.M.	
M	♐ 5;3	correct for about $\phi = 27$ [17]	
full moon	♈ 2;42	Sept. 27 4^h A.M. ♈ 5;30	—3

The case of Venus is of interest since the text must be based on tables which allow an elongation of 51° where Ptolemy's tables (Almagest XII, 10) would only lead to 46;30 (in agreement with the above given result). A maximum elongation of 50° is assumed by Theon of Smyrna (p. 296 Martin), by Chalcidius with reference to Heraclides Ponticus (p. 206 f. Mullach) and by Cleomedes (p. 226 Ziegler).

No. L 463

CCAG 8,4 p. 224,21 to p. 225,5

TRANSLATION

Nativity of a non-viable.[1, 2]

This was the child of a king and died at 5⅙ months [3] in Byzantium. Year of Diocletian 179, month April 25, 7th hour of the day. Sun Taurus 3, moon Capricorn 23, Saturn Aquarius 26, Jupiter Virgo 17, Mars Aries 1, Venus Aries 15, Mercury Taurus 26, Horoscopos Leo 26, Midheaven Taurus 17, ascending node Aquarius 24,

---

[3] The correct sign of the text (♐) was mistakenly emended to Taurus by the editors.

[4] Should be 26;26 because $\lambda_H + (\lambda_☾ - \lambda_☉) =$ ♒ 23;30 + 7^{s}2;56° = ♍ 26;26. The error is caused by dittography of the following number.

[5] From ♒ 23;30 — 7^{s}2;56° = ♋ 20;34.

[6] Counting simply signs M and IMC would be ♏ (♀) and ♉ (♄ and ☾) respectively.

[7] *Cf.* the preceding note.

[8] The sun in ♎ 5 is 41;30° distant from Δ = ♌ 23;30 thus in the second 30-degree section following a center. By signs, however, ♎ is the third sign from ♌.

[9] Aquarius is a house of Saturn.

[10] Saturn is the ruler of this triangle at daytime, as is the case in this nativity.

[11] Because Saturn and moon are in ♉, a house of Venus.

[12] We are following here Delatte p. 67 note 5 in retaining the κειμένων of the manuscript while the published text has κειμένην.

[13] A translation of the whole horoscope is given by Delatte-Stroobant [1] p. 65 to 70.

[14] P. 222,15 and p. 224,6. Correct for both Systems A and B in clima I.

[15] P. 224,9 f.; correct for System A, clima I.

[16] P. 222,23.

[17] Corresponding to the birthplace of Pamprepius, Panopolis in the Thebais.

No. L 463

[1] *CCAG* 8,1 p. 77, F. 24^v: "Est ultimum cap. operis Rhetorii."

[2] *Cf.*, e.g., *CCAG* 8,2 p. 61 ff. περὶ ἀτρόφων (from *Hephaestion*). *Cf.* also No. L 482, note 10.

[3] The text has μηνῶν ε′ ς′ which only can mean 5⅙ months (= 5 months 5 days). The editor refers to p. 221.7 where ἐτῶν μδ′ ἕκτον means, of course, 44⅙ years (= 44 years 2 months). From this he then seems to conclude that ἕκτον always means "menses duo"! Once being on the wrong track he further suggests that our nativity concerns a child of Leo I and Verina, born 464 April 25, which died at the age of 2 years and 2 months! The text, however, says only 5⅙ months and concerns the date 463(!) April 25.

(preceding) full moon at the 6th hour of night Libra 24;36, (Lot of) Fortune Taurus 16,[4] Daimon Sagittarius 6,[5] Exaltation (of the nativity) Leo 11.[6]

Explanation: The triangle rulers ($\jupiter$) of the Horoscopos ($\leo$ $\sagittarius$ $\aries$) and ($\venus$ and the moon) of (the triangle of) the luminary of the sect[7] (the moon) were in apoklima (namely the moon) and opposing themselves (being in opposition to their own houses: $\jupiter$, $\venus$, and the moon) and under the aspect of Mars (the moon in $\capricorn$).

COMMENTARY
Diagram: pl. 27

For Diocletian 179 = A.D. 463 April 25 1 P.M. one finds

Text	Computed	Text—comp.
$\sun$ $\taurus$ 3	$\taurus$ 6	—3
$\moon$ $\capricorn$ 23	$\capricorn$ 24	—1
$\saturn$ $\aquarius$ 26	$\aquarius$ 28	—2
$\jupiter$ $\virgo$ 17	$\virgo$ 19	—2
$\mars$ $\aries$ 1	$\aries$ 3	—2
$\venus$ $\aries$ 15	$\aries$ 10	+5
$\mercury$ $\taurus$ 26	$\taurus$ 26	0
H $\leo$ 26 7^h day	1 P.M.	
M $\taurus$ 17	$\taurus$ 18;20[8]	
$\node$ $\aquarius$ 24	$\aquarius$ 26;30	—2;30
full moon: $\libra$ 24;36 6^h night	April 19 11^h A.M.	
	$\libra$ 25;40	—1

No. L 474
CCAG 1 p. 102

TRANSLATION

57.[1] Katarche of a (voyage by) ship.

A person left Caesarea,[1a] 191st year of Diocletian, month October 1, fifth [hour] of day[time].[2] Sun in $\libra$ [...], $\moon$ $\scorpio$ 27;36, $\saturn$ in $\cancer$ 27, $\jupiter$ in $\virgo$ 16;30, Mars in $\aries$ [3] 17;30, $\venus$ in $\leo$ 23;30, $\mercury$ in $\libra$ 25;30, Horoscopos

---

[4] From $A_F = \leo\ 26 + (\capricorn\ 23 - \taurus\ 3) = \taurus\ 16$.
[5] From $A_D = \leo\ 26 - (\capricorn\ 23 - \taurus\ 3) = \sagittarius\ 6$.
[6] The exaltation of the sun is $\taurus$ 19; thus the sun in $\taurus$ 3 is 16° behind its exaltation and the exaltation of the nativity should be in $\leo$ 26 — 16 = $\leo$ 10 and not in $\leo$ 11.
[7] We are dealing with a daytime birth, thus the day-ruler of the triangle in question ($\capricorn$ $\virgo$ $\taurus$) is chosen, i.e., the moon (in $\capricorn$).
[8] Using Theon's tables for Byzantium; Clima 5 would lead to $\taurus$ 19;20. H = $\leo$ 25, instead of $\leo$ 26, would give for Byzantium M = $\taurus$ 17.

No. L 474
[1] From Palchus. Cf. also the special discussion of ships by Demetrius CCAG 8,3 p. 98 note 1.
[1a] Presumably in Palestine.
[2] This is the obvious interpretation of the text and confirmed by the position of sun and M, as well as by the longitude of the moon.
[3] So correctly in the text; the edition gives Gemini.

$\sagittarius$ 6, Midheaven $\virgo$ 22, $\node$ $\cancer$ 12;30. Because the $\moon$ happened to be in the terms of $\saturn$ and in the (12th) locus of Bad Genius and in slow motion[4] it made the (ship have) storm and delay in the voyage. And the diametrical aspect of $\mars$ (in $\aries$) to $\mercury$ (in $\libra$) made the oars(?) of the boat suffer (damage) and made quarrels and fights occur in the ship. But [this aspect] did not harm the Katarche because of the ruler of the Horoscopos ($\jupiter$ [5]). And $\venus$ being found in a dry sign ($\leo$)[6] and [the moon[7] (in $\scorpio$)] in aspect with $\saturn$ (in $\cancer$) and $\mars$ (in $\aries$)[8] [................]. Thrown on land he was loved by a woman because the quartile aspects (of the moon in $\scorpio$) were dry and beneficent ($\venus$ in $\leo$).[9] And the bicorporeal sign of the Horoscopos ($\sagittarius$) and of $\jupiter$ (in $\virgo$) made him embark on another ship at Abydos[10] and when the $\moon$ came into $\libra$ in diameter to $\mars$ (in $\taurus$) according to computation[11](?), while $\mars$ in $\taurus$ was in diameter to the $\moon$ (in $\scorpio$) according to computation[11](?), the exit from equi-zonal (signs, $\libra$ and $\taurus$) and also the entrance into equi-zonal signs ($\scorpio$ and $\aries$) took place.[12]

COMMENTARY
Diagram: pl. 27

For the given date, Diocletian 191 = A.D. 474[13] October 1, 5th hour of daytime one finds:

Text	Computed	Text—comp.
$\sun$ $\libra$ [...]	$\libra$ 10	
$\moon$ $\scorpio$ 27;36	$\scorpio$ 30	—2;30
$\saturn$ $\cancer$ 27	$\cancer$ 30	—3
$\jupiter$ $\virgo$ 16;30	$\virgo$ 18	—1;30
$\mars$ $\aries$ 17;30	$\aries$ 19	—1;30
$\venus$ $\leo$ 23;30	$\leo$ 26	—2;30
$\mercury$ $\libra$ 25;30	$\scorpio$ 2;30	—7
H $\sagittarius$ 6	about 11^h A.M.	
M $\virgo$ 22	$\virgo$ 23	
$\node$ $\cancer$ 12;30	$\cancer$ 15;20	—2;50

---

[4] The equation of center is found to be about —5°.
[5] Sagittarius is a house of Jupiter.
[6] According to CCAG 1 p. 165,16 all signs from $\aries$ to $\scorpio$ are "dry" with the only exception of $\cancer$.
[7] For the importance of the moon cf. CCAG 1 p. 100,21 to 23 and CCAG 8.3 p. 98 note 1.
[8] The meaning of this passage is perhaps: both Venus and moon are in dry signs, Venus in aspect with Mars, the moon in aspect with Saturn.
[9] The ὑπό inserted by Kroll should be eliminated.
[10] At the Hellespont; cf. the Commentary.
[11] πῆξις occurs in Vettius Valens for "composition, computation" of a table (e.g., p. 24,18 and p. 295,3 and 4 "computation of Table I").
[12] "Equi-zonal signs" are defined in CCAG 1 p. 153,27. Cf. also P. Mich. 149, XVI where we find the same zonal arrangement explained as based on the planetary houses; thus $\libra$ and $\taurus$ are associated, and $\scorpio$ and $\aries$. What is meant is probably that the moon moved from $\libra$ into $\scorpio$ while Mars had retrograded from $\taurus$ into $\aries$.
[13] Incorrectly stated as A.D. 475 in CCAG 1 p. 102 note.

Note that all planetary positions are given in the text with an accuracy of ½ degree. Of interest is the determination of M. Using the Handy Tables of Theon one finds for M in clima 4 (Rhodes, $\phi = 36$) ♍ 20 and for clima 5 (Hellespont, $\phi = 40;56$) ♍ 23;30. Since Abydos on the Hellespont has a latitude of $\phi = 40$ the value ♍ 22 can be explained as the result of interpolation (accurately ♍ 22;50).

No. L 475

CCAG 1, p. 103,1 to 30

TRANSLATION

58.[1] Inquiry about a Katarche concerning fear for a journey to Athens.

Year of Diocletian 19[1], month Epiphi (XI) 22, first hour of the day. The sun 21 of Cancer, the moon 21 of Gemini, Saturn [........] 17 of Leo, Mercury 6 of Cancer, [....] the ascending node [...] of Gemini. Investigating first of all the Lot of Fortune, I found it in the 12th [locus] which is a very watery sign (♋). And you will find this configuration set forth in the 7th chapter of the *καταρχαί*.[2]

[There follows a discussion of astrological details from which we quote:] but since the lords of the hour (☿ [3]) and of the moon (☉ [4]) were in apoklima, they indicated that the inquiry was about foreign (places [5]). I said that they suggested danger and loss and damage because Mars and Saturn were in the Horoscopos. And again [......] because of the presence of Venus (in ♌) and because of the moon contacting (the quartile of ♃) toward Jupiter, they had fallen into trouble and, given 15 [days ?], they were saved. [In the following discussion, which also quotes Dorotheus, it is shown what the ship must have carried, e.g., four-footed animals—and indeed it brought camels.] Observing the diametrical and the quartile position of the moon,[6] to be such that the sun would come to its own house (♌, the place of) the Horoscopos, I said when the moon would be in Sagittarius or Pisces the ship would arrive. And when the moon came into Sagittarius they arrived. If one peruses the chapters of the ancients one will find not a few excellent suggestions.

[1] From Palchus.

[2] Dr. Weinstock suggests chapter 7 of the same compilation, published *CCAG* 5,3 p. 126 f. = *CCAG* 11,1 p. 270 f.

[3] The day in question is a Wednesday; thus Mercury rules the first hour of the day.

[4] The sun is in ♋, the house of the moon.

[5] See for this significance of apoklima Bouché-Leclercq, *AG* p. 455.

[6] After 15 days the moon would be in ♐, in quartile to Jupiter (in ♍). The diametrical position would be reached in ♓.

COMMENTARY

Diagram: pl. 27

The date of this horoscope is given in the text as Diocletian 190 Epiphi (XI) 22 which corresponds to A.D. 474, July 16.[7] The given positions, however, require a date one year later; thus one must read Diocletian 191 [8] Epiphi 22 which corresponds to A.D. 475 July 16, a Wednesday.

The text contains many lacunae. Obviously there is something missing between the mention of Saturn and that of Mercury.[9] Probably one can restore " Saturn [in ♌ ..., Jupiter in ♍ ...., Mars in ♌ ...., Venus] in ♌ 17." After Mercury we can supply " [Horoscopos in ♌ ...]." The elongation of the moon from the sun is 30° and therefore the Lot of Fortune must be located in ♌ ± 30, thus either in ♋ or in ♍. From line 7 we know that the Lot of Fortune fell in the 12th locus which is ♋, a humid sign.[10] Thus the Lot of Fortune was computed according to the formula

$$\Lambda_F = \lambda_H + (\lambda_{☾} - \lambda_{☉})$$

as is to be expected for daylight (*cf.* p. 8).

For 475 July 16 one obtains the following positions:

Text			Computed	Text—comp.
☉	♋	21	♋ 24	—3
☾	♊	21	♊ 23;30	—2;30
♄	♌	[..]	♌ 4	
♃	[♍	..]	♍ 27	
♂	[♌	..]	♌ 17	
♀	[♌]	17	♌ 22	—5
☿	♋	6	♌ 16	—10
H	[♌	..]	about 8 A.M.	
☊	♊	[..]	♊ 30	

No. L 478

CCAG 6, p. 64,26 to p. 65,21

TRANSLATION

Another [1] Katarche, concerning the lost linen of a (slave) girl.

Year of Diocletian 195 Thoth (I) first day [to] second, 4th hour of night. The sun in Virgo 14 degrees, the moon in Pisces 9, Saturn in Virgo 11, Jupiter in Sagittarius 26, Mars in Libra [2] 20, Venus in Virgo 24,

[7] Not 475 July 17 as stated *CCAG* 1, p. 103 apparatus to line 3 [read 2].

[8] A similar error — 190 for 195 — happened in No. L 479, cod. Vindob.

[9] This lacuna is, however, not indicated by the editors.

[10] *Cf.* Bouché-Leclercq, *AG* p. 155; *cf.* also Vettius Valens p. 8,23 ff.

No. L 478

[1] Referring to No. L 487. From Palchus; chapter 164 in Vindob.

[2] Error for Gemini; *cf.* note 11.

Mercury in Leo 18, Horoscopos in Taurus 28, Midheaven in Aquarius [14],[3] ascending node in Aries, Lot of Fortune in Scorpio.[4] The Lot of Fortune falling in the setting sign (♏) indicated the inquiry (to be) about something out of sight; and that the luminaries were in aspect with the Horoscopos,[5] (as was) also its ruler (♀ [6] in ♍), indicated that the thief was in the household and not from outside; and that the moon at full moon [7] (in ♓) was in contact with (the diameter of) Saturn (♍ — ♓) at evening setting [8] indicated that the thief was an old man. Since then Saturn was in the house of Mercury (♍) (it indicated) that the thief was an educated person and a rascal and frustrated in intercourse. Again the terms of the moon (♀ [9]) showed that the stolen article belonged to a woman, and the ruler of the terms of the moon and of the Horoscopos (♀) being in depression (♍) showed that the loser was a humble person or slave and that the lost article was old and wretched because it (♀) was in evening phase.[10] And the following configurations caused the finding of it. First the moon moving toward the ruler (♂ in ♊) of the receiving house (♈ [11]) and in aspect to the ruler of the Lot of Fortune (♏), Mars,[12] and its likeness (?); and the sun (in ♍) being in the aspect of trine to the Horoscopos (♉) and in sextile to the (Lot of) Fortune (♏) and in company with the ruler of the Horoscopos (♀ [6]), and diametrical to the moon (in ♓), by every scheme made the finding quick. And because the ruler of the Horoscopos (♀) happened to be in its own triangle (♍ ♑ ♉) and the house ruler [13] (♃) of the moon (in ♓ [14]) was in its own house (♐), the finding took place.

COMMENTARY

Diagram: pl. 27

In several codices [15] there is preserved a group of six horoscopes [16] collected by Palchus. Their dates in chronological order are

(II)	478 Aug. 29	theft
(VI)	479 July 14	ship from Alexandria to Smyrna
(III)	483 July 8	taming of a little lion
(IV)	484 July 18	coronation of Leontius
(V)	486 March 17	Theodoros, praefect of Alexandria
(I)	487 Sept. 5	incorrect prediction concerning a letter

where the roman numbers in ( ) indicate the arrangement in a manuscript in Vienna.[17]

The date of the earliest one, translated above, is given as Diocletian 195 Thoth (I) 1 to 2 = 478 August 29, 4th hour of night. For it one finds

Text		Computed	Text—comp.
☉	♍ 14 [18]	♍ 7	+7 [18]
☾	♓ 9	♓ 11	−2
♄	♍ 11	♍ 14	−3
♃	♐ 26	♐ 28	−2
♂	[♊] [19] 20	♊ 22	−2
♀	♍ 24	♍ 28	−4
☿	♌ 18	♌ 19	−1
H	♉ 28	about 11 P.M.	
M	♒ 14	correct for Syene	
☊	♈	♈ 29;40	

No. L 479

CCAG 1 p. 103/104, including variants from Cod. Vindob. 1, fol. 300v/301r (CCAG 6 p. 14)

TRANSLATION [1]

59.[2] Another [3] inquiry, in Smyrna regarding fear about a ship; for it was expected long before to arrive from Alexandria and had not arrived.

Year of Diocletian 195 [4] Epiphi (XI) 20, 3rd hour of the day, Saturday. Sun Cancer 19, moon Scorpio 16, Saturn Virgo 16, Mars Virgo 18, Jupiter Aquarius 8, Venus Gemini 6, Mercury Leo 4,[5] retrograde, Horo-

[3] From diagram in Vindob. (unpublished).

[4] From $\Lambda_F = \lambda_H - (\lambda_{☾} - \lambda_{☉}) = \text{♉}\,28 - 5^s25° = \text{♐}\,3$ it follows either that only signs were counted ($\Lambda_F = \text{♉} - 6^s = \text{♏}$) or that $\lambda_{☉} = \text{♍}\,14$ is an error for ♍ 4 (thus $\Lambda_F = \text{♏}\,23$) as is suggested by computation (*cf.* note 18).

[5] The sun (in ♍) in trine, the moon (in ♓) in sextile.

[6] Taurus is a house of Venus.

[7] The moment of the Katarche falls only a few hours before full moon. At full moon Saturn in ♍ 11 is in close "contact" with the line from the sun (in ♍ 4 or 14) to the moon (in opposition).

[8] Saturn in ♍ 11 (with sun in ♍ 4 or 14) is obviously past evening setting but moving toward morning rising.

[9] The moon in ♓ 9 is in the terms of Venus.

[10] Venus was indeed evening star not very long after evening rising.

[11] The moon is moving from ♓ toward ♈ which is a house of Mars and Mars was in ♊, and hence the next planet to be met by the moon.

[12] The moon in ♓ is in quartile to Mars in ♊.

[13] The text has again "ruler of the receiving house."

[14] A house of Jupiter is ♓.

[15] *Cf. CCAG* 6 p. 63 note 1

[16] As a seventh example is added the horoscope of Hadrian (No. L 76), from an older collection.

[17] *Cf. CCAG* 6 p. 14 (F. 292 v.).

[18] *Cf.* note 4; the emendation ♍ 4 would lead to the expected deviation of −3.

[19] Text and diagram give ♎ but the astrological discussion uses correctly ♊.

No. L 479

[1] In general we follow the text published in *CCAG* 1 (F).

[2] From Palchus (*cf. CCAG* 1, p. 53 F. 193). He is not mentioned in the manuscript V which assigns the number 164 to this chapter.

[3] This is a reference to No. L 475.

[4] V, incorrectly, 190. *Cf.* also No. L 475.

[5] Text and diagram (unpublished) in V give ♌ ♂ where the

scopos Virgo 2,[6] Midheaven Taurus 26,[7] ascending node Aries 11, {conjunction ♋ 11},[8] Lot of Fortune Sagittarius 29.[9, 10] Taking the ruler of the day (and of the first hour) and the ruler of the (3rd) hour — Saturn and Mars [11] — and again having observed Saturn and Mars to be in the Horoscopos (♍) and the moon moving toward Saturn,[12] hence I said that the ship had met with a great storm, but that it was saved because Venus (in ♓) in Midheaven [8] and moon (in ♏) were in aspect to Jupiter (in ♒) and the moon's [13] (position on the) seventh day was moving away from [13] Jupiter [14] and its [15] dodecatemorion was also moving away from [15] Jupiter.[16] And that the problem of the Katarche concerned a ship out of sight, observe that the Lot of Fortune was in Sagittarius, with which Argo rises;[17] and furthermore, the ruler (♃) of it (Argo)[18] was in a very watery sign (♒). Having observed that the Horoscopos was in a bicorporeal sign (♍) {and that the ruler of the Horoscopos (☿)[19] was retrograde, and the ruler of Midheaven (☿),[20] Venus, was in a bicorporeal sign (♓)} [21] and the ruler (♃)[18] of the (Lot of) Fortune (in ♐) was in a bicorporeal sign [22] and (the planet) which receives the ruler of the Lot of Fortune [23] was in a bicorporeal sign {and the house ruler of the moon (♂) was in a bicorporeal sign (♍)} [21] I said that they would change from ship to ship. And because Virgo [24] was a winged sign and also [25] Sagittarius [26] I said they were bringing some feathered things with them. And because the moon was in the {house of Mars (♏) and} [21] terms of Mercury I said they were probably bringing books and papyrus with them and some bronze objects because of Scorpio. Having noted that Asclepius [27] was rising with the moon (in ♏) I said that they were bringing medical implements with them. And (as to) when it ought to come, I said the moon would be in Aquarius {or Pisces; in ♒ because of its (the moon's position on the) seventh day [28] and because the Lot of Fortune (in ♐) was moving [29] toward Jupiter (in ♒); in ♓ because of its motion toward its own house ruler.[30] And they arrived on the 8th day (the moon being) in ♒} [31] and being asked they spoke {about the delay and said that there was an upheaval at sea,} [32] and, as the sea parted, the rudder [33] struck a rock and was broken and they were greatly driven by the storm. Having reached a harbor [34] they transferred the cargo to another ship on which they came, bringing, indeed, ostrich feathers and plain papyrus, because Mercury was retrograde, and cooking implements, because of Scorpio, and a shipload of medical supplies, because of Asclepius and Hygeia.[27]

COMMENTARY

Diagram: pl. 28

For Diocletian 195 Epiphi (XI) 20 — A.D. 479 July 14 [35] 9 A.M. one finds

	Text	Computed	Text—comp.
☉	♋ 19	♋ 22	—3
☽	♏ 16	♏ 19	—3
♄	♍ 16	♍ 20	—4
♂	♍ 18	♍ 10	+8
♃	♒ 8	♒ 10	—2
♀	♓ 6	♓ 8	—2
☿	♌ 4 retrogr.	♌ 10 retrogr.	—6

---

editors of *CCAG* 1 p. 104,2 assume a lacuna and then read δυ(τικὸς).

[6] V, incorrectly: 12; cf. note 9 and note 7.

[7] The diagram of V gives ♉ 7, the text 7 only, leaving a blank space for the zodiacal sign. For H = ♍ 12 (cf. the preceding note) one would obtain M = ♊ 8;30 but never ♉ 7.

[8] Omitted in F.

[9] From Λᵧ = ♍ 2 + (♍ 16 — ♋ 19) = ♍ 2 + 117 = ♐ 29.

[10] V adds here the sentence "Midheaven falls either in the 9th locus (counting by degrees) or in the tenth" (counting by signs).

[11] This is correct as is seen from the fact that the day in question is a Saturday, which means that Saturn is the ruler of the first hour, to be followed by Jupiter, Mars, the sun, etc.

[12] Saturn was in ♍, Jupiter in ♒; thus the moon in ♏ is approaching ♃, not ♄.

[13] Following the text of V: αὐτῆς ἀπὸ instead of αὐτῶν πρὸς in F (*CCAG* 1 p. 104,8); cf. note 15.

[14] Seven days later (July 21) the moon is in about ♒ 17, thus past Jupiter in ♒ 8.

[15] V has here αὐτῆς πάλιν φέρεσθαι ἀπὸ for αὐτῶν πρὸς in F (p. 104,9); cf. note 13.

[16] The dodecatemorion of the moon (in ♏ 16) is ♉ 28, thus beyond Jupiter (in ♒ 8).

[17] Cf. No. L 118.

[18] Argo rises with Sagittarius and Sagittarius is a house of Jupiter.

[19] Virgo is a house of Mercury.

[20] Taurus is a house of Venus.

[21] Omitted in V.

[22] This is an error because ♒ is not a bicorporeal sign. It is the Lot of Fortune itself which is located in a bicorporeal sign (♐).

[23] The ruler of the Lot of Fortune, Jupiter, is in ♒, a house of Saturn. Thus Saturn is the planet which receives Jupiter, and Saturn is in ♍, a bicorporeal sign.

[24] Following V; F has here Venus.

[25] V: καὶ; F: κατὰ.

[26] For this classification cf. *CCAG* 1 p. 166,31 f.

[27] Both Asclepius and Hygeia are equivalent names for Ophiuchus. Cf., e.g., Gundel, *HT* p. 65,18 f.: "Ab octavo usque ad decimum (gradum Scorpionis) Ophiuchus et Aesculapius et Sanitas et duo Dracones perplexi." Furthermore Gundel in *RE* 18,1 col. 661,10-50 and 662,56-663,13.

[28] Cf. note 14.

[29] With increasing elongation $\lambda_{\mathrm{C}} - \lambda_{\odot}$, the Lot of Fortune moves in the direction of increasing longitudes.

[30] The Lot of Fortune is in ♐, a house of Jupiter, and it moves (cf. the preceding note) toward Jupiter (in ♒) whose second house is ♓.

[31] Omitted in F.

[32] Following V; F has a slightly different version.

[33] "The rudder of the ship": V.

[34] "Stopping place" (ἔνδιον): V.

[35] In *CCAG* 1 p. 103 incorrectly given as A.D. 480(!) July 15(!).

	Text	Computed	Text—comp.		Text	Computed	Text—comp.
H	♍ 2 3ʰ day	about 9 A.M.		♃	[♈] 28	♈ 30	−2
M	♉ 26	correct for Byzant.		♂	♓ 29	♈ 1	−2
☊	♈ 11	♈ 12;47	−2	♀	♉ 15	♉ 18	−3
prec. conj.	♋ 11	July 5: ♋ 13	−2	☿	♉ 20	♓ 19	+1
				H	♍ 6	about 5 P.M.	
				M	♓ 2	correct for Hellesp.	

No. L 482

CCAG 8,1 p. 240,17 to 28

TRANSLATION

As:[1] sun 29;43[2] of Pisces, terms of Saturn, moon 6 of Libra, terms of Saturn,[3] Saturn 25 of Libra, terms of Venus, Jupiter 28 of Aries,[4] terms of Saturn, Mars 29 of Pisces, terms of Saturn, Venus 15 of Taurus, terms of Jupiter, Mercury 20 of Pisces, terms of Mars, Horoscopos 6 of Virgo, terms of Mercury, Midheaven 2 of Gemini, terms of Mercury, Lot of Fortune 12 of Pisces,[5] terms of Venus.[6] We find Saturn receiving, in its terms only, sun and moon; and in its terms and triangle and exaltation sun (in ♓), Mars (in ♓), Mercury (in ♓); in triangle the Horoscopos.[7] Then Saturn and Venus will be house rulers, that is Saturn ruling and Venus ruling with him, and the two according to the horimaia[8] each one forecast death. And this is in the 4th chapter[9] on non-viables.[10]

COMMENTARY

Diagram: pl. 28

For 482 March 21 one finds

	Text	Computed	Text—comp.
☉	♓ 29;43	♈ 2	−2
☾	♎ 6	♎ 8	−2
♄	♎ 25	♎ 29	−4

---

[1] From Rhetorius, in a chapter (No. 19) entitled "On nativities having no house ruler, according to Valens."

[2] The edition has 20;43 but the photo of P shows clearly 29;43.

[3] Libra 6 is the boundary between the terms of Saturn and Mercury; cf. the commentary to No. 46.

[4] The edition has Taurus but both computation and "terms" require Aries. The photo of P shows a sign which is definitely not ♉ but ♈.

[5] From Λ_F = ♍ 6 + (♎ 6 − ♓ 29;43) = ♓ 12;17 ≈ ♓ 12. The edition has incorrectly ♓ 2.

[6] Venus omitted in the edition though given in text. Pisces 12 is the upper boundary of the terms of Venus; cf. note 3.

[7] The text seems confused. Probably it was stated that ☉ ♂ ☿ (in ♓) were in a triangle (♓ ♋ ♏) of Venus and in the sign of the exaltation of Venus (♓) and that the Horoscopos (in ♍) was also in a triangle (♍ ♑ ♉) of Venus.

[8] Cf. Tetrabiblos II,10 Robbins p. 280 = III,11 Boll-Boer p. 132,18; Vettius Valens p. 146,32 and p. 342,12; Bouché-Leclercq, AG p. 413 f.

[9] Cf. the note of the editors. Still-born or non-viables are mentioned in Vettius Valens I,24 (p. 53,25). Cf. also Cumont, Lux perpetua, p. 309 ff.

[10] That is, infants who live less than one year (cf. Tetrabiblos III,9 ed. Robbins p. 265 = III,10 ed. Boll-Boer p. 126).

No. L 483

CCAG 6 p. 65,22 to p. 66,15

TRANSLATION

About a small lion, whether he will be tamed.[*]

Year of Diocletian 199, Epiphi [1]4,[1] 1st hour of the day. The sun in Libra[2] 18, moon in Aquarius 10, Saturn in Scorpio 11, Jupiter in Gemini 19, Mars in Libra 6, Venus in Leo 7, Mercury in Cancer 18, Horoscopos in Leo 2, Midheaven in Aries 24, ascending node in Capricorn 23, (preceding) full moon in Capricorn 11, Lot of Fortune in Aquarius 29.[3] The Horoscopos (♌) indicated the kind of animal, also the dodekatemoria of the Horoscopos,[4] sun (in ♋[5]) and moon (in ♒) and Mars (in ♎) falling in four-footed signs[6] showed that the Katarche was about animals. And Venus being in the Horoscopos (♌) and Jupiter (in ♊) in the (11th locus) Agathos Daimon, about to receive the contact of the moon[7] indicated tameness; and especially the third(?) of the moon and the seventh(?) from Venus — all this showed that it would be tamed and brought up with man; and that the moon and the Lot of Fortune happened to be in the setting sign (♒) showed that it would go abroad on a ship because the setting sign was watery. And that the moon (in ♒) had left contact with Venus (in ♌)[8] to make contact with Jupiter (in ♊)[7] joined the fortunes of the sender and the receiver;

---

[*] From Palchus.

[1] The text has επιφι δ instead of επιφι ιδ. The year is given only by P.

[2] Error for Cancer; also in the (unpublished) diagram in V. The emendation Cancer follows e.g., from the position of the full moon, the hour, etc.; cf. also note 3.

[3] From Λ_F = λ_H + (λ_☾ − λ_☉) = ♌ 2 + 6°22° = ♒ 24; the number of the text, 29, is perhaps caused by using mistakenly ♌ 9, the place of Venus, instead of H = ♌ 2. For the sun the correct longitude ♋ 18, was used.

[4] Because it is ♌ 26, thus again in Leo.

[5] Cf. note 2.

[6] According to CCAG 1 p. 166,10 the "four-footed" signs are ♈, ♉, ♌, and ♐. Thus, neither the sun (in ♋) nor the moon (in ♒) nor Mars (in ♎) is located in a four-footed sign. The dodekatemoria fall in ♒ for the correct solar position, in ♉ for the erroneous position (♎ 18), in ♊ and ♐ for moon and Mars respectively.

[7] The quartile of Jupiter (♓) will be "connected" (cf. p. 157) by the moon.

[8] "Contact" is in the case of Venus understood as contact with the diameter from ♌ 7 (♀) to ♒ 7; for Jupiter cf. note 7.

for the sender was to become consul and the receiver was consul.

COMMENTARY

Diagram: pl. 28

For Diocletian 199 Epiphi (XI) [1]4 — A.D. 483 July 8 one obtains

	Text	Computed	Text—comp.
☉	[♋] 18	♋ 16	+2
☾	♒ 10	♒ 14	−4
♄	♏ 11	♏ 4	+7
♃	♊ 19	♊ 20	−1
♂	♎ 6	♎ 8	−2
♀	♌ 7	♌ 12	−5
☿	♋ 18	♋ 24	−6
H	♌ 2 1h day	about 7 A.M.	
M	♈ 24	correct for Alex.	
☊	♑ 23	♑ 25;44	−2;40
full moon	♑ 11	July 6 2h A.M. ♑ 14	−3

The consul of the year 483 was Anicius Acilius Aginatius Faustus (*cf. RE* 6,2 col. 2094, No. 20) in the West, Flavius Trocondus in the East (Grumel, *Chron.* p. 352). For related horoscopes *cf.* No. L 478.

No. L 484

CCAG 1 p. 107 f. and CCAG 6 p. 66,16 to p. 67,7

TRANSLATION [1]

{Katarche [2] of Leontius crowned at Antioch.[3]

This person taking a Katarche from two astrologers was crowned and immediately expelled from kingship and fortune.} [3] Year of Diocletian [200] month Epiphi (XI) 24, beginning of the first hour of daylight,[4] sun in Cancer 23,[5] moon in Scorpio [6] 7, Saturn in Scorpio 15, Jupiter in Cancer 5,[7] Mars in Cancer 20,[7] Venus in Gemini 26,[8] Mercury in Leo 19, Horoscopos in Cancer 23, Midheaven in Aries ...,[9] ascending node [10] in Capricorn 14, conjunction in Cancer, Lot of Fortune in Scorpio 17,[11] Daimon in Aries,[12] the exaltation {of the nativity} [3] in Aries.[13] Those who gave the Katarche were much misled by the sun and Jupiter and Mars being in the Horoscopos (♋) and Mercury in epanaphora (♌) and the moon (in ♏) being in (the 5th locus) Agathe Tyche towards Saturn (in ♏) and Jupiter [14] (in ♋). But they did not turn (their attention) first to (the fact that) Mercury (in ♌), the ruler of day and hour,[15] had fallen into passivity, for it had its greatest elongation from the sun [16] and was in aspect only [17] to Saturn (in ♏). And this indicates violent death. For Venus (in ♊) alone [3] was not able to relieve (the passivity), for the sun (in ♋) radiated in between. And further I observed [18] that the moon, being ruler of the sun and of the Horoscopos and of Jupiter and Mars and of the (preceding) conjunction (all in ♋ [19]), was in depression (in ♏) and in (the 12th locus) Bad Daimon.[20] In order that none of these schemata should be effective it did not suffice(?) that sun and Horoscopos and Jupiter were together.[21]

COMMENTARY

Diagram: pl. 28

For Diocletian 200 Epiphi (XI) 24 — 484 July 18, Wednesday, 6 A.M., one finds

	Text	Computed	Text—comp.
☉	♋ 23	♋ 26	−3
☾	♏ 7	♏ 11	−4
♄	♏ 15	♏ 14	+1
♃	♋ 5	♋ 17	−12 [or +3]
♂	♋ 20	♋ 7	+13 [or −2]
♀	♊ 26 [22]	♊ 28	−2
☿	♌ 19	♌ 23	−4
H	♋ 23	about 6 A.M.	
☊	♑ 14(!)	♑ 5;48	+8
prec. conj.	♋ —	about July 8: ♋ 18	

[1] Following, in general, *CCAG* 6. *Cf.* also the translation in Bouché-Leclercq, *AG* p. 514 ff. The Vindob. gives a diagram (unpublished).

[2] *CCAG* 1: § 88 in Palchus.

[3] From *CCAG* 1.

[4] Correct in *CCAG* 1, meaningless "beginning 11th day" in *CCAG* 6.

[5] *CCAG* 1: 2 (β) and 6.

[6] *CCAG* 1 gives erroneously Cancer (as for the sun).

[7] Computation suggests the interchanging of these two numbers; *cf.* the commentary.

[8] *CCAG* 1: 27.

[9] About ♈ 9 would be correct for Rhodes.

[10] *CCAG* 1 erroneously: descending node.

[11] Error for 7 since Λ_F = ♋ 23 + (♏ 7 − ♋ 23) = ♏ 7. If sun and H coincide then also the moon (in ♏ 7) and the Lot of Fortune must fall together. And one cannot change the moon's position to ♏ 17 since the Daimon would then be moved to ♓ 29 (*cf.* note 12).

[12] From $\Lambda_D = \lambda_H - (\lambda_☾ - \lambda_☉)$ = ♋ 23 − 104 = ♈ 9.

[13] If sun and H coincide the exaltation of the nativity is always in ♈.

[14] Jupiter (in ♋) is (after Saturn) the next planet to be met by the moon, at which time Venus would have passed Jupiter. For the importance of these "contacts" for a Katarche *cf.* Bouché-Leclercq, *AG* p. 247 and above No. L 483.

[15] *Cf.* Bouché-Leclercq, *AG* p. 515 note 2; indeed 484 July 18 was a Wednesday.

[16] This is indeed almost exactly the case.

[17] Ignoring the moon.

[18] *CCAG* 1: they did not observe.

[19] The house of the moon.

[20] Counted with respect to the Lot of Fortune which coincides with the moon.

[21] There follows, in support, a quotation from Dorotheus, translated in Bouché-Leclercq, *AG* p. 516.

[22] or 27.

The comparison of the text with the computed longitudes of Jupiter and Mars suggests the interchange of the degree numbers.

Cumont, in his commentary (*CCAG* 1 p. 107) says that Leontius was crowned in Tarsus (not in Antioch as our horoscope incorrectly says) "anno 483, ut videtur." In his original article concerning this text [23] he discusses the "incroyable confusion dans la chronologie des événements." In view of the date of the coronation, 484 July 18, only 484 seems possible for the reception in Antioch.[24] Bouché-Leclercq, *AG* p. 514, following Krumbacher, accepts 484 June 27. Our computation fixes the date as 484 July 18.

For related horoscopes *cf.* No. L 478.

No. L 486
CCAG 1, p. 100 to 101 [*]

TRANSLATION

31.[1] Katarche when Theodoros the augustalian prefect entered Alexandria.

He began well and he caused prosperity, he was not a thief and he was truthful and in good repute with the city, but was soon dismissed in disgrace and under punishment for thievery. Year Diocletian 202, month Phamenoth 25,[2] Monday, first hour, the sun in 26 of Pisces, the moon in 27 of Capricorn, Saturn in 11 of Sagittarius, Jupiter in 27 of Leo, Mars in 25 of Aries, Venus in 12 of Pisces, Mercury in 23 of Pisces, Horoscopos in 26 of Aries, Midheaven in 10 of Capricorn, the ascending node in {32(!) of}[3] Sagittarius, {the descending node in 20 of}[3] Virgo, the (Lot of) Fortune in 27 of Aquarius.[4] Observe that Mars is in the Horoscopos (in ♈) and in its own house[5] and in triangular aspect to Jupiter (in ♌) and Saturn (in ♐) and that Venus (in ♓ 12) and the (Lot of) Fortune (in ♒ 27) are in Agathos Daimon [6] and that the moon is in Midheaven (♑) and that Jupiter (in ♌) and Venus (in ♓ 12), the rulers of the city,[7] are in epanaphora with respect to the Horoscopos (in ♈). And you will find the first part of his government excellent and efficient and worthy of all praise. But looking for the house ruler of the (Lot of) Fortune (in ♒) and of the moon (in ♑), you will find Saturn [8] in apoklima (in ♐) and fallen under the effect of eclipses,[9] which caused the overthrow with disgrace. And that he was soon succeeded was caused by Mars and the Horoscopos (in ♈) and the moon (in ♑), (all) happening to be in tropical signs and signs of short rising time.[10] And that he was punished was caused first by the ruler (♄) of the (Lot of) Fortune being in apoklima; {and also Mars of the city caused it, and Jupiter and Venus of the city being in epanaphora to the Horoscopos; and he was obstinate and insolent because of the nature of Mars in the Horoscopos by sect.}[11]

COMMENTARY
Diagram: pl. 28

This horoscope, contained in a compilation by Palchus, belongs to the same group as No. L 478. It concerns an otherwise unknown praefectus augustalis Aegypti Theodoros. Cumont [12] drew attention to its historical interest, conjecturing as its date 487 March 23 but adding "les positions sidérales données dans la suite du text permettront de contrôler ce chiffre hypothétique." As usual this caution has been forgotten and consequently Theodoros appears as prefect of Egypt for 487 in our handbooks [13] without a question mark.[14] In fact, the horoscope fixes the date at 486 (= Diocletian 202), Monday, March 17:

	Text		Computed	Text—comp.
☉	♓	26	♓ 28	—2
☽	♑	27	♑ 29	—2
♄	♐	11	♐ 13	—2
♃	♌	27	♌ 29	—2
♂	♈	25	♈ 28	—3
♀	♓	12	♓ 5	+7
☿	♓	23	♓ 16	+7
H	♈	26	about 7;30 A.M.	
M	♑	10	} *cf.* commentary	
☊	♐	32(!)		
☋(!)	♍	20		

---

[23] Cumont [12] p. 10 ff.
[24] Theophanes gives June 27, 483 (year) or 484 (indictio). Brooks in the *Cambr. Med. Hist.* I p. 478 note 1 substitutes July for June.

No. L 486
[*] For variants *cf. CCAG* 6 p. 63 note 1, next to the last paragraph.
[1] From Palchus.
[2] *Sic*, instead of 21.
[3] Omitted in the Vindob., in the text as well as in the diagram (unpublished).
[4] From $\lambda_F = \lambda_H + (\lambda_☽ - \lambda_☉) = ♈ 26 - 59 = ♒ 27$.
[5] Because Aries is a house of Mars.
[6] The 11th locus is here computed accurately by degrees, not simply by signs.
[7] *Cf.* Bouché-Leclercq, *AG* p. 513 note 5.

[8] Saturn's houses are ♒ and ♑.
[9] Saturn is close to the ascending node of the moon. Furthermore, there occurred within two months from the Katarche the total solar eclipse of 486 May 19, visible in Alexandria, framed by two lunar eclipses (485 December 7 and 487 April 23).
[10] Signs with rising times less than 30° are the six signs from ♑ to ♊.
[11] This concluding section is obviously corrupt. Wünsch (*CCAG* 1 p. 175) suggests deleting the mention of Jupiter and "omnia intelligentur" without saying how. For variants *cf. CCAG* 6 p. 63 bottom. A consistent emendation of the whole passage was suggested by Bouché-Leclercq, *AG* p. 514.
[12] Cumont [12] p. 9 ff.
[13] E.g., in *R.E.* V,A col. 1903 (No. 85). Also Kroll [3] p. 567.
[14] In fact even for Cumont's date the suggested reading Phamenoth 29 is wrong and should be Phamenoth 27, which is a Monday (*cf. CCAG* 1, p. 100, apparatus to line 3).

The date given in the text, Phamenoth (VII) 25, must be corrected to Phamenoth 21. If ♈ 26 is rising, one finds for Alexandria ♑ 17 culminating; the text gives ♑ 10. Furthermore: if ♈ 26 is rising and the sun is in ♓ 26, one finds for Alexandria an interval of 1;23ʰ since sunrise (or 1;24 in seasonal hours). Consequently the text should have said "second hour" instead of "first hour."

The data concerning the nodes are obviously wrong: ♐ 32 is in itself impossible and ♍ 20 is not diametrically opposite to any point in ♐. The correct value for the ascending node would be ♐ 3;40, thus ♊ 3;40 for the descending node. Probably ♍ 20 does not refer to the descending node but concerns the preceding full moon, because for 486 March 6 one finds the moon in ♍ 20 and the sun almost in opposition (♓ 17). The text of the Vindobonensis gives only ☊ ♐ ♍ which shows the existence of a lacuna, while the diagram gives only ☊, without number, in ♐, and nothing at all in ♍.

No. L 487

CCAG 1 p. 106/107 and CCAG 6 p. 63,2 to p. 64,25

TRANSLATION [1]

Year of Diocletian 204, month Thoth 7,[2] Saturday, first hour of daytime,[3] there were brought distressing letters to a certain person with everything contrary to the expectation of the receiver. The Katarche was found thus: the sun in Virgo degrees 10, the moon Libra degrees 4, Saturn in Sagittarius degrees 5;56, Jupiter in Libra degrees 7;55, Mars Capricorn 8, Venus in Leo 8, Mercury in Virgo 25, Horoscopos Libra 0;41,[4] Midheaven Cancer 0;41,[4] Lot of Fortune in Libra 25,[4ᵃ] the ascending node[5] in Scorpio 2;24. And we found Jupiter in the Horoscopos by degree[6] and Venus (in ♌) {as ruler[7] — namely}[8] of the Horoscopos {and of the moon and of Jupiter and of the Lot of Fortune — in Agathos Daimon}[9] and rising in the morning and adding to its numbers[10] and in superior aspect to (the position of) the moon on the third (day)[11] and we found that its[12] dodekatemorion happened to be Scorpio. The associate ruler of the Horoscopos,[12ᵃ] Jupiter, in the Horoscopos, was also moving forward. And I was misled by these configurations and said that the letter had a fine appearance, but (in fact) they meant the opposite. For afterwards investigating the Katarche accurately, I found that Venus (in ♌), house ruler[7] of the Katarche, had no claim[13] with respect to the Horoscopos (♎); (nor) the moon and Jupiter and the Lot of Fortune, because they were intercepted[14] by Mars[15] (in ♑), according to what Dorotheus says. And I found Mars (in ♑) and Saturn (in ♐) hemming in Mercury (in ♍) and the Horoscopos and the moon and Jupiter[16] according to what Antigonus teaches. And besides Mars (in ♑) was diametrical to Midheaven (in ♋) and destructive. Therefore, it is necessary in every Katarche to observe the interception of the sun and the stars and the hemming in of the Horoscopos and of the moon and of the star that has a claim to the houserulership of the Katarche. For not only were the Horoscopos and the moon (in ♎) hemmed in by Saturn (in ♐) and Mars (in ♑) but also Venus which had the greater claim in reference to the Katarche; for Mars (in ♑) was diametrical to Cancer[17] (M) and Saturn (in ♐) was in quartile to Virgo (☉ and ☿). Thus nearly all the stars were hemmed in by the two maleficent ones. This configuration is mentioned by Antigonus in the sixth nativity of the third book. If you investigate the figure you see Mercury (in ♍) injured by Saturn (in ♐) and also located in the 12th locus.

COMMENTARY

Diagram: pl. 28

For Diocletian 204, Thoth 7 = 487[18] September 5, Saturday, first hour of the day, one finds

Text		Computed	Text—comp.
☉	♍ 10	♍ 13	−3
☽	♎ 4	♎ 6;40	−2;40

---

[1] We follow the better version, CCAG 6, quoting significant variants in the notes. The Vindob. gives a diagram (unpublished). CCAG 1 gives as title "87. An efficacious Katarche in which I failed and went astray and afterwards finding the reason I marveled at its force. On letters brought." The number refers to excerpts from Palchus.

[2] CCAG 1 incorrectly 17.

[3] Daytime omitted in CCAG 1.

[4] CCAG 1 gives 1 (degree).

[4ᵃ] From Λ♀ = ♎ 0;41 + (♎ 4 − ♍ 10) = ♎ 24;41 ≈ ♎ 25.

[5] CCAG 1 has here the symbol for the descending node (misprint?).

[6] Since Jupiter is about 6° distant from H the meaning of this term is not clear.

[7] Because ♎ is a house of Venus; cf. note 9.

[8] CCAG 1 has here a meaningless κατὰ.

[9] We are following here CCAG 1 which states correctly that H, ☽, ♃ and the Lot of Fortune are in ♎ and thus ruled by Venus, which is in ♌, the 11th locus or Agathos Daimon. According to CCAG 6 the moon, Jupiter, and the Lot of Fortune would be in the 11th locus.

[10] Venus was indeed morning star and in direct motion.

[11] The moon, after 3 days, would be in ♏, thus in quartile to Venus in ♌. Cf. for the significance of the 3rd day No. L 40.

[12] This might be either Venus (whose dodekatemorion is ♏ 14) or the moon (dodekatemorion ♏ 22). CCAG 1 has αὐτῆς, CCAG 6 αὐτοῦ thus referring to Jupiter, but this is wrong since Jupiter's dodekatemorion is in ♑.

[12ᵃ] We follow here CCAG 1. Jupiter is "associate ruler" of the Horoscopos because it is in the same sign (♎), the "ruler" being Venus; cf. note 7.

[13] λόγος.

[14] Cf. Bouché-Leclercq, AG p. 251 f.

[15] CCAG 1: by Jupiter.

[16] Omitted in CCAG 1.

[17] CCAG 6 has here Libra which is wrong. This whole passage is confused in CCAG 6.

[18] Not 488 as stated in CCAG 1 p. 106.

	Text	Computed	Text—comp.
♄	♐ 5;56	♐ 18(!)	— [2]
♃	♎ 7;55	♎ 10	— 2
♂	♑ 8	♑ 8	0
♀	♌ 8	♌ 11	— 3
☿	♍ 25	♍ 30	— 5
H	♎ 0;41 or 1	about 7 A.M.	
M	♋ 0;41 or 1	correct [19]	
☊	♏ 2;24	♏ 5;14	— 2;50

Probably 5;56 for Saturn is a scribal error for 15;56.

The hour cannot be the 1st hour [20] of daylight since for a solar longitude ♍ 10 and for ♎ 1 or 0;41 as Horoscopos one finds a time of at least 1;30 seasonal hours [21] after sunrise for any geographical latitude from Alexandria to Byzantium. Thus " 2nd hour " would be required.

No. L 488
CCAG 8,1 p. 235,1 to p. 237,24

TRANSLATION

Another [1] (and) clearer (example) to the same. Clima 6.[2] Sun 18;20[3] of Aries, moon 17 of Gemini, Saturn 11 of Capricorn, Jupiter 19 of Scorpio, Mercury 25 of Aries, Venus 2 of Taurus, Mars 25 of Virgo, Horoscopos 2 of Leo, Midheaven 14 of Aries. Starter then is the sun and receives as chronocrators first Mercury, then Venus, then Saturn in trine, then Jupiter at diameter, then Mars in trine, then the moon, then as destroyer Saturn at diameter.

Midheaven 14 of Aries, sphaera recta 103 time-degrees. Sun 18;20 of Aries, sphaera recta 107. Excess 4 time-degrees. Mercury 25 of Aries, sphaera recta 113. Excess relative to Midheaven 10 time-degrees. Subtract the excess of the sun, 4 time-degrees, remainder 6. But since we ought first to take the (star) which took the nearest lead to the sun, it is evident that the moon had the near(est) lead, by virtue of sextile, for 7 years 6 days.

Then Mercury takes over the (next) period. Hours of daytime 16;40. Divide these into the 10 of the excess, it makes half of a seasonal hour [4] which Mercury is distant from Midheaven.

Again, Venus 2 of Taurus; sphaera recta 120 time-degrees. Subtract from these the 103 time-degrees of Midheaven, there remains as excess 17. Hours of daytime for (the place of) Venus 17;7. Divide these into the excess, making about 1 seasonal hour, which it (Venus) is distant from Midheaven. (Multiply) [5] the 0;36 [6] seasonal hours [7] of the distance of Mercury from Midheaven by the 17;7 time-degrees; it makes 10;16. Subtract these from the 17 time-degrees of the excess, remainder 6;47 [8] time-degrees. Thus Mercury will be the ruler of the period for 6 years and 7 months,[9] that is from the beginning of the 7th year to 11 years 7 months, and then he will transfer (rulership) to Venus.[10]

Again, the trine of Saturn (in ♑ 11) 11 of Taurus. Sphaera recta 129 [11] time-degrees. Subtract from these the 103 of Midheaven; there remains as excess 26. [Hours of daytime of the trine of Saturn 17;39. Divide these into the excess] making 1;30 [12] seasonal hours. Hence the triangle of Saturn is distant from Midheaven 1 and one half seasonal hours. Now again (multiply) the distance of Venus from Midheaven, that is 1 hour, by the hours of daytime of the trine of Saturn, (namely) 17;39. And it makes 17;39 time-degrees. Subtract this from the excess of Saturn, namely 26. Remainder 8;21. Thus Venus which took over the period from 11 years 7 months will rule 9 years,[13] that is until 20 years 7 months.

Again, the diameter of Jupiter (in ♏ 19) 19 of Taurus. Sphaera recta 136;28. Subtract the 103 time-degrees of Midheaven. There remains as excess 33;28. Hours of daytime of the diameter of Jupiter 18;7. Divide (these into) the excess, i.e., 33;28,[14] it makes 1;11 [15] seasonal hours. Then the diameter of Jupiter is distant from Midheaven 1;11 seasonal hours. Again (multiply) the 1;30 seasonal hours of the distance of Saturn from Midheaven by the hours of daytime of the [diameter

---

[19] Since H practically coincides with the autumnal equinox it is clear without computation that the arc HM must be very close to 90°.

[20] Written out: ἡμερινὴ πρωτ(η).

[21] Equinoctial hours make the discrepancy slightly greater.

No. L 488

[1] Referring to No. L 401 which directly precedes this horoscope. Both are from Rhetorius.

[2] Following Rome's emendation. The manuscripts have Clima 4 as in No. L 401, but clima 6 is required by the values quoted for seasonal hours.

[3] Following Rome's emendation; the manuscripts have 18;5 but later on 18;20.

[4] Accurately 0;36 and this value is used in the next paragraph.

[5] The editors restore incorrectly [divide]: cf., however, their own note in which they state that 0;36 × 17;7 = 10;16,12 (note the misprint: 1°16′12″).

[6] The editors read here 0;37 though they use correctly 0;36 in the note (cf. our note 5). The text has also the correct value ο λϛ′; the critical apparatus ad locum being wrong.

[7] Cf. note 4.

[8] The codices have here χρόνοι ϛ′ ἡμέραι ζ′ which the editors change to the equally meaningless χρόνοι ϛ′ μῆνες ζ′ obviously because also the subsequent lines give a siglum for " day " where month is expected. It is, however, not difficult to restore the text because 17 — 10;16 = 6;44. Thus we have to read χρόνοι ϛ′ μδ′ where μ′ = 40 was misinterpreted as μῆν. month. ζ′ is an error for δ′. Indeed codex V has here " signum trigoni " △ which represents, however, δ = 4 and not trine.

[9] 6;47 taken as years (or better 6;44) would be 6 years 9 months.

[10] Cf. the astrological commentary (below p. 152).

[11] Should be 128;29.

[12] Accurate would be 1;28.

[13] The text has time-degrees instead of years. But why 9 years instead of 8;21 years = 8 years 4 months?

[14] The words " into hours [of daytime] " make no sense here.

[15] Should be 1;47.

of] Jupiter (namely) the 18;7. It makes 27;10 [16] time-degrees. Subtract these from the excess of Jupiter (namely) 33;28. Remainder 6;18.[17] Thus the trine of Saturn, beginning at 20 years 7 months, will rule 6 years 5 months 17 days,[18] i.e., until 26 years 11 months [17] days.

Again, the trine of Mars (in ♍ 25) 25 of Taurus. Sphaera recta 142;34. Subtract the 103 time-degrees of Midheaven. There remains as excess 39;34. Hours of daytime of the trine of Mars 18;25. Divide these into the excess; it makes 2;9 [19] seasonal hours. Thus the trine of Mars is 2;9 seasonal hours distant from Midheaven. Now the diameter of Jupiter was 1;11 seasonal hours distant from Midheaven. Multiply these by the daylight time-degrees of the trine of Mars, i.e., 18;25. It makes 21;48 [20] time-degrees. Subtract these from the excess, namely 39;34. Remainder 17[;46] time-degrees (that is) [17 years] 11 months 9 days.[21] But the diameter of Jupiter which began at 26 years [11 months 17 days] will extend until 44 years 10 months 26 days.

Again, the moon in Gemini 17. Sphaera recta 165;50. Subtract from them the 103 time-degrees of Midheaven. There remains as excess 62;50. The hours of daytime (of the position) of the moon: 19;13. Divide these into the excess; it makes 3;21 [seasonal] hours. Thus the moon is 3;21 seasonal hours distant from Midheaven. But the trine of Mars was 2;9 seasonal hours distant from Midheaven. Multiply this by the daylight time-degrees (of the position) of the moon, namely 19;13. It makes 41;19.[22] Subtract this from the excess of the moon, namely 62;50. Remainder 21;31 time-degrees. Thus the trine of Mars, which began to rule from the year 45, will end at 66 years 7 months 5 days.

Again, the diameter of Saturn (in ♑ 11) 11 of Cancer. Sphaera recta 192. Subtract 103; excess 89.[23] Daylight time-degrees of the diameter of Saturn 19;16. Multiply the distance of the moon from Midheaven, 3;21 hours, by 19;16. It makes 64 time-degrees, 6 months 17 days.[24]

Thus the moon which took over the period from 66 years [7 months] 5 days will rule until 26 more years 3 months 12 days, that is up to a total of 92 years 10 months 17 days. Here the diameter of Saturn was destructive and destroyed.

ASTRONOMICAL COMMENTARY
Diagram: fig. 19

It can be shown that there is no possibility of placing this horoscope before the third century. Similarly dates after 500 can be excluded, which is consistent with the authorship of Rhetorius. But also within this interval

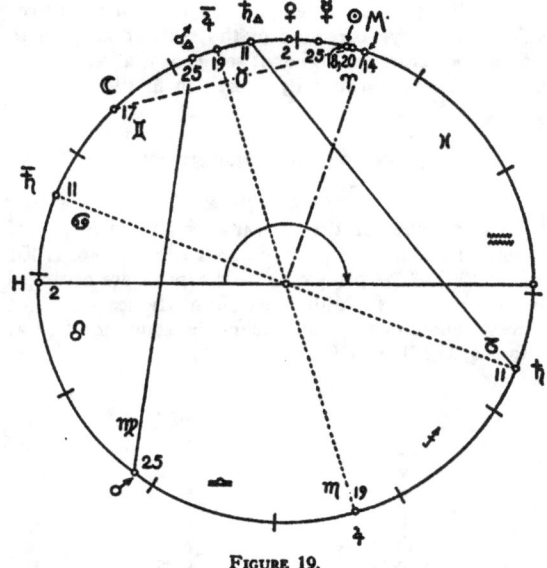

FIGURE 19.

no good solution can be found. Partial agreement is reached for 488 April 2:

Text		Computed	Text—comp.
☉	♈ 18;5 or ♏(!) 18;20	♈ 15	+3
☾	♊ 17	♊ 19	−2
♄	♑ 11	♑ 6	+5
♃	♏ 19	♏ 4(!)	
☿	♈ 25 or ♏(!)	♈ 5(!)	
♀	♉ 2	♉ 4	−2
♂	♍ 25	♉ 24(!!)	
H	♌ 2	about noon	
M	♈ 14	clima 6: ♈ 14	

Jupiter and Mars can be placed more correctly for October 11 in the same year (488):

---

[16] Accurately 27;10,30. Codex V gives 27 only.
[17] Codex V gives 6;28 in consequence of the value quoted in note 12.
[18] One expects either 0;18 years = 3;36 months = 3 months 18 days instead of 5 months 17 days; or (with the values quoted in the notes 4 and 5) 0;28 years which would be 5 months 18 days. The codices have 5 months 17 days which the editors replace by 4 months 17 days in order to obtain the total given in the text.
[19] Accurately 2;8,54, ...
[20] Accurately 21;47,35.
[21] Expected: 0;46 years = 9;12 months = 9 months 6 days.
[22] Accurately 41;18,57.
[23] Text: 79, but 89 is used in the next step.
[24] Accurately 64;32,36. Taking this as years one finds 64 years 6 months 16 days. In fact the result should be 64;33° and then be subtracted from the excess 89 (cf. note 23). The result 24;27° counted as years would give 24y5m12d for the rulership of the moon. The final total, given in the text, presupposes an interval

of 26y3m12d. The sentence which mentioned the length of the interval has fallen out.

☾	♊ 14;40	but	☉	♎ 20	instead of	♈ 18;5
♃	♏ 13		♄	♑ 1		♑ 11
♂	♍ 27		♀	♏ 21		♉ 2
			☿	♎ 4		♈ 25.

If one accepts the variant ♏ for the longitudes of ☉ and ☿ one finds for 488 November 7

☉	♏ 17	but	♂	♎ 14	instead of	♍ 25
♄	♑ 3		♀	♏ 10		[♏] 2
♃	♏ 18		☿	♏ 5		♏ 25
			☾	♊ 10		♊ 17.

Since all longitudes are extensively used in the astrological part, any correction of the text would lead to serious contradictions. Thus it seems as if at least some of the data were arbitrarily chosen to fit the example.

ASTROLOGICAL COMMENTARY

Diagram: fig. 19

This example for the computation of the length of chronocratorship follows the same pattern as No. L 401 (*cf.* p. 134). The planets or their aspects are arranged in their relation to Midheaven since Midheaven is the nearest center to the sun. Using the same notation as with No. L 401 we have

n		λ_n
0	Sextile of ☾	♈ 17
1	☉	♈ 18;20
2	☿	♈ 25
3	♀	♉ 2
4	Trine of ♄	♉ 11
5	Diameter of ♃	♉ 19
6	Trine of ♂	♉ 25
7	☾	♊ 17
8	Diameter of ♄	♋ 11.

In principle the sun is the "starter" but, the sextile of the moon being still nearer to M = ♈ 14, the moon is given a first period of 7^y6^d without explaining how. Furthermore, the length of this period contradicts the subsequent statement according to which the rulership of Venus ($n = 3$) begins at 6^y7^m(!) and ends after 7^y at 11^y7^m(!). Obviously the whole text up to the discussion of Venus is corrupt. In general it must be observed that the replacement of degrees and their fractions by the equivalent number of years and months is in a chaotic stage and, similarly, the adding up of these intervals is almost always wrong.

It is surprising to see that while the trivial additions of years and months are mostly wrong, the complicated formula (1) of p. 134 is correctly used from $n = 3$ on. All values are taken from Theon's Handy Tables.

No. L 497

CCAG 1, p. 171,2 to 14 and CCAG 4, p. 106,6 to p. 109,4

TRANSLATION

First Version (CCAG 1)

Photo: pl. 32

The example of the theme is as follows: let it be the year 214 from the beginning of the reign of Diocletian which is the year 821 from the death of Alexander the Macedonian, first (day) of the Alexandrian month Athyr (III), seasonal hour for Alexandria in Egypt 7;12 of the day. For this time and the said locality, that is Alexandria, we computed accurately the positions of the stars and the centers and put them down without change because thus it seems right to the godly Ptolemaios. Thus the positions of the seven planets and the centers and the lots are as follows:[1] ☉ ♏ 4;22,[2] ☾ ♉ 18;10, ♄ ♈ 24;28, ♃ ♍ 4;12, ♂ ♌ 17;35, ♀ {♏ 21;58, ☿ ♏ 14;32,[3] Horoscopos[4] ♒ 2;12, Midheaven[4] ♏ 19;22, (Lot of) Fortune ♌ 16, Daimon[4]}[5] ♋ 18;24,[6] full moon[7] ♉ 3;7, ascending node[4] ♈ 16;4.

☉ ♏ 4;22,[8] inclined towards the equator[9] on the circle[10] which passes through its pole towards the south 13;6[11] degrees; it descends to the south in the third step,[12] occupying of this(?) step ¼ ¹⁄₃₀.[13] Near to it is the fixed star which is between the claws of ♏, of magnitude 4. It occupies in longitude ♏ 4;46; in latitude it occupies in the circle through the middle of the zodiacal (signs) towards the south 1;30. Thus it is

---

[1] The second version (*CCAG* 4, p. 106) replaces this whole introduction by one sentence "The following is the investigation of the seven planets." For the different manuscripts *cf.* below p. 188.

[2] Second version: 4;8,50.

[3] In the second version this is the last entry preceding the next paragraph which again begins with the sun.

[4] Represented in the manuscript by a siglum.

[5] This whole passage (one line) omitted by the editor. Checked on photo.

[6] From $\Lambda_F = \lambda_H + (\lambda_☾ - \lambda_☉) =$ ♒ 2;12 − 166;12 = ♌ 16 and $\Lambda_D = \lambda_H - (\lambda_☾ - \lambda_☉) =$ ♒ 2;12 + 166;12 = ♋ 18;24 respectively.

[7] The manuscript has the abbreviation ☾π̇ which the editor wrongly interpreted as σελήνη πρὸς.

[8] The excerpt given in *CCAG* 1 ends at this point. What follows is based on a photostat of the manuscript, kindly supplied to us by the Laurentian Library.

[9] Written with ligature; *cf. CCAG* 4, p. 106,10 apparatus.

[10] The ζ′ printed in *CCAG* 4, p. 106,11 which follows κύκλου makes no sense; perhaps abbreviation for μεγίστου. Similarly p. 107,2.

[11] This value from the second version; the first version has 3;6 only.

[12] "Step" (βαθμός) is in the first version always abbreviated by β which is underlined by an s-shaped line.

[13] Second version: "occupying of the 4th step 30" which makes no sense and is probably a mistake.

distant from the ♂ in longitude towards following (longitudes) 0;24, in latitude 1;36. House (of) ♂, terms according to Ptolemy and the Egyptians (of) ♂. Triangle (of) ♀, partner ♂ {and third (partner) ☾. In the exaltation of none, in the depression (of) ☾. In the first decan,}[14] in the countenance (of) ♂; monomoiria of the moon.

☾ ♉ 18;10 being distant in latitude from the circle through the middle of the zodiacal (signs) towards north 2;42 north. [For the continuation *cf.* the following summary.]

SUMMARY OF CONTENTS

Cf. also p. 188

In a similar fashion a detailed discussion of the positions of all planets is given, followed by a less complete list for the centers and other significant points. In the following these data are rearranged in groups of equivalent elements.

A. Positions

1.	☉	λ = ♏ 4;22	δ = −13;6
2.	☾	♉ 18;10	β = + 2;42
3.	♄	♈ 24;28	− 3;5[15]
4.	♃	♍ 4;12[16]	+ 0;35
5.	♂	♌ 17;35	+ 1;5
6.	♀	♏ 21;58	− 0;39
7.	☿	♏ 14;32	− 1;38

B. Fixed stars. The planetary positions are compared with nearby fixed stars for which name, magnitude, longitude and latitude are given and also the differences Δλ and Δβ between star and planet. We also compare these data with those given in the Catalogue of Stars of the Almagest.[17]

Text	Ptol.
1. Between the claws of ♏	The southern one[18] (of the 3 in the claws)
λ = ♏ 4;46 β = −1;30 *m* = 4	λ = ♏ 1;10 β = −1;30 *m* = 4
2. On the base of the northern horn of ♉	same[19]
λ β *m* ♉ 19;16 −0;15 4	λ β *m* ♉ 15;40 −0;15 4
3a. The last of (the stars) in the rhomboid quadrangle of ♈ and (also) at the tip of the nostril of Cetus [♈] 6;6[21] −7;45 4	at the end of the nostril[20] ♈ 17;40 −7;45 4
3b. The first of the three in the tail of ♈ ♈ 27[23] +1;40 4	same[22] ♈ 23;50 +1;40 4
4. At the tip of the southern and left wing of ♍ ♍ 2;36 +0;10 3	same[24] ♌ 29 +0;10[25] 3
5. The southernmost of the two in the buttocks of ♌ ♌ [1]9;56[27] +9;40 3	same[26] ♌ 16;20 +9;40 3
6a. The last of the three in the body[29] of ♏ ♏ 18;6 −5;30 3	same[28] ♏ 14;30 −5;30 3
6b. In the right and posterior knee of Ophiuchus ♏ 24;46 +7;30 3	In the right knee[30] ♏ 21;10 +7;30 3
7. The middle and reddish one in the body[29] of ♏ called Antares ♏ 16;16 −4;0 2	same[31] ♏ 12;40 −4 2

The comparison with the data from the Almagest shows identity of latitudes and magnitudes in all cases whereas the longitudes are consistently 3;36° greater than in the Almagest. Using Ptolemy's value of one degree per century for the constant of precession we obtain a time interval of 360 years. Since Ptolemy's catalogue refers to the beginning of the reign of Antoninus Pius in A.D. 137, we find that the positions in our text correspond to the year 137 + 360 = 497 which is exactly the year of the horoscope.

In all cases the text gives the differences Δλ and Δβ in longitude and latitude between the fixed star and the planet (or sun or moon) nearby. The numbers are:

[14] The second version has here: "In the first decan and third the moon, in the exaltation of none, in the depression of the moon."

[15] Omitted in second version.

[16] Erroneously given as 2;12 in the second part of the first version; correctly 4;12 in the first part and in the diagram.

[17] The same comparison was made by Boll, *CCAG* 4 p. 100 f.

[18] Heiberg II p. 108/109,12.

[19] Heiberg II p. 88/89,8.

[20] Heiberg II p. 130/131,3. *Cf.* also Gundel, *RE* 11 col. 367,25 ff.

[21] Error for 21;16. The second version is also confused; *cf.* *CCAG* 4, p. 107,6 apparatus.

[22] Heiberg II p. 84/85,10.

[23] Error for 27;26 in both versions.

[24] Heiberg II p. 102/103,7.

[25] Heiberg II p. 103,7 gives 0;20 but Peters-Knobel *PCS* p. 38 No. 501 give 0;10 (*cf.* also p. 24 Table V first group and Heiberg's apparatus).

[26] Heiberg II p. 100/101,2.

[27] The text has here 9;56 only (both versions).

[28] Heiberg II p. 110/111,8 and 6.

[29] Both versions write στόματι (in the mouth) instead of σώματι.

[30] Heiberg II p. 68/69,12.

[31] Heiberg II p. 110/111,7.

case	1	2	3a	3b
Δλ	+0;24	+1;6	−3;12 [32]	+2;58
Δβ	1;36	2;57	4;40 [32]	4;45

	4	5	6a	6b	7
	+1;36	+2;21	−3;52	+2;48	+1;44
	0;45	8;35	4;51	8;9	2;22 [33]

where Δλ is called positive, "towards following (longitudes)," when the longitude of the fixed star is greater than the longitude of the movable body, whereas the differences in latitudes are only given as absolute values. These differences allow us to check the given data; they either agree with the text or confirm the restorations which were made independently on the basis of Ptolemy's values corrected for precession. In the case of the sun the difference in "latitude" is actually the sum of latitude of the star and declination of the sun.

Following longitude and latitude (respectively declination) of sun, moon, and planets the text gives the number of the "step" — either "upgoing" or "downgoing" — and the fraction of this step which is occupied by the planet. Such "steps" are defined by Theon in his Commentary to the Handy Tables.[34]

For the sun each quadrant, bounded by solstices and equinoxes, is divided in six 15° segments, or "steps," counted from 1 to 6 in each quadrant. In the present text the sun is in ♏ 4;22 or 34;22° beyond the autumnal equinox. Thus the sun is 4;22° inside the 3rd step, "descending" toward south. Since 4;22° represent a fractional part of 0;17,28 of one step of 15° the text is quite correct in saying that the sun has occupied ¼ ¹⁄₃₀ i.e., 0;17 of the 3rd step.

For the moon the situation is similar. We know that the ascending node is at ♈ 16;4, the moon itself at ♉ 18;10, thus 32;6° beyond the node. Again operating with "steps" of 15° each, we can say that the moon is "ascending" in the 3rd step of which it occupies 2;6° or 0;8,24. The text gives ¹⁄₁₀ ¹⁄₂₄ or 0;8,30.

For the planets one "step" is defined as ⅙M if M denotes the extremal latitude toward north or toward south, and these steps are counted from 1 to 6 beginning with the nodes or with the points of extremal latitude. The text assigns the following data to the planets

♄ in 1st step upgoing, occupying ¹⁄₃₀ of this step
♃ in 3rd step upgoing, occupying ½ ⅕ of this step
♂ in 2nd step upgoing, occupying ½ of this step
♀ in 1st step downgoing, occupying ⅓ ¹⁄₁₅ of this step
☿ in 3rd step downgoing, occupying ⅕ ¹⁄₁₀ of this step.

---

[32] Second version −3;12,10 and 4ω' = 4⅔? respectively.
[33] Second version 22 only.
[34] This is the so-called "small" commentary of Theon, published by Halma in vol. I of his edition of the *Tables manuelles* (Paris, 1822). For the sun cf. p. 54, 22-29, for the moon p. 55, 16, for the planets p. 59, 4-23.

The extremal values M of the planetary latitudes are given in the Handy Tables and in Theon's Commentary to these tables,[34a] and thus also the values of the "steps," namely ⅙M, are known.[34b] For our purposes the following values are needed:

	M	⅙M	⅓M	β
♄	−3;6	−0;31		−3;5
♃	+2;3	+0;20,30		+0;35
♂	+4;23	+0;43,50		+1;5
♀	−8;56	−1;29,20		−0;39
☿	−4;18	−0;43	−1;26	−1;38

If we compare the given planetary latitudes β with the values of the steps we see that ♄ is 0;1° above the southernmost point. Since one step amounts to 0;31 the part $\frac{0;1}{0;31} \approx 0;1,56$ is occupied by the interval from M to β. The text gives instead the fraction ¹⁄₃₀ = 0;1,12. If we proceed in the same fashion for all planets we find for the occupied fractions

♃ 0;42,26   text: ½ ⅕ = 0;42
♂ 0;29   text: ½ = 0;30
♀ 0;33,48   text: ⅓ ¹⁄₁₅ = 0;24
             [emend ½ ¹⁄₁₅ = 0;34]
☿ 0;16,48   text: ⅕ ¹⁄₁₀ = 0;18.

Thus, except for the scribal error in the case of Venus, the agreement between the accurate values and the approximation by unit fraction is reasonably good.

We have still to check the numbers given to the steps. From the comparison of β with the multiples of ⅙M one obtains always two possibilities, depending on approach or retreat of the planet with respect to the endpoints of the four quadrants. Thus we find

♄	6th or 1st step	text: 1st upgoing
♃	5th or 2nd step	text: 3rd upgoing
♂	5th or 2nd step	text: 2nd upgoing
♀	6th or 1st step	text: 1st downgoing
☿	4th or 3rd step	text: 3rd downgoing.

Except for Jupiter the counting of the text does not contradict the results obtained by using β only. To find the trend of the motion we should know whether it took into consideration only the deferent or the epicycle or both. Our material seems not to suffice to distinguish between these possibilities.

The information concerning the positions of the planets is followed by remarks about their motion. We

---

[34a] Halma, Tab. Man. III p. 1-10 and I p. 59, 14-23. Better than Halma's printed text are the manuscripts Cod. Vat. 1291 fol. 78ʳ—82ʳ and Vat. Gr. 208 fol. 84ʳ 30-35 respectively (both unpublished).

[34b] The values of the steps are given at the foot of the tables of the planetary latitudes in the Handy Tables. These numbers are badly garbled in Halma's edition but correctly given in Vat. Gr. 1291 fol. 78ʳ—82ʳ.

quote them here as given in the text and refer to the Commentary for discussion.

♄ retrograding from the first station, visible in the evening, being in opposition [35] for many days

♃ not progressing, being in the first station,[36] visible in the morning

♂ not progressing, being in the first station, visible in the morning

♀ not progressing, being in the first station, visible in the evening, having made its evening rising many days before [37]

☿ not progressing, being in the first station, invisible in the evening, tending towards evening rising [38] within 4 days.

About "the preceding full moon" it is said that it happened "according to the Alexandrian Phaophi (II) 30, seasonal [39] hours of daylight 6;20."

The relative positions with respect to fixed stars are only given for sun, moon, and planets. The astrological data, houses, terms, etc., are mentioned also for the preceding full moon, for H and M, and for the Lots of Fortune and Daimon. Table 12 gives the whole list. An

		House	Terms Pt.	Terms Eg.	Triangle	Ex.	Depr.	Dec.	Pros.	Mon.
☉	♏ 4;22	♂	♂		♀ ♂ ☽	—	☽	1	♂	☽
☽	♉ 18;10	♀	♃		♀ ☽ ♂	☽	(—)	(2	☽)	♃
♄	♈ 24;28	♂	(♂)		☉ ♃ ♄	☉	♄	3	♀	☿
♃	♍ 4;12	☿	☿		♀	♃	♀	1	☉	♂
♂	♌ 17;35	☉	♀	♄	☉ ♃ ♄	—	—	2	♃	☽
♀	♏ 21;58	♂	☿	♃	♀ ♂ ☽	—	☽	3	♀	♂
☿	♏ 14;32	♂	♃	☿	♀ ♂ ☽	—	☽	2	☉	♂
Full ☽	♉ 3;7	♀	♀		♀ ☽		—	1	☿	♄
H	♒ 2;12	♄	♄	☿	♄ ☿ ♃	—		1	♀	♂
M	♏ 19;22	♂	♀	☿	♀ ♂ ☽	—	☽	2	☉	♄
Lot of F.	♌ 16;0	☉	♀	♄	☉ ♃	—	—	2	♃	♀
Lot of D.	♋ 18;24	☽	☿		♀ ♂ ☽	♃	♂	2	☿	☉

TABLE 12.

empty space indicates omission by the text whereas "—" indicates a special remark in the text that no planet has its exaltation or depression in this sign.[40]

The "houses" are all correct. The "terms" are given both acording to Ptolemy and according to the Egyptians. Full agreement is found for this second system if one counts ♏ 19;22 of M as belonging to ♏ 19, which is the boundary between the terms of ☿ and ♄, and assigns it still to ☿. The terms of Ptolemy are all correct if one accepts the version of the edition of the *Tetrabiblos* by Boll-Boer [41] but not the version of Robbins.[42]

The triangle rulers are correct if one uses the scheme of Vettius Valens (*cf.* p. 13) which gives three rulers.[43] Exaltations and depressions are correct with the only exception of Jupiter which is said to be in its own exaltation, which is ♎, however, and not ♍ (which is the exaltation of Mercury). The "decans" are simple numbers, 1, 2, or 3, and correct. The "Prosopa" or planetary faces are also in agreement with the standard list. The "Monomoiria" are found by means of the method explained on p. 10.

The text turns finally to some general remarks about the careful consideration of all astrological data and ends with a quotation from Homer. Then follows a crude diagram which we reproduce in figure 20. Below it is written *τέλος τοῦ θέματος* "end of the theme."

NOTES TO THE DIAGRAM (FROM F)

The title is *τὸ δε σχῆμα τοῦ θεάματ(ος)* (*sic*). Following a counterclockwise motion we begin at the left with *ā ωρ(οσ)κ(οπος)* ♒ *β ιβ′*. To each of the following zodiacal signs is given a number of degrees and minutes such that the quadrants between the centers are divided in as nearly as possible equal parts:

			differences	
H:	♒	2;12		
	♓	7;55	35;43	
	♈	13;39	35;44	
IMC:	♉	19;22	35;43	total: 107;10
	♊	13;39	24;17	
	♋	7;55	24;16	
Δ:	♌	2;12	24;17	total: 72;50

and similarly on the other semicircle. This shows that the different sectors of the diagram represent the twelve loci. Their numbers and names are given as follows

	1	ωρ(οσ)κ(οπος)
2	2	βίος
		θέα
4(?)	4	υπ(ογειον)
	5	ἀγαθὴ τυχ(η)
6	6	κακὴ τύχ(η)
	7	δυνω(ν)
	8	ἀργὸς
9	9	θεὸς
10	10	μ(εσου)ρ(ανημα)
	11	αγαθ(ος) δαίμο(ν)
	12	κακος δαίμο(ν).

The number of the third locus is omitted and the mean-

[35] ποιησάμενος τὴν ἀκρονυχίαν.

[36] προ(σ)θέτης μηδέπω ποιησάμενος τὸν ᾱ στηριγμόν.

[37] ποιησαμένη τὴν ἑσπερίαν ἀνατο(λὴν) πρὸ ἡμέρων πολλῶν.

[38] ἑσπέριος ὕπαυγος μέλλων ποιεῖσθαι τὴν ἑσπερίαν ἀνατολήν.

[39] The scribe wrote instead *καὶ ρικ ᾱς*, thus taking *καιρικὰς* as "and 100 10 20 1⅓."

[40] Data in ( ) indicate occurrence in the second version only. No information is available from the second version beyond Mercury.

[41] P. 51 f., identical with Bouché-Leclercq, *AG* p. 211.

[42] P. 106/107 for ♂, Lot of Fortune, and M.

[43] They are enumerated in the text as *τρίγων*, *μετοχή* and *τρίτου* (or *τρίτης*).

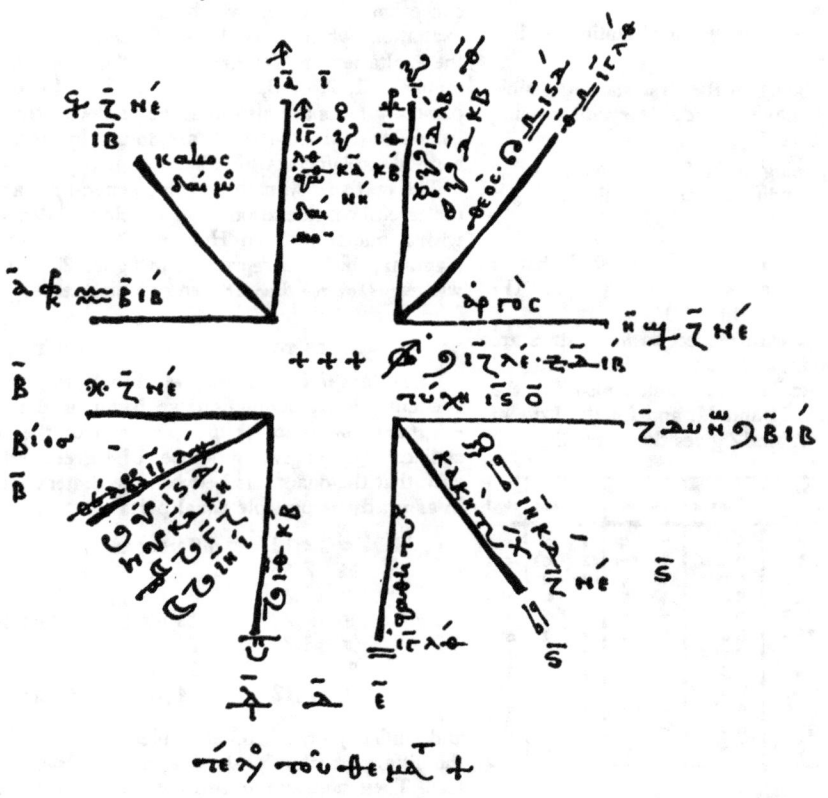

FIGURE 20.

ing of the first 4(?) is doubtful since it is a ligature of δ and ι (as in διά). The symbol for ♈ on the division line between second and third locus is also strange.

In the third locus we find the ascending node, Saturn, full moon, and moon. The 6th locus contains the Lot of Daimon (♌ 18;24) while the Lot of Fortune is in the 7th locus (♌ 16;0), together with Mars and Jupiter. The 9th locus is occupied by the descending node, the sun and Mercury, the 10th by Venus.

ASTRONOMICAL COMMENTARY
Diagram: fig. 20 and pl. 29

The date of this horoscope is 497 October 28, which is the equivalent of the date given in the text, Diocletian 214, Athyr (III) 1, and is confirmed by computation.[44]

[44] We have here used P. V. Neugebauer, *Tafeln zur astronomischen Chronologie* II.

		Text		Computed	Text—comp.
☉	λ =	♏	4;22	♏ 7;0	−2;38
	δ =	−	13;6	− 13;12	0;6
☽	λ =	♉	18;10	♉ 21;15	−3;5
	β =	+	2;42	+ 2;45	0;3
♄	λ =	♈	24;28	♈ 26;46	−2;18
	β =	−	3;5	− 2;43	0;22
♃	λ =	♏	4;12	♏ 7;21	−3;9
	β =	+	0;35	+ 1;10	0;45
♂	λ =	♌	17;35	♌ 20;9	−2;34
	β =	+	1;5	+ 1;47	0;42
♀	λ =	♏	21;58	♏ 25;38	−3;40
	β =	−	0;39	− 0;29	0;10
☿	λ =	♏	14;32	♏ 18;2	−3;30
	β =	−	1;38	− 1;33	0;5
☊	λ =	♈	16;4	♈ 18;41	−2;30

If we assume that $-2;30°$ is the correction for precession then we see that the errors of the planetary positions are as much as about 1° only for Venus and Mercury in longitude whereas the latitudes deviate ¾° only for Jupiter and Mars and are otherwise quite small.

The further statement (cf. p. 155) that Saturn is retrograde and near opposition is obviously correct. All four remaining planets are said to be in their first station. In fact, Jupiter is about two months before this point, Mars almost twice as much. Mercury is said to be shortly before evening rising. The scribe should have realized that this implies greatest direct motion and thus excludes being stationary. In fact, Mercury is not 4 days but about 12 days before evening rising (Nov. 10).[45] Venus is evening star, about three weeks after evening rising (about Sept. 19).

As culminating point the text gives ♏ 19;22 for H = ♒ 2;12. The same result is found from Theon's tables.

The preceding full moon is said to have occurred on Phaophi (II) 30 at 6;20 in seasonal hours of daylight, in ♉ 3;7. The moment in question (Oct. 27, noon) precedes the moment of the horoscope by only 25 hours. For this interval the text implies a solar motion from ♏ 3;7 to ♏ 4;22, thus of 1;15°, and a lunar motion from ♉ 3;7 to ♉ 18;10, thus of 15;3°. Modern computation for Oct. 27 noon gives for the sun a longitude of ♏ 5;55 which corresponds to a motion of 1;5°, and for the moon ♉ 7;58 corresponding to 13;17° motion. This also shows that we should place the opposition about 4 hours earlier.

No. L 516
CCAG 8,1 p. 231,5 to 31

TRANSLATION

Horoscopos 12 of Taurus,[1] Midheaven 23;30 of Capricorn, clima 5. The sun 12 of Taurus, the moon 30 of Libra, Mars 18 of Taurus, Mercury 21 of Taurus, Venus 20 of Libra.[2] I investigate the starting of the moon,[3] which is in Libra 30, and when it contacts each of them as follows: Lower Midheaven Cancer 23;30. Sphaera recta 205;26 time-degrees. The moon in 30 of Libra, sphaera recta 297.[4] Excess 91;34. Seasonal hours of the night (of the position) of the moon 16;43. Divide the 91;34 by the 16;43. It makes about 5½ seasonal hours.[5] So much is the moon distant from the lower center.

One must now find when the moon makes the connection with the diameter of the sun. The diameter of the sun is 12 of Scorpio. Rising time of sphaera recta 309;28. Lower Midheaven Cancer 23;30. Sphaera recta 205;26. Excess 104;2. Time-degrees (of the diameter) of the sun 17;21. Multiply this 17;21 by the 5;28 seasonal hours. It makes 94;37 time-degrees.[6] Subtract these from the 104;2. Remainder 9;25. I then say that the moon connected with the sun in diameter at 9 years and about 2 months.[7]

Again, Mars in Scorpio 18 according to diameter. Sphaera recta 315;28. Excess over Lower Midheaven (205;26) 110;2. Time-degrees of daylight[8] (of the position) of Mars 17;38. This multiplied by the 5½ seasonal hours of the distance of the moon (from IMC) makes 97 time-degrees.[9] Subtract this from the 110;2. Remainder 13. I then say that the moon connected with Mars and after these 13 years Mars took from it (the moon) the period and became chronocrator.

Again, the diameter of Mercury 21 of Scorpio. Sphaera recta 318;29. And with Lower Midheaven are culminating 205;26 time-degrees. Excess 113;3. Time-degrees of daylight[8] (of the position) of Mercury 17;46. This multiplied by the 5½ hours. It makes 97;43.[10] Subtract this from 113;3. Remainder 15;20. Declare then that the moon contacts with the Diameter of Mercury at 19 years 4 months.[11]

ASTRONOMICAL COMMENTARY
Diagram: pl. 29

From the incomplete data of this horoscope one can nevertheless deduce as a plausible date 516 May 1, in agreement with the time of Rhetorius.

Text			Computed	Text—comp.
H	♉	12	about sunrise	
M	♑	23;30	clima 5: ♑ 23;30	
☉	♉	12	♉ 12	0

---

[45] Using the Tafeln zur Berechnung der jährlichen Auf- und Untergänge der Planeten, by P. V. Neugebauer, AN 264 (1938) col. 313-322.

No. L 516

[1] This is the correct reading of codex V. Codex P: Aries. Cf. also other fragments of the same horoscope:

CCAG 8,1	p. 223,21 f.	H = ♉ 11;13	M = ♑ 23
	p. 227,19 f.	H = ♉ 11	M = ♑ 23
	p. 229,13	H = ♉ 12	M = ♑ 23
	p. 230,11 f.		IMC = ♋ 23

The author is Rhetorius.

[2] This is obviously impossible if the sun is in Taurus. The fragment p. 227,19 f. gives ♓ 21 which is in trine to ♎ 2[1]. One could also think of the diametrical position ♈ 20. Computation confirms neither one of these possibilities (resulting in ♈ 0°). It is also suspicious that Venus is omitted from the astrological discussion.

[3] This means: the moon is the "starter."

[4] Sic, instead of 297;50. Note that in the related example CCAG 8,1 p. 230,11 the position of the moon is given as ♎ 29 for which α = 296;53.

[5] Accurately: 5;28,36, ... The value 5;28 is used in the next step. If we had replaced 297 by 297;50 (cf. note 4) we would have obtained here 5;31,38, ... ≈ 5;32.

[6] Actually 94;50,48.

[7] One would expect 5 months.

[8] Time-degrees of daylight for a certain position is the same as time-degrees of night for the diametrically opposite position.

[9] 17;35 · 5;30 = 1,36;52,30.

[10] Exactly equal 17;46 · 5;30.

[11] Sic, instead of 15 years 4 months.

158 LITERARY SOURCES: No. L 516— No. L 621

	Text		Computed		Text—comp.
☾	♎ 30		♎ 26		+4
☊	♉ 18		♉ 19		−1
☿	♉ 21		♉ 18		+3
♀	[♈][12] 20		♈ 0(!)		
IMC	♋ 23;30				

The omission of Jupiter and Saturn deprives us of the most important element for a definitive date. The incomplete example of *CCAG* 8,1 p. 229,14 which also uses clima 5, H = ♉ 12, M = ♑ 23 places Jupiter in ♋ 11. In May 516 Jupiter was, however, in ♓ 25. Thus we must assume that either the two examples are independent, or that we are dealing with a fictitious example, as probably is the case also with Nos. L 401 and L 488 from Rhetorius.

ASTROLOGICAL COMMENTARY
Diagram: fig. 21

We have here the same procedure for determining the duration of chronocratorship as in No. L 401. We begin again with the moon. This time, however, the next

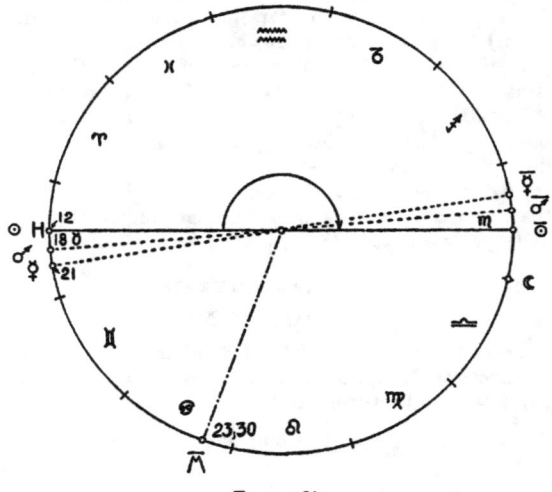

FIGURE 21.

center to be reached by the moon in its daily rotation is Lower Midheaven (IMC = $\bar{M}$). Consequently, one now defines as "distance" of the moon from $\bar{M}$

$$(1) \quad \Delta_1 = (a(\lambda_1) - a(\bar{M})) \frac{1}{h(\lambda_1)}$$

where $\bar{h}(\lambda) = h(\lambda + 180)$ represents the length of a seasonal hour of the night for a solar longitude λ or, which is the same, the length of a seasonal hour of day-

[12] The text gives the diametrically opposite sign ♎.

light for a solar longitude $\bar{\lambda} = \lambda + 180$. For the positions which follow the moon ($n \geq 2$) we have

$$(2) \quad \Delta_n = (a(\lambda_n) - a(\bar{M})) - (a(\lambda_{n-1}) - a(\bar{M})) \frac{\bar{h}(\lambda_n)}{\bar{h}(\lambda_{n-1})}$$

(cf. No. L 401 p. 134).

In our specific example we have

n		λ_n	$a(\lambda_n)$	$\bar{h}(\lambda_n)$
1	☾	♎ 30	297(!)	16;43
2	Diameter of ☉	♏ 12	309;28	17;21
3	Diameter of ☊	♏ 18	315;28	17;38
4	Diameter of ☿	♏ 21	318;29	17;46

The values for a and $\bar{h}$ agree with Theon's Handy Tables except for $a(\lambda_1)$ where 297 is an error for 297;50, which, however, affects all subsequent results.

The transformation of the final results from degrees and minutes into years and months is again very carelessly handled—a fact which is equally striking in Nos. L 401 and L 488 though nothing is involved but the multiplication of the minutes by 12 in order to obtain the number of months.

No. L 621

Stephanus, Usener Kl. Schr. III p. 273,10 to p. 275,9

TRANSLATION

[After a long introduction on the philosophy of astrology Stephanus reports about the visit of his friend Epiphanius, an Arabic merchant, who told him about a prophet Μωάμεδ who promises to his followers "victory in war, rulership over their enemies, and the pleasures of paradise."[1] Cause enough for alarm and for astrological investigation.] The aforesaid merchant came to us in the month Thoth, according to the Egyptians, {5, day 3, hour 3}.[2] And we find the sun occupying 9;5° in Virgo. Referring this to the astrolabe we found (for the)[3] 3rd hour of day that the 20th degree of Libra was rising and the 22d degree of Cancer in Midheaven. Similarly the 20th degree of Aries was setting and in Lower Midheaven was the 22d degree of Capricorn.

[1] This is indeed a perfect description of the teaching of early Islam. Also about Muḥammad's life Stephanus is well informed. He says (Usener p. 272,10) that the prophet was of the Quraysh tribe and that he appeared in the desert of Yathrib in the month Pharmouthi or April of the year 932 of the era Philip (= A.D. 609). Islamic tradition places Muḥammad's first revelation exactly one year later (Ramadan 610). Usener (p. 263) assumes (incorrectly, as it seems to us) that the marginal note in which the year 932 is called ἀπ' ἀρχῆς Φιλίππου is an error for the Seleucid era (S.E. 932 = A.D. 621). As name of Muḥammad's father our text gives Ismael, instead of 'Abd Allah; cf. Theophanes, *Chronographia* I p. 512,1 ed. Classen (Bonn 1839): οὗτος μιᾶς γενικωτάτης φυλῆς κατήγετο ἐξ Ἰσμαήλ, υἱοῦ Ἀβραάμ.
[2] Cf. Commentary and note 12 there.
[3] The text gives "we found the 3rd hour of day." But obviously solar longitude and hour are the data from which H and M were determined.

Oh the misery! alas the change of things! Libra is rising, (being) the exaltation and day-triangle[4] of Saturn; Saturn itself being in Midheaven, approximately in the very degree of Midheaven and in the terms of Venus,[5] and Venus itself in Agathos Daimon,[6] together with the ascending node, and in the sign of Leo. The horoscopic premise then indicates a nation of Saturnian type and a state akin to Venus, rising quickly and mastering many nations; for rising Libra[7] brings servitude to all men. Further, Saturn being actually in the South and Venus likewise in the South[8] indicate a nation from the South, outstanding in great strength and unrestrained speed and constant motion and of enduring possession, with epic battles and most brilliant deeds of valor and distinguished by a new type of state. And further, because Saturn was high in Midheaven this (nation) has an unrestrained spirit and irrepressible enterprise. And it opposes the nations (belonging to) Jupiter and Mars until it comes to the dominion of the sun; for only the nations (belonging to) the sun are exempt from the yoke of such a nation [etc., making, e.g., Venus responsible for observing the sixth day in Islam.]. It prevails over all Egypt and Libya — Egypt as belonging to Mercury and Libya to Mars — and also Palestine and Syria as belonging to the moon and Mercury. It drives the Roman nation from all Syria and the other countries, and subdues, slaughters, and imprisons Cilicia; and it will completely devastate Cappadocia, because of Mars. But it will not overwhelm the kingdom of the Romans itself, because of the sun and because of Mars being near Midheaven. And further, because of Mars being in its own depression, that is in Cancer, the nations belonging to Mars will be in depression in the face of this people; but because of Mars itself being in Midheaven, the kingdom of those people under it (Mars) will remain, that is the (kingdom) of the Romans and the Turks and the Khazars and the Bulgars and the like.[9] Further also because of Jupiter being in his own depression, namely in Capricorn, the kingdom of the eastern peoples and of the Persians themselves will be entirely suppressed by this nation [etc.]. And all Persia will be subjugated as far as the Indians; and further, because of Jupiter in his own depression and in Lower Midheaven, Armenia and Iberia and Colchis and parts of Scythia, because of Mars. [There follows a long discussion of the loci and of the "starters," ending in detailed "predictions" of the events up to the Abbāsid period (A.D. 775)[10] which is obviously the time at which the horoscope was actually written. The predictions which follow about the duration of Islamic power are dictated by wishful thinking and completely wrong.]

COMMENTARY

Diagram: pl. 29

The elements of this horoscope are badly preserved; they come from a diagram in two families of manuscripts, A to M and V, the latter giving rounded-off figures of the first.[11] At least two errors (for ♀ and the Lot of Fortune) are common to both.

Computation shows that the horoscope of Islam was taken as of 621 September 1, that is the beginning of the Byzantine year in which the beginning of the first year of the Hijra falls (July 16, 622). The equivalent of 621 September 1 in the Alexandrian calendar is Thoth 4 but the text seems to give either Thoth 5 or 3.[12] The following comparison shows that September 1 is the only possibility which agrees with the lunar position:

	Text	Computed	Text—comp.
☉	♍ 9;5	♍ 11	−2°
☾	♑ 12;16	♑ 10	+2
♄	♋ 23;30[13]	♋ 26	−2½
♃	♑ 20;39	♑ 23	−2¼

[10] Usener, p. 259. *Cf.* also Bouché-Leclercq, *AG* p. 371.

[11] For the diagram of the manuscripts A to M (Usener's notation p. 289) *cf.* Usener p. 321. In his figure Usener gives ☋ ♌ 19;50 and again ☋ ♒ 19;50. The text says correctly that the ascending node was in ♌. Unfortunately the modern use of the symbols ☊ and ☋ is exactly opposite to the meaning of these sigla in mediaeval Greek manuscripts. Bouché-Leclercq, *AG* p. 371 fig. 39 follows the figure of Usener p. 321 with two exceptions: he writes ☊ in ♒ 19;50 (which makes both nodes wrong, assuming modern notation) and he omits in the middle field the word ἡμέρας (*cf.* above note 4). The diagram of V is given in Usener p. 322 and again in Daremberg-Saglio, *Dict.*, art. Zodiacus [Cumont], vol. V, p. 1055, fig. 7594. This diagram contains the following data:

☉	♍ 9	♂	♋	H	♎ 20
☾	♑	♀	♌	M	♋ 22
♄	♋ 24	☿	♍	☊	♌
♃	♑			κλ.τ.	♑

The ascending node is given by V correctly in Leo while Usener and Daremberg-Saglio substitute the wrong symbol (☋) for it, at least if one follows modern usage.

[12] Usener takes as text "Thoth 5, day 3, hour 3" and quotes from V in the apparatus "month Thoth, which is September, day 3, hour 3." Cedrenus (Unger p. 257) quotes "September 3 day 5"; this shows that the text was already corrupt in the 11th century. Similarly Const. Porphyrog., Usener p. 257/258 note.

[13] "Terms of Venus": wrong in the "Egyptian System" but correct in the "Chaldean System" and in the system of Ptolemy (*Tetrabiblos* I,20,21 [Robbins]; 21,22 [Boll-Boer]).

---

[4] This must be the meaning and not "rising at daytime" as Usener's text has. The diagram of M .(Usener p. 321) gives correctly τρίγωνον ἡμέρας ♄. [Usener retained the siglum "6" for ἡμέρα]. V says only "Libra is rising, Saturn's exaltation and triangle."

[5] *Cf.* Commentary. V says "house of Venus" which is wrong.

[6] Correct: Leo is the 11th locus from H = ♎.

[7] This is a pun, the Greek for "Libra" meaning literally "yoke."

[8] Saturn is in the meridian and Venus shortly before it.

[9] This fits exactly the political situation under Leo III (717 to 741).

160   LITERARY SOURCES: No. L 621

	Text	Computed	Text—comp.
☌	♋ 2	♋ 3	—1
♀	♌ 25;6	♋ (!)26	[—1]
☿	♍ 9;5	♍ 27(!)	?
H	♎ 20 [14]	about 9 A.M.	
M	♋ 22	clima 5	
☊ [15]	♌ 19;50	♌ 23;40	—3;50
Lot of F.	♑ 22;9	♒ (!) [22;9]	

Venus and the Lot of Fortune were probably entered in the wrong fields of the diagram; ♍ 9;5 for ☿ is dittography from the solar position. The Lot of Fortune has the right longitude if one emends the longitude of the Horoscopos from ♎ 20 to ♎ 18;58. This emendation is confirmed by the fact that H = ♎ 20 gives M = ♋ 23;30 for clima 5 (or ♋ 24 for Byzantium) whereas H = ♎ 19 yields M = ♋ 22 for clima 5 [16] as given in the text (♋ 23 for Byzantium). The inaccuracy in the numbers of H and M might be caused by the fact, mentioned in the text,[17] that the position of these points was determined on the astrolabe from the given hour.

---

[14] "Exaltation of ♄": correct; "terms of ♀": wrong for Ptolemy as well as in the "Chaldean System" and in the Egyptian System as given in the *Tetrabiblos*, according to which the terms of Venus extend from ♎ 21 to 28, but correct according to Vettius Valens I,3 (p. 16,16 ff. Kroll) who gives the limits ♎ 19 and 26; "day-triangle of ♄": correct; "decan of ♄": extending from ♎ 10 to 20 (cf. Bouché-Leclercq, *AG* p. 228).

[15] Modern notation; cf. note 11.

[16] According to Theon: clima 5 ♎ 19: $\alpha = 204;4$. Sphaera recta ♋ 22: $\alpha_0 = 203;50$.

[17] Cf. above p. 158.

III. GENERAL COMMENTS

A. CHRONOLOGICAL DISTRIBUTION

At present we know of 16 Babylonian horoscopes, the earliest of which concerns the date of — 409 April 29 (time of Darius).[1] The latest cuneiform horoscope is cast for S.E. 243 = — 68 (cf. fig. 22[2]). Practically contemporary is the coronation horoscope of Antiochus I of Commagene (— 61), being part of the famous monument on the summit of the Nimrud Dagh. This is the horoscope, as far as known to us, is cast for the coronation of the Sasanian King Khosro Anōsharwān (531 August 18).[4] The latest Greek horoscope discussed here professes to be cast for A.D. 621 in order to predict the future of the Islamic empire but was in fact written about 150 years later when the development of Islamic astronomy under the Abbāsids begins. We have excluded from this study horoscopes cast during the later

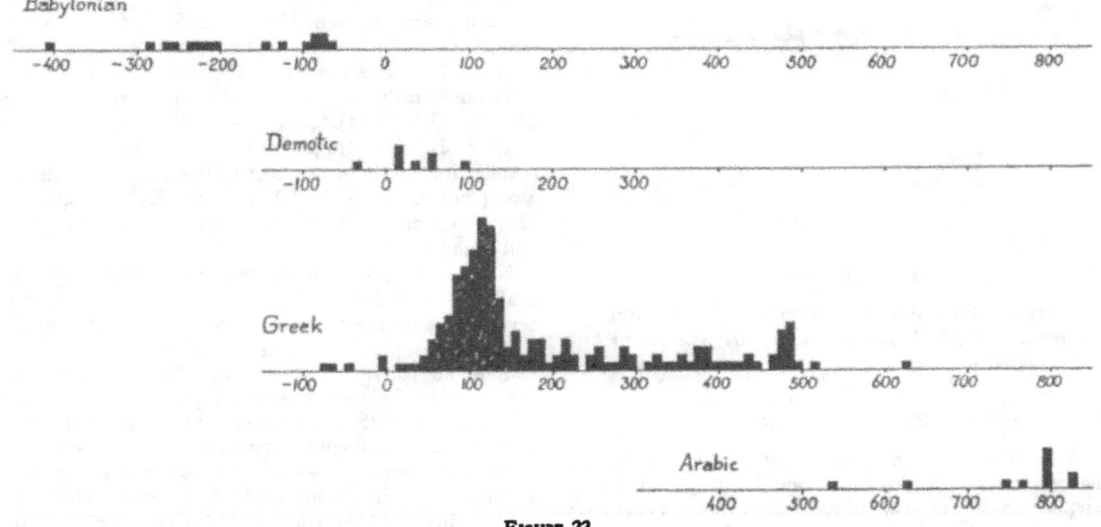

FIGURE 22.

earliest extant Greek horoscope. The earliest horoscopes preserved in Greek literary sources are dated to — 71 and — 42. The date of the first Greek horoscope written on papyrus is — 9. All Egyptian horoscopes known so far, nine in number,[3] fall into the interval from — 37 to + 93. The bulk of about 180 Greek horoscopes belong to the first five centuries A.D. The earliest Arabic

Byzantine period, since Islamic sources then become of primary importance.

The unevenness in the time distribution of Greek horoscopes is primarily due to the large collection of horoscopes by Vettius Valens (second century) and then, secondarily, to the examples contained in the writings of Palchus, Rhetorius, etc. (fifth/sixth century).[5] This is clearly shown by the distribution in figure 23 in which the literary sources are separated from the preserved originals (mostly papyri, plus a few graffiti and one ostracon).

It is clear that we are limited to accidental finds of texts and to a random preservation of astrological treatises. Nevertheless, the combined graphs of figure

---

[1] For the texts known in 1952 cf. Sachs [2]. Several more horoscopes were found by Dr. Sachs in the British Museum. The dates which he established were put at our disposal and are included in the graph of fig. 22. Cf. now Pinches-Sachs, LBAT p. XXXIV.

[2] Fig. 22 gives the number of horoscopes per decade, each little square represents one horoscope (duplications not counted).

[3] The earliest Demotic horoscope, probably cast for — 37 April 21, was only recently discovered by Professor J. Černý in a collection of ostraca in Oxford. For five more Demotic ostraca and one horoscope on a coffin lid cf. Neugebauer [1]. Two horoscopes, from the ceiling of a tomb, were published by Petrie, Athribis, London 1908.

[4] Taqizadeh [1], p. 128 ff.

[5] The only Latin horoscope, known to us, comes from Firmicus Maternus and was cast for 303 March 14. This date has been established by W. Koch in the Astrolog. Rundschau 23,6 (1931) and again (1935) by Neugebauer [2].

22 illustrate clearly the fact of the overwhelming role of the Hellenistic period in the development of astrology. It is equally clear that the original invention belongs to Babylonia; but it is hardly an accident that we have less than 20 cuneiform horoscopes from a period which has given us well over 1,800 astronomical tablets. With Greek horoscopes the situation is just the reverse. We have barely 20 astronomical documents as compared with about ten times as many horoscopes. It is also clear that Vettius Valens had no material at his disposal which was essentially older than represented by the statistics, however fragmentary, of our papyri. This

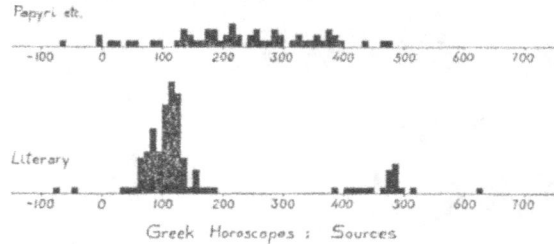

FIGURE 23.

supports our hope that the chronological distribution of the material available to us gives a fairly correct picture of the actual facts.

B. THE HOROSCOPES IN GENERAL

Were it not for an extensive astrological literature, the original horoscopes alone would hardly reveal their purpose to foresee the future of a person or coming events from the initial configuration of the planets. In No. —3 we find an explicit reference to the future: "There are dangers; take care for 40 days because of Mars" and a similar warning for 17(?) months ahead is expressed in No. 138/161. A very vague prediction concludes No. L 142. No. 345 makes a statement about the significance of the data in question for the well-being of the father. All other horoscopes list only factual data or data which are necessary for the astrological investigation but are in definite numerical relation with purely astronomical elements. In No. 95 the horoscopic section is continued in a kind of astrological treatise (eventually using the Coptic language) which contains many "ifs" and "whens" but obviously leaves any definite conclusion for the individual in suspense. Many of the literary horoscopes give an extensive analysis of the causal connection between the astronomical data and the recorded events but only *ex post facto*.

The absence of predictions in the literary horoscopes is not surprising since these horoscopes serve as illustrations or empirical confirmations for the doctrines of the general text. It is, however, of interest to note that men like Vettius Valens must have had large collections of case histories at their disposal in order to be able to select suitable examples in support of the doctrines in question. In several instances we find Vettius Valens using the same initial horoscope for the discussion of different phases in the life of the patient.[6] And it cannot be doubted that many of the isolated data quoted by Vettius Valens and in the later astrological literature are taken from actual horoscopes. As an illustration may serve a sequence of examples quoted by Rhetorius which all use H = ♉ 11, M = ♑ 23.[7]

Perhaps we can recognize even among the papyri the existence of collections of horoscopes. The largest group is formed by the horoscopes of PSI 22, 23, and 24 which concern years between 338 and 385 (376 and 381 in duplicate) and which might be interpreted as a family archive. The same could hold for P. Oxy. 596 (139 and 149) but is rather unlikely for the four horoscopes Nos. 217, 219 II,1, 219 II,12, and 244 of the magical papyrus Warren 21. The fragment P. Aberd. 126 is perhaps a collection of three horoscopes from the first century A.D. (Nos. 50,1; 50,2; 50.3). Nos. 258 and 260 (P. Oxy. 1563 and 1476) were written by the same hand and found together.

No. 137 exists in duplicate plus a variant, and the graffiti in Dura from 219 reproduce the same horoscope seven times on the same wall, for reasons unknown.

The geographical distribution of our sources is very simple: all papyri, one ostracon (No. 207), and one graffito (No. 353) come from Egypt.[8] The graffiti Nos. 176, 219 I, and 250,1 were found in Dura on the upper Euphrates. The monument which gave us the famous Greek horoscope, No. — 61, crowned the summit of the Nimrud Dagh in Commagene. The codices which contain the literary horoscopes are, of course, of Byzantine origin, now scattered among the great European libraries. Very little is known about the authors though it may be safely assumed that they lived in Alexandria or the Hellenistic cities of the eastern Mediterranean world.

THE NAMES OF THE PLANETS

With very few exceptions the horoscopes use the familiar names for the planets κρόνος, ζεύς etc. The older names

φαίνων:	♄	φωσφόρος:	♀
φαέθων:	♃	στίλβων:	☿
πυρόεις:	♂		

occur in only five of the papyri (Nos. 46, 81, 95, 137c and No. 250.2 — the latter presumably written in the third century) and in one graffito from Dura (No. 176). Among the literary horoscopes only the poem of

---

[6] Four times in L 180 and 120,II; three times in L 75; twice in L 74,XI 82 83 113,VII 122,XII. Hephaestion also uses the same data (L 380) three times.

[7] *CCAG* 8,1 p. 220 ff.

[8] Nothing is known about the origin of the parchment No. 326, now in the Louvre.

Manetho (No. 80) uses the old names for Saturn and Jupiter, and Kypris for Venus. The coronation horoscope of Antiochus of Commagene (No. —61) represents also the archaic terminology.[9] Associated with the older planetary names are usually epithets for sun and moon such as " ruler of the world " etc. (Nos. 46, 81, 137c).

SYMBOLS

Only three symbols occur in the papyri: the ligature of ω and ρ for ὥρα as well as for ὡροσκόπος, the crescent ☾ for the moon and ☉ for the sun.[10] No symbols are ever used in the papyri for the planets or for the zodiacal signs. A quite different situation prevails in the codices through which the literary horoscopes are preserved and which were written between five and ten centuries later than the latest extant horoscopes on papyrus. In these Byzantine codices we find not only the above-mentioned symbols in frequent use plus symbols for the planets and the zodiacal signs — though full writing of planetary and zodiacal names occurs at all periods — but also an ever increasing number of astrological symbols, e.g., for the aspects, lots, etc. The modern editions of these texts show a complete disregard for such details. It is, e.g., perfectly possible to find a passage like ♄ ἐν ♋ κζ̄ rendered by Κρόνος Καρκίνου κζ´. Furthermore, if one realizes that the existence of diagrams in the codices is usually not even mentioned in the modern editions, then it is clear that a study of astronomical and astrological symbols could be undertaken only by beginning anew with the manuscripts themselves, a task which would have added years to our work. The following is therefore taken exclusively from the papyri and graffiti, all of which, with the exception of a few papyri which have been lost since their first publication, we have checked in photographs.

The symbols for sun and moon occur only in P. Warren 21 (horoscopes Nos. 217 219 II,1 219 II,12 and 244) though even this text writes ϲελήνη once. A crude lunar crescent is drawn in the graffiti 219,I in Dura. All other texts spell out in full, or occasionally abbreviate, the names of sun, moon, and planets.

The ligature ᾧ for the Horoscopos is first attested in a Michigan papyrus from Karanis, concerning the year 182. Unfortunately, the beginning is destroyed so that we do not know how ὥρα was written. In Nos. 125 and 190 we find ᾧ used for ὥρα while ὡροσκόπος is written out. The next instance of the use of the ligature ᾧ for the Horoscopos is probably No. 207 (Wilcken, Ostracon 1602) but a check is impossible since the text is now lost. Beginning with P. Warren 21, concerning the years 217, 219, and 244, our symbol is frequently used, often in the same text, for ὥρα and for ὡροσκόπος. In No. 260 we have only the symbol for ὥρα but ὡροσκόπος written out, while Nos. 277 and 283 use the symbol for the hour but call the Horoscopos ὥρα. In No. 293 VIII ὥρα is written for hour as well as for Horoscopos while Nos. 326 and 338 give the spelling ὡρασκοπος for the latter. From No. 345 on, the symbol ᾧ is used commonly for both words.

In No. 478 the monogram χ + ρ might have occurred at the beginning of the horoscope if we may rely on the printed text. Unfortunately, the original is lost.

DIAGRAMS

Diagrams are very rare. Only one papyrus (P. Oxy. 235 — No. 15/22) gives a crude circular diagram. Otherwise, only the graffiti (Dura: 176, 219,I, 250,1 and Abydos: 353) have drawings which represent the circle of the zodiac and its division. In the Byzantine codices diagrams become more frequent (though consistently ignored in *CCAG*) and are of three types as indicated in figure 24. For a photograph of an example of the

FIGURE 24.

first type *cf.* figure 20. The second and third types are very common in Latin and Islamic astrological literature.

The invocation ἀγαθῇ τύχῃ at the beginning of a horoscope is fairly common in papyri from No. 227 on. Before then we find it only at the end of No. 81 and at the beginning of No. 137a. In the duplicate of the latter text (No. 137b) ἀγαθῇ τύχῃ εὐτύχει is written at the end. In No. 81 διευτύχει ends the horoscope proper (end of column VII); similarly in Nos. 138/161, 277 and 283.

C. ARRANGEMENT OF THE BASIC DATA

With the exception of a few elaborate horoscopes the text either consists of or begins with an enumeration of the positions of the celestial bodies. The arrangement of the planets in this enumeration follows definite patterns which we shall describe presently. Before doing so, however, we must draw attention to a habit which prevails in the majority of texts which give only zodiacal signs for the planetary positions, that is, no degrees and minutes. If, then, two planets fall in the same sign, they are mentioned together, regardless of any scheme of

---

[9] *Cf.* p. 162.
[10] The modern symbol ☉ for the sun seems first to appear in Renaissance printed books, e.g., Fracastoro. *Homocentrica* (Venice 1538) p. 50. Copernicus uses it rarely in his manuscript of *De Revolutionibus* (e.g., fol. 90 to 94, 127ᵃ, 143ᵇ); ordinarily he writes *sol*.

planetary arrangement. For example, we find in No. L 85,II the arrangement

$$(\odot \, \male \, \female \, \tauri)(\moon \, \jupiter) \, \saturn \, H$$

where all planets in one group (...) fall in the same sign. The underlying pattern to which this text belongs is nevertheless

$$\odot \, \moon \, \saturn \, \jupiter \, \male \, \female \, \tauri \, H.$$

In the following discussion we disregard all variations of the basic scheme caused by the accidental grouping in the same zodiacal sign.[11]

It is of interest to remark that the Babylonian horoscopes also modify their standard order of enumeration

$$\jupiter \, \female \, \tauri \, \saturn \, \male$$

whenever two planets belong to the same zodiacal sign.[12] The Babylonian order itself, however, is no longer reflected in Greek horoscopes, except, probably, the horoscope of Antiochus of Commagene.[13]

The few Demotic horoscopes known follow a different principle. After sun and moon the centers and related places are enumerated in a fixed order. The planets are mentioned whenever they happen to coincide with one of the places mentioned.[14]

1. ORIGINAL SOURCES

In enumerating the positions of the seven celestial bodies and of the Horoscopos the papyri follow two major schemes: one is determined by the planetary order

$$\saturn \, \jupiter \, \male \, \female \, \tauri$$

the other follows the order of the zodiacal signs.

The first class is represented in about two-thirds of our texts and can be divided into two subgroups according to the position of sun and moon at the beginning or at the end.[15] The arrangement

$$\odot \, \moon \, \saturn \to \tauri \, H$$

is the rule for the earlier horoscopes (before 150); the order

$$\saturn \to \tauri \, \odot \, \moon \, H$$

is common between 171 and 277; from 320 on a modification appears in which the Horoscopos is placed at the beginning, thus leading to

$$H \, \saturn \to \tauri \, \odot \, \moon.$$

The habit of beginning with the Horoscopos seems in general to be a characteristic later development.

The earliest text which shows the zodiacal order is No. 150 which begins with the sun but puts H at the end, out of the zodiacal order. Nos. 217 and 258 begin with Saturn. All the remaining examples of the zodiacal arrangement belong to the interval from 245 to 394 and start with the Horoscopos (13 cases).[16]

In the papyri the enumeration of the positions of sun, moon, and planets is only rarely amplified, except by the addition of the Horoscopos. The four centers are mentioned in some of the earlier texts: Nos. — 3, 15/22, 46, 95 and 137a,b. Horoscopos, Midheaven and Lot of Fortune are given in No. 81. Nos. 137a,b mention, besides the centers, the "first" and "second" Lots. Four Lots (Fortune, Daimon, Eros, Necessity) occur in Nos. 138/161 and 338, in the latter followed by the date of the ἀποκρουσις.[17] Horoscopos plus Lot of Fortune are given in No. 182 and in six out of seven horoscopes of the group Nos. 351 to 385.[18] This represents a total of 15 cases among about 60 texts.

2. LITERARY SOURCES

The L-horoscopes show a much greater uniformity than the papyri. Except for a few special cases where the doctrine under discussion requires a particular selection of data we find the arrangement

$$\odot \, \moon \, \saturn \, \jupiter \, \male \, \female \, \tauri \, H$$

consistently followed (about 120 cases). Among those only a few minor deviations can be observed: the moon is out of place, generally near the end of the list, 9 times; Jupiter is out of order 4 times, Saturn and Mars twice. Quite irregular is the arrangement in Nos. L 516 and 621 but none of these variants reflects a consistent system.

The zodiacal order, which is quite common among the later papyri, is represented only once among the literary horoscopes (No. L 132, with the moon as "starter"). The two earliest horoscopes (Nos. L —71 and L —42) begin with H, but probably only on the basis of a diagram.

All four centers are mentioned only once, characteristically in one of the earliest horoscopes, No. L 40 (from Antigonus). Midheaven is noted in Nos. L —71 and L —42, then four times by Vettius Valens, but becomes the rule in all late horoscopes (from No. L 380 on [19]). On the other hand, the Lots (particularly the

---

[11] Occasionally a text follows strictly the planetary order, disregarding identity of signs (No. L 50 is an example).
[12] Cf. Sachs [2] p. 64. Sun and moon, or moon and sun, precede the planets.
[13] Cf. the commentary p. 14.
[14] Cf. Neugebauer [1].
[15] Quite exceptional is No. —9 which begins with the moon: $(\moon \, \saturn)(\odot \, \tauri) \, \male . . . .$

[16] Fotheringham's statement (*The Oxyrhynchus Papyri* XII p. 232) "that the horoscopes always name them in order, either beginning or ending with the Horoscopos" needs modification: the zodiacal order is observed only in about one-fourth of all cases and the Horoscopos is only once put at the end (but out of the zodiacal order) and twice the lead is given to Saturn.
[17] Cf. the commentary to No. 338 (p. 66).
[18] Omitted in No. 366.
[19] The only exceptions are the two special examples which are additions to Vettius Valens (Nos. L 419 and 431).

Lot of Fortune) are much more frequently quoted in the literary horoscopes than in the papyri. In Vettius Valens almost a third of all examples quote the Lot of Fortune and this ratio increases to one-half of all cases with the late authors.

In about one-third of all cases in which the Lot of Fortune is mentioned the Daimon is also given. The "Exaltation of the Nativity" is listed in Nos. L 95,V,18 and L 440; 463; 484. The ascending node of the moon is never mentioned in a papyrus, ostracon, or graffito. It occurs in No. L —71, but in Vettius Valens it is given 3 times (Nos. L 74,IV; 75,1; 115,II) but 13 times between No. L 412 and L 497. In 10 out of 16 cases in which the ascending lunar node is mentioned the preceding conjunction or opposition is also given. A syzygy alone, without giving also the position of the node, occurs only 3 times (Nos. L 419; 431; 440).

The preceding statistics concern exclusively the initial enumeration of data. In several of the papyri and in the majority of the L-horoscopes additional data may be mentioned in the astrological discussion. An enumeration of these data would be of little interest since their selection depends on the contents of the chapter to which the horoscopes are appended as examples.

D. CALENDARIC DATA

As far as can be seen at the present state of preservation only two horoscopes on papyrus were not dated: No. 217, one out of a collection of four horoscopes (P. Warren 21) and No. 46, assuming that not a whole column is missing, which may or may not be the case. No date is given on the graffito No. 353 (Abydos) and the ostracon No. 207 mentions only month and day but no year.

In sharp contrast herewith not a single date is given in the about one hundred horoscopes in Vettius Valens.[20] The majority of the late literary horoscopes, however, are dated[21] (in the era Diocletian) with the exception of four horoscopes from Rhetorius (Nos. L 401, 440, 482, 488, the first and last of which are suspect of being artificially made up), of No. 516 and of the fictitious horoscope of Islam (No. L 621).

1. YEARS AND ERAS

The following is the complete list of references to specific years,[22] found in our material.

---

[20] We disregard here the group of about 40 partial horoscopes listed in table 15.

[21] We disregard here partial versions; for example No. L 380 first version gives the date Diocletian 97 Athyr 30. The second version gives Athyr 30 only, the third version no date at all.

[22] A parenthesis ( ) in connection with the word "year" means use of the abbreviation for ἔτους. In all other cases it indicates an addition made by us.

No.	
—9	year 20 of Caesar (Augustus)
—3	year 27 of Caesar (Augustus)
15/22	[the ...] year of Tiberius
32	year 18 of Tiberius
81	third year of the Divine Titus
138/161	[year ...] of Antoninus Caesar the Lord
137a,b	(year) 1 of Antoninus Caesar the Lord
137c	1(st year) of Antoninus Caesar the Lord
139	2 of Antoninus Caesar the Lord
145	eighth of Antoninus
149	13 of Antoninus Caesar
150	13(th year) of Antoninus Caesar the Lord
161	1(st year) of Antoninus [and] Verus, the Lords Augusti
176	487 (Seleucid era)
177	17(th year of Marcus Aurelius)
184	(year) 24 (of Commodus)
187	year twenty-seven (of Commodus), Mesore 6, according to the Greeks, according to the Egyptians (year) 28, Thoth 23
190	[30th (year)] of Commodus
208	(year) 15 (of Septimius Severus and Caracalla)
219,I,b	year 530 (Seleucid era)
219,II,1	(year) 2 (of) Antoninus the Lord(?)
219,II,12	(year) 2 of the same
220	year [..]
227	year 7 of Severus Alexander
244	(year) 1 of Philip
258	6(th year) of our Lords, Valerian [and Gallienus] and Cornelius Saloninus, Augusti
260	1(st year) of Macrinianus and Quietus, Augusti
277	2(nd year) of Probus
282	1(st year) of Carinus
283	1(st year) of Carinus
284	in the blessed reign of Diocletian; [(year)] 1 of Diocletian
293,VIII	9(th year of Diocletian) and 8(th year of Maximianus) and 1(st year of Constantius and Galerius) completed
316	33(rd year) of Diocletian
320	year [...] 38 of Diocletian
326	42 (years) from Diocletian
338	55(th year) of Diocletian
345	(year) 61 (of Diocletian)
351	67 of Diocletian
366	[82nd] (year) of Diocletian, 9(th) indiction, presidency of Isidorus, son of Pamenes
370	86 of Diocletian, 13(th) indiction
373	89 of Diocletian
376,a	6(th) indiction ... 93 of Diocletian, Hermogenes, son of Theotimos, president
376,b	6(th) indiction, in the time of the Night-

No.	
	watch commander Macarius, son of Hermias Gesdos, 93 of Diocletian
L 380	in the 97(th) year from the reign of Diocletian
381,a,b	97 of Diocletian
385	101 from Diocletian, in the presidency of Serenus, (son of ?) Artemidoros Phanousire
394	(year) 111 (of Diocletian)
L 419	Reign of Valentinian .... year of Diocletian 135
L 428	(year) of Diocletian 145
L 431	of Diocletian year 147
431	148 of Diocletian
L 463	year of Diocletian 179
465	from Diocletian 181
L 474	year of Diocletian 191
L 475	year of Diocletian 191
478	year of the blessed reign of Diocletian 194
L 478	year of Diocletian 195
L 479	year of Diocletian 195
L 483	in the year of Diocletian 199
L 484	year of Diocletian [200]
L 486	year of Diocletian 202
L 487	year of Diocletian 204
L 497	year 214 from the beginning of the reign of Diocletian which is the year 821 from the death of Alexander the Macedonian.

This list shows that from 284 on the era Diocletian is used consistently. The Seleucid era appears only in the graffiti from Dura (Nos. 176 and 219,I). The date of No. 187 is a clear example of the association of regnal years and calendar years in Egypt. The era Augustus is not attested in the horoscopes though it is used by Vettius Valens in his rules for the determination of the weekdays and of planetary positions [23] and similarly in the excerpts from Balbillus for the nodes.[24] In the Greek planetary tables P. Tebt. 274, years of Titus are quoted beside the regnal years of Trajan.[25] The era "from the death of Alexander" appears only in No. L 497 and in the form of "from the beginning of Philip" as a marginal note of one manuscript in the introduction to No. L 621.[26] The fact that this era was used in the Handy Tables of Ptolemy and Theon makes it surprising that it is so rare in astrological context.

2. MONTHS

The Macedonian names of the months occur in our material only in the graffiti from Dura.[27] For the two instances in which we are able to establish the corresponding Julian dates (Nos. 176 and 219,I) we find that the Hyperberetaios corresponds roughly to the month of October. As is common in Greek calendars, the months are either civil or accurately lunar. Consequently, some dates are given in both versions, plain (i.e., civil) and "according to the moon."

The Roman calendar occurs in No. 81 in a triple dating in Alexandrian, Egyptian, and Roman fashion ("calends of April"). Roman counting is suggested by our restoration of No. 171. Roman month names are used in Nos. L 412,[28] L 463 and L 474 but the days are counted simply as ordinals (April 17, April 25, and October 1 respectively). This seems to be the earliest known instance of this simplified method of counting.[29]

The honorific month name "Hadrian" is used in Nos. 137a, b, and c, in the latter version described as "according to the Greeks." All three versions give the Egyptian equivalent (Tybi).

The Egyptian names of the months are used [30] in all other cases (more than 50), in 37 of which the Alexandrian calendar is meant. No. — 3 speaks about Phaophi 5 κατὰ καίσαρος whereas Nos. 161, 284, and 287 say "according to the Greeks." Normally, however, no nearer description is given and only the astronomical data make it possible to distinguish between the Alexandrian and the unreformed Egyptian calendar. We have found no instance later than 207 among the horoscopes [31] for a date in the old Egyptian calendar.

This old Egyptian calendar was used, perhaps, without further explanation, in two papyri, horoscopes Nos. 139 + 149, and 177 but in all three cases the dating causes great difficulties and our assumption of the Egyptian calendar might be wrong.[32]

The "old calendar" is explicitly quoted in four cases of double dates: Nos. 15/22, 81, 137a and b, 161. In two other horoscopes (Nos. 137c and 187) dates "according to the Greeks" and "according to the Egyptians" are given beside each other. No. — 9 and No. 145 have only a date "according to the ancients" and No. 207 (ostracon) "according to the Egyptians," being the latest instance for this style of dating among all horoscopes.[33]

---

[23] Vettius Valens I, 10 and 20 (Kroll p. 26, 12 f. and p. 33, 31 f. etc.).
[24] CCAG 8,4 p. 239,16. The era Augustus is also used by Ptolemy in his Canobic Inscription (opera II p. 152,1 Heiberg).
[25] Cf. Neugebauer [4] p. 240 ff.
[26] Cf. p. 158 note 1.

[27] Nos. 176; 219,I; 250,1.
[28] Written 485/486.
[29] Cf. Ideler, Chron. II p. 191; Ginzel, Chron. III p. 117. It is worth noticing that the "Ephemeris for 467 A.D." uses the Roman style of denoting the days of the Roman months; cf. Curtis-Robbins [1].
[30] Their list is given on p. 3.
[31] In the strictly astronomical literature the old Egyptian calendar remains in use in Greek and in Islamic works and then again in Western Europe during the Renaissance (Copernicus).
[32] This holds particularly for No. 139; cf. the commentary to this horoscope.
[33] Of the Demotic planetary tables only the earliest one, covering the years from Augustus 14 (= —16) to 41 (= A.D.

3. DAYS; NIGHT AND EPOCH

The days are usually denoted only by their ordinal number. Four times the names of the weekdays are added. The earliest instance is No. 219,I,b, a graffito from Dura. The three other ones come from Palchus: Nos. L 479, L 486, L 487.[34]

For the day numbers the following rules can be observed:

(a) If the horoscopic moment is a moment of daylight, then only a single number is given as date.

(b) If we are dealing with an hour of the night, then the date is given in the form "n to[35] $n+1$."

The rule (a) is followed without exception (35 cases). The rule (b) is not strictly obeyed but among 26 dates associated with an hour of night there are 6 single dates[36] against 19 cases[37] which follow rule (b) and one ambiguous case (No. 125) where the restoration of a double day number is possible but not certain.

The difference in notation for daytime and night poses a problem which is related to the intricate problem of "epoch" in the counting of days in the Egyptian and in the Alexandrian calendar. Before entering upon the discussion of this wider question we shall make some preliminary remarks.

First: one must carefully distinguish between "double day numbers" as occurring in the above stated form (b) for the reference to an hour of night and "double dates" which relate two different calendars. Secondly: it is convenient to illustrate our discussion by means of schematic figures in which the time axis is divided in sections, alternately representing daylight (lightly drawn) and night (heavily drawn). Finally: one must realize that modern chronological tables (e.g. Schram, *Taf.*) count their Julian days in modern fashion from midnight to midnight. An equivalence obtained from such a table between an ancient and a Julian date may actually be correct only for a part of the day. For example, the relation between an Egyptian day with morning epoch and the corresponding Julian date is represented in figure 25.

We now can turn to the evidence concerning double day numbers. From the fourth century B.C. to the late fifth and early sixth century A.D. we find the rules (a) and (b) followed, namely single day numbers for daytime (without exception) and preferably double day numbers for the night.

The earliest occurrence of the double day number, known to us, is the "dream of Nectanebo," presumably referring to — 342 July 5/6, preserved in a Serapeum papyrus written about two centuries later.[38] Several other reports on dreams from the same group of texts do quote single day numbers only, but we do not know whether daytime or night is implied.[39] A double day number is given, however, for a dream in — 158 February 10[40] and in a contemporary complaint about a robbery.[41] About a century earlier two Petrie papyri furnish us with double day numbers again from reports

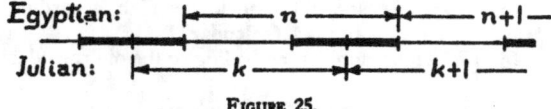

FIGURE 25.

on night robbery.[42] These examples suffice to prove that the use of double day numbers for the time of night is independent of any astronomical practice and also precedes by centuries the Augustan reform of the Egyptian calendar.

The next large group of double day numbers for the time of night comes from astronomical observations quoted by Ptolemy in the Almagest. Smyly was fully justified when he said[43] that Ptolemy "merely adopted a system which had been in popular use for centuries before his time in Egypt." The Almagest contains more than 60 instances of double day numbers, referring to observations ranging from — 720 to + 141.[44] These double day numbers are never associated with any month name except the Egyptian ones (Thoth, Phaophi, etc.) and in almost all cases it is explicitly stated that this is a month "according to the Egyptians."[45]

---

11), uses the Egyptian calendar. The Stobart tables, preserved for Vespasian 3 (= A.D. 70) to Hadrian 17 (= 132) are based on the Alexandrian calendar. *Cf.* below p. 173. For Greek papyri in general, A.D. 237 seems to be the latest attested date for the use of the "old calendar" (P. Grenf. II, 67) though the question is unsettled as to which calendar is meant in documents which give no double dates or give dates without specifications.

[34] In No. L 486 the day in question is a Monday; the three other cases are Saturdays.

[35] εἰς. The same expression in P. Ryl. 27, 59.

[36] Nos. 284, 320, 431 certain. In Nos. 145, 207, 394 the critical passage is partly destroyed but a restoration to a double day-number is made very unlikely by the spacing or shape of the text.

[37] Here are included Nos. 138/161 and 177 where the restoration of a double day number is practically certain.

[38] Wilcken, *UPZ* I No. 81 (p. 370) col. II,1: φαρμουθι κα εις την κβ̄.

[39] *Cf.*, e.g., Wilcken, *UPZ* I No. 77.

[40] Wilcken, *UPZ* I No. 78, p. 359 line 2: τυβ[ι] ῑβ εις [τ]ην ῑγ.

[41] P. Tebt. 958 (162 B.C.?): τηι κθ τηι νυκτι [τηι φερουσηι] εις την λ [του παχων]. Obviously one has to restore P. Tebt. 804 (112 B.C.?) according to the same scheme: [τηι δ τηι νυκτι τηι φερο[υση]ι εις την ε του φαμενωθ. *Cf.* Hipparchus on the autumn equinox of — 146 September 26/27: τῇ γ′ τῶν ἐπαγομένων τοῦ μεσονυκτίου τοῦ εἰς τὴν δ′ φέροντος (Almagest, Heib. 204,5). Similarly P. Reinach 17 for — 108 December 1: τηι νυκτι τηι φερ[ου]ση̄ς εις τ͞ε του αθυρ του θ (ετους).

[42] P. Petrie III p. 59: νυκτος τηι κ̄ε εις τη[ν κ̄η] του παχων (about — 260) and p. 72: τηι νυκτι [ᾱ ε]ις την β̄ του παχων (about — 240).

[43] Smyly [2] p. 87.

[44] The double day numbers are always given in the form of, e.g., Φαμενωθ ιθ′ εἰς τὴν ιθ′ (Heib. II, 263,24).

[45] Examples of omission: Heib. II, 311, 12 f. ("21 of Mesore to the 22th") or Heib. I, 195, 12 f. ("Mesore 30 at sunset").

168 GENERAL COMMENTS

Many cases concern double dates, i.e., comparison with other calendar systems. Among these some of the earliest dates [46] are of interest since single and double day numbers occur simultaneously, e.g. " in the 36th year of the Calippic period, Poseideon 25, (i.e.) Phaophi 16, beginning of the 10th hour (of night) .... in the year 454 of Nabonassar Phaophi 16 to 17, 3 seasonal hours after midnight." This, and several similar instances, show clearly that " Phaophi 16 " includes the whole night which is otherwise denoted as " Phaophi 16 to 17." This agrees with the well-known fact of a morning epoch of the Egyptian day.

The earlier Greek horoscopes give the impression that the double day numbers belong to the Egyptian calendar in contrast to the reformed Alexandrian calendar. We have the following passages:

No. —9: Mesore the 24th to the 25th, according [to the old (calendar)]
No. 15/22: Phaophi 1, according to the old calendar Phaophi 11 to [12]
No. 81: the sixth day (επιφωςκουση εκτηι), the third hour of the night; as the Romans reckon, the calends of April; according to the old (calendar), Pachon the first (νεομηνια) to the second.

In the later horoscopes, however, double day numbers are given for the hours of night though only the Alexandrian calendar is used. The same holds for the dates in Vettius Valens with one exception [47] which is, however, explicitly stated as "according to the Alexandrians," though all other dates are also based on the Alexandrian calendar.

The implication that double day numbers reflect an Egyptian terminology is not supported by Egyptian sources. The evidence for this statement has been assembled by Sethe in his *Zeitrechnung* [48] and need not be repeated here. We can add only the dates of two Demotic horoscopes of the year Tiberius 4 (A.D. 17/18) [49] which run as follows

year 4, (month) 2 of Akhet, day 9, the 11th hour of the night [50] (= A.D. 17, September 26 about 5 A.M.)

and

year 4, (month) 4 of Peret, (day) 1, the 4th hour of the night [50] (= A.D. 18, February 25 about 10 P.M.).

Though both horoscopes (ostraca) use the old Egyptian form of dating, with three seasons of four months each, the astronomical investigation shows that the first one is using the old Egyptian calendar, based on the Egyptian year of 365 days, the second one, however, the reformed calendar which is based on the Julian year. And neither one uses double day numbers for the night.

In Greek sources the double day numbers are not restricted to the Ptolemaic and Roman period. We find them still in use in our horoscopes Nos. L 380, 428, 431 and 478 [51] as well as in observational records from the years 498, 503, and 509 made in Alexandria by Ammonius and Heliodorus.[52]

After having established the continuous use of double day numbers for the time of night during a period of about eight centuries we can now turn to the thorny question of epoch, at least as far as reflected in our material.

It is the generally accepted rule to consider the epoch for the days in the Egyptian calendar and in the Alexandrian and Roman calendar as identical, at least for the Roman imperial period which concerns us here. This means one assumes morning epoch for all those calendars [53] and a relation to our norm of the Julian day which is illustrated by figure 25 p. 167. All astronomical data in the Almagest agree with this norm. Some of the horoscopes, however, show a less simple picture.

We first mention Nos. 366 and 373, both written by the same hand. In No. 366 a moment somewhere between sunset and about 4 seasonal hours after sunset is called

τυβι οψε εις ιβ̄ " Tybi, late to 12(th)"

whereas No. 373 says

τυβι η ωρα οψε εις θ " Tybi 8, late hour to 9(th)"

for a moment which could fall anywhere between 2 hours before or after sunset. Since No. 366 refers, in all probability, to an hour of night (our computation would require at least one hour after sunset), it seems plausible to assume also for No. 373 a moment after sunset (which would agree with our computation).

The simplest explanation of this terminology would be, of course, a midnight epoch. This would also motivate the use of double day numbers " from n to $n + 1$ " for the time of night. This conclusion was indeed drawn by Smyly [54] though he was forced to declare the dating

---

[46] All from observations by Timocharis in Alexandria.
[47] Nos. 23-26 in the tables 15 and 16 p. 180 f.; Kroll p. 26,21.
[48] Particularly p. 130 ff. The only case of a date which resembles the Greek double day numbers was communicated to us by Dr. H. Goedicke from a Bucheum inscription concerning the year —144: "month IV šmw (XII) 27, the 11th hour of night, as the morning of the 28th day dawned" (Mond-Myers, *Buch.*, II p. 7 and III pl. 41 A).
[49] Neugebauer [1] p. 116 ff.
[50] Literally "evening" (rhwy).

[51] From Hephaistion, Rhetorius, add. to Vett. Val., Palchus respectively.
[52] First published by Bouillaud in 1645, then discussed by Delambre, *HAA* I p. 518 (1817). Place and authors were determined by Tannery (1884), *Mém. Sci.* II p. 125 (the date 502 is an error for 503).
[53] Ginzel, *Chron.* III, p. 6; I p. 162 ff. Sethe, *Zeitr.* p. 137.
[54] Smyly [2] p. 86. He quotes Pliny *NH* II,188 in support of his thesis, but the same passage ascribes, e.g., a morning

of No. 81 as an error,[55] because this text is incompatible with a midnight epoch but requires an evening epoch both for the Alexandrian and for the Roman calendar by its equating (for A.D. 81 March 31 about 9 P.M.)

Egypt. IX 1/2 with Alex. VIII 6 — Roman April 1

(cf. figure 11 p. 25). If one accepts this relation as correct — and No. 81 is one of our best written and most elaborate documents — then one may perhaps see in this double date the outcome of a combination of the original Greek evening epoch with the traditional Egyptian morning epoch. This would again motivate single day numbers for daytime, but double day numbers for the night (cf. figure 26). Otherwise one must follow

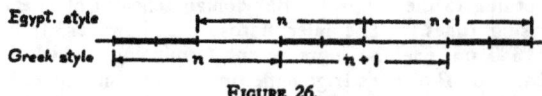

FIGURE 26.

Smyly and declare the double dates of No. 81 as an error for

Egypt. IX 1/2 corresponds to Alex. VIII 5 — March 31

as would be the case for common morning epoch. Under no circumstances can No. 81 be an evidence for a midnight epoch.[56]

The material presented here seems to lead to the following conclusion: The standard relations between Egyptian, Alexandrian, and Julian calendar can be applied safely wherever we are dealing with a single date which concerns daytime, or with a double day number referring to the night from n to $n+1$. If, however, a single date includes the night, one must be cautious. In case of the Egyptian calendar it seems to be safe to assume morning epoch throughout. For Alexandrian and Roman dates, however, an evening epoch may have existed much later than usually assumed.

4. HOURS

It is natural to assume that numerals in connection with hours of day or night are to be taken as ordinals, denoting the current hour. This is confirmed by the explicit occurrence of ordinals in Nos. 15/22, 81, 145, and 190. The same follows from cases like No. 244 where Horoscopos and sun are both located in ♑ and the time is given as ὥρα β ἡμέρας. Since even an extremal distance of H and ☉ cannot correspond to more than 2 seasonal hours after sunrise, it is clear that β can only denote the 2nd hour and not 2 completed hours.

It is often impossible to distinguish between equinoctial and seasonal hours, particularly if only the zodiacal signs of Horoscopos and sun are given. But there exist limiting cases where a decision is possible [57] and it then falls in favor of seasonal hours. For example, No. 220 associates the 9th hour of day with a position of the Horoscopos at the "end of ♊" and a solar longitude in ♒. From this it follows that the time elapsed after sunrise can only be at the most 7;56 equinoctial hours but between 8;33 and 9;11 seasonal hours. Similarly in No. 260: in this case the exact positions of H and sun are known and the numbers of the text agree only with seasonal hours.

In No. 15/22 the hour is said to be the fourth hour of night. Computation shows that the time was at most 3^h after sunset (both in equinoctial and seasonal hours since the date would correspond to the time of equinox). The text belongs, however, to so early a period that we must assume that the computation was not carried out with trigonometric tables but with the arithmetical schemes of System A or B for the rising times. Using system A one would obtain as upper boundary for the time after sunset $3;6,24^h$, and $3;10^h$ with system B. This may just suffice to explain the "fourth hour."

In all other cases from original sources where a check is possible the hour given by the text lies within the computed limits. Twice [58] the hour is called "approximately" (περί), in several cases [59] the "beginning" of an hour occurs.

It is probably mere carelessness if No. 385 only says "hour of the day" without numeral. The numeral alone is used in Nos. L 380 L 419 L 486 always indicating an hour of daylight. The word "hour" is omitted in No. 338.

The horoscopes from the earlier group of literary sources, i.e., before A.D. 190, never mention either date or hour. Nothing is known for two centuries until the later authors (from 380 onward). Two-thirds of these late horoscopes give date and hour, 10 of which can be checked numerically from the given longitudes of H and ☉. Five times the statement of the text agrees both with seasonal and with equinoctial hours.[60] In 4 cases [61] only seasonal hours agree with the text.[62] The only possible evidence for the use of equinoctial hours could be No. L 380 but the data of the text show internal discrepancies which destroy the basis for such a conclusion.[63]

---

epoch to the Babylonians, in flagrant contradiction to the facts known from the cuneiform texts (evening epoch).

[55] Smyly [2] p. 88.

[56] Wilcken, apparently not realizing that the astronomical conditions of No. 81 determine the hour as one before midnight, assumed (Ostr. I p. 793) as part of the Augustan reform the introduction of the Roman midnight epoch into the Alexandrian calendar.

[57] Nos. 190, 220, 260.

[58] Nos. —3 and 219,I,b.

[59] Nos. 137a,b 190 227 282 L 419 L 428.

[60] In two additional cases (Nos. L 486 and L 487) the text gives the wrong hour (2nd instead of 1st hour) regardless of what type of hour had been used.

[61] Nos. L 428, L 463, L 479, L 483.

[62] This agrees with Ginzel's general experience (cf. Chronol. III, p. 297).

[63] Cf. the commentary to No. L 380 (p. 132).

To summarize the investigation of our whole material we may say that with the exception of a very small number of cases all references to hours can be interpreted as meaning current seasonal hours.

E. ASTRONOMICAL DATA

1. If one looks for explicit astronomical data in the horoscopes, beyond the position of the celestial bodies and the "centers," then one finds only little in the papyri, with the exception of No. 81. In the literary sources the situation is different. The direct enumeration of the data is as limited as in the papyri but the fact that the literary horoscopes serve as examples in astrological treatises has preserved for us many deductions from the initial data in the course of which additional astronomical elements are mentioned. This gives us at least some glimpse of the astronomical techniques and concepts which were at the disposal of men like Vettius Valens, Palchus, and Rhetorius. Though our material can only represent a minute fraction of the astrological literature of antiquity, one obtains the impression that the later authors required a greater numerical accuracy — at least formally — than Vettius Valens in the second century. The papyri, however, seem to show the opposite trend, the more elaborate horoscopes being the earlier ones. The two texts which are of greatest interest astronomically are four centuries apart: No. 81 (P. London 130) and No. L 497 (from Eutocius), both giving detailed planetary positions with respect to fixed stars, a procedure which is reminiscent of the procedure followed by certain classes of cuneiform texts [64] of the last centuries B.C.

In the following we give a summary of astronomical data found in the horoscopes. For details the commentaries to the single texts must be consulted.

2. The rising times of the zodiacal signs are only once mentioned in a papyrus, namely, in the statement of No. 95 [65] that "the governing Horoscopos was fixed in complete conformity (with the rules of) the rising times," which probably means that from the given hour the Horoscopos was computed with the use of rising times.

In the literary sources the rising times are mentioned frequently in astrological context (e.g., for the periods of life) — about 30 times in Vettius Valens. Usually they are taken from the arithmetical schemes of System A or System B which are derived from the corresponding Babylonian schemes. In the late astrological treatises, however, the much more accurate tables of Ptolemy or Theon were used and detailed computations are carried out which also involve the tables for sphaera recta and the tables of seasonal hours for given latitudes (e.g., Nos. L 401, L 428, L 488, etc.).

3. All cases which operate with rising times contain some geographical data, usually in the form of quoting the number of the corresponding "clima." Occasionally, however, a purely geographical position is given such as "clima of Spain" (No. L 419) or of Alexandria (No. L 74,IV or No. L 497). No. 81 says "computed in Hermopolis where the horizon has the ratio seven to five (of longest to shortest daylight)" which is the standard norm for Lower Egypt.

The concept of "clima" is never mentioned in any of the horoscopes on papyrus but is quite common in the literary horoscopes. In the earlier cases, all of which come from Vettius Valens and thus belong to the first and second centuries A.D., the climata are defined according to the originally Babylonian schemes of linear rising times.[66] The later horoscopes, however, from L 380 on, use the more recent terminology which is based on Ptolemy's trigonometric table of rising times.[67]

The details of the distributions are as follows. Of the literary horoscopes of the first two centuries A.D. only five are not from Vettius Valens. None of them mentions the clima, but No. 113,IV (from Antigonus through Hephaestion) uses the arithmetical rising times of System A or B for Alexandria. Hephaestion himself refers in No. L 380, first version, to clima 3 of Lower Egypt and uses tables which agree with Theon's Handy Tables.

Rhetorius gives two examples from clima 4 [68] and two from clima 5;[69] twice (L 482 and L 463) no clima is mentioned, while in No. L 440 rising times for Alexandria of the old type, System A, are used implicitly.[70]

The additions to Vettius Valens Nos. L 419 and L 431 mention the "clima of Spain" and "clima 4," respectively. The latter corresponds to $\phi = 36$ which is also the latitude of southern Spain.

No reference to the clima is made by Palchus (8 horoscopes from L 474 to L 487). Eutocius (L 497) expressly gives "Alexandria in Egypt" as locality. The last of our horoscopes, L 621, seems to have been computed for clima 5.

4. Fixed stars play in general a minor role in astrological works. A doctrine involving the heliacal rising of Sirius (on Epiphi 25 — July 19) is used in the first version of No. L 120,II. Perhaps merely for poetical reasons, Centaur is mentioned in Manetho's horoscope (No. L 80). In Hadrian's horoscope (No. L 76), the moon "was about to come in conjunction with a certain

---

[64] Cf. Sachs [1].
[65] III, 59 f.

[66] Cf. the tabulation in table 3 (p. 5) left side.
[67] Cf. p. 5 table 3 right side.
[68] Nos. L 401 and L 488, though the latter is actually computed for clima 6.
[69] Nos. L 428 and L 516.
[70] In the introduction to CCAG 8,1 p. 220 ff. Cumont remarks on the higher frequency of clima 5 or Byzantium. In fact, however, the number of examples is greatly reduced if one observes that the same data are repeatedly used for different purposes. Then one is left with 2 cases for clima 3 (Alexandria), 2 for clima 4, and 3 for clima 5. Obviously no conclusion can be based on such limited evidence.

bright fixed star" in ♒ 20°, which we could not identify. In No. L 479 the rising of Asclepius (Ophiuchus) with the moon is noted.[71]

In No. 81 the positions of the planets are not only defined by their ecliptic coordinates but also in relation to constellations or individual stars — the distances being expressed in "fingers" and in "lunar diameters." The identification of the stars mentioned causes various difficulties which are discussed in our commentary to No. 81.

A much simpler situation is met in No. L 497. Here the coordinates λ and β of nine fixed stars are given such that there can be no doubt about their identification. Their names agree closely with the names in the Almagest, and not, as one perhaps would have expected, with the Handy Tables.[72] The coordinates and magnitudes are exactly as with Ptolemy except for the necessary correction of λ for precession. For details cf. p. 153.

"Simultaneously rising stars" (συνανατέλλοντα)[73] occur in No. 137c, unfortunately, only in very fragmentary context.

5. In No. 137c the longitude of the sun is said to be "calculated according to the lowest and the greatest [of velocity]," which might be a reference to the arithmetical methods of Babylonian origin. The declination of the sun is described in No. L 497 as "inclination toward the equator on the circle which passes through its pole." Ordinarily the solar longitude is the only numerically given coordinate, or, in other cases, the solar position is given with more sexagesimal digits than the planetary positions.

In No. 81 it is mentioned that the spring equinox is past, the sun having reached ♈ 14;6. For the question which longitude is associated with the vernal equinox cf. p. 182.

A solar eclipse (484 January 14) which took place shortly before the death of Proclus is mentioned in his horoscope (No. L 412), together with a predicted solar eclipse (486 May 19). Probably this latter eclipse in combination with the preceding and following lunar eclipse (485 December 7 and 487 April 23) are made responsible for some ill effect on Theodoros in his career as augustalian prefect in Alexandria (No. L 486).

6. The lunar nodes, which play an important role in Islamic astrology, are of only secondary importance in our material. They are never mentioned in any of the original documents but the ascending node occurs in No. L —71, three times in horoscopes from Vettius Valens (Nos. L 74,IV L 75,1 L 115,II) and 13 times in the late horoscopes (Nos. L 412 ff.).[74] The lunar latitude was mentioned in the now destroyed context of No. 95. Otherwise, the moon's latitude is only given in No. L 497.

The moon is said to be in slow motion in No. L 474. Its mean motion is used in No. L 380. The moon is called "rising" in No. 137a,b though the elongation from the sun (which is actually rising) is almost — 50°. In No. 81 the moon is called "waxing in crescent" (ἀπ' ἀνατολῆς φερομένη ὀξυγώνιος) at an elongation of about 29°.

7. In planetary motion retrogradation is mentioned six times[75] against two references to direct motion.[76] Opposition is mentioned twice,[77] the second stationary point is referred to in No. 81 (♃) and in No. 137a,b (♄, ♃, ♂) whereas No. L 497 places ♃, ♂, ♀, and ☿ in the first station. Planets are said to be "rising" in Nos. 81, 137a,b, L 40, L 76, L 487, and L 497; reference to a phase 7 days later is made in Nos. 81 and L 76. Setting is mentioned only once (♀ in No. 250,2). Mars is said to be "dim," being close to the sun (No. L 112,VII).

The latitudes of the planets are given only in No. L 497. The perigee of Mercury is recorded in No. 81. It is of interest that the position (♈ 10°) agrees with Ptolemy's value for the apogee of the eccenter of Mercury for about A.D. 140. In very obscure and doubtful context an "anomaly" of Mars is mentioned in No. 345.

A "magnitude" or "weight" (ὄγκος) is assigned to the planets in No. 81 but we do not know exactly what is meant since this concept is not attested elsewhere in any astronomical treatise known to us.

For the planetary periods which occur in astrological context in various disguises cf. the remarks in the technical glossary s.v. "Periods" (p. 10).

8. The tabulation of the deviations between the longitudes of the celestial bodies given by the later literary horoscopes (almost all of which belong to the fifth century) and found by modern computation reveals some interesting facts. A glance at the graphical representation of the differences "text — computation" (figures 37 and 38 p. 188 f.) shows that they remain essentially constant except for the irregular fluctuations which are caused by minor errors and inaccuracies. This is particularly evident in the case of the slow and regular motion of the sun and of the ascending lunar node. This graph stands in marked contrast to the graph of the solar longitudes obtained from Vettius Valens (figure 33 p. 182): whereas the later literary horoscopes show a

---

[71] The Pleiades are mentioned in astrological context in No. L 87,I.
[72] Halma, Tab. man. III p. 44 ff.
[73] For the importance of this concept in Greek astrology cf. Boll, Sphaera.
[74] The descending node appears in one version of No. L 486 but certainly by mistake.
[75] Nos. 46, 250,2 260 and Nos. L 75 112,VII 479.
[76] Nos. 95 and L 487.
[77] Nos. 260 and L 40. This concerns only mention as astronomical element; in the form of an "aspect" opposition is often mentioned in the literary horoscopes, then called "diameter."

practically constant deviation, the deviations in the texts from Vettius Valens decrease continually at a rate corresponding to precession. This shows that the longitudes in Vettius Valens are sidereal longitudes whereas the later authors operate with tropical longitudes, obviously following the norm adopted by Ptolemy and Theon.

From the graphs, figures 37 and 38, a deviation of about 2° or 3° between the textual longitudes and

No.	☉	☾	♄	♃	♂	♀	☿
46	0	+3	+8	-4		+5	-6
81	+5	+2	+5	+5,30	+3	+8,40	-1
137a,b	+1,30	+3,30	+6,30	+1,30	-3	-2	
137c	+4,30						
258	0						-9
260	+2	-1	+6	-0,30		-3	-4
284	0		+6		-2		
338			-1		-1	0	+7
478	-2	0	-1	-4	+4		

TABLE 13.

the modern ones results. Let us assume that longitudes were correct at the time of Ptolemy (+150) and that his constant of precession of 1° per century had been used until the fifth or sixth century. Then in the middle of the fifth century the resulting longitudes would be about 1° too small, owing to the

is illustrated by the case of Nos. 137a,b and No. 137c where a difference of 3° for the solar longitude appears in two versions of the same horoscope. The literary horoscopes give us a more significant picture not only by their greater number but also by their uniformity.

9. It is of interest to note that some early Arabic horoscopes continue where the Greek literary horoscopes end. The table given here (table 14) shows almost exact agreement between the differences found in the Greek horoscope No. L 621 of the Islamic state (written about 775) and in the Arabic counterpart of A.D. 622 March 21 given by al-Kindī (who died about 873).[79] The same trend is shown by the next six horoscopes from a work [80] of Māshā'allāh (who died about 815/820) which was communicated to us by Professor F. Carmody.[81] The explanation of this close agreement between Greek and Arabic literary horoscopes lies, of course, in the use of the same type of tables, most likely the Handy Tables of Ptolemy or Theon.

10. It is, of course, plausible to assume that the later Greek horoscopes were also based on the tables of Ptolemy or Theon. No. L 497 states explicitly that its author, presumably Eutocius, follows the rules of Ptolemy, though without giving further details. Very little is known, however, about the methods of computation of the earlier horoscopes. "Perpetual Tables"

Date	☉	☾	♄	♃	♂	♀	☿
622 March 21	-3,0	-3,20	-3,0	-3,0	-3,0	-2,50	
791 Febr. 13	-4,10	-4,30	-5,10	-5,40	-6,40	-3,40	-1,10
" March 22	-3,40	-1,20		-6,50	-6,0	-1,20	-0,50
" April 11	-4,10	-4,10	-6,40	-5,40	-6,30	-5,20	-6,30
" April 21	-4,40	-4,40	-7,0	-5,20	-6,30	-4,0	-10,40
" May 23	-3,40	-5,30	-5,30	-3,50	-6,30	-3,20	-0,40
794 Dec. 1	-5,0	-4,10	-4,10	-5,0	-7,20	-4,30	+2,40

TABLE 14.

defect in the value of Ptolemy's constant of precession. Actually, however, we find values 1° or 2° smaller. Since Ptolemy's fixed star longitudes for the zodiacal belt have longitudes about ¼° too small,[78] we are left with a deviation of $1 \pm \frac{1}{4}°$ for which we cannot offer any plausible explanation.

The material from the papyri is much too small to allow for similar statistics since the majority of papyri give zodiacal signs only for the longitudes of the planets and for sun and moon. The few cases which can be compared numerically with modern computation are tabulated in the following table (table 13). The extent to which individual differences might influence the data

are quoted in No. 81 but no further details are given. Only from the numerical values assigned to the longitudes of the planets it seems to follow that these tables gave at least two or three sexagesimal places.

For the horoscopes which give zodiacal signs only, a plausible procedure can be reconstructed. We do have Demotic as well as Greek planetary tables which list the dates of the entries of a planet into a sign (both in direct or in retrograde motion). Given the date of a birth these tables allow us to say without any further

---

[78] Peters-Knobel, PCS, p. 17.

[79] Published Loth [1].
[80] *De receptione planetarum sive de interrogationibus.* Several printed versions and many manuscripts are extant.
[81] No dates are given by Mā'shāllāh for these horoscopes but the astronomical determination causes no difficulties.

computation in which sign the planets were at the given day.

The above-mentioned tables are preserved for the following years:[82]

P: Demotic Pap. Berlin 8279: from Augustus 14 (— — 16) to 40 (— A.D. 10)

S: Demotic Stobart Tablets: from Vespasian 4 (— 71) to 10 (— 78)
from Trajan 9 (— 105) to Hadrian 3 (— 119)
from Hadrian 11 (— 126) to 17 (— 132)

T: Greek Pap. Tebtunis 274: from Trajan 10 (— 106) to 18 (— 115)

L: Greek P. Lund. Inv. 35b: Hadrian 4 (— 120).

The dates in P are understood to refer to the Egyptian calendar while S, T, and L, are based on the Alexandrian year. The fact that T overlaps a section of S permits us to establish the existence of some small discrepancies in the dates given by the two texts.

Unfortunately, only two horoscopes on papyrus belong to an interval covered by the preserved planetary tables, namely, the two earliest ones, Nos. —9 and —3. In both cases one finds agreement between the zodiacal signs obtained from the Demotic tables and the positions given by the horoscopes.

From Vettius Valens, however, we have a relatively large amount of material at our disposal which can be compared with the planetary tables; more than 30 horoscopes can be checked in this fashion. A detailed investigation of all these cases reveals 6 instances where the Stobart tables would have given different results. On the other hand, there are about 25 cases in which the horoscopes give the expected positions at a date which is quite close to the moment of transition from one sign to the next as listed in the planetary tables. The Demotic tables agree with the text of the horoscope several times when modern computation shows a different zodiacal sign.[83] This is, of course, due to the difference in the definition of longitudes between modern computation on the one hand and the ancient tables and the horoscopes on the other. On the whole, the hypothesis of making tables of the Stobart type the basis of the determination of the zodiacal signs for the planets in the horoscopes finds some support in the existing material, in spite of some undeniable contradictions. And it must also be remembered that No. L 75 exists in three versions, two of which give zodiacal signs only, while one has degrees and minutes of longitude. Thus simple omission of accurately computed longitudes is an ever present possibility.

We find 3 cases in which Saturn is differently located from what the Stobart tables would indicate. In No. L 110,III Saturn is placed on March 12 in ♓ 1;25 while the Demotic tables give March 20 as the day of entering ♓. This case, however, is not significant for our problem since it is *a priori* unlikely that positions with degrees and minutes are based on tables which give entries in signs only. In No. L 110,XII, however, we find Saturn in ♒ on December 15 while the Demotic tables (Stobart C_1) give November 20, the Greek table T November 19, as date of entering ♓. In No. L 114,VII Saturn is placed in ♈ on July 26 though it was in ♉ from June 28 to August 29 according to the Stobart tables.

Jupiter and Venus show no discrepancies. Mars in ♊ in No. L 78 is 2 days earlier in this sign than according to the Stobart tables. In No. L 108,XI Mars should not be in ♍, following the Stobart tables, since it entered ♎ five days earlier. Finally, No. L 109 assigns to Mercury a position in ♊, two days before the date of entry according to the tables.

The following cases may suffice as examples for narrow agreement with the dates given by the Demotic tables. In No. L 107 Venus is in ♈ one day before entering ♉. No. L 113,VII places Venus in ♈, one day before entering ♉ according to the Stobart tables; similarly Mercury is in ♊ only one day after entering this sign. In No. L 117,VI Jupiter is in ♋ only seven days before leaving this sign, Mars in ♉ only two days before its exit, and Mercury in ♌ two days after entry. In view of such good results one must not overlook, however, the fact that the same agreement is reached by modern computation which is certainly not the equivalent of the ancient procedure.

11. Of instruments the water clock is mentioned in No. 95 and the occurrence of the measure cotyla in connection with a time of night in No. 171 points in the same direction. The astrolabe was used for the determination of the Horoscopos from given date and hour in No. L 621, a horoscope which was actually written much later (A.D. 775) and by an author who was probably already influenced by Islamic astronomy.

F. ASTROLOGICAL DATA

Since the primary purpose of this work is to make the horoscopes available for the investigation of their astronomical content, we do not make an attempt to give a systematic discussion of the astrological doctrines and practices which are reflected in the horoscopes. On the other hand, for whatever use one wishes to make of these texts a great deal of astrological doctrine and technical terminology must be known to the reader. For this purpose we have compiled the " Glossary of Astro-

---

[82] For P, S, and T cf. Neugebauer [4], for L cf. Knudtzon-Neugebauer [1]. Neugebauer suggested these texts might be the "perpetual tables," criticized by Ptolemy (cf. above p. 24). Van der Waerden [2] (1947) argued for a deviation of all these texts from Babylonian methods. This would, however, exclude eccenters and epicycles, contrary to Ptolemy's description.

[83] Examples: ♄ in Nos. L 112,VII and L 112,VIII or ♃ in No. L 111,IV.

logical and Technical Terms" which concludes the Introduction (p. 2 ff.). In the following section we shall discuss a few specific cases of astrological practice which often can be used to verify or restore purely astronomical elements.

1. THE LOT OF FORTUNE

From the longitude λ_H of the Horoscopos the "Lot of Fortune" is derived by adding or subtracting from λ_H the elongation $\lambda_{\mathbb{C}} - \lambda_{\odot}$ of the moon from the sun:

(1) $\lambda_F = \lambda_H \pm (\lambda_{\mathbb{C}} - \lambda_{\odot})$.

In the literary horoscopes, without exception, the rule holds that addition has to be used in (1) by day, subtraction by night. This statement is based on almost 50 cases from Vettius Valens and 13 cases from the late authors.

In the papyri, however, the Lot of Fortune occurs only rarely. The earliest example, No. 81, uses the formula (1) with the plus-sign though it concerns a night birth; and the remark is added that "ignorant people" would use the opposite sign. In view of the above-quoted evidence it seems as if the scribe of No. 81 was violating the rule and not the other people.

In No. 137a a "first" and a "second" Lot of Fortune are determined but we are not able to explain the procedure. It seems doubtful that the term here means the same as in the other cases.

Four "Lots," Fortune, Daimon, Eros, and Necessity, were determined in Nos. 138/161 and 338 but in neither case are we able to determine all details of the procedure because of lacunae or errors in the texts.

In No. 182 the Lot of Fortune alone is mentioned but not enough is preserved to let us see the method of computation.

There remain only six cases, two of which are partially based on restorations, in which the definition (1) is applied. All six examples belong to the same late group of horoscopes.[84] Thus the papyri alone would not give us an adequate picture of the role which the concept "Lot of Fortune" plays in the astrological literature.

The importance which the literary texts attach to the Lot of Fortune leads to associating with it a "Midheaven," that is to say, a point in the ecliptic which would culminate when the Lot of Fortune was rising, thus taking the place of the Horoscopos. This "Midheaven with respect to the Lot of Fortune" is mentioned several times in the horoscopes from Vettius Valens[85] and appears in use in No. L —42 (cf. p. 78). The numbering of the Loci can be reckoned from the Lot of Fortune[86] instead of from the Horoscopos.

---

[84] Nos. 351, 370, 373, 376a/b, 381a/b, 385.
[85] Nos. L 50 72 74,XI 82 83 85,II 97,XI 123,I. In the last-mentioned example also a "Midheaven with respect to Daimon" is used.
[86] E.g., No. L 105 or No. L 484.

2. SYZYGIES

Various astrological doctrines require the position of the preceding conjunction and opposition. One from Antigonus,[87] four from Vettius Valens,[88] and eleven from late authors[89] mention a syzygy. In all these cases the choice between conjunction or opposition follows the rule that the syzygy that is nearer to the day of the horoscope has to be taken.

A subsequent conjunction is only once taken into consideration, namely, in No. L 114,V. Since it falls nine days later than the horoscope, the preceding opposition would have been much closer to the date of the horoscope.

All these cases come from the literary horoscopes. Among the papyri only one doubtful instance exists (No. 338) in which perhaps the preceding opposition is meant by the term ἀπόκρουσις.[90]

Eclipses are mentioned in the horoscope of Proclus (No. L 412) and in No. L 486. The full moon near conception is considered in No. L 74,IV.

3. THE FIRST TABLE IN VETTIUS VALENS VIII

The horoscopes No. L 419 and No. L 431 determine the length of life by making use in a somewhat modified form[91] of a table which is given at the end of Book VIII of the Anthology of Vettius Valens.[92] It was this practical application which led us to the discovery of the method by which this table was originally constructed. Since this procedure illustrates very well the combination of serious astronomical concepts with arbitrary manipulation of numbers in order to obtain new complex astrological rules we shall explain here the essential steps in the building up of the whole table.[93]

The starting point is a table of rising times of System A for Alexandria with the provision that the equinox is placed at ♈ 8°, thus following System B of Babylonian astronomy. From the 8th degree of one sign to the 8th degree of the next, linear interpolation is used. It is characteristic for the latitude of Alexandria that the ratio of longest to shortest daylight is given by $M:m = 7:5$. This suffices for computing the whole table of rising times and from it a table of the length of daylight for each single degree of solar longitude. As usual these values are expressed in "time degrees"; consequently one assumes $M = 3,30$ at ♋ 8° and $m = 2,30$ at ♑ 8° respectively. Presently we shall give the values for all signs at 8°.

A table of this kind is of purely astronomical interest

---

[87] No. L 40.
[88] Nos. L 74,IV 75,1 115,II 123,VII.
[89] Out of 17 horoscopes between L 412 and L 497.
[90] The reading is not certain, however.
[91] Cf. the commentary to No. L 419 (p. 136).
[92] Kroll p. 321 to 324.
[93] We did not succeed, however, in explaining all details in the second table (p. 325 to 328 Kroll).

ASTROLOGICAL DATA

and indeed an indispensable tool for all computations of rising times, culminating degrees and similar problems which make use of the arithmetical methods of Hellenistic astronomy. This table is now arbitrarily combined with a scheme of the even numbers from 2 to 30 which has no astronomical significance whatever. These numbers are arranged in five groups as follows

(1)
$$\begin{array}{ccccc} 26 & 20 & 14 & 8 & 2 \\ 28 & 22 & 16 & 10 & 4 \\ 30 & 24 & 18 & 12 & 6 \\ 2 & 26 & 20 & 14 & 8 \\ 4 & 28 & 22 & 16 & 10 \\ 6 & 30 & 24 & 18 & 12. \end{array}$$

The increase in each column is 2 (mod. 30) between consecutive numbers, and $-6 \equiv +24$ (mod. 30) horizontally. These 30 numbers are now associated with the 30 degrees of each zodiacal sign in a fashion which may be illustrated in the case of ♈: The degrees from ♈ 1 to 6 are associated with the group 14, 16, ..., 24; then, for the six degrees from ♈ 7 to 12 the group 8, 10, ..., 18 is used, then follows for the next six degrees the group 2, 4, ..., 12 then 26, 28, ..., 6 and finally 20, 22, ..., 30. Exactly the same principle is followed for the other signs. Within each sign the groups follow each other in the order of the scheme (1), read cyclically. The whole scheme of associations between the 360 degrees of the ecliptic and the 12 times repeated numbers (1) is therefore known if we indicate the numbers associated with the first degree of each sign. The law which governs this construction is very simple: one begins at ♎ 1° with 2 and adds 12 mod. 30. Thus one obtains for the first degree of all signs the following numbers:

(2)
$$\begin{array}{cccccccccccc} ♎ & ♏ & ♐ & ♑ & ♒ & ♓ & ♈ & ♉ & ♊ & ♋ & ♌ & ♍ \\ 2 & 14 & 26 & 8 & 20 & 2 & 14 & 26 & 8 & 20 & 2 & 14 \end{array}$$

and from it, using (1) the numbers for all other integer degrees, e.g. for Libra:

(3)
$$\begin{array}{cccccccccc} ♎ & 1° & 2 & 3 & 4 & 5 & 6 & 7 & 8 & \text{etc.} \\ n = & 2 & 4 & 6 & 8 & 10 & 12 & 26 & 28 & \text{etc.} \end{array}$$

All these numbers are tabulated in the column headed by ἀριθμοί.

Having now all the numbers n at our disposal we can proceed to the columns called "years, months, days" which are used for the astrological prediction. These time intervals are obtained as follows. Let λ be an integer degree of longitude, e.g., $\lambda = $ ♎ 8. With this longitude are associated (a) the number n and (b) the length of daylight $c(\lambda)$ — in our example $n = 28$ and $c = 3,0°$. We now form the products

$$t = c \cdot 0;n$$

where $0;n$ indicates that we assign to the numbers n the place value of sixtieths. The resulting integers of t are then counted as years, $0;5$ as one month, $0;2,30$ as 15 days. Thus we obtain for our example

$$t = 3,0 \cdot 0;28 = 1,24 = 84 \text{ years}$$

and indeed 84 is the entry in the table under "years" for ♎.[94]

In the same fashion all entries for t in the whole table can be obtained though the fractions are usually rounded off to a nearby multiple of $0;5$ i.e. "months." In details many scribal errors can be detected, e.g., omission of

8°	$c \cdot 0;n = t$			Text			Notes
♈	3,0	· 0,10 =	30	30	30ʸ	0ᵐ	
♉	3,16;40	· 22	1,12;6,40	1,12;10	72	2	
♊	3,26;40	· 4	13,46,40	13,45	13	9	
♋	3,30	· 16	56	56	56	0	
♌	3,26;40	· 28	1,36;26,40	1,36;30	96	6	
♍	3,16;40	· 10	32;46,40	32;25	32	5	(a)
♎	3,0	· 28	1,24	1,24	84	0	
♏	2,43;20	· 10	27;13,20	27;10	27	2	
♐	2,33;20	· 22	56;13,20	56;10	56	2	
♑	2,30	· 4	10	10	10	0	
♒	2,33;20	· 16	40;53,20	40;40	40	8	(b)
♓	2,43;20	· 28	1,16;13,20	1,15;35	75	7	(c)

TABLE 14a.

Notes to Table 14a:

[a] One would expect $32;45 = 32$y. 9m. (cf. ♓). Confusion of ϵ and θ explains this error.

[b] The 8 is dittography from the preceding line ($\equiv$ ♒: 39y. 8m.). Correct would be 40y. 9m.

[c] Expected: $1,16;10 = 76$y. 2m. (cf. ♏ and ♐). The 7 is dittography from the preceding line (♓ ♒: 70y. 7m.).

lines and resulting incorrect alignments, dittographies etc. but it would now be an easy task to reconstruct the whole table in its original form. As an example the results for the 8th degrees of all signs are listed in the table 14a.

The table of the text contains an additional column in which a planet is associated with a group of lines. Since this column is not used in our material we need not discuss a possible restoration of this entry which was particularly maltreated by successive scribes and editors.

---

[94] Kroll p. 321, IV, 11.

IV. THE AUTHORS OF THE LITERARY HOROSCOPES

One might arrange the "literary horoscopes" either according to their dates or according to the literary source. Both classifications lead to almost exactly the same division of our material. About 130 horoscopes are older than A.D. 190; about 20 are later than A.D. 380; no literary horoscopes are preserved from the two centuries in between. In the older group all but seven come from the *Anthology* of Vettius Valens. Two come from Balbillus, three concern Hadrian and his family and are taken from Antigonus of Nicaea, the sixth is the astrologer Manetho's horoscope (No. L 80), the seventh the horoscope of the rhetor Aristides (No. L 117,X). The later group contains the horoscope of Proclus (No. L 412), 8 examples from Palchus, 7 from Rhetorius, one each from Hephaestion and Eutocius, plus two of unknown origin.[1] Hence we have two major groups: the older authors Balbillus, Antigonus, Vettius Valens, and through him perhaps Critodemus, all from the first and second century A.D.; then the later sources, Hephaestion, Rhetorius, and Palchus, essentially of the fifth century.[2] The literary horoscopes conclude with the horoscope of Islam, allegedly cast in the first year of the Hijra but actually composed in A.D. 775, the year of al-Manṣūr's death, the ruler under whom Islamic astronomy took over the lead from the Greeks.

A. VETTIUS VALENS

1. The importance of the *Anthology* of Vettius Valens for our subect can be illustrated by the following figures. With its about 130 (partial or complete) horoscopes it contains twice as many examples of Greek horoscopes as all papyri combined. Without Vettius Valens (whose examples range from A.D. 37 to 188) we should have only five examples of "literary" horoscopes before A.D. 380.

In 1899 Bouché-Leclercq wrote[3] "J'ai renoncé à exploiter à fond cet ouvrage, tout en casuistique sans idées et en problèmes d'arithmétique que l'incertitude des sigles et des chiffres rend le plus souvent inintelligibles." Since then Kroll's edition appeared (1908) and some scattered fragments were published in the twelve volumes of the *CCAG*. The *Anthology*, during the Middle Ages, was a much used work and consequently the text, as we have it, is in very bad shape. It may suffice to quote two instances from the present material: the horoscope L 61,X in III,8 appears once more in a slightly different form in VIII,8 (where it does not belong at all) and the horoscope L 118, published in *CCAG* 2 from an anonymous *De planetis*, is merely a version of a horoscope of which three other versions are preserved in the *Anthology*. Only a prolonged and detailed study will lead to a better understanding of the text, its composition, and its — undoubtedly great — influence upon later treatises. The horoscopes provide, as we shall see, an important help for the establishment of an internal chronology of the work itself and of its sources. On the other hand we gain valuable information of the astronomical, astrological, and historical character of a comparatively large and uniform group of horoscopes. It is for these reasons that we must discuss the horoscopes of the *Anthology* as a separate unit.

2. All horoscopes in Vettius Valens serve a specific purpose, namely to provide examples for certain doctrines which are discussed in the main body of the work. *A priori* it would be perfectly permissible to assume that such examples were arbitrarily made up in order to demonstrate by concrete configurations how conclusions should be drawn from them according to theory. If this were the case we should have very little chance to find such a configuration actually occurring, say, during the first two centuries of our era. The fact that every one of these horoscopes can be shown to be astronomically correct for a date in the first or second century A.D. is therefore proof that Vettius Valens was using empirical material exclusively, collected either by himself or by his predecessors.[4] A more detailed analysis of these data will show us that a large proportion of this material was gathered during the decades from about 140 to 170. Thus, during these years Vettius Valens himself was systematically collecting and analyzing a large amount of statistical material of birth data, life histories, and deaths in order to confirm or to modify the theoretical structure of astrology. This shows that Hellenistic-Roman astrology was still under development near the end of the second century A.D. Ancient propaganda as well as insufficient criticism of modern scholarship has often succeeded in obliterating such facts by postulating an origin of astrology in the far remote past.

3. The *Anthology* contains some 40 explicit references to years of the Roman imperial period and about 100 horoscopes which do not quote their dates but which can be dated astronomically. Both groups provide a

---

[1] Preserved as additions to Vettius Valens, perhaps taken from Rhetorius.

[2] No. L 380 is the earliest horoscope in this second group. The year 379 is the epoch for the list of stars of the "Astrologus anni 379" published *CCAG* 5,1 p. 196 ff. Paulus Alexandrinus supposedly wrote in 378 (*l.c.* p. 196).

[3] *AG* p. XIII.

[4] This confirms his repeated claims of drawing on his own experience. *Cf.*, e.g., No. L 122,XII first version.

total of 77 cases where either events are quoted which happened at a certain age of the subject or where the age is mentioned at which the person died. By adding these years to the date of birth we obtain dates which are earlier than, or at the most equal to, the date of the writing of the section in question. Since it happens that these totals lie, usually, within very narrow limits for the single chapters, we can be sure that the date of writing practically coincides with the latest mentioned dates.

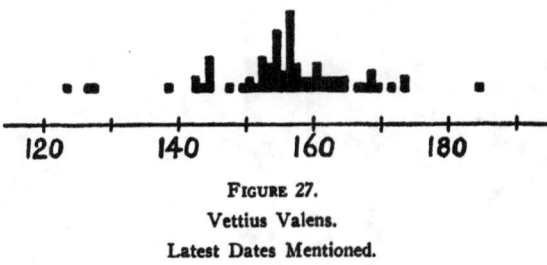

FIGURE 27.
Vettius Valens.
Latest Dates Mentioned.

Figure 27 shows the distribution of these latest recorded dates which range between 123 and 184. Of the 77 cases in question, 71 belong to the interval from 140 to 170 (or 55 to the years from 150 to 165). This shows

question. Some of these references are found in book I (11 cases) as shown in line "b." So far all dates concern birth only. Almost all the following cases give not only the date of birth but also the age at death. This is indicated in figure 29 by lines which connect the first and last year of life.[5] It is obvious from this graph that chapter VIII,6 must have been written in the years 156/158 (with one late addition in 174) whereas VIII,7 was completed about ten years later.

The same sort of situation is found from the investigation of the much greater mass of astronomically dated horoscopes (figure 30). The early books contain dates of birth only, almost all belonging to the interval between A.D. 50 and 120 though the very latest horoscope, for 188, is among them (II,26). Then follow horoscopes which either mention the length of life of the person (indicated by a closed line in figure 30) or investigate events during lifetime to a certain age (open lines in figure 30). Some of these horoscopes are used repeatedly for different purposes. Such repetitions are indicated in figure 30 by little empty squares. One sees clearly how certain chapters are based on closely contemporary material. Chapter V,10, for example (line "n" in figure 30), was completed essentially in 156/157, chapter VII,5

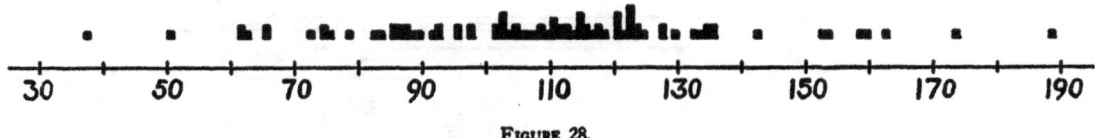

FIGURE 28.
Vettius Valens.
Total of Horoscopes.

that the gathering of material extended over a period of at least 25 or 30 years with perhaps a few borrowings from older sources and a few later additions up to A.D. 184.

Essentially the same picture is obtained from the dates of birth gathered from all horoscopes which can be astronomically dated (figure 28). A few isolated cases belong to the period from 140 to 188. The majority of subjects were born between 100 and 140 and met important events or death to be recorded some decades later. We shall return to this age distribution presently. Figure 28 suffices to show that some older material must have been taken from other sources, though not to a very great extent and only a few decades earlier.

4. By the same method one can even obtain some insight into the chronology of the composition of the *Anthology*. This can best be illustrated by means of figures 29 and 30. The first figure concerns the 40 cases where a date of birth is explicitly quoted. Their distribution is shown in line "a" where each little square represents one case belonging to the year in

(line "u") in 164, whereas III,16 ("k") and VII,3 ("s") reach to 169/170 and even 173 respectively. Thus all evidence agrees that Vettius Valens must have worked for at least twenty years, from 154 to 174, on the composition of the *Anthology*, even if one disregards the isolated latest additions.

5. Since there is very little statistical material available for the life expectancy in antiquity, we give here in figure 31 a graphical representation of the ages of life reached by individuals whose horoscopes were discussed by Vettius Valens. The graph gives the totals for consecutive decades. Thus the first point indicates that 6 children died before having completed ten years.[6] We realize, of course, that 42 cases cannot have a real statistical value though the individuals whose lives Vettius

---

[5] Lines not ending in a black square indicate that the final year was investigated for its astrological significance but that we do not know whether it is the year of death or of some important event during lifetime.

[6] For a much more extensive statistical material *cf.* Hombert-Preaux [1].

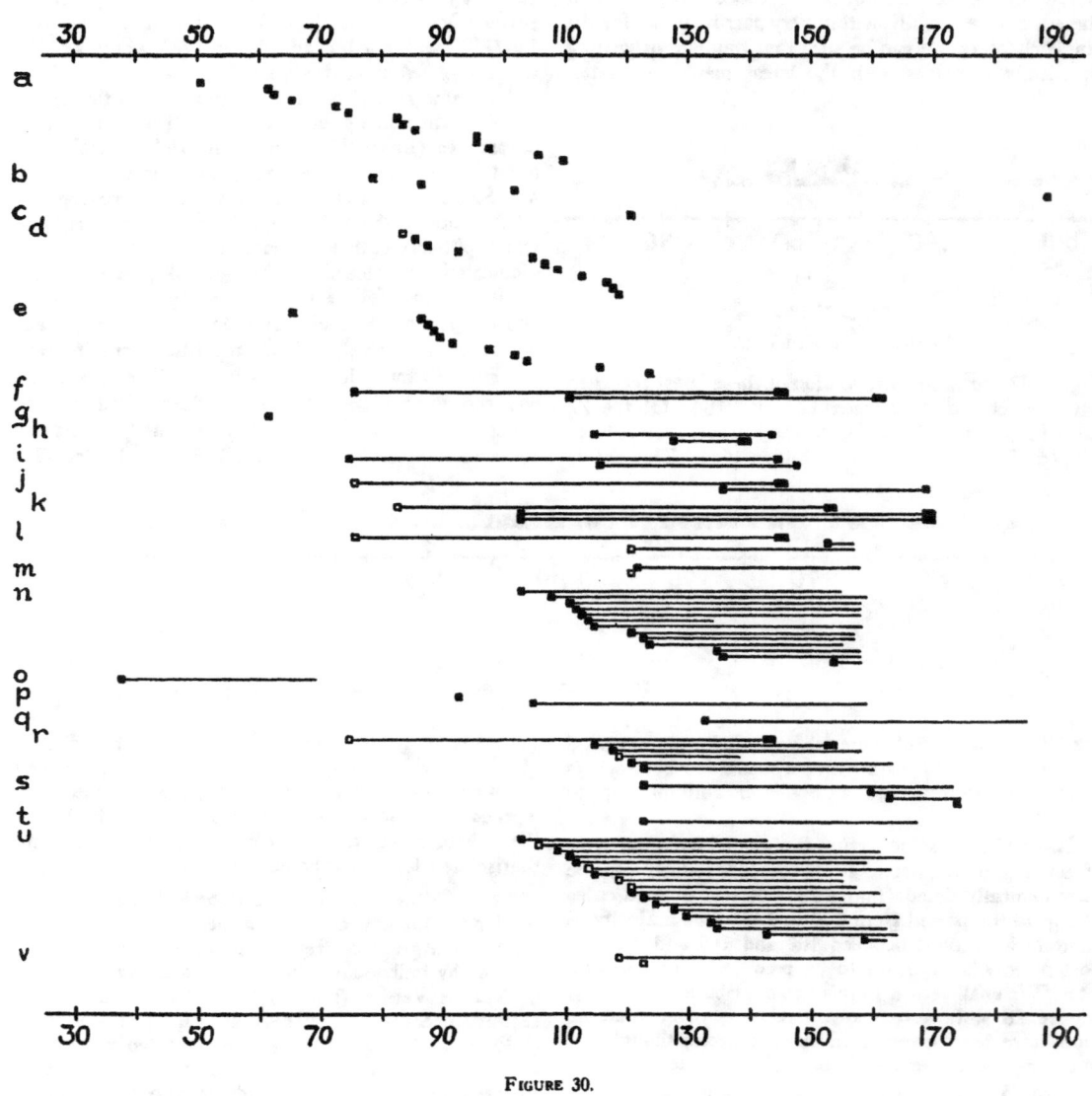

Figure 30.

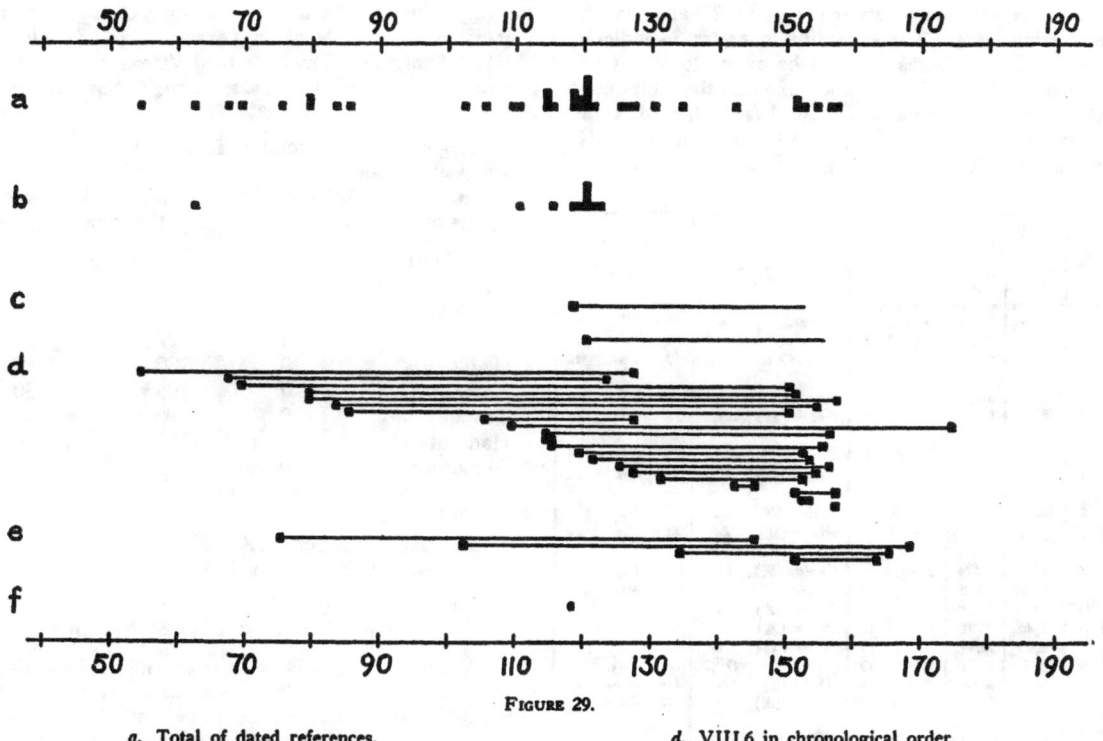

FIGURE 29.

a. Total of dated references.
b. Total from Book I.
c. III,11 and V,7.
d. VIII,6 in chronological order.
e. VIII,7 in chronological order.
f. IX,19.

Valens recorded seem not to be restricted to any particular group of people.

A similar graph is figure 32 which gives, again for consecutive decades, the ages at which persons were still alive when Vettius Valens wrote about them. As usual, most of the events recorded in astrological literature are "crises," either of health or of daily life. This explains the slight interest in the lowest age group.

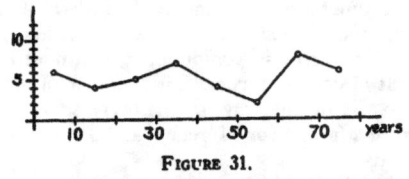

FIGURE 31.

6. We may now turn to astronomical questions in connection with those cases where explicit dates are mentioned by Vettius Valens. These are listed in chronological order in table 15. In the majority of cases only the positions of sun and moon are given, together with the Horoscopos and the hour.[7] Thus it is possible to check these data by computation. This leads to several interesting results.

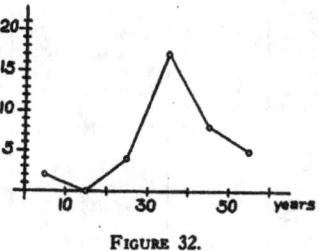

FIGURE 32.

A. In all cases the longitudes of the sun given by Vettius Valens are greater than the longitudes found by modern computation[8] (cf. table 16 and figure 33).

[7] In some cases, which are marked in the tables by an asterisk, we also have the corresponding complete horoscope which can be found under the L-number for the corresponding date.— All dates are understood to be Alexandrian dates.

[8] All conversions of Alexandrian to Julian dates were obtained by means of the tables of Schram and then checked by means of the lunar positions. All differences in longitudes are rounded off to full degrees. In the fourth case two points are plotted corresponding to the difference in the solar position as given in VIII,7 and in the horoscope No. L 75.

The deviations vary between 7° and 2° and show a decreasing trend. The skew line in figure 33 indicates how much the decrease would be owing to precession. It seems to be admissible to assume that the tables upon which these computations were based showed a difference of about $+5°$ in A.D. 50 and of about $+3\frac{1}{2}°$ in A.D. 160. This seems in general agreement with the deviation observed in the Demotic planetary tables which belong to the period from -16 to A.D. 131.[9]

B. Whenever the hour is an hour of daylight only one date is given (example No. 4: Epiphi 22 5^h day-time). Whenever the hour is an hour of the night a double day number is given (example No. 3: Thoth 14/15 11^h night). Thus Vettius Valens follows the general usage for quoting dates at night time in Hellenistic and Roman Egypt.[10]

C. Regnal years are counted according to the Alexandrian calendar and are in agreement with the canon given in I,19 (p. 32,24 ff. Kroll). The following table gives the attested correspondences to this canon:

Thoth (I) 1 of			Julian year	
Nero	year	0	53	August 29
Vespasian		0	67	30
Titus		0	77	29
Domitian		0	80	29
Nerva		0	95	30
Trajan		0	96	29
Hadrian		0	115	30
Antoninus		0	136	29

Examples:

Domitian 2 Pachon (IX) 20
= 80 + 2 + 1 May 15 = 83 May 15.

Domitian 5 Athyr (III) 24
= 80 + 5 November 20 = 85 November 20.

7. We are now in a position to extend our investigation of the counting of longitudes to all horoscopes in Vettius Valens. The results are of necessity less accurate since deviations between ancient and modern computation involve not only the difference in the definition of the zero point of longitude but also the inaccuracies of both the ancient and the modern computation. In the case of the sun these inaccuracies are much smaller than in the case of the planets with their much higher irregularity of motion. In fact the interest of the following investigation lies less in the confirmation of a difference in counting of longitudes than in a possibility of obtaining some insight into the accuracy of the planetary tables used by Vettius Valens.

From the following comparison between ancient and modern computation we exclude, of course, those cases where the ancient text is entirely wrong, either, perhaps, because of an error in computation, or, more likely, as the result of copyist errors. The number of such gross deviations is, by the way, surprisingly small. We find 9 cases for Mars, 4 for Mercury, and 3 each for Jupiter and Venus.

The deviations which are of interest are of a different character. In seven horoscopes we find positions given by degrees, some even by degrees and minutes. Here a direct comparison with modern computation is possible. Almost all cases show longitudes greater than according to modern computation. The corresponding differences are entered in the following tables (table 17). As ex-

No.	Kroll ch.	Kroll p.	Text year		Text month		Julian Date	
1	VII,6	305,3	Nero	1	Athyr (III)	2	54 Oct.	29
2	I,23	50,25	"	8	Mesore (XII)	6/7	62 July	31
3	VII,6	309,33	"	14	Thoth (I)	14/15	67 Sept.	13
4	III,6	307,9	Vespas.	1	Epiphi (XI)	22	69 July	16
5*	VII,7	317,15	"	7	Epiphi (XI)	25/26	75 July	19
6	III,6	305,26	Titus	1	Pham. (VII)	20/21	79 March	16
7	"	311,26	"	2	Choiak (IV)	1	79 Nov.	28
8	VII,6	310,34	Domit.	2	Pachon (IX)	20	83 May	15
9	"	311,11	"	5	Athyr (III)	24	85 Nov.	20
10	IX,7	316,24	Trajan	6	Choiak (IV)	1/2	102 Nov.	28
11	VII,6	308,18	"	8	Pham. (VII)	26	105 Apr.	21
12*	"	310,11	"	12	Payni (X)	8	109 June	2
13	I,20	35,15	"	13	Pham. (VII)	18	110 March	14
14*	I,23	52,16	"	17	Mesore (XII)	2	114 July	26
15	VII,6	306,8	"	17	["	17/18	114 Aug.	10
16	"	307,21	"	17	Thoth (I)	14/15	114 Sept.	12
17	"	306,24	"	18	Payni (X)	14	115 June	8
18	II,11	149,1	Hadr.	3	Athyr (III)	27	118 Nov.	23
19	I,19	32,6	"	3	"	28	118 Nov.	24
20*	II,14	163,5	"	3	"	30	118 Nov.	26
21	VII,6	313,5	"	3	Pham. (VII)	29/30	119 March	25
22	I,20	36,4	"	4	Athyr (III)	30	119 Nov.	27
23*	I,4	20,9	"	4	Mechir (VI)	13	120 Febr.	6
24*	I,10	26,21	"	4	"	"	"	"
25*	I,20	36,15	"	4				
26*	V,7	215,30	"	4				
27	I,16	29,30	"	4	Pham. (VII)	19	120 March	15
29	VII,6	312,17	"	5	Pachon (IX)	23/24	121 May	18
30	"	308,32	"	9	Pham. (VII)	28/29	125 March	24
31	I,23	51,19	"	10	Mechir (VI)	13/14	126 Febr.	7
32	VII,6	306,35	"	12	Athyr (III)	1	127 Oct.	28
33	"	309,16	"	15	Epiphi (XI)	16	131 July	10
34	VII,7	317,32	"	18	Pham. (VII)	2	134 Febr.	26
35	VII,6	307,27	Anton.	5	Tybi (V)	28/29	142 Jan.	24
36	"	312,5	"	14	Mechir (VI)	23	151 Febr.	17
37	VII,7	318,7	"	15	Athyr (III)	25/26	151 Nov.	23
38	VII,6	302,36	"	15	Tybi (V)	12	152 Jan.	8
39	II,11	149,10	"	17	Pham. (VII)	11	154 March	7
40	X,7	215,31	"	20	Phaophi (II)	10	156 Oct.	7
41	VII,6	308,9	"	21	Athyr (III)	28/29	157 Nov.	24

TABLE 15.

---

[9] Cf. above p. 173.

[10] Cf. above p. 167 ff.

TABLE 16.

pected they are mostly near + 4° or + 5° with the exception of only No. L 127,XI which shows even negative deviations. By far the best agreement is found for sun and moon. Larger deviations in some cases for the planets must be due to a lower accuracy of planetary tables and to the increased chance for errors in computing.

Here the results are less regular (figure 35), especially for Mars and Mercury, where deviations of less than + 3° must be interpreted as caused by real deviation in the tables, probably due to special situations as retrogradations or stations. Obviously Mars is by far the worst case.

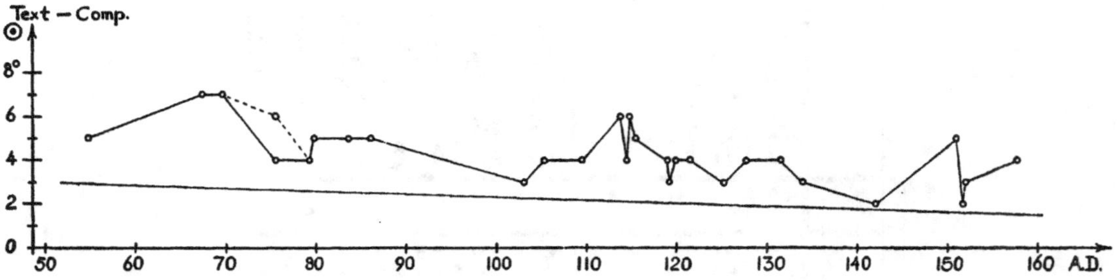

FIGURE 33.

No. L	☉	☾	☊	☌	☿	♀	♂	♃	♄
61 X						≦ + 5	wrong	+ 5	+ 4
74 IV	+ 4	+ 4	+ 4	+ 4					
75	+ 5;30*)	+ 2			+ 2;30	+ 7;30	− 1;30	wrong	+ 8
110 III	+ 2	+ 2			wrong	+ 1	+ 8	+ 4	+ 8;30
114 V	+ 5	+ 5		+ 3;30	+ 5	+ 1	+ 7;30	+ 4;30	+ 5
115 II	+ 3;30	> + 4	+ 5	+ 4					
127 XI	[+ 2]	− 1;30			− 9	+ 4	+ 0;30	− 3	− 2;30

*) Variant: + 3;30.

TABLE 17.

For the majority of horoscopes no such direct numerical comparison is possible since all positions are given by zodiacal sign only. In many cases, however, we find that modern computation leads for one or two planets to a position near the end of a sign (e.g., ♉ 28) where the text gives the next sign (♊). Such formal deviations are marked in our commentaries by (!). In fact, however, one must expect exactly this phenomenon if ancient longitudes are numerically greater than modern ones. Thus we have here a possibility of saying that the ancient longitudes are at least so much greater than the modern ones if we register all differences up to the boundary of a sign. Figure 34 shows all these lower limits [11] which again agree very well with an actual deviation of 4° or 5°.

Similarly one can look for upper limits by collecting all cases where modern longitudes fall, say, between 25° and 30° without producing a change of sign in the text.

[11] The connecting lines between the single points are, of course, of no significance whatever.

8. On several occasions Vettius Valens tells us the "clima" where the subject of a horoscope was born. This gives us some indication of the wide range of persons with whom the author had come in contact. Since the climata are in almost all cases mentioned in connection with the rising times of zodiacal signs, we obtain in this way also some astronomical information together with an insight into the astrological practice which appears here as rather unconcerned with the accuracy of its data.

We have 53 cases in which Vettius Valens mentions the "clima" in which a person was born. The distribution of cases is shown in figure 36.[12] Clima 1, Alexandria, clima 2, Syria and Palestine, and clima 6, Rome and Italy, furnish the majority of cases with 11, 17, and

[12] In one case, clima 2, the date of the horoscope is not certain (L 75,1). In two cases, L 105 and No. 32 in the list of table 15 (127 Oct. 28), the clima is doubtful: clima 6 or 3 in the first case (Kroll p. 308,18 and 25), clima 5 or 1 in the second case (p. 306,35 and 307,7). The dotted line in figure 36 corresponds to the second alternative.

15 persons respectively, whereas only 10 cases are left for the four remaining climata. Such a distribution is in good accord with Alexandria as locality where Vettius Valens [13] and his predecessors collected their evidence; it also confirms our previous conclusion that we are dealing with real cases and not with artificially composed horoscopes.

In several horoscopes astrological doctrines are discussed for which the rising times of the zodiacal signs play a role, e.g., in the determination of the length of life or of critical years. Since the rising times depend upon the geographical latitude, Vettius Valens gives in these cases the "clima" in which the person was born. His definition of the seven climates is found in I,7 where he states that for the first clima the total of the rising times from ♋ to ♐ amounts to 210° and 4° more for each clima up to 234° for the seventh clima. Since $a_4 + \cdots + a_9 = M$, the longest daylight, his statement is equivalent to saying that $M = 3,30$ in the first clima, 3,34 in the next, etc. A glance at table 2, p. 4, shows that he is operating with the sequence (a) which is based on Alexandria. Since he furthermore assumes constant differences for the rising times of a given clima, he has chosen System A.[14]

This clear theoretical statement in Book I is, however, not in accord with his practice. If one investigates all

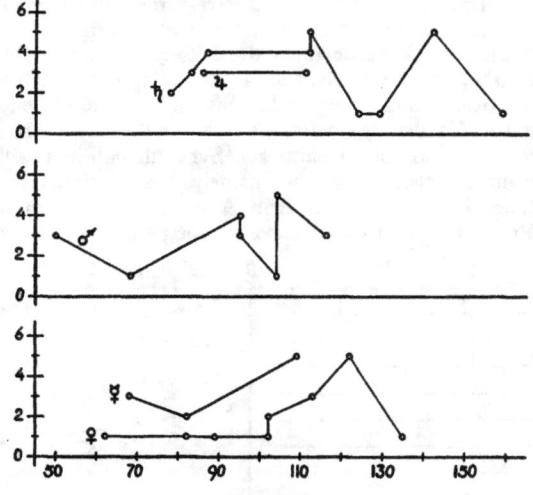

FIGURE 34.

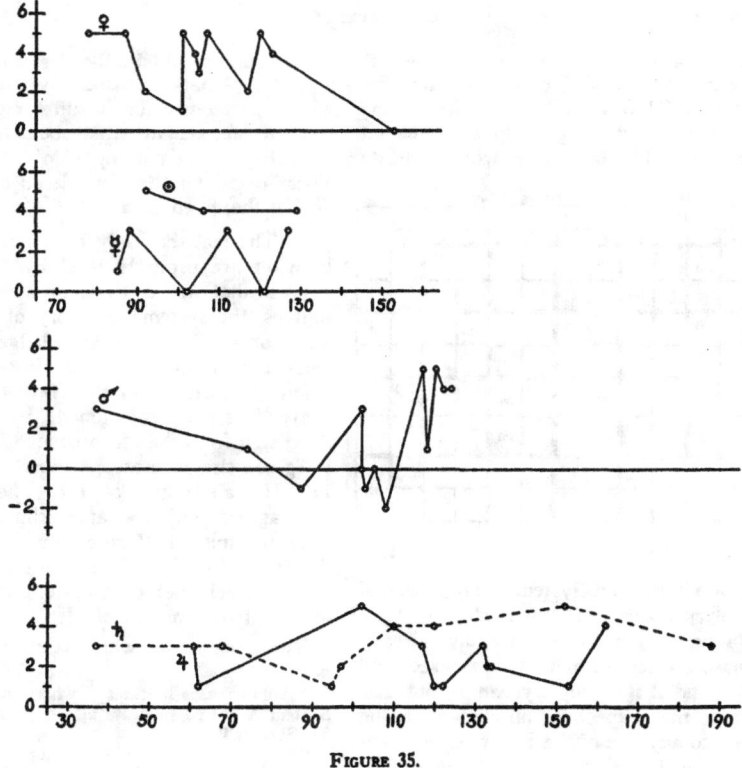

FIGURE 35.

---

[13] That he reached Egypt after much travel he reports himself in IV,11 (p. 172,4; p. 173,21 ff. Kroll).

[14] Cf. p. 12.

cases in which climata and rising times are stated, one finds some interesting results. "Clima 1" means in all (14) cases Alexandria, System A, since the numbers agree exactly with our table 2 p. 4 col. 1,a. "Clima 2" is always (23 cases) Babylon (col. 1,b in table 2) and System A, with one exception (2 cases in No. L 120,XII) where System B was used. So far one is simply dealing with the two historically most important and oldest lists of rising times: Systems A and B for Babylon, System A for Alexandria. But for the remaining climata almost complete arbitrariness is apparent. This can be best illustrated by means of the following table 18 in which an x-mark indicates a certain case clima, whereas our table shows that the use of the proper clima is not the rule but the exception. The majority of all values can be explained as belonging to the clima 2,b (Babylon) and System B,[15] regardless of the clima in which the subject is said to have been born.

Thus we have obtained the following result. Only for the first clima, Alexandria, does Vettius Valens use the rising times which he has instructed us to compute in I,7. He already violates his rules for the second clima, for the region of Palestine and Syria, though he is still using a consistent scheme, namely, the classical rising times for Babylon,[16] System A and, once, System B. From this point on he chooses rising times as he sees

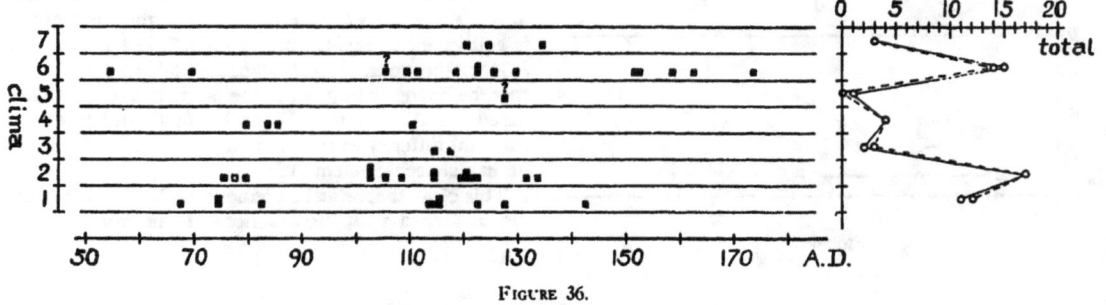

FIGURE 36.

whereas (x) means that either System A or System B is possible. For example, $a_{10} = 27$ occurs in clima 7,a in System A or in clima 2,b System B. If Vettius Valens had followed the rules which he gave in I,7, then all cases would be represented in the squares which indicate

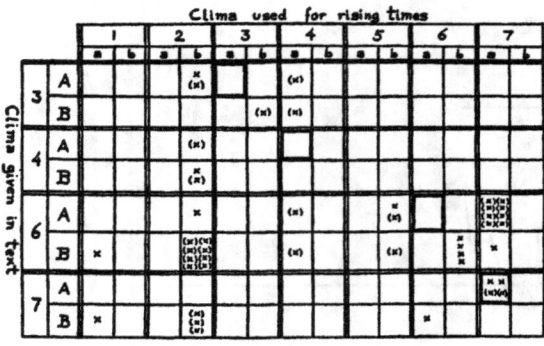

TABLE 18.

the first half (a) of a clima and System A (represented in table 18 by a black frame). But it is clear that we cannot expect so much strictness since we have already seen that for clima 2 the second half (b) was used, and for both climata 1 and 2 not only System A but also System B. Thus we may expect for all climates rising times which belong to any one of the four possible combinations of A and B with (a) and (b). We ought not find, however, rising times which belong to a different

fit in order to obtain the desired results in his astrological explanations. Both Systems A and B are used with a preference for B and for the clima of Babylon. But the fact remains that about one-third of the given rising times have nothing to do with the clima in question. These cases must be considered as deliberate modification of the given data.

9. The fact that only the climata of Alexandria and Babylon are properly used by Vettius Valens might suggest that in his time, or at the time of his main sources, the systematic theory of the "seven climata" was not yet common knowledge. Also the accurate determination of the culminating point (M) is not yet of any importance in the *Anthology*. Though "Midheaven" is often mentioned, it is generally simply the third sign from the Horoscopos. In only three cases are we given the accurate position of M (Nos. L 75 75,1 110,III), all from book III. The result in the second case agrees with the approximate method of Paulus Alexandrinus, the third case with accurate computation, the first with neither one. Thus we are not in a position to say which method was followed by Vettius Valens for the determination of M.

The lunar nodes which later on, especially in Hindu

---

[15] All the cases in clima 6 where values (x) from clima 7,a System A are possible are also explicable as values from clima 2,b System B.

[16] In fact the only climates not designated by number are Alexandria (in No. L 74,IV) and Babylon (reference to 79 Nov. 28 = No. 7 in table 15 p. 180).

astrology, obtain the same significance as the planets, play a very secondary role in Vettius Valens. The ascending node is mentioned three times only, again book III (Nos. L 74,IV; 75,1; 115,II), though rules for its computation are given in I,16. Since a good mean value for the nodal motion is easily obtainable, it is not surprising that we find good agreement with modern computation.

In spite of the great extent of astrological doctrine contained in the *Anthology*, one receives the impression that the development of the theory had not yet reached its climax in the second century A.D. when both Vettius Valens and Ptolemy wrote their compendia. And both authors were satisfied by using simple arithmetical schemes, e.g., for the rising times, which belong to a period of astronomical theory which had been long surpassed at that time. The cliché which is so popular in histories of astronomy about the stimulating influence of astrology on exact astronomy is nowhere born out where we are able to control the details.

B. CRITODEMUS

The determination of the dates of the horoscopes which are contained in the *Anthology* also puts the problem of the age of the sources of Vettius Valens in a new light. Since almost all the dates in the earlier books (*cf.* figure 30) belong to the second half of the first century A.D., one has no definite proof of the use of essentially older sources. This raises a new difficulty in the dating of Critodemus who is frequently quoted by Vettius Valens and seems to be one of his most important sources besides the traditional Nechepso-Petosiris.[17] Cumont [18] considers Critodemus as "un des premiers auteurs qui révélèrent aux Grecs les arcanes de l'astrologie babylonienne." Though the factual basis for such a role is, to say the least, exceedingly slim, the mere fact that Critodemus precedes Vettius Valens makes him one of the earliest known authors on astrology and hence the determination of his date becomes a problem of some interest.

The existing evidence outside of Vettius Valens is easily summarized.[19] Pliny quotes him among his authorities for Book II and Book VII [20] and mentions him together with Berossus in Book VII in connection with fictitious Babylonian chronological schemes.[21] Thus Critodemus could be at the latest, a contemporary of Pliny, in the first century A.D.[22] No real information about the contents of his writing flows from this source.

After Vettius Valens not much more is available. Firmicus (fourth century) mentions Critodemus in the introduction to Book IV [23] as a known authority beside the traditional mystical names. Hephaestion (who died later than 382 [24]) quotes him verbatim in II,10, a chapter on stillborn children.[25] This is straightforward astrological doctrine of the ordinary type. Finally, Theophilos [26] (eighth century) ascribes to Critodemus, with Valens, Dorotheus and Timocharis, the use of ♈ 0° for the counting of longitudes, a statement which obviously has no merit of historical accuracy.

Turning now to Vettius Valens we must distinguish two types of references to Critodemus: (*a*) quotations from and remarks about a work of Critodemus called ὅρασις "Vision" [27] and (*b*) excerpts concerning specific astrological doctrines.

It is well known how Vettius Valens criticizes the ὅρασις for obscurity and artificiality of style. All this has been described excellently by Boll [28] and Cumont [29] and need not be repeated here. One is probably right in saying that the obscurity of writing in the ὅρασις agrees with its being an early astrological work of a more general character.

Be this as it may, no particular obscurity can be detected in those chapters of the *Anthology* of Vettius Valens which are directly ascribed to Critodemus or can be assigned to him with good reason. These chapters contain also several horoscopes, whose dates, now established, all point to the decades around 100 A.D., thus to one or two generations before Vettius Valens, but definitely after Pliny. We now present the evidence for these excerpts from Critodemus.

(A) Vettius Valens III,7 is called "On dangerous loci and starters from (the writings of) Critodemus." [30] This chapter is followed by III,8 and VIII,8 (which again are concerned with dangerous loci) preserved in two slightly different versions, as shown in our discussion [31] of the horoscope No. L 61,X which is contained in this chapter. Critodemus is again quoted in connection with the doctrine of the "starter" in V,11 [32] and V,12.[33]

---

[17] *Cf.* the Index in Kroll p. 374.
[18] *Lux perpetua*, p. 311.
[19] A nearly complete list of fragments is given by Schnabel, *Berossos*, p. 18 to 120. As we shall see, his fragments No. 14 and 15 are actually identical; *cf.* note 25.
[20] Ed. Ian-Mayhoff vol. I p. 14,36 and p. 22,44 respectively.
[21] Ian-Mayhoff, vol. II p. 68,4 (VII,193).
[22] Schnabel (*Ber.* p. 118) guesses that this whole passage comes from Posidonius, the convenient repository for all sources which cannot be localized.

[23] Kroll-Skutsch, vol. I p. 196,24.
[24] Book II contains the horoscope No. L 382.
[25] The first half of this long quotation is published *CCAG* 8,2 p. 64,24 to p. 65.4. The same text plus a second part of about the same length is preserved in a cod. Vindob., quoted *CCAG* 6 p. 14 F. 302ʳ (checked on photo). This is Schnabel's fragment No. 14. The same text, once more, F. 303ʳ and again F. 341ʳ (Schnabel's fragment No. 15).
[26] *CCAG* 1 p. 130,19 ff.
[27] Vett. Val. III,12 (p. 150,11 ff.) and IX, pref. (p. 329,18 ff.) = *CCAG* 1 p. 79,15 ff. and *CCAG* 5,2 p. 48,30 ff.
[28] *RE* 11,2 col. 1928(4) ff.
[29] *Lux perpetua* p. 311 ff.
[30] Kroll p. 142,21 ff. to p. 143,4 = *CCAG* 1, p. 36, F. 303ʳ; *CCAG* 2, p. 73, F. 318ʳ; *CCAG* 5,2 p. 6 (III,7).
[31] *Cf.* p. 82 ff.
[32] Kroll p. 233,22.
[33] Kroll p. 234,31 f. *Cf.* for the Latin version of this chapter Gundel, *HT* p. 39,24 and p. 335 f. Only the horoscope No.

These two chapters contain the horoscopes Nos. L 37, L 92,II, and L 104,VII. The investigation in No. L 37 is extended to the year 68.

(B) Chapter III,12 in which the first reference to the ὅρασις occurs,[34] is followed in III,13 by examples concerning the doctrines about the length of life. The same problem (χρόνοι ζωῆς) is discussed in IX,8 which contains two references to Critodemus;[35] as mentioned before, the introduction to book IX provides us with the second reference to the ὅρασις.[36] Chapter III,13 contains three closely related horoscopes: L 74,IV for the clima of Alexandria and recording a death in May of 144; L 115,II for a person who died in July 147; and L 75,1 with the age of 76 years at death.

(C) The anonymous author of a section in cod. Paris. 2425[37] has given us a short table of contents from some work of Critodemus, the title of which is not given. At the end the scribe intended to mention the date of Critodemus but failed to insert it.[38] Cumont realized that these chapters correspond to Vettius Valens IV,17 to 24. This is supported by the fact that Critodemus is given as author by Vettius Valens for IV,26[39] dealing with the length of life. Unfortunately, no horoscopes are included in these chapters.[40]

(D) Chapter II,41 on "violent death" contains a series of eleven horoscopes, dates of which range from 65 to 123.[41] Nine of these horoscopes (now ranging from 65 to 115) are also contained in a Vatican codex,[42] but now placed at the end of Book IX though with the remark[43] "end of Book II" as is indeed the place of II,41. The title of this chapter is now[44] "On the yielding of chronocratorship of the planets; (chapter) 2: the assignments. From the (books) of Critodemus." There is also one more horoscope given at the beginning of this version than given in II,41, namely, No. L 112,VIII.[45] It is found in Kroll's edition of Vettius Valens at the end of II,36,[46] which is its proper place since it does not concern a case of violent death but a case of sickness, the subject of II,36. The evidence of such disorder in the Paris version combined with the fact that its title does not fit the cases of "violent death" of the horoscopes discredits the attribution to Critodemus. On the other hand a doctrine on violent death is ascribed to Critodemus by Rhetorius.[47]

(E) Gundel identified in the "Liber Hermetis Trismegisti" which he had discovered, two sections which contain enlarged versions of chapters assigned by Greek texts to Critodemus. The first is Vettius Valens V,12, mentioned above in (A).[33] The second is a discussion of the "terms," the Greek version of which is published in CCAG 8,1;[48] its greatly enlarged version is chapter 25 of the Liber Hermetis.[49] No horoscopes are contained in these chapters.

Summarizing the impression one obtains from all these passages which are ascribed to Critodemus, one would hardly discover any significant difference between them and the rest of the Anthology. The dates of the horoscopes connected with these sections would point to an author perhaps fifty years older than Vettius Valens. All this is in flagrant contradiction to a date before Pliny and to the stylistic character of the ὅρασις. The obvious solution of this dilemma, however, the assumption of two persons called Critodemus, finds no support whatever in the way in which "Critodemus" is quoted by Vettius Valens and everywhere else. The chronology of early astrology is still full of unsolved problems.

C. ANTIGONUS OF NICAEA

The date of this author is determined[50] by the fact that he refers in his discussion of the horoscope of Hadrian (No. L 76) to the death of this emperor (A.D. 138), while his writings are quoted by Porphyry in his "Introduction" to the Tetrabiblos[51] late in the third century.[52] No sharper limits seem to be obtainable from the astrological literature.[53]

Kroll assumed[54] identity of the astrologer with a

L 104,VII would fall within the portion covered by the Latin version of V,12, but it is omitted.

[34] Kroll p. 150,11 f.
[35] Kroll p. 348,9 = CCAG 5,2 p. 124,24 and Kroll p. 350,31 = CCAG 5,2 p. 126,34.
[36] Cf. note 27.
[37] CCAG 8,3 p. 102 (cf. also p. 91); CCAG 8,4 p. 38.
[38] CCAG 8,3 p. 102,16 and note 2.
[39] Kroll p. 202,26 f. (omitted in his index p. 374) = CCAG 5,2 p. 7 (IV,26).
[40] The earlier parts of Book IV belong undoubtedly to Vettius Valens himself and were written in 154/156.
[41] Cf. the list given in the introduction to No. L 65,V.
[42] CCAG 5,2 cod. 15.
[43] CCAG 5,2 p. 10 (IX,20).
[44] CCAG 5,2 p. 120,28 f.
[45] CCAG 5,2 p. 120,30-32.
[46] Kroll p. 114,16-18.

[47] CCAG 8,4 p. 199,15-202,10 (cf. also p. 117 next to last paragraph). The same in CCAG 2 p. 43 F. 103ʳ and in CCAG 8,1 p. 75, F. 15.
[48] CCAG 8,1 p. 257,21-261,2. Cf. also CCAG 2 p. 6 F. 204ʳ; 5,1 p. 40 F. 160ʳ and p. 45 F. 184ʳ; 5,2 p. 76 F. 88ʳ; 8,1 p. 105 F. 138.
[49] The parallel passages are listed by Gundel, HT, Index p. 357.
[50] Cumont in CCAG 8,1 p. 242 note 1. There are three articles on Antigonus in Pauly-Wissowa: vol. 1, col. 2422 (No. 24) by Riess, Suppl. 1 col. 90 by Cumont, Suppl. 5 col. 2 by Kroll.
[51] Ed. Boer-Weinstock, CCAG 5,4 p. 223,18.
[52] Porphyry died between 301 and 305.
[53] Aside from the three horoscopes (Nos. L 40 L 76 L 113,IV) in Hephaestion and the passage in Porphyry only the following references are known: from Palchus, CCAG 1 p. 80.24 = CCAG 5,1 p. 205,14; CCAG 1 p. 107,4,13 = CCAG 6 p. 64,14,23; from Rhetorius, CCAG 2 p. 188,23; CCAG 8,1 p. 242,15 f.; from Lydus, CCAG 1 p. 81,8; from an anonymous author who wrote near the year 1000, CCAG 2 p. 44 F. 111.
[54] RE Suppl. 5 col. 2.

physician of the same name, who probably belongs to the time of Hadrian. We see no gain in such hypothetical constructions.

Kroll also ascribed the fourth version of No. L 118 to Antigonus without realizing, however, that there existed three more versions of the same horoscope in the work of Vettius Valens. Thus No. L 118 has no bearing on the problem of dating Antigonus.

D. THE LATE AUTHORS

The gap in our material of about two centuries between Vettius Valens and Hephaestion is probably caused by the accidents of preservation. With Palchus and Rhetorius, however, as Cumont has pointed out,[55] we come near to a period in which astrological practice was severely suppressed until its revival in the eighth century. The horoscopes of the fifth century are therefore probably the latest texts which are directly connected with the origin and development of Hellenistic astrology.

1. HEPHAESTION

The horoscope No. L 380 seems to be the only contemporary one found in the writings of Hephaestion.[56] It is contained in Book II in the first chapter "On conception and childbirth" and is partially repeated in the next chapter and in II,11.[57] It is the sole basis for the date of Hephaestion.[58] In *CCAG* 8,2 p. 86,21-23 three longitudes of stars are quoted from Book II,18. They compare as follows with the longitudes given in the Almagest:

	Hephaestion	Almagest
Regulus:	♌ 5	♌ 2;30
ν Aquarii:	♒ 20	♒ 17;40
α Tauri:	♉ 15	♉ 12;40

An increase of about $2\frac{1}{4}°$ between Ptolemy (A.D. 140) and Hephaestion would mean a time difference of about 250 years if Hephaestion had derived these longitudes from Ptolemy by using his constant of precession, counting longitudes from the vernal point. The closeness of the resulting date, about 390, to the date of our horoscope seems to support the above-mentioned relationship to the Ptolemaic catalogue of stars and a date near A.D. 400 for the life of Hephaestion.

2. PALCHUS

The lifetime of Palchus is known to us within comparatively narrow limits from horoscopes in which he refers to statements made by himself, correct ones as well as an incorrect prediction.[59] The earliest of these references is found in No. L 475, the latest in No. L 487. With the exception of No. L 474 all other horoscopes preserved in the writings of Palchus belong to this interval.[60]

From No. L 479 it seems that Palchus was in Smyrna at the time of the arrival of the ship from Alexandria about which the "inquiry" was made. No. L 474 concerns a ship sailing from Caesarea (in Palestine) to the Hellespont, No. L 475 deals with a journey to Athens, No. L 484 with an event in Antioch. Only No. L 486 concerns events in Alexandria, though without a direct indication of the presence of the writer.[61]

The fact that three horoscopes concern travel by ship, while No. L 487 discusses the prediction of the contents of a letter, is of interest since these topics correspond to two chapters in the work of Palchus which he took from a much earlier treatise which goes under the name of Zoroaster: Περὶ ἐπιστολῆς ἀναδόσεως and Περὶ πλοίων ἀναγωγῆς.[62] Apparently ancient doctrines were illustrated by up-to-date examples.

Except for the first (No. L 474) and the last (No. L 487) horoscope, all planetary positions are given in degrees only. In No. L 474 half-degrees occur for the planets, minutes for the moon. In No. L 487 degrees and minutes are given for the longitudes of Saturn and Jupiter but their correctness is not free from suspicion.

The comparison with modern values of the longitudes given by Palchus, particularly for sun, moon, and nodes, shows deviations which place his zero point of longitudes close to ♈ 3° of the modern norm (cf. figure 37).

3. RHETORIUS

Cumont used a list of stellar positions [63] for the determination of the date of Rhetorius and found [64] A.D. 500.[65] This agrees well with the dates Diocletian 145 — A.D. 428 and Diocletian 179 — A.D. 463 given in two of the seven horoscopes which are preserved under the name of Rhetorius. To the same period belong two astronomically dated horoscopes, No. L 482 concerning a "non-viable" child, and No. L 440 for the grammarian Pamprepius, who died 44 years old, i.e., in 484. There remain three more horoscopes of very uncertain date, namely, Nos. L 401, L 488, and L 516.

---

[55] Cumont [11] p. 53 f.; Boll [4] p. 105.

[56] Hephaestion's work contains in addition the three early horoscopes Nos. L 40, L 76 (Hadrian), and L 113,IV (Pedanius Fuscus) which were taken from Antigonus of Nicaea.

[57] Cf. the second and third versions, respectively.

[58] Boll in *RE* 8 col. 309,60 ff. and *CCAG* 8,4 p. 246 note.

[59] Cf. Cumont [11]; for Palchus in general cf. Cumont [12].

[60] Nos. L 478, 479, 483, 484, and 486.

[61] Cumont [12] p. 3 and p. 6 states that Palchus was an Egyptian according to his own statement. Cumont himself, however, realized later on that the passage in question belongs to the "Anonymous of the year 379" (*CCAG* 5,1 p. 194 f. and p. 204,18). Cumont also states that the name Palchus points to Egyptian origin. This may be so but the Egyptological proof is lacking.

[62] Published by Bidez-Cumont, *Mages* II, pp. 209-224, and assigned to an early period of astrological literature because of the use of the older planetary names, Φαίνων, Φαέθων, etc.

[63] *CCAG* 5,1 pp. 219-226.

[64] Cumont [11] p. 43; also Boll, *Sphaera*, p. 15.

[65] Another list leads to A.D. 484; cf. Cumont [11] p. 43 and p. 47.

188   THE AUTHORS OF THE LITERARY HOROSCOPES

The horoscope No. L 516 refers to clima 5 (Hellespont), while Nos. L 401 and L 488 give clima 4 (Rhodes) as geographical latitude, though in fact No. L 488 uses values which belong to clima 6.[66] No. L 440, the horoscope of Pamprepius, who was born in Panopolis in the Thebais[67] is based on a latitude between clima 2 and 3 as should be the case for the latitude of Thebes. Nos. L 482 and L 516 are based on clima 5, while No. L 463 was probably computed for Byzantium. Cumont collected [68] all the evidence for an Egyptian origin of Rhetorius but even the seemingly most solid argument, the use of rising times valid for Alexandria,[69] actually supports strongly Cumont's suspicion that the whole passage is lifted from Antiochus.[70] The rising figure 38), ignoring, of course, gross deviations caused by some error of computation or copying.

4. EUTOCIUS

As far as we were able to establish from *CCAG* the text of the horoscope No. L 497 occurs in three manuscripts

V: Vindob., Cod. philos. gr. 179 — *CCAG* 6 Cod. 3
(cf. p. 32)

M: Mutin., Cod. 85 — III, C, 6 — *CCAG* 4 Cod. 11
(cf. p. 32)

F: Florent., Plut. 28, cod. 34 — *CCAG* 1 Cod. 12
(cf. p. 69).

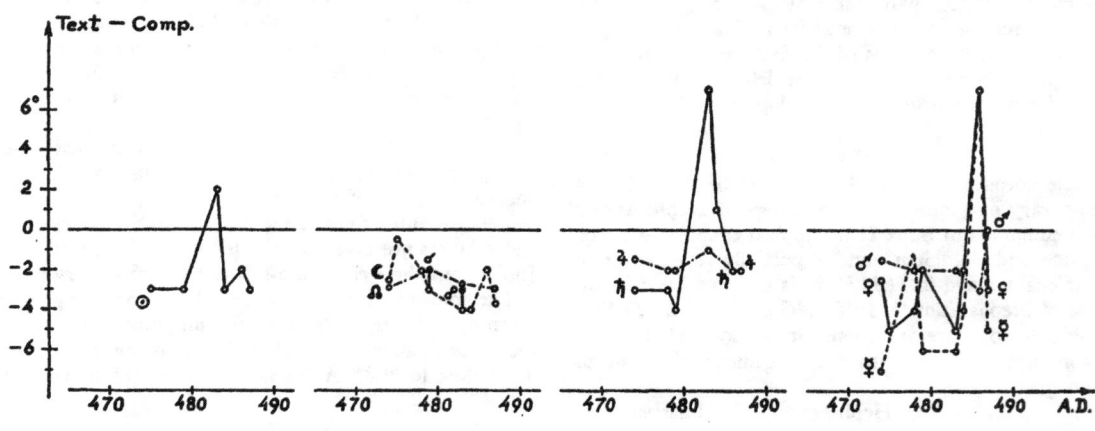

FIGURE 37.
Palchus.

times in question follow the primitive arithmetical scheme of System A, whereas Rhetorius operates in all his horoscopes with the accurate values of the Handy Tables.[71]

The three earlier horoscopes (Nos. L 401(?), 428, 440) give the longitudes in degrees and minutes, while the later ones (Nos. L 463, 482, 488(?), 516(?)) use degrees only, with the exception of degrees and minutes for the sun in Nos. L 482 and L 488, and for the full moon in No. L 463. The deviations between ancient and modern computation are about the same as in the horoscopes of Palchus, i.e., between $-2°$ and $-3°$ (cf.

Cumont, who did not know that F was a parallel to V and M, used the two latter manuscripts to establish a text which we call here the "second version" in contrast with F which is our "first version." Cumont's text is published in *CCAG* 4 pp. 103 to 110 with a long introduction (pp. 99 to 103). Of F only the beginning is published (*CCAG* 1 p. 170 f. F. 140^v), the main part being available to us in a photostat (cf. pl. 32).[72]

In V our horoscope forms part of "Chapter 2" of a treatise called *Astronomical Investigation of Julianus of Laodicaea*. In M, however, Paulus Alexandrinus is given as author and Julianus appears as author of one chapter only (F. 83, corresponding to Chapter 3 in V) which is taken verbatim from the *Tetrabiblos*.[73] Cumont

---

[66] Middle Pontus; cf. p. 5.
[67] Cf. *RE* 18,3 col. 409,56 ff.
[68] Cumont [11] p. 44 f.
[69] *CCAG* 1 p. 163,4-14.
[70] Cf. Cumont [11] p. 44 note 7.
[71] Cf. also Rhetorius' own words, *CCAG* 1 p. 152,9-16 where he opposes the rising times of Ptolemy, which are "accurate to minutes," to the "cruder" schemes "of the Egyptians and Valens."

[72] To complicate matters, the folio numbers are no longer the same as those given in *CCAG* 1 p. 69 but one higher. Our references give the printed, not the secret, numbers. The number 142a visible on pl. 32 corresponds to F. 141^r of *CCAG* 1.
[73] Cf. *CCAG* 4 p. 109.

emended M and published a combined text [74] under the title as it appears in V. He realized, however, that the longitudes of the fixed stars require a date near 500 whereas he assigned Julianus a date 120 years earlier.

author. In all probability Julianus (in V) as well as Paulus (in M) has nothing to do with the chapter which contains the horoscope and one may only hope that the reference to Eutocius is more trustworthy. One

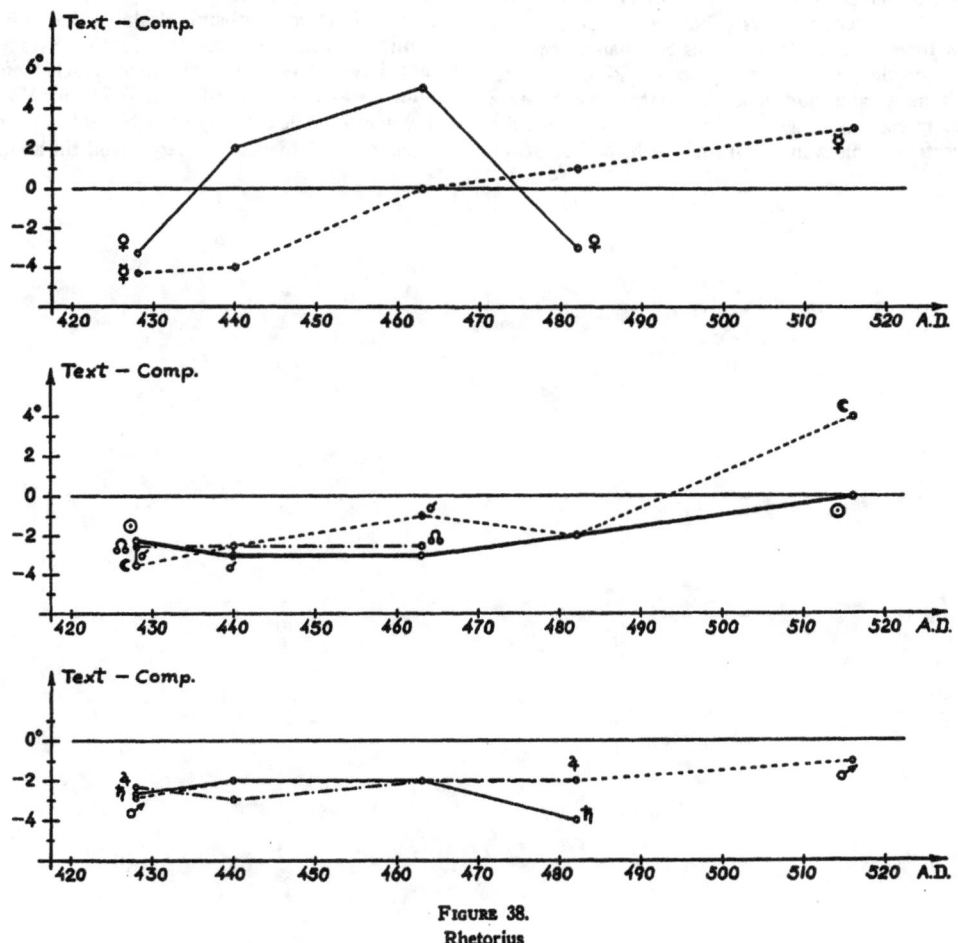

FIGURE 38.
Rhetorius

The version of F not only gives the exact date, Diocletian 214 — Philip 821 [75] — A.D. 497, fully confirmed by the astronomical elements, but it quotes Eutocius as

could then at least conclude that Eutocius had not died in 497.

In other words, the date of the horoscope sheds no new light on the difficult question of the lifetime of Eutocius as derived from the dedication of his commentaries to Archimedes and Apollonius.[76]

---

[74] Using also Angelicus 29 (*CCAG* 5, 1 Cod. 2; *cf.* p. 32) which seems not to contain the section with the horoscope. A parallel is the Florent., Plut. 28 cod. 33 = *CCAG* 1 Cod. 11, F. 247 (p. 56).

[75] Kroll ([5] p. 128), who first drew attention to this horoscope, thought he had found some discrepancy between these two dates but was corrected by Cumont (Kroll [5] p. 192 but read "Zu S. 128 f. 140"). Unfortunately, Kroll expressed himself inaccurately by saying that 498 was the corresponding date in our era (instead of 497/498) whereas Cumont (*CCAG* 1

p. 170 note 1) incorrectly equated it with 498/499. Cumont [11] p. 47 note 1 gives correctly 497 as date of the horoscope, associating it, however, with Julianus.

[76] *Cf.* Tannery, *Mém. Sci.* 2 p. 118 ff. He assumes (p. 121) that Eutocius was born about 465. Hultsch in his article, *RE* 6,1 col. 1518 uses simultaneously the contradictory assumptions of Heiberg and Tannery.

5. STEPHANUS

Usener, who edited the treatise which contains the horoscope No. L 621, assumed [77] that its author, Stephanus, was the neoplatonist Stephanus of Alexandria who was a contemporary of the emperor Heraclius (610 to 641). He realized, of course, that the horoscope itself could not have been written by this Stephanus, since it reveals a knowledge of events up to A.D. 775, and considered it as a later addition. Cumont, however, was undoubtedly right in rejecting this hypothesis since the whole treatise forms a unit [78] from which the horoscope cannot be separated. He therefore identified the author with a " Stephanus the Philosopher " of whom an astrological treatise " On the mathematical art " is, at least in part, preserved and the author of which came from Persia to Constantinople.[79]

Since this later Stephanus declares that this art was transmitted from the Romans to the Saracens,[80] he cannot have made this statement much before 800. Baghdad was not founded until 762, and Māshā'allāh, with whom Islamic astrology can be said to have become the heir of Hellenistic astrology, lived till about 810.

---

[77] *Kleine Schriften* III p. 247-322 (Bonn 1880).
[78] *CCAG* 2 p. 181.
[79] *CCAG* 2 p. 182,2.
[80] *CCAG* 2 p. 182,29 f.

GLOSSARY

A. GREEK

Numbers, separated by a colon, indicate text-number and line, as published here in Chapter I (p. 14 ff.). Numbers preceded by L refer to literary sources, translated in Chapter II (p. 76 ff.). In these cases reference is made to the original publication of the Greek text, particularly to *CCAG* (quoted as C, followed by volume number, page and line) and to Kroll's edition of Vettius Valens (quoted as V, followed by page and line). Example: L 135 I V 156:17 means: Horoscope No. L 135 I, Vettius Valens p. 156, line 17, ed. Kroll.

Square brackets [ ] indicate doubtful or supplied readings.

The glossary is complete for the original sources. From the literary horoscopes selective references are made only to technical and unusual words. Additional indices are found in *CCAG* and in Vettius Valens, ed. Kroll. Occasionally we have listed words which do not occur in the passages translated here; then no L-number is given.

For the explanation of technical terms *cf.* Introduction p. 2 ff.

1. General Greek Glossary

ἀ see ὁς
ἀγαθοδαιμονέω be in Agathos Daimon L 62 V 87:2
ἀγαθός 81:212; 137a:[1]; 137b:43; 227:1; 277:1; 282:1; 284:1; 316:1; 320:1; 326:1; 338:1; 345:2
αγαθυ 376b:31
ἀγαθὸς δαίμων 95:72
ἀγαθοτυχέω be in Locus Good Fortune L 50 V 83:20
ἄγαν 81:137
ἀγαπητός 15/22:[2]
ἄγνοια 81:181
ἄγω 81:43
ἀγωγή 138/161:5
'Ἀδριανός (month) 137a:4; 137b:[3]; 137c:8
αθυρ 284:6, verso [2]
αἰγόκερως 95:[70]; 137a:16, 36; 137b:16, 35; 137c:19; 138/161:4; 177:[4]; 219 II 12:7; 220:7; 250,2:21; 258:[9]; 277:6; 283:7; 284:12; 293 XII:5; 326:10; 353: diagr.; 366:8; 431:6; 478:8
αγογερω 182:1, 3
αγωγερω 182:7
αιγ 176; 219 I a; 219 I b: diagr.; 219 I c: diagr.; 219 I d: diagr.; 219 I e; 250,1: diagr.
αιγο 15/22: diagr.
αιγογαιρω 207:3
αιγογερως 46:2, 14; 338:6, 10, 12; 351:17; 370:34; 373:7; 376a:16; 376b:37
αιγοκ 217:2; 219 II 1:7; 260:[12]
αιγοκερο 149:5
αἰγύπτιος 81:1; 137c:9; 187:5; 207:9
αἵρεσις configuration (of stars) L 412 Marinus (Cousin) 61:8
— διά division by L 123 I V 231:15
— division (of periods) L 135 I V 156:17

— method V 237:35
— sect L 50 V 83:16
αἱρετικός of the sect L 463 C 8,4 225:4
αἰσθητός perceptible, actual L 428 C 8,1 222:8
αἰτιατικὸς τόπος responsible locus L 123 I V 231:6
αἰώνιος 81:13
ἀκίνητον (ζῳδιακόν) fixed zodiac L 428 C 8,1 222:7
ἀκριβῶς 81:17
ἀκρονυχία opposition L 497 note 35
ἀκρόνυχος in opposition L 87 VII V 129:13
ἀκρώνυκτος 260:5
ἀκτινοβολέω project rays, radiate L 113 IV C 8,2 86:5
ἀκτινοβολία projection of rays, radiation L 75 V 142:4
ἀμαρτύρητος not in aspect L 78 V 93:30
ἀμαυρός dim, faint L 112 VII V 229:35
ἀναβαίνων ascending node L 478 C 6 65:2
ἀναβιβάζω 81:87, 128
ἀναβιβάζων ascending node L 74 IV V 154:3
ἀναγκαῖος 15/22:1
ἀνάγκη
  αναγκη 137c: frag. 3
  — (κλῆρος) 138/161:8; 338:16
ἀναδίδωμι 284:2
ἀναίρεσις destruction, destructive influence L 75 V 142:6
ἀναιρέτης destroyer, destruction L 87 VII V 129:11
— (κλῆρος) Lot of destroyer L 440 C 8,4 221(b):12
ἀναιρετικός destructive L 488 C 8,1 237:24
ἀναιρέω be destructive, destroy L 76 C 6 68:8
ἀναλογία 95:60
ἀναμφίβολος 81:29
ἀνάπαλιν 81:179
ἀναποδίζω subtract L 114 V V 147:17
ἀναποδισμός 46:9
ἀνατέλλω 137c: II 9
ἀνατολή 81:74; 137a:21; 137b:[21]; L 497 notes 37, 38
ἀνατολικὸν κέντρον rising center, Horoscopos L 412 Marinus (Cousin) 64:35
ἀνατολικός 81:103, 147; 137a:9; 137b:[8]
ἀνατρέχω 81:109
ἀναφέρω 95:80
ἀναφορά rising time L 75 V 156:11
ἀναφορᾶς, ἐπί in epanaphora L 72 V 85:5
ἀναφορικός 95:60
ἀνήρ 81:[2]
ἀνοικοδεσπότητος without house ruler L 75 V 141:32
ἀντίκειμαι be in opposition L 61 X V 143:15
ἀντιπαράδοσις succession (of chronocratorship) L 122 I 22 V 230:27
ἀνωμαλία 345:11
ἀνωφερῶς upwards (in direction of daily rotation) L 74 IV V 154:4
ἄοιμος 95:113
ἅπας 138/161:10
ἀπεργάζομαι effect L 40 C 8,2 85:9
ἀπέχω 81:155
ἀπλανὴς ἀστήρ fixed star L 497 C 4 106:12
ἀπλῶς 81:23
ἀπό 81:51, 74; 95:100, 103, 126, 134; 137c:2, 3, 5; 138/161:15; 326:3; 385:[21]; 465:1
ἀπόκλιμα apoclima L 83 V 84:15
ἀποκλίνω be in apoclima L 61 V V 86:23
ἀπόκλισις apoclima L 75 V 169:1
ἀπόκρουσις 338:17

GLOSSARY

ἀπολείπω 81:14
Ἀπόλλων see Στίλβων
ἀπόλυσις
  απολυε 95:115
  — outcome L 123 I V 231:22
ἀπορρέω leave contact with L 483 C 6 66:12
ἀπόστασις elongation L 484 C 6 66:26
ἀπόστροφος turning away L 75 V 168:4; L 87 VII V 129:12; L 91 V 130:20; L 101 III V 94:8; L 114 V V 147:21; L 123 VII V 128:37
αποτ[137c:II 14
ἀποτέλεσμα prognostic, portent L 102 XII 14 V 284:1; L 118 C 2 168:27
ἀποτελεσματικός 95:[82]
ἀποτελέω have effect; have influence; forecast; portend 95:111; L 40 C 8,2 85:7; L 106 V 113:6; L 114 XI V 231:3; L 134 XI V 228:8; L 482 C 8,1 240:27
ἀποτέμνω 81:152, 167
Ἀπρίλειος 81:44
ἀργὸς τόπος locus of trouble L 497 notes to diagr., C 8,2 68:26
Ἄρης —9:[5]; —3:[8], 17; 15/22:11, [12], 15, diagr.; 32:7; 46:7, 11, 18; 50,1:4; 50,2:[3]; 81:59, 118, 126, 169; 95:[28], 42, 54, 67, 74, 76, 78, 111; 125:9; 137a:13, 16, 38, 41; 137b:13, 16, 36, [39]; 138/161:[5], 8; 139:7; 149:7; 150:5; 177:4; 182:2; 187:9; 190:7; 207:3; 217:5; 219 I b:diagr.; 219 I c:diagr.; 219 I d:diagr.; 219 I e; 219 II 1:6; 219 II 12:6; 220:5; 244:2; 245:7; 250,2:10; 258:4; 260:7; 277:5; 283:5; 284:14; 293 VIII:6; 293 XII:[5]; 316:10; 320:8; 326:8; 338:10, 21, 23, 25, 29, 30; 345:8, 9; 351:[15]; 353:diagr.; 366:10; 370:37; 373:8; 376a:16; 376b:37; 381a:26; 381b:34; 385:25; 431:8; 465:6; 478:9
ἀριθμός 95:44; 250,2:[24]
ἀρρενικός 81:61
  αρσενικος 15/22:8, [10], 12, 14
ἀρχαῖος —9:[2]; 15/22:6; 81:45; 137b:3; 137c:3; 145:2
  αρχαι̊ 137a:4
  αρχεος 161:6
ἀρχή 95:113; 137b:4; 219 I b:diagr. (?); 227:4; 282:3; 293 VIII:8
  αρχ 137a:5; 245:5
ἄρχω 190:2
ἀσελγής vicious, lascivious L 40 C 8,2 84:26; L 116 V 113:23
ἀστήρ 81:97, 109, 122, 127, 132, 139, 154, 158; 95:82; 137a:43; 137b:42; 137c:II 7, II 10; 138/161:11, 12; 250,2:[21]; 284:9
ἀστροθεσία configuration of the stars L 102 IV b V 157:24
ἀστρολάβος astrolabe L 621 Usener III 273:12
ἀστρολογία 81:27
Αὐδυναῖος
  αυδεναιου 219 I d:1
  αυδιναιου 219 I b:1
  αυδυναιου 219 I c:1; 219 I f:1
αὐξιφωτέω wax L 40 C 8,2 85:8
αὐτός 81:15, 94; 137a:42; 137b:41; 137c:5, 15, frag. 5; 219 II 12:1; 345:14
αφ.ερουμε 95:96
ἀφαιρέτης retrograde C 4 107:13
αφαιρετι[κο]ς 260:6
ἀφαιρέω 250,2:[23]
ἄφεσις start; starting point; beginning (first chronocratorship) L 75 V 142:3; L 110 III V 142:13; L 516 C 8,1 231:7
ἀφέτης starter; initiator; first chronocrator L 61 X V 144:9; L 75 V 142:1; L 110 III V 142:12
ἀφετικός starting L 110 III V 142:14
ἀφθόνως 81:[11]
ἀφίστημι be distant from L 37 V 233:29
Ἀφροδίτη —3:[9]; 15/22:[8], [9], 11, 13, 16; 32:[3]; 46:5, 13, 15; 50,2:[4]; 50,3:[2]; 81:81, 102, 138, 182; 95:29, 32, 38, 39, 48, 57, 63, 67, 68, 73, 79, [80], [83], 93, [103], [112]; 137a:[7], 15, 20, 25, 27, 32, 34, 43; 137b:6, 15, 20, 24, 27, 32, 34, 37, 40, 41; 138/161:1, [7]; 149:4; 171:[5]; 177:[4]; 182:3; 187:[10]; 190:7; 207:4; 217:3; 220:6; 245:6; 250,2:2, [16]; 258:5; 260:9; 277:7; 282:5; 284:15; 293 VIII:5; 316:11; 320:[9]; 326:10; 345:7, 14; 353:diagr.; 373:7; 431:5; 478:[10]
αφρο 219 I c:diagr.; 219 I e
αφρο̇δ 81:195; 137a:38, 41
αφροδ 150:6; 219 I a; 219 I b:diagr.; 219 I d:diagr.; 219 II 1:7; 219 II 12:7; 244:2; 283:8
αφροδει 137a:32, 35
αφροδειτ 184:8
αφροδιδη 465:[7]
αφροτιδη 338:11, 24, 26, 28; 351:[16]; 366:7; 370:36; 376a:15; 376b:36; 381a:25; 381b:33; 385:27
αφροτιτη 125:10

βαθμός step 95:[25]; L 497 C 4 106:12
β(αθμός) L 497 note 12
βάλλω send (rays), radiate L 110 III V 142:15
βασιλεύς 137c:4
— (τόπος) Locus of King L 107 V 229:3
βασιλία 284:5; 478:3
βασιλικὸς τόπος Locus of King L 134 XI V 228:9
βάσις basis L 78 V 93:29
βίβλος 137c:2
βίος 137c frag.:2, 9
βλέπω be in aspect to L 478 C 6 65:16
Βορειάς 81:62
βορειότερος 81:122
βραβεῖον 138/161:13

γαμοστολικός (τόπος) Locus of marriage L 111 IV V 228:29
γαμοστόλος 137a:28; 137b:28
γάρ 81:27; 95:114; 345:12
γείνομαι see γίγνομαι
γενέθλη nativity L 80 Manetho III:739
γένεσις 15/22:2; 95:[72], 79, 107; 137a:1, 42; 137b:1, 41; 138/161:[17]; 150:1; 177:1; 227:[2]; 277:1; 282:2; 283:1; 316:1; 320:[1], 11; 326:2; 338:2; 345:2; 351:[11]; 366:[1]; 370:29; 373:1; 376a:11; 381a:21; 381b:30; 385:20
γῆ see ὑπὸ γῆν
γίγνομαι 81:2; 95:106, 108, 111; 219 I b:3; 219 I d:3; 219 f:2; 284:3; 338:18
γεινεται 138/161:15
γεινησεται 95:[110]
γνησίως 81:3
γνώμων (ἡλιακός) (solar) gnomon L 419 V 365:34
— (ὡροσκοπικός) (horoscopic) gnomon L 419 V 365:34
γνῶσις 81:15
γονειςιος 95:109
γονεύς 95:109; 137c: frag. 8
γράφω 219 I:2; 250,1:[2]
γυνή 95:112

δαίμων (κλῆρος) 138/161:4; 338:14
  ἀγαθὸς δαίμων 95:72
Δαίσιος
  δαισιον 219 I:1
δάκτυλος 81:123
δέ 15/22:4, [5]; 81:43, 46, 73, 96, 108, 126, 138, 157, 165, 173, 177; 95:[21], [26], 34, [40], 45, 59, 105, 106, 108, 111; 137a:4; 137b:3; 137c:4, 9, 15, fragm. 4; 138/161:11; 161:6; 187:5; 284:9; 345:14
δέκα 81:112, 160
δέκα ἕξ 81:[129], 141
δέκα ὀκτώ 81:168
δέκα τέσσαρες 81:54
δέκα τρεῖς 81:78

GLOSSARY

δεκανός 81:66, 89, 171; 95:[29], 35, [41], [47], [52], 58, 65
δεκατεύω be in superior aspect L 487 C 1 106(b):11
δέκατος 81:57
δεσπότης ruler (of place in zodiac) C6 69:17
δεύτερος 81:47, 66, 89, 119; 95:115; 250,2:[14]
δή 81:113, 143
δι see δίδυμοι
διά 81:12, 162; 95:59
διάθεμα configuration; nativity 81:164 L 76 C 6 68:4
διάθεσις configuration L 412 Marinus (Cousin) 61:6
διακατέχω hold rule L 158 V 284:23
διακρατέω rule (as chronocrator) L 488 C 8,1 236:9
διαμετρέω be at diameter; be in opposition to L 61 X V 143:11
διάμετρος diametrical L 61 X V 143:18
διανοητικὸς τόπος Locus of intellect L 123 I V 231:30
διάστασις interval L 37 V 233:25
διάστημα distance L —71 C 8,4 237:9
διασφάλλω 81:198
διατεκμαίρομαι determine; mark out L 80 Manetho III:750
διαυγάζω 81:70
δίδυμοι 15/22:diagr.; 32:4; 46:19; 95:43; 137a:28; 137b:28; 139:5; 177:[3]; 180:8; 190:8; 207:1; 217:5; 219 II 12:8; 220:9; 244:[3]; 293 VIII:[8], 9; 293 XII:4; 316:9; 326:7, 11; 338:14, 18; 345:7; 353:[diagr.]; 376a:18; 376b:[39]; 381a:27; 381b:35; 385:[24]; 465:[6]
διδ 176; 219 I a; 219 I b:diagr.; 219 I c:diagr.; 219 I d:diagr.; 219 I e; 250,1:diagr.
διδημης 431:10
δίδωμι 15/22:3; 137c:7
διέπων ruler of the hour; associate ruler L 479 C 1 104:4; L 487 C 1 106(b):13
διευτυχέω 81:184; 138/161:[16]; 277:10; 283:9
διμ(ερισμός) 338:19 to 30
διοδεύω pass through L 75 V 167:22
διοικέω 81:10
διπλοῦς 345:10
διπ 81:199
δισμυροχιλεξακοσιοστός 81:114
δίσωμον (ζώδιον) bicorporeal (sign) L 124 V 281:19
δοξοτη see τοξότης
δορυφορέω attend; accompany; escort L 40 C 8,2 85:3; L 101 III V 94:5
δοῦλος 85:105
δούλων, τόπος περὶ Locus of slaves L 111 IV V 228:25; L 112 VII V 229:18
δρόμος 137c:[15], II 12
δυάς second interval L 37 V 233:29
δυνάστης 81:51
δύνω —3:13; 15/22:15; 46:[19]; 250,2:[11]
δύο 81:123, 156
δυσικός 95:70
δύσις 250,2:[25]
δυστύχημα 138/161:[10]
δυτικόν setting point; setting sign L 478 C 6 65:3
δωδεκατημόριον 81:68, 91; 95:[19], [27], 33, [39], [46], [50], 56, 62
δωδέκατος 137c:14
δωρέω 81:181

ε[15/22:2; 250,2:20
ἐάν 95:105, 106; 137c:fragm. 5; 345:14
ἐαρινός 81:51
ἑαυτόν 345:10
ἑαυτός itself, i.e., its own house L 123 I V 231:31
ἑβδομαία seventh day (of moon); seventh day (from Venus) L 479 C 1 104:8; L 483 C 6 66:9
ἔγγιστος
ενγιστωι 46:7
ἐγώ 15/22:[3]; 81:11, 33; 137c:3, 5, 7; 258:1; 284:2
εἰ 137c:fragm. 6

εἶδος figure L 487 C 1 107:14
εἴδωλον 81:136
[εικοσι]εκτ[α] 187:1
εἰκοστός 81:129
εἰμί (see also τουτέστι) 81:114, 145, 180; 95:72, 105; 137c:6, 13, 16; 284:9; 345:13
εἰς 81:178
(a) εις (b) (in dates) —9:2; 15/22:6; 81:47; 138/161:[18]; 139:3; 149:2; 150:3; 177:[2]; 184:2; 219 II 1:2; 219 II 12:2; 338:4; 366:[4]; 370:31; 373:3
εἴσειμι 345:12
εἰσχωρέω 95:[110]
εἶτα —3:[13]
ειχθυες see ἰχθύες
ἐκ 95:60
ἔκαστος 81:16, 25
ἐκδέχομαι συναφήν receive contact L 483 C 6 66:8
ἐκλάμπω 81:95
ἐκλειπτικόν eclipse L 486 C 1 101:5
ἐκπαραπίπτω be unfavorably located L 78 V 93:30
ἐκπίπτω 95:61
ἕκτος 81:40
εκχθρος see ἐχθρός
ἐλάχιστα, τὰ minimum period L 75 V 156:10
ἐλάχιστος 137c:11
ἐλευθερόω 95:[105]
ἐλλήγω
ευληγω 95:50, 56
Ἕλλην 161:4; 187:4; 284:6
Ηλληνων 137c:9
ἐμβολιμός intercalary (day) L 380 C 8,1 146:18
ἐμπεριέχω hem in C6 71:9; L 122 XII V 268:20
ἐμπεριοχή hemming in L 487 C 1 107:7
ἐν —9:3, 4, 5; —3:4 to 8, [9], [10]; 15/22:8 to 12, [15]; 32:4, [5] to [8]; 46:[2], 4, 6, 8, 9, 11, 12, 14, 17; 81:27, 120, 133, 152, 201; 95:[17], 19, [23], [27], 31, 33, 37, [39], 43, 46, [49], 50, 54, 57, 61, 62, [66], [68], [70], [73], 75, 77, 80; 137a:6, 10, 11, 12, 13, 17, 20, 23, 24, 26, 28, 30, 33, 36, 39; 137b:[5], [7], [9], [10], 12, 13, 17, [20], 23, 24, 26, 28, 30, 33, 35, 38; 137c:17, [22]; 150:3, 4, 5; 177:[3], 5; 187:7, 8, 9, [10]; 219 I b:4; 250,2:7, [27]; 293 XII:1 to 5
νε 187:11, 12, 13
ἐναντίος in opposition L 61 X V 143:9
ἐναντιόομαι be in opposition L 62 V 87:3
— ἑαυτῷ be in opposition to own house L 463 C 8,4 225:5
ἐναντίωσις opposition L 124 V 281:8
ἔνατος
εννατη[c] 190:2
ενγιστος see ἔγγιστος
ἐνχρονίζω 81:26
ἔνθα 81:202
ἐνιαυτός 208:3
ευληγω see ἐλλήγω
εννατος see ἔνατος
ἕξ 81:56, 99, 111
εξ[.]..... 293 VIII:10
ἑξάγωνος sextile (aspect) L 120 XII V 267:32
— πλευρά side of hexagon L 117 VI V 269:16
ἐξαλείφω 353:3
ἑξηκοντάς 137c:13
ἑξηκοστός 81:100
ἐξουσίας, τόπος περὶ Locus of property L 123 I V 231:13
επ.[208:4
ἐπαγόμεναι 293 VIII:[2]
επαγὸ 184:1
ἐπαναφέρομαι be in epanaphora L 50 V 83:15
ἐπαναφέρω 95:111
ἐπαναφορά epanaphora L 113 IV C 8,2 86:6
επειφ 244:1; 345:3; 478:4

ἐπέμβασις approach (to chronocratorship) L 120 II V 216: 2
ἐπέχω 81:54, 110
ἐπί 81:40, 63, 86, 104, 148; 137c:7; 250,2:[20]; 345:10; 376b: 32; 385:21
ἐπιβλέπω be in aspect to L 478 C 6 65:14
ἐπιγιγνώσκω 81:6
ἐπίκεντρος 258:[4]; 260:[7]
ἐπίκοινος (κύριος) co-ruler L 65 X V 86:8
ἐπικρατέω rule (a period of life); be in power 138/161:11; L 102 XII 14 V 283:17; L 142 V 288:27, 31
ἐπιλέγω 81:25
ἐπιμαρτύρησις be in aspect C 6 70:24
ἐπιμαρτυρία aspect L 75 V 168:18
ἐπιμερισμός apportionment L 75 V 167:25
ἐπίσκεψις 81:23
ἐπιτελέω portend L 122 I 22 V 230:30
ἐπιτίθημι 95:[107]
ἐπιτολή rising L 120 II V 101:22
ἐπιτόπως in own domain L 50 V 83:18
ἐπιτρέχω 81:158
ἐπίτριτος 81:106
ἐπιτυχία 345:16
ἐπιφώσκω 81:39
ἕπομαι 208:[4]
ἐποχή position L 497 C 1 171:8
ἑπτά (see also εἰκοσιεπτά) 81:7, 161, 205; 137c:1
ε..ρ... 95:105
Ἑρμῆς — 9:[4]; —3:[10]; 15/22:[11], diagr.; 32:[3]; 46:16; 50,2:[4]; 81:59, 83, 131, 151, 157, 169; 95:20, 31 to 34, 38, 40, 43, 44, [47], [51], 53, 54, 69, 74, 78, 79, 93; 125:13; 137a: 9, 11a, 23, 29, 31, 35; 137b:8, 11, 23, 29, 31, 34; 137c:II 6; 138/161:3, 12; 139:6; 149:4; 150:4; 171:4; 177:5; 182:4; 184:6; 187:11; 190:3; 207:5; 217:5; 219 I b: diagr.; 219 I c: diagr.; 219 I d: diagr.; [219 I e]; 219 II 1:7; 220:3; 244:3; 245:[3]; 250.2:[26]; 258:7; 260: 10; 277:8; 282:6; 283:8; 284:16; 293 VIII:[10]; 316: 6; 320:5; 326:9; 338:12, 20, 21, 27, 29, 30; 345:7; 351: 16; 353:diagr.; 366:8; 370:35; 373:6; 376a:19; 376b: 40; 381a:25; 381b:34; 431:4; 465:8
ερμ 81:192; 219 II 12:5, 7
ἔρρωσο see ῥώννυμι
ἔρχομαι 345:9
ἔρως (κλῆρος) 138/161:6; 338:15
ἐρώτησις inquiry (of Katarche) L 475 C 1 103:1; L 479 C 1 103(b):1
ἑσπέρα 137a:24
ἑσπέριος 137a:24; 137b:24
ἔσχατος 137a:16; 137b:16; 138/161:13; 220:8
ἐτάξω 137c: fragm. 6
ἔτι 81:79, 112
ἔτος — 9:1; —3:1; 32:1; 15/22:5; 81:37; 95:[90], [103], 184:2; 187:[1]; 219 I b:1; 219 I c:1; 219 I d:1; 219 I f: [1]; 219 I g; 220:[1]; 227:3; 320:2; 478:3
 L (= ἔτους) 95:84, 117, 126, 134, 135; 137a:3; 137b:[2]; 137c:7; 138/161:[17]; 161:2; 176; 177:2; 184:1; 187:6; 190:[1]; 208:1; 219 II 1:1; 219 II 12:1; 244:1; 258:1; 260:1; 283:2; 284: verso [1]; 345:3; 394:1
 other symbols for ἔτους 150:1; 277:2; 282:2; 293 VIII:1; 316:3; 326:3; 338:3; 366:2
εὑρίσκω 46:1, [4]; 95:[16], [22], 30, [36], [42], 48, [53]; 137c: 12; 138/161:2, 4, 6, 8
εὐσεβής 478:3
εὐτυχέω 137b:43; 182:9
ευτυχηϲτατοϲ 284:4
ἐχθρὸς ἀστήρ hostile planet L 61 X V 143:5
εκχθρος 138/161:9
ἔχω — 3:15; 81:[203]; 138/161:12; 219 I; 284:8
ἑῷος 81:103, 137, 147; 137a:21; 137b:21; 250,2:25
 ἑῷος φάσις morning phase L 76 C 6 68:19
ἕως — 3:16; 95:[90], 117, 126, 135

Ζεύς — 61; —3:[7]; 15/22:10, 11, diagr.; 46:2, 8, 12; 50,2: 2; 81:60, 101, 108, 132, 146, 178; 95:36, [50], [57], 62, 78, 125, 127; 125:7; 137a:7, 12, 19, 22, 24, 27, 34, 38, 41; 137b:[6], 12, [19], 21, 24, 27, 33, 36, 39; 137c:14, [15]; 138/161:9; 139:8; 149:6; 150:4; 171:3; 177:3; 182:1; 184:7; 187:8; 190:6; 207:2; 217:2; 219 I a; 219 I b: diagr.; 219 I c:diagr.; 219 I d:diagr.; 219 I e; 219 II 12:5; 220:5; 245:[2]; 250,2:9; 258:4; 260:8; 277:5; 282:4; 283:6; 284:13; 293 VIII:[9]; 293 XII:[4]; 316:9; 320:7; 326:8; 338:9, 22, 23, 28; 345:4; 351:[14]; 353:[diagr.]; 370:32; 373:8; 376a:18; 376b:[39]; 381a: 24; 381b:33; 385:27; 394:6; 431:7; 465:5; 478:[8]
ζευ 219 II 1:5; 244:2; 250,1:diagr.
ζόη 95:82
ζυγός —3:4; 15/22:8, 11, diagr.; 32:[8]; 81:181; 95:[68]; 149:8; 184:8; 187:[7]; 217:1; 245:4; 258:6, [7]; 260: 11; 293 VIII:5; 316:[6]; 320:[7]; 326:12; 345:1, 4; 351:[12;] 353:diagr.; 373:5, [10]; 376a:19; 376a:40; 381a:24, 28; 381b:[33], 36; 385:29; 431:3, 4, 5
ζυγ 176; 219 I a; 219 I b:diagr.; 219 I c:diagr.; 219 I d: diagr.; 219 I e
ζωδιακῶς by, or according to, sign L 61 X V 144:11
ζώδιον 15/22:8, 9, [10], 12, 14; 81:58, 62, 80, 85, 101, 116, 130, 145, 169, 178; 137c:12, 22; 208:4; 250,2:[28]
 — animal L 483 C 6 66:4
ζωνη[137c:II 2
ζῷον sign L 80 Manetho III:740; L 483 C 6 66:5

ἤ 95:110, 111
ἡ see ὅς
ἡγέομαι 15/22:1
ηγροχοος see ὑδροχόος
ἦθος 138/161:5
ἡλιακὸν ζῴδιον solar sign L 103 V 129:27
ἥλιος — 9:[4]; —3:4; 15/22:[7], 16, diagr.; 32:3; 46:1; 81:50, [193]; 95:[16], 47, 55, [69]; 137a:6, 14; 137b: [5], [14]; 137c:[11]; 139:4; 149:4; 150:3; 182:5; 207:7; 245:1; 260:11; 277:8; 282:6; 283:4; 284:10; 293 VIII:4; 293 XII:[1]; 316:6; 320:9; 326:9; 338:6; 345:8; 351:17; 353:diagr.; 366:8; 370:34; 373:7; 376a: 19; 376b:40; 381a:25; 381b:[33]; 385:28; 431:3; 465: 9; 478:[5]
η^λ 219 I b: diagr.; 219 I c: diagr.; 219 I d: diagr.
η^λ 219 I e
ηλιc 125:7; 171:4; 177:8; 184:4; 187:12; 190:3; 220:3; 250,1:[diagr.]; 258:[6]
ꙋ 217:4; 219 II 1:5; 219 II 12:5; 244:3
κλιον 338:19
Ἕλλην see Ἕλλην
ἡμέρα 32:[2]; 81:210; 95:[15], [90], [117], 126, 134, 135; 137a:5; 137b:4; 137c:10; 187:[3]; 190:[2]; 220:2; 227:4; 250.2:verso; 258:2; 260:2; 277:3; 283:3; 293 VIII:3; 316:5; 326:4; 345:3; 376a:12; 376b:34; 381a: 22; 381b:[32]; 385:23; 465:[2]; 478:[4]
 symbol: L 621 note 4
ημερ 244:1
ημερ^L —3:3; 95:126
ημερα(c) 161:5
ημερω(ν) —3:16
 — Κρόνου Saturday 219 I b:2; L 479 C 1 103(b):4
 — ϲεληνηϲ 219 I:3
ἡμερινοὶ ὡριαῖοι χρόνοι time degrees of daylight L 516 C 8,1 231:21
ἡμερινός 219 I b:3; 219 I c:2; 219 I d:2; 219 I f:2
ἡμιόλιος 81:124
ἥμισυς middle of L 158 V 284:18
ημιϲι 137c:[14]
Ἡρακλῆς see Πυρόεις

θανατικὸς (τόπος) Locus of death L 65 V V 130:3

GLOSSARY

θάνατος 95: [77]
— Locus of death L 89 V 130: 13
θέμα theme (of nativity); horoscope L 110 XII V 282: 16
θεάματ(ος) L 497 notes to diagr.
θεματ(ος) L 497 diagr.
θεμάτιον nativity L 76 C 6 68: 10
θεός divine 81: 37; 95: 7(?)
— moon 81: 73
— planet 81: 8; 137c: 1; L 61 X V 143: 11
— sun 137c: 11
— θεοῦ, τόπος περί Locus of god L 62 V 87: 1
— θεοῦ καὶ βασιλέως, τόπος περί Locus of god and king L 107 V 229: 3
θεωρέω be in aspect L 65 V V 130: 4
θηλυκός 15/22: 9; 81: 83
θωθ 161: 6; 187: 6; 258: 2; 282: 2; 316: 4; 431: 2

ἰδιοθρονέω be in own house L 486 C 1 100(b): 10
ἴδιος 46: 12; 81: 82, 146; 95: [20], 40, 49, 71; 137a: 11, 18, 19; 137b: [10], [18], 19, 22; 250,2: [23]
ἰδιοτοπέω be in own domain L 65 X V 86: 7
ἵνα 81: 24, 196
ἰνδικτίων
 ιντικτιονος 366: 2; 370: 30; 376a: 12; 376b: 32
ἰσανάφορον equality of rising times L 158 V 285: 3
ἰσανάφορες of the same rising time L 61 X V 143: 13
ἰσημερία 81: 52
ἰσημερινός equator L 497 C 4 106: 10
ἰσόμοιρον of equality of degrees L 75 V 142: 5
ἰσοσκελὲς τρίγωνον equilateral triangle L 440 C 8,4 222: 6
ἵστημι 137c: fragm. 7
ἰχθύδιον 219 I b: 4
ἰχθύες —3: 5; 15/22: diagr.; 32: [7]; 46: 19; 50,2: 5; 81: 97, 140; 125: 8, 13; 137a: 33; 137b: 33; 139: 4; 149: 5; 150: 4; 190: 4; 207: 7; 220: 5; 245: 7; 250,2: [8]; 258: 3; 277: 7; 283: 8; 284: 14; 326: 8; 338: 9; 351: [14]; 353: diagr.; 376a: 17; 376b: 38; 381a: 25; 381b: 34; 385: 27
 ειχθυες 171: 3
 ιςχυει 171: 5
 ιχθ 176; 219 I a; 219 I b: diagr.; 219 I c: diagr.; 219 I d: diagr.
 ιχθησι 465: 4
 ιχθυ 219 I e
 ιχθυς 81: 105; 182: 4, 5, 6

κ[137c: fragm. 8
κα[95: 114; 250,2: 29
κα..κα...α 95: 109
καθυπερτερέω be in dominant aspect; prevail L 75 V 168: 3; L 85 XI V 113: 14; L 89 V 130: 14; L 142 V 288: 33
καθυπερτεροῦμαι be in inferior aspect L 97 II V 129: 4; L 101 III V 94: 11
καθυπερτέρησις dominant aspect L 110 XII V 282: 22
καί —9: 3, [4]; 81: 6, 10, 20, 21, 22, 50, 55, 61, 73, 78, 84, 102, 111, 117, 125, 142, 178, 179, 196; 95: 74, [76], 78, 79, 93, 106 to 109, 112, 114, [115]; 137c: 4, 5, 6, 11, 16, fragm. 2; 138/161: 5, 6, 7, 9, [12], 15; 161: [2]; 258: 1; 260: 1; 293 VIII: 1; 316: [6]; 345: 15, 16; 385: 20
καιρικὴ ὥρα seasonal hour L 401 C 8,1 232: 10
Καῖσαρ (see also κορυ') —9: 1; —3: 1; 137a: 3; 137b: 2; 137c: 8; 138/161: [17]; 139: [1]; 149: 1; 150: 2
 καισαρ —3: 2
κακοδαιμονέω be in Locus Bad Daemon L 474 C 1 102(b): 7
κακοδαιμόνημα Bad Daemon L 380 C 8,2 68: 8
κακοπαθέω 95: 114
κακὸς δαίμων Bad Daemon L 497 diagr.
καλάνδαι 81: 44; 171: [1]
καλέω 81: 67, 90, 134
καλῶς 95: 108
κανών table 81: 12; L 37 V 233: 25

καρκίνος —3: [7]; 15/22: diagr.; 32: [6]; 46: 8; 81: 110; 95: 61; 137b: 38; 138/161: 2; 150: 5; 187: 8, 9; 244: 3; 258: 4; 282: 4; 283: 6; 316: 7; 320: [8]; 353: [diagr.]; 373: 4; 381a: 23; 381b: [32]; 385: 25; 478: 5, [10]
 καλκινω 345: 8
 καρ 219 I a; 219 I b: diagr.; 219 I c: diagr.
 καρκ 176; 208: 2; 219 I d: diagr.; 219 I e
 καρκεινω 137a: 39
κατά —9: [2]; —3: [2]; 15/22: 4, 5; 81: 45; 137a: 4; 137b: 3; 137c: 7, 9, 11, II 3; 145: 2; 161: [4], 6; 187: 4, 5; 284: 1, 6
 κατα σεληνην 219 I: 2; 219 I b: 2; 219 I d: 2
καταβιβάζω 81: 105
καταβιβάζων descending node L 486 C 1 100(b): 9
καταλαμβάνω 345: [15]
καταλήγω fall in; be located in L 50 V 83: 24
καταρχή Katarche L 474 C 1 102(b): 1
κατοπτεύω be in dominant aspect L 106 V 113: 8
κατορθόω 81: 31
κατωφερῶς downward; backwards; in direction opposite rotation L 115 II V 154: 14
κέντρον 95: 80
κεντροῦμαι be in a center C 6 69: 4
κέντρωσις center L 428 C 8,1 221: 14
κιν[137c: 23
κινδυνεύω 95: [113]
κίνδυνος —3: 15
κινέω 250,2: [7], [19]
κίνησις 81: 8; 95: [16], [22], 30, 36, [42], 48, 53
κλεψύδρα 95: 60
κληρονομέω be destined L 65 X V 86: 6
κληροῦμαι τὴν τύχην be located in the Lot of Fortune L 82 V 85: 14
κλῆροι Lots (Fortune and Daimon) L 65 X V 130: 3
κλῆρος 81: 183; 95: [72], [75], [77], [104], [115]
— Lot of Fortune L 62 V 86: 34
— ἀνάγκης 138/161: [8]
 κλ αναγκης 338: 16
— ἀναιρέτου Lot of Destroyer L 440 C 8,4 221(b): 12
— δαίμονος 138/161: 4
 κλ δαιμονος 338: 14
— ἔρωτος 138/161: [6]
 κλ ερωτος 338: 15
— τύχης 81: 177; 137a: 36, 39; 137b: 35, 38; 138/161: 2; 182: 8; 338: 13; 351: 19; 370: 38; 376a: 20; 376b: 41; 381a: 28; 385: 29
 κλ τυχης 373: 10; 381b: 36
κληρόω allot L 37 V 233: 28
κλίμα clima L 74 IV V 154: 7
κλιμακτήρ crisis; critical point L 61 X V 143: 9
κλιμακτηρικός 138/161: 15
κλιον see ἥλιος
κοίτη 95: [109]
κολλάω contact L 401 C 8,1 233: 4; L 516 C 8,1 231: 8
κόλλησις contact; approaching conjunction C 6 70: 24
κορυ' 219 II 1: 1
κορυφή (κατὰ κορυφὴν τόπος) midheaven C 6 69: 10, 15
κοσμοκράτωρ 46: [1]
— chronocrator V 170: 36
κόσμος
 κοσ⟨μος⟩μου 95: 110
κοτύλα 171: 2
κουριος see κύριος
κρατέω rule L 102 XII 14 V 283: 25
κριός 15/22: diagr.; 46: 4; 81: 53, 65, 158; 95: 54, 66, 75; 125: 9; 137b: [12]; 138/161: 8; 149: 6; 150: 4; 171: 4; 207: 4; 217: 3; 219 II 1: 6; 219 II 12: 6; 220: 7; 245: 2; 283: 4; 316: 10; 351: 13; 366: 9; 376a: 20; 376b: 41; 385: 28
 κρειος 137a: 12; 184: 7; 260: 5; 353: diagr.
 κρι 176; 219 I a; 219 I c: diagr.; 219 I d: diagr.; 219 I e
 κριο 219 I b: diagr.

GLOSSARY

κρονικός saturnine L 621 Usener 273:20
Κρόνος — 9: [3]; —3: [6]; 15/22:10, [14], diagr.; 32:6; 46:6, [10]; 50,2:1; 81:96, 130, [194]; 95:30, [51], 55, 68, 70, [78], [84], 112; 125:5; 137a: [9], 10, 14, 18, [37]; 137b:8, [9], 14, 18, 36; 138/161:1, [5]; 139:6; 149:6; 150: [5]; 171: [3]; 177:3; 184:5; 187:7; 190:5; 207:1; 208:2; 217:1; 219 I a; 219 I c: diagr.; 219 I d: diagr.; 219 I e; 220:4; 245: [4]; 258:3; 277:4; 282:7; 283:7; 284:12; 293 VIII:8; 293 XII: [3]; 316:8; 320:6; 326:7, 11; 338:8, 19, 20, 24 to 27; 345:6; 351: [13]; 353: [diagr.]; 366:6; 370:33; 373:8; 376a:17; 376b:38; 381a:26; 381b: 34; 385:25; 394:5; 465:4; 478: [7]
ημερα κρονου 219 I b:2
κρο' 219 II 1:4; 219 II 12:4; 244:2
κρονον 260:5
κρωνος 431:6
κρούω 81:175
κρυμός
κρυεμες 95:110
κρύσταλλος 81:150
κρωνος see Κρόνος
κύκλος cycle (of planet) L 104 VII V 237:26
κυριεύω 95:107
κύριος 125: [1]; 137a:3; 137b:2; 137c:5, 8; 138/161:17; 150: 2; 161:3; 258:1; 478:2
κοριῳν 139:2 (see also κορυ')

λ see λεπτόν
λαγών 81:64
λαμπρός 81:121, 154; 95: [21], [26], [40], [45], 55, 64; 250,2: [22]
λαμπ 95:51
λε see λεπτόν
λε.. 95:77
λείπω 81:99
λεπτόν 81:20, 56, 112, 142; 95: [20], 28, [37], [43], 66, [73], 75; 137a:17; 137b:5, 7, [9], [12], 17, 20, 23; 137c: [13], II 13; 260:8
λεπτα τριτα 81:112
λ 207: [7]; 338:8, 10
λε 338:7, 16, 18; 478:5 to 8, [9]
λεπ 260:5, 12
λεπτο 137a:6, 8, 10, 12
λεπτον (for λεπτῶν) 137a:21, 24
λέων —9:4, 5; —3: [12]; 15/22: [15], diagr.; 95:19, 46; 176; 177:4; 182: [2]; 187: [10], [11], 12; 217:6; 258:5; 260: 7; 282:5; 284: [17]; 293 VIII:10; 338:5; 351:18; 353: diagr.; 366:5; 370:37; 394:4; 465:7, 8
λεοντ 150:5
λεω 219 I a; 219 I b: diagr.; 219 I c: [diagr.]; 219 I d: diagr.; [219 I e]
λεωντι 125: [12]; 431:8
λόγος 81:203
— claim L — 71 C 8,4 237:7; L 487 C 1 106(b):17; C 1 107: 7, 10
λοξόω incline, slant L 497 C 4 106:10

Μάιος 171: [2]
μάλιστα 137c:4
μαρτυρέω be in aspect to L 75 V 168:21
μέγεθος 81:156; 137c:11
μέγιστος 137c: [15]
μεγιστοτατος 81:49
μελεοκοπούμενον ζώδιον misfortune-causing sign L 113 IV C 8,2 86:5
μέλλουσα σύνοδος next (following) conjunction L 114 V V 147:16
μέν 81:49
μερίζω count; allot; apportion 95: [85], [86]; L 110 XII V 282: 25; L 117 VI V 269:21

μέρος 81:57, 79, 99, 113, 144; 137c: [13], II 5
μεσεμβολέω intercept, radiate between L 484 C 6 67:2
μεσεμβολία interception L 487 C 1 107:6
μεσημβρινός
μεσηβρινος 81:174
μεσορη —9: [1]; 161:5; 187: [3]; 465:2
μεσουρανέω —3: [12]; 15/22: [13]; 46:19
— κατὰ τὸν κλῆρον culminate with respect to the Lot of Fortune L 82 V 85:15
— τῷ κλήρῳ culminate with respect to the Lot of Fortune L 74 V 84:32
μεσουράνημα 95: [66], [81]; 137a:30
μεσουρ 15/22: diagr.
μεσουρανημα 137b:30
μεσουράνησις culmination L 428 C 8,1 222:14
μέσσος οὐρανός Midheaven L 80 Manetho III:748
μετά
μεθ 138/161:14
μετοχή partnership (as triangle rulers) L 497 C 4 106:17
μέτρον 138/161:3
μεχειρ 149:2; 208:1; 219 II 1:2; 219 II 12:2; 220: [1]; 326: 4; 381a:22; 381b:31
μή 81:24, 198; 353:3
μηδείς 345:11
μήκος 95: [44]
μήν 15/22:5; 95: [90], 93, [99], 117, 126; 137a:4; 137b: [3]; 137c:8; 138/161:14; 219 I b:1; 219 I c: [1]; 219 I d:1; 219 I f:1; 219 I g; 250,1: [1]; 394:2; 478:4
μινι 284:6
μήτηρ 46:3
μιμνήσκω
μνηςθ[η] 219 I:2
μινι see μήν
μιρ see μοῖρα
μοῖρα 46:2, 4, [7], [9], 11, 14, 17; 81:19, 53, 58, 77, 80, 98, 100, 111, 113, 128, 140, 144, 159, 167; 95: [20], [28], 31, [37], 39, [43], 46, [49], 51, [54], 57, 61, 63, 66, 70, [73], 75, 77; 137a:6, [8], 10, 12, 16, 20, [23], 26, 28, 30, 33, 37, 40; 137b:5, 7, [9], 12, 16, 20, 23, 26, 28, 30, 33, 35, 38; 137c:12, [14], 16, II 4, II 11; 138/161: [1], 2, 4, 6, 8; 250,2: [8]; 284:10 to 16, [17]
μ 182:1, 2, 4 to 7
μιρ/ 207: [2], 8
μο 478:5 to 8, [9]
μοι 258:3, [4], 5 to 8; 260:3, 4, 12
μοιρ- 95:23; 207: [2], 8
μυρ 338:5 to 16, 18
μοιρικός (reckoned) by degree L 61 X V 143:19
μονάς first interval V 234:1
μονομοιρία monomoiria L 497 C 4 106:19
μόριον degree L 412 Marinus (Cousin) 61:12
Μουνυχιών Munychion (Athenian month) L 412 Marinus (Cousin) 61:26
μυρ see μοῖρα

νε see ἐν
νεομηνία 81:46
νέος 184:2
νή 353:3
νοητὸν ζωδιακόν ideal or theoretical zodiac L 428 C 8,1 222:6
νότιος
νοτεις 81:148
νυκτερινοὶ ὡριαῖοι χρόνοι time degrees of night L 516 C 8,1 231:10
νυκτοστράτηγος 376b:32
νύξ —9:2; 15/22:7; 81:41; 125:4; 138/161: [18]; 139: [3]; 145:3; 149:3; 150:3; 171: [2]; 177:3; 184:3; 219 II 1: 3; 219 II 12:3; 284:7, verso 3; 338:4; 351:12; 370:31; 394:3; 431:2
νῶτον
νωιτωι 81:121

GLOSSARY

ξένης, τόπος περί Locus of travel L 118 V 269:8; L 122 I 22 V 230:30
ξενιτεύω 95:114

ὁ, ἡ, τό — 9:1, 2; —3:3, 17; 15/22:3, [4], [5], 7; 46:1, 3, 6, 8, 11, 13, 16; 81:1, 4, 7, 9, 14, 22, 27, [36], 39, 41, 47, 48, 50, 64, 68, 71, 73, 86, 88, 91, 94, 96, 104, 108, 119, 120, 126, 133, 136, 138, 148, 151, 152, 157, 164, 165, 173, 175, 177, 183, 196, 199, 204, 205, 208; 95:[15], [16], [19], [21], 22, [25], [26], [27], 30, 33, 34, 36, 39, [40], [42], 44, 45, [46], 48, [50], 51, 53 to 56, 59, 60, 62, 64, 72, 77, 79, 80, 82, 107, 113, 114, 115, 125, [134]; 125:1, 4; 137a:3, 4, 5, 11, 13, 17, 36, 39, 42, 43; 137b:2, 3, 4, [10], 13, 17, 35, 38, 41; 137c:4, 5, 7 to 11, 13, 17, fragm. 4, 8, 9; 138/161:2 to 10, 12 to 15, 17; 139:2; 150:2; 161:3, 6; 171:2; 177:[2]; 184:1, 2; 187:[2]; 207:9; 208:3; 219 I:2; 219 II 1:2; 219 II 12:1; 250,2:[5], [14], [17], [27]; 258:1, [2]; 284:1, 4, 8; 338:17; 345:9, 15; 353:3; 385:20; 478:2
των as rel. pron.(?) 95:35, 41, [47], 57, 65
ὄγδοος 145:1
ὄγκος 81:107, 124
ὅθεν 81:16, 48
οἶαξ 81:165
οἰκεῖος own L 76 C 6 68:21
οἰκείως in own domain; favorably located L 76 C 6 68:21; L 86 VIII V 93:23
οἰκητήριον 15/22:14
οἰκοδέκτωρ ruler of receiving house; house ruler L 478 C 6 65:14, 20
οἰκοδεσποτεία 138/161:13
οἰκοδεσποτέω 15/22:16; 81:163
οἰκοδεσπότης 95:[79]; 137a:42; 137b:41; 138/161:10, [15]
— τῆς καταρχῆς house ruler of the Katarche L 487 C 1 106(b):16
οἶκος 15/22:8, 9, [10], 11, [12], 13, 15, 16; 46:12; 95:[20], [28], 31, 33, 37, 40, 43, 47, 49, 51, [54], 57, 63, 67, [68], 70, 73, 76, 78; 137a:7, [9], 11, [13], 18, 22, 24, 27, 29, 31, 34, 37, 40; 137b:[6], 8, 10, 13, 18, 21, 24, 26, 28, 31, 33, [35], [38]; 137c:14; 138/161:1, 2, 4, 6, 8; 250,2:[9]; 338:19 to 30
οκωι 95:61
ὀλιγαναφορος of short rising time L 486 C 1 101:7
ὀλιγοδρομέω be in slow motion L 474 C 1 102(b):7
ὅλος 81:9, 50, 136, 165; 95:59
ὁμόζωνος equi-zonal (sign) L 474 C 1 102(b):18
ὅμοιος 81:86, 149; 95:[115]
ὁμοίως 95:115
ὁμόκεντρος in same center C 6 69:20
ὁμόλογος 81:31
ὁμονύμως 81:135
ὅμως 81:182
ὄνειρος 95:113
ὄνομα 95:34, 35, 41, 45, 52, 56, 67, [71]
ονομ- 95:26, 64, 69
ὀξυγώνιος 81:76
ὁποῖος 95:[107]
ὁποῖουν 95:108
ὅπου 95:110
ὁράω look upon; be in aspect L 40 C 8,2 85:1; L 118 C 2 169:5
ὄργανον table L 61 X V 143:11
ὀρθός 81:173
ὁρίζων 81:204
ὁρικὸς κύριος ruler of term L 61 X V 143:15
ὅριον 46:2, [5], 7, 10, 12, 15, 18; 81:59, 82, 101, 116, 132, 151, 168; 95:20, [29], 32, 38, 40, 43, 44, 47, 50, [51], 55, 57, 62, [67], [69], 70, [74], [76], 78; 137a:7, 9, 11, 15, 19, 22, 25, 27, 29, 32, 35, 38, 41; 137b:6, 8, [10], 15, [19], [21], 24, 27, 29, 32, 34, 36, [39]; 138/161:1, [3], [5], 6, [9]; 338:19 to 30
ὅρος 138/161:15

ὁροσκόπος see ὡροσκόπος
ὅς, ἥ, ὅ 81:68, 91, 113, 143, 163, 180; 95:[19], 26, [27], 33, 34, 35, 39, 41, 45, [46], [50], 52, 56, 62, 64, 67, 69, [71], 72, 112, 113(?); 137c:6, 13, 16; 138/161:14
ὅσος 95:107
οὐ 345:12; 353:3
οὐδείς
οὐδέν 137a:17; 137b:17; 345:13
οὐράνιος 81:4
οὗτος see also τουτέστιν 15/22:4; 95:[111]; 138/161:3, 5, [7], [9]
οὕτως 81:27; 284:8
ὀψέ 366:4; 373:3

παθητικὸν ζῴδιον sign indicating weakness L 116 V 113:24
πάθος 95:[111]
— passivity L 484 C 6 66:26
παλαιός 81:2
πάλιν 81:93, 183
Πάναμος (month)
παν 176
πανσεληνιακὴ φάσις full moon C 6 70:18
πανσεληνιακὸν ζῴδιον sign of (preceding or following) full moon L 97 II V 129:5
πανσέληνος full moon L 75 V 167:12
— (abbreviation) L 497 note 7
— (γένεσις) (nativity) of full moon type L 114 V V 147:15
παντοδαπής 138/161:9
παρά 15/22:2; 207:9
παραδίδωμι 137c:2
— yield, give over (chronocratorship) L 104 VII V 237:29
παράδοσις yielding, transfer (of chronocratorship) L 104 VII V 237:31; L 111 IV V 228:28
παραιρέτης destroyer or reducer of life period C 6 70:31
παραιρετικός destructive L 487 C 1 107:5
παραιροῦμαι destroy; take away C 6 70:29
παραλαμβάνω take over (chronocratorship) L 75 V 167:31
παράληψις taking over (of chronocratorship) L 75 V 168:18
παρανατέλλω rise with L 479 C 1 104:11
παραπίπτω be unfavorably placed L 92 XI V 113:36
παρέρχομαι 138/161:14
παρθένος —3: [8], [10]; 15/22:diagr.; 46:18; 95:31, 33, 37; 125:6; 137a:30; 137b:30; 184:4, 5, 6; 190:6; 260:8, 9, 10; 282:6; 284:13; 293 VIII:4; 293 XII:2; 316:11; 338:13; 353:diagr.; 370:32; 385:26; 465:3, 9; 478:7
παρ 219 I b:diagr.; 219 I c:diagr.; [219 I e]
παρθ 176; 219 I a; 219 I d:diagr.; 244:[2]
παροδεύω pass by V 234:6
πάροδος passage V 234:30
πᾶς 46:3; 138/161:11
πατήρ 345:15
παῦνι 32:2
παχών 81:46
πειράω 15/22:3
πέμπτος 250,2:18
πέντε 81:206
πεντηκαιδέκατος 81:143
περί —3:3; 81:4, 15, 71, 93; 95:54, 61, 77, 82, 112; 138/161:5, 7, 9, 10; 219 I b:2; 219 I d:2
περίγειος 81:160
περίοδος period (of moon) L 142 V 288:28
περίπατος orbit L —42 C 8,4 236:19; progression (of moon) L 401 C 8,1 232:1
περιποίησις Locus of accomplishment L 61 V V 86:22
περιποιητικὸν ζῴδιον sign of (Locus of) accomplishment L 50 V 83:22
Περίτιος (month) 250,1:1
πῆξις computation L 474 C 1 102(b):17

πλανώμενοι, οἱ ἑπτά planets L 497 C 1 171 : 10
πλάτος 95 : 24 ; 137c : 21
πλευρά, ἑξάγωνος side of hexagon; sextile aspect L 117 VI V 269 : 16
πλευρᾶς, ἐκ τετραγώνου in quartile aspect L 117 X Aelius Arist. (Keil) 440 : 27
πληρόω 81 : 98, 140 ; 260 : 3 ; 293 VIII : 1
πλουςι[95 : 106
πλούσιος 95 : 106
πλουσιώτερος 95 : [106]
ποιέω 81 : 162 ; 95 : [16], [23], 31, 36, 42, 49, 53, 108 ; 137c : fragm. 4 ; 219 I : 1 ; 345 : [14]
ποηςι 345 : 10
— παρά multiply by L 431 V 365 : 26
πολεύων ruler (of day) L 478 C 1 104 : 4
— καὶ διέπων ruler of day and hour L 484 C 6 66 : 25, 26
πόλος pole L 497 C 4 106 : 11
πολύς 137c : 2
πολλαν 293 VIII : 7
πρακτικοί χρόνοι period of life activities L 75 V 168 : 7
πρακτικὸν κέντρον operative center C 6 70 : 3
— τόπος Locus of activity, Locus of affairs L 123 I V 231 : 14
πρᾶξις affairs (political) L 61 V V 86 : 19
πράσσω 345 : 15
πρό 81 : 161 ; 171 : [1] ; 345 : 9
προαναφέρομαι precede (in the direction of daily rotation) L 40 C 8,2 85 : 4
προγίγνομαι 338 : 17
προεδρεύω 366 : 3 ; 385 : 22
πρόεδρος 376a : 14
προήγησις lead L 488 C 8,1 235 : 13
πρόκειμαι 81 : 190
προρρητικός
προρητικος 81 : 28
πρός 15/22 : 3 ; 81 : 19, 22, 62, 173, 206 ; 95 : 60
προσδέω 95 : 59
προσέτι 81 : 55
προσθέτης in direct motion C 4 108 : 17
προσθετικὸς τοῖς ἀριθμοῖς adding to its numbers L 487 C 1 106 (b) : 11
προστάσσω 81 : 125
προστίθημι 95 : [44]
πρόσωπον countenance L 497 C 4 106 : 19
πρῶτος 81 : 71 ; 95 : [72], [83], [102], 114 ; 219 I c : 2 ; 219 I f : 2 ; 250,2 : [5]
πτερωτὸν ζῴδιον winged sign L 479 C 1 104 : 17
Πυρόεις 81 : 126
πυ 176
πυροεις ηρακλ[εους] — 61
πυρωις 46 : 10 ; 95 : 42

ρ[137c : fragm. 6
ῥάχις 81 : 87, 175
Ῥωμαῖος 81 : 43
ῥώννυμι 81 : 32

ς[137c : 21
σαπρός 345 : 13
σεβαστός 161 : 4 ; 258 : 2 ; 260 : 2
σελαγίζω 81 : 65
σελασφόρος 81 : 73
σελήνη — 9 : 3 ; — 3 : 5 ; 15/22 : 9 ; 32 : 5 ; 46 : 3 ; 81 : 74, 116, 170 ; 95 : [22], 50, 62, 64, [74], 95 ; 125 : 12 ; 137a : 8, 40 ; 137b : [7], 39, 138/161 : [2] ; 139 : 5 ; 149 : 5 ; 150 : 6 ; 171 : [4] ; 182 : 6 ; 184 : 9 ; 187 : 13 ; 190 : 4 ; 207 : 6 ; 219 I : 2 ; 219 I b : 2 ; 219 I d : 2 ; 219 II 12 : 8 ; 220 : 7 ; 245 : [5] ; 258 : 8 ; 277 : 5 ; 282 : 4 ; 283 : 6 ; 284 : 11 ; 293 VIII : 7 ; 293 XII : [2] ; 316 : 7 ; 320 : 10 ; 326 : 12 ; 338 : 7 ; 345 : 5 ; 351 : 18 ; 353 : diagr. ; 366 : 9 ; 370 : 32 ; 373 : 5 ; 376a : 17 ; 376b : 38 ; 381a : 27 ; 381b : 35 ; 385 : 26 ; 431 : 9 ; 465 : [10] ; 478 : [6]
ημερα ϲεληνηϲ 219 I : 3
ϲελ[η] 260 : 12
ϲελην- 15/22 : diagr. ; 95 : 28
☾ 217 : 2 ; 219 I b : diagr. ; 219 I c : diagr. ; 219 I d : diagr. ; 219 I e ; 219 II 1 : 7
σεληνιακός 81 : 155
σημαίνω 95 : 107 ; 138/161 : 3, 5, [7], 9
σημεῖον point L 428 C 8,1 222 : 28
σινωτικός 138/161 : 14
σκέπτομαι 137c : 2
σκορπίος — 3 : [9], [11] ; 15/22 : 12, 15 ; 81 : 70, 92, 166 ; 95 : 27 ; 184 : 9 ; 190 : 5 ; 219 II 1 : 4 ; 219 II 12 : 4 ; 220 : 4 ; 245 : 5 ; 277 : 4 ; 284 : 10, 15, [16] ; 293 VIII : 6 ; 338 : 7, 11 ; 345 : 5 ; 353 : diagr. ; 366 : 6 ; 370 : 33, 38 ; 376a : 14 ; 376b : 35 ; 431 : 9 ; 465 : 5
ϲκορ 219 I e ; 394 : 5
ϲκορπ 15/22 : diagr. ; 219 I d : diagr.
ϲκρο 219 I a
ϲκροπι 219 I c : diagr.
ϲκροπιον 219 I b : diagr.
ϲκωρ' 150 : 7
σός
ση 284 : 3
σοφός 137c : 3
σπόριμος 81 : 208
στ[137 : 18
στάσις conflict; disagreement L 76 C 6 68 : 14 ; L 142 V 288 : 36
— position; aspect L 40 C 8,2 84 : 19 ; L 102 XII 14 V 283 : 34 ; L 124 V 281 : 5
στερεός 81 : 84, 159
στηριγμός 137a : 11, 13, 17 ; 137b : [10], [13], 17 ; 137c : II 8 ; L 497 note 36
στηρίζω 81 : 119
Στίλβων 46 : 16 ; 81 : 157 ; 95 : 53 ; 250,2 : [27]
ϲτιλ[βων 46 : 15
ϲτιλβ 176
ϲτιλβων απολλωνοϲ — 61
στοιχεῖον 81 : 60
στρωφάω revolve, turn L 80 Manetho III : 749
σύ 15/22 : 2 ; 81 : 19
ση 284 : 3
συμμαρτυρέω be in common aspect with C 6 70 : 17
συμμερίζω also allot; divide L 122 VI 30 V 274 : 10
συμμεσουρανέω culminate with L 428 C 8,1 222 : 21
συμμεσουράνησις simultaneous culmination L 428 C 8,1 222 : 16
συμμετέχω share; be coruler (of period) L 120 II V 287 : 8
συμπαρατυγχάνω be present with; be situated with C 6 70 : 13
συμπάρειμι combine periods (to the same total) L 110 XII V 282 : 26
συμπαρουσία being in company with L 478 C 6 65 : 17
συμπλοκή association, combination L 37 V 233 : 26 ; L 134 VI V 281 : 30
συναιρετίστης co-ruler L 101 III V 94 : 7
— of the same sect L 74 XI V 84 : 29 ; L 132 V 253 : 17
συναινατέλλω 137c : [15]
συναποκατάστασις common return (to same distance) L 37 V 233 : 25 ; L 104 VII V 237 : 27
συνάπτω be in conjunction; contact L 76 C 6 68 : 18 ; L 475 C 1 103 : 16 ; L 478 C 6 65 : 6
συναφή conjunction; contact, approach to conjunction L 76 C 6 69 : 2 ; L 102 XII 14 V 283 : 32 ; L 120 IX V 290 : 13 ; L 483 C 6 66 : 7
σύνδεσμος node L 621 Usener III 273 : 19
συνεδρεύω
         η
ϲυνϋδρευϲαν 137c : 5

GLOSSARY

συνέχω 81:9
συνοδική μοίρα degree of conjunction L 127 XI V 147:26
— φάσις conjunction C 6 70:18
σύνοδος conjunction L 40 C 8,2 84:21; L 78 V 94:1
συνοικοδεσπότης co-house-ruler (of nativity) L 482 C 8,1 240:26
σύστασις 95:109; 138/161:7; 345:16
σφαίρος
  εφηρος 219 I:1
σφόνδυλος 81:72
σχήμα configuration 81:21; L 113 IV C 8,2 86:4; L 121 V 210:12
— diagram L 497 notes to diagr.
σώμα 95:[106]
σωματικοί χρόνοι physical life periods L 75 V 167:15

τ[137c: fragm.9; 208:3
τάξις 81:197; 250,2:[6], [15], [18]
ταπεινοῦμαι be in depression L 478 C 6 65:11
ταπείνωμα 81:117
  ταπιν- 137a:14
  ταπινωματι 137a:19, 32, 35, 38, 41
  ταπινωιματι 137b:18, 31
  ταπινωμα 95:[28], 32, 34, 36, [38], 55, 69, 78; 137b:14, 39; 345:12
τάσσω 81:18
ταῦρος —3:[6], [13]; 15/22:9, 13, diagr.; 32:[5]; 81:76, 88; 95:[49], 63, [73]; 138/161:6; 139:8; 207:6; 217:4; 244:[2]; 245:3; 283:5; 284:11; 293 XII:3; 338:[16]; 351:15; 353:[diagr.]; 366:10; 381a:26; 381b:35; 385:28; 478:9
  ταυ 176; 219 I a; 219 I b: diagr.; 219 I d: diagr.
  ταυρ 219 I c: diagr.; 219 I e
τ[αχ...]ε 137c:12
τε 81:4, 10, 19, 21, 124; 138/161:10
τείνω 81:24
τέκνον 95:[108]
  τέκνων, τόπος περί Locus of Children L 107 V 229:5
τερπνός 95:105
τέσσαρες 81:142
τέταρτος 15/22:7
τετραγωνίζω L 487 C 1 107:11
τετραγωνικῶς in quartile L 40 C 8,2 85:1
τετράγωνος πλευρά side of quartile L 75 V 142:3
  ἐκ τετραγώνου πλευρᾶς quartile aspect L 117 X Aelius Arist. (Keil) 440:27
τετράπους four footed (sign) L 483 C 6 66:5
τις 81:180; 95:108; 137c: fragm. 5
τοξότης 15/22:10, diagr.; 46:6, 17, 20; 95:39, 57; 137a:6, 20, 23, 26; 137b:[5], [20], 23, 26; 137c:12; 149:7; 190:[7]; 220:6; 282:[7]; 293 VIII:7; 293 XII:1; 316:13; 338:8, 15; 353: diagr.; 366:7; 370:33, 35; 376a:15; 376b:36; 478:6
  δοξοτη 187:13; 373:6
  τοξ 176; 219 I a; 219 I b: diagr.; 219 I c: diagr.; 219 I d: diagr.; 219 I e
τοπική μοίρα longitudinal degree L 428 C 8,1 222:5
τόπος 81:94
τουτέστιν 81:30, 56; 137c:3
τρέχω 81:77
τριακοντάς thirtieth (of a sign); degree L 61 X V 144:8
τριάς triad; third interval L 37 V 233:30
τριγωνίζω 95:112
τρίγωνος 81:131, 170, 179; 137c:14; 250,2:[2], [10], [22]
τριταία third (of moon) L 487 C 1 106(b):12
τρίτη third (of moon) L 483 C 6 66:8
τρίτος 81:37, 41, 112; 95:77, [115], 125; 145:3
τροπή 81:36
τροπικὸν ζῴδιον tropical sign L 486 C 1 101:7

τρόπος 81:29; 138/161:5
τυβι 137a:5; 137b:[4]; 137c:10; 190:1; 277:2; 320:4; 351:[12]; 366:4; 370:31; 373:3
τυγχάνω
  τυγχανω 250,2:[6], [15], 18
  τυνχανω 15/22:4, 7
τύχη 81:213; 137a:[1]; 137b:43; 137c: fragm.: 2, 9; 138/161:3; 227:[1]; 277:1; 282:1; 284:1; 316:1; 320:1; 326:1; 338:1; 345:2; 376b:31
— (κλῆρος) 81:177; 137a:36, 39; 137b:35, 38; 138/161:2; 182:8; 338:13; 351:19; 370:38; 373:[10]; 376a:20; 376b:41; 381a:28; 381b:36; 385:29

ὑγίεια
  υγειαν 353:2
ὑδροχόος —3:14; 50,3:[2]; 81:127; 137b:[7], [9]; 171:[3]; 177:[5]; 219 II 12:[5]; 245:[1], 6; 258:[8]; 465:10
  ηγροχ[ο]ω 125:11
  υγρηχωι 46:11
  υδρ [176]; 219 I b: diagr.; 219 I c: diagr.; 219 I d: diagr.
  υδρηχο 326:9
  υδρηχοος —9:3, [4]; 95:51, 77; 150:6; 190:[3]; 277:8, 9; 282:3; 326:5, 6; 345:6; 351:16, 19; 353: diagr.; 370:36; 373:9
  υδρηχου 260:4
  υδρηχω 316:8
  υδριχω 207:2, 5, 8
  υδρο 219 I a; 219 I e
  υδροχους 219 I b:4
  υδροχοω 15/22:14, diagr.
  υδροχω 137a:8, 10; 220:3
  υδροχωι 139:7
  υδροχῶ 219 II 1:5
  υδρυχωου 431:7
υἱός 137c:6
ὑπάρχω be effective L 484 C 6 67:6
ὕπαυγος invisible, close to sun L 40 C 8,2 84:20; L 497 note 38
ὑπέργειον μεσουράνημα upper culmination L 40 C 8,2 84:11
ὑπερέχω 95:109
ὑπό 95:108, 113; 284:3
— γῆν —3:14; 15/22:[15], diagr.; 46:20; 95:68; 137a:33; 137b:33
ὑπόγειος lower culmination L 40 C 8,2 84:12
ὑποδέχομαι receive (in house, terms, triangle, or sign of exaltation) L 482 C 8,1 240:22
ὑποδεχόμενος receiving (star); ruler of receiving house L 479 C 1 104:14
ὑπομένω be destined to; have to expect L 75 V 168:18
ὑπόστασις 138/161:15
ὕψος 81:109
ὑψόω 81:63
ὕψωμα exaltation 81:81, 102, 146; 95:32, 38, 49, 54, 63, 68, 74; 137a:14, 18, 31, 34, 37, 40; 137b:14, 18, 31, 33, 36, 39; L 482 C 8,1 240:23
  υψ 95:73
— exaltation of the nativity L 65 X V 86:8; L 95 V 14 V 86:2
— κατὰ τὴν γένεσιν exalt. of the nativ. L 123 I V 231:23
— τῆς γενέσεως — — — — L 50 V 83:22
ὑψωματικόν (ζῴδιον) sign of exaltation L 111 IV V 228:19
ὑψωτικόν (ζῴδιον) — — — L 111 IV V 228:20

Φαέθων —61; 46:[8]; 81:108; 95:36; [176]
Φαίνων 46:[6]; 81:96; 95:30; 137c: II, 1
  φαιν 176
φαμενωθ 139:[2]; 283:2
φαμεθ 150:2
φαντασία prestige L 61 V V 86:19

φαρμουθι 81:38; 177:2; 207:[9]; 385:[23]
  φαρμουθ 125:2
  φαρμουτι 145:2
φάσις phase 81:21, 161; L 76 C 6 68:19
φαωφι — 3:2; 15/22:5, 6; 260:2; 316:4; 376a:11; 376b:34
φέρω 81:52, 75
φθάνω overtake, extend L 120 II V 174:12; L 380 C 8,1 146:17
φιλία 138/161:7
  φιλίας, τόπος περί Locus of Friendship L 122 I 22 V 230:29
φιλοπονέω 81:5
φίλτατος 81:33; 284:3
φρουρέω accompany; attend L 87 VII V 129:10
φυλάττω — 3:16
φυσικός 95:[79]
φύσις 95:109
φυτεία 95:[108]
φῶς luminary L 76 C 6 68:18
φωστήρ luminary C 6 70:31
Φωσφόρος 81:138; 250,2:17
  φωςφ 176
  φωςφορους 95:48
  φωςφωρος 46:13

Χαλδαικός 137c:3
χάριν — 3:17
χειμερινός
  χιμεριν.[137c:20
χιλιοστός
  χιλιοστ̊ 81:79
χλαμύς 81:133
χοιακ 338:4; 384:2
  χυακ 227:4
χρ 478:1
χρηματίζω 95:[98], [102], 117, 125, 127, 134
χρηματιστικός operative; effective 95:[82]; L 74 XI V 84, 29
χρονογραφία period of time L 75 V 168:14; L 122 X 11 V 268:25
χρονοκρατέω be chronocrator L 401 C 8,1 233:16
χρονοκράτωρ chronocrator; ruler of period of life L 152 V 170:24
χρόνος 15/22:4, 6; 81:35, 209; 95:125, 134; 137c:7; 284:2
  — time degree L 401 C 8,1 232:23
χρυσός 81:85
χρῶμα 95:110

ψηφίζω 81:17, 188, 200; 137c:11
  — παρά multiply L 431 V 365:12
ψῆφος 138/161:11
ψόγος 95:113
ψυχρός 95:112

ὥρα hour; see also ὡροσκόπος —9:2; 15/22:7; 32:2; 81:42;
  137a:5; 137b:[4]; 137c:10; 138/161:18; 139:3; 145:3;
  149:2; 150:3; 161:5; 171:5; 177:[2]; 184:3; 187:[2];
  220:2; 227:4; 250,2:verso; 258:[2]; 293 VIII:2; 316:
  5; 320:[4]
  φ 125:3; 190:2; 219 I b:3; 219 I c:2; 219 I d:2; 219 I f:
  2; 219 II 1:3; 219 II 12:3; 244:1; 260:3; 277:3; 282:3;
  283:3; 284:7, verso 3; 326:4; 345:3; 351:[12]; 370:31;
  373:3; 376a:12; 376b:34; 381a:22; 381b:32; 385:23;
  394:3; 431:2; 465:2
  ωρ∟ — 3:3
ὡριαία διαστολή hourly distinction L 61 X V 143:28
ὡριαῖον μέσον δρόμημα hourly mean motion L 380 C 8,1 146:12
ὡριαῖος χρόνος time degree L 401 C 8,1 232:8; L 516 C 8,1 231:16
ὡριμαία horimaia L 482 C 8,1 240:27; cf. L 482 note 8
ὡρονομέω be in horoscope L 483 C 6 66:7
ὡρονόμος 95:59

ὡροσκοπέω — 3:[11]; 15/22:13; 46:18; 150:7
ὡροσκόπος 15/22:diagr.; 32:8; 81:165; 95:59; 125:5; 137b:
  26; 139:[9]; 149:8; 177:[3]; 187:14; 220:8; 258:9;
  260:4; 284:[17]; 316:12
οροσκοπος 137a:26; 190:8; 353:[diagr.]
  φ 182:7; 207:8; 217:6; 219 I b:4; 219 II 1:4; 219 II 12:
  7; 244:[3]; 245:[1]; 282:3; 326:6; 345:4; 351:12;
  366:5; 370:32; 373:4; 376a:14; 376b:35; 381a:23;
  381b:[32]; 385:[24]; 394:4; 431:10; 465:3
  ωρα 277:9; 283:4; 293 VIII:4; 320:5
  ωρασκοπος 326:5; 338:5
  ωρκ L 497 notes to diagr.
  ωροσκοποι, οι λς λαμπροι 95:21, [26], 34, 40, [45], 52, 56, 64
ὡς 81:42, 189; 137c:2
ὥσπερ 137c:5

2. Greek Words, incomplete

]α 95:108   ]ϲϲκο[.] 138/161:13
]αιτη 137c:fragm. 4   ]κο 137c:19
]αν 95:114   ]ρω 137c:17
]δεια 250,1:1   ]ϲ 95:2
]δεδει 138/161:14   ]ϲης 95:1
]ειϲ 95:8   ]ϲτο 95:5
]ετα 137c:fragm. 5   ]τερο[137c:fragm. 3
]..η.[138/161:13   ]τερον 137c:fragm. 6
]ηρης 50.1:1   ]το..ν.[137c:fragm. 1
].λϛ 250,2:1   ]τος 138/161:18
]λλ...[137c:24   ]τω 137c:18
]λο 95:10   ]..χο 95:9
]ναι 15/22:1   ]ω 137c:16; 138/161:10
]ν..ιον[137c:25   ]ωι 95:14
]ονται 137c:fragm. 7   ]ωρα 95:6
ο]υϲ 95:114

3. Names of Persons and Places

Ἄβυδος L 474 C 1 102(b):16
Ἀθῆναι L 440 C 8,4 221(b):3; L 475 C 1 103:1
Αἰγύπτιος 81:1; 137c:9; 187:5; 207:9; L 497 C 4 106:17
Αἴγυπτος L 621 Usener III 274:17
Ἀλεξάνδρεια L 479 C 1 103(b):2; L 486 C 1 100(b):2
Ἀλέξαν(δρος) Μακεδών 219 I b:3
Ἀλέξανδρος, Σεουῆρος 227:3
Ἀνουβίων 137a:1; 137b:[1]
Ἀνοῦπ 478:2
Ἀντίγονος L 40 C 8,2 84:2; L 487 C 1 107:4 = C 6 64:14
  — ὁ Νικαεύς L 76 C 6 67(b):6
Ἀντιόχεια L 484 C 1 107(b):2 = C 6 66:16
Ἀντωνῖνος 137c:3; 137b:[2]; 137c:7; 138/161:17; 139:1;
  149:1; 150:1; 161:2
  Αντωνεινου 145:1
  Αντωνιν 219 II 1:1
Ἀπολλωνία 385:20
Ἀπολλώνικος 219 I b:4
Ἀρμενία L 621 Usener III 275:8
Ἁρποκρατίων 161:1
Ἀρτεμίδωρος 353:1; 385:22
Ἀσκληπιός 137c:6
βασιλεύς (i.e. Necho) 137c:4
Βησᾶς 353:3
Βούλγαροι L 621 Usener III 275:2
Βυζάντιον L 440 C 8,4 221(b):4; L 463 C 8,4 224:22
Γαλληνός 258:[1]
Δίδυμ(ος) 217:1
Διοκλητιανός 284:5, verso 1; 316:3; 320:3; 326:3; 338:3;
  351:[11]; 370:30; 373:2; 376a:13; 376b:34; 381a:21;
  381b:31; 385:21; 431:1; 465:1; 478:3

GLOSSARY

Διακλητιανου 366:2
Διοκλητιανου βασιλεία L 380 C 8,1 146:4
— ἔτος L 419 V 365:30; L 431 V 365:3; L 475 C 1 103:2
Διονυσία 244:1
Δωρόθεος L 475 C 1 103:20; L 487 C 1 107:2 = C 6 64:12
Ε..... 345:2
Ἐθρίβων, ἔρημος τῆς L 621 Usener III 272:10
Ἕλλην 161:4; 187:4; 284:6
  Ἑλλήνων 137c:9
Ἐπιφάνιος ὁ Ἀραβικὸς ἔμπορος L 621 Usener III 271:24
Ἑρμείας 376b:33
Ἑρμειόρη 351:11
Ἑρμείων 373:1
Ἑρμῆς (god, astrologer) 137c:6
Ἑρμησίων 338:2
Ἑρμογένης 376a:13
Ἑρμούπολις 81:201
Ἕρμων 81:33
Εὐτόκιος L 497 C 1 170(c):1
Ἡρακλείδης 316:2
Ἡφαιστίων Θηβαῖος L 380 C 8,1 141:1
Ἥφηστος 137c:6
Θεόδωρος ὁ Αὐγουστάλιος L 486 C 1 100(b):1
Θεότιμος 376a:14
Θεωνῖνος 277:1
Θηβαῖος L 440 C 8,4 221(b):2; see also Ἡφαιστίων
Ἰβηρία L 621 Usener III 275:9
Ἰμούθης 137c:6
Ἰνδοι L 621 Usener III 275:7
Ἰσίδωρος 366:3
Ἰσμαήλ L 621 Usener III 272:11
Ἰωάννης 376a:11; 376b:31
Καισάρεια L 474 C 1 102(b):2
Καππαδοκία L 621 Usener III 274:21
Καρῖνος 282:2; 283:2
Κιλικία L 621 Usener III 274:21
Κολχίς L 621 Usener III 275:9
Κόμμοδος 190:[1]
Κοπρίας 326:2
Κορασιανῶν, φυλὴ τῶν λεγομένων L 621 Usener III 272:10
Κορνήλιος Σαλωνεῖνος 258:1
Κριτόδημος L 61 X V 143:6; cf. C 5,2 120:29
Κυιῆτος 260:1
Κυρίλλους 370:29
Κῦρος 381a:21; 381b:30
Λεόντιος L 484 C 1 107(b):1 = C 6 66:16
Λιβύη L 621 Usener III 274:17
Μακάριος 376b:33
Μακεδών 219 I b:3
Μακριανός 260:1
Μωάμεδ L 621 Usener III 272:11
Νεχεῦς 137c:4
  βασιλεύς 137c:4
Νησία 385:20
Νικαγόρας ἄρχων L 412 Marinus (Cousin) 61:25
Νιλάμμων 366:1
Ὄκις 219 I f:3
Ὄκκας (?) 219 I d:3
Οὐαλεντινιανὸς βασιλεύς L 419 V 365:29
Οὐαλεριανός 258:1
Οὐῆρος 161:3
[Ὀ]φέλλιος 138/161:17
Παλαιστίνη L 621 Usener III 274:18
Παμένης 366:[3]
Πέργαμον L 117 X Aelius Arist. (Keil) 440:22
Περσίς L 621 Usener III 275:7
Περσῶν βασιλεία L 621 Usener III 275:4
Πετοσίρις 137c:[4]

Πιτήνιος 81:186
Πιχίμε 283:1
Πλάτων L 117 X Aelius Arist. (Keil) 440:20
Πρόβος 277:2
Πρώταρχος 177:1
Πτολεμαῖος L 428 C 8,1 221:13; L 497 C 4 106:17
— ὁ θεῖος L 497 C 1 171:10
Πτολεμαΐς 227:2
Ῥωμαῖος 81:43
  Ῥωμαίων βασιλεία L 621 Usener III 274:22
— ἔθνος L 621 Usener III 274:19
Σαλωνεῖνος 258:1
Σαραπάμμων 260:13
Σεουῆρος Ἀλέξανδρος 227:3
Σερῆνος 385:21
Σκυθία L 621 Usener III 275:9
Σμύρνα L 117 X Aelius Arist. (Keil) 440:21; L 479 C 1 103(b):1
Σπανίων, κλίμα διὰ L 419 V 365:29
Συρία L 621 Usener III 274:18
Τιβέριος 15/22:5; 32:1
Τίτος 81:38
  Τίτος Πιτήνιος 81:185
Τοῦρκοι L 621 Usener III 275:1
Τρύφων 15/22:2
Φανουσίρε 385:22
Φίλιππος 244:[1]
Φιλόη 150:1
Χάζαροι L 621 Usener III 275:1
Ψανσνῶς 137a:2; 137b:1
Χαλδαϊκός 137c:3
Χρ 478:1

4. Constellations and Star Names

(except for zodiacal signs, for which see the General Glossary, and for decans, for which see the Coptic Glossary)

ἀντάρης L 497 C 4 108:25
ἀργώ L 118 C 2 168:32; L 479 C 1 104:11
ἀσκληπιός L 479 C 1 104:21
βέλος 137c:18
γανυμήδης 81:135
ἰχθῦς, νότειος 81:148
— χελιδοναῖος 81:104
καρκίνος (bright star at the back of) 81:110, 120-122
κένταυρος L 80 Manetho III:749
  βέλος (i.e. κένταυρος) 137c:18
κῆτος (star at nostril of) L 497 C 4 107:5
κριός (flank of) 81:64
— (star in the rhomboid quadrangle of) L 497 C 4 107:4-5
— (star at tail of) L 497 C 4 107:9
κύων (Sirius) L 120 II V 101:22
λέων (back of) 81:174
— (star in the buttocks of) L 497 C 4 107:30
λίνοι 81:153
οὐρανείσκος 137c:17
ὀφιοῦχος (star at right knee of) L 497 C 4 108:14
παρθένος (star at the back and right wing of) L 497 C 4 107:20
πληιάς L 87 I V 111:37
σκορπίος (star between claws of) L 497 C 4 106:13
— (star in the body of) L 497 C 4 108:10, 25
ταῦρος (back of) 81:87
— (star at northern horn of) L 497 C 4 106:23
ὑγεία L 497 C 1 104:30
ὑδροχοῦς (bright star at 20th degree of ♒) L 76 and commentary note 20 C 6 68:27
— (star in the cloak of) 81:133-134
χελιδονιαῖος
  χελειδονιαιον 81:104

B. COPTIC

For the Coptic section of No. 95 cf. Černý-Parker [1], p. 93 ff.

αρου⟩ 95: 45, 52	ϲⲉⲛⲧⲁχορ 81: 68
αρωι αρωι 95: 41	ϲιϲρωι 95: 71
αρωθ 81: 91	ϲρωι 95: 67
βικωτ 95: 35	ϲτωμα 95: 69
ετϥⲉ 95: 34	]τι 95: 29
θουμουθ 81: 172	φουτϥⲉ 95: 41
κατ/ ⲕⲟⲩⲁⲧ/ 95: 58	χⲉⲛ 95: 21
κνουμⲉ 95: 65	χⲉⲛⲧⲁ[95: 56
ⲟⲩⲉϲϣⲏ βικωτ 95: [35]	χω[95: 52
ρⲉⲙⲉⲛⲁⲁⲣⲉ 95: 47	ϩⲣⲁⲕⲛⲟⲩμ 95: 65
ρⲉⲙⲉⲛⲁχ.[....] 95: 65	ϭⲉϭⲭⲟϲ 376b

C. ENGLISH

For references see the Greek Glossary

accompany δορυφορέω; φρουρέω
accomplishment, locus of περιποίησις; περιποιητικὸς τόπος
activity, locus of πρακτικὸς τόπος
adding to its numbers προσθετικὸς τοῖς ἀριθμοῖς
affairs (political) πρᾶξις
allot κληρόω; μερίζω
animal ζῴδιον
anomaly ἀνωμαλία
apoclima ἀπόκλιμα; ἀπόκλισις
—, be in ἀποκλίνω
apportion μερίζω
apportionment ἐπιμερισμός
approach (to chronocratorship) ἐπέμβασις
ascending node ἀναβαίνων; ἀναβιβάζων
aspect ἐπιμαρτυρία; στάσις
— see also diameter; quartile; sextile; trine
—, be in βλέπω; ἐπιβλέπω; ἐπιμαρτύρησις; θεωρέω; μαρτυρέω; ὁράω
— —, along with συμμαρτυρέω
— —, not in ἀμαρτύρητος
—, superior δεκατεύω; καθυπερτερέω; καθυπερτέρησις
associate ruler διέπων
association συμπλοκή
attend δορυφορέω; φρουρέω

Bad Daemon κακοδαιμόνημα; κακὸς δαίμων
— —, be in κακοδαιμονέω
basis βάσις
be with συμπαρατυγχάνω; συμπαρουσία
bicorporeal sign δίσωμον ζῴδιον

center κέντρον; κέντρωσις
—, be in a ἐπίκεντρος; κεντροῦμαι
—, in same ὁμόκεντρος
children, locus of τόπος περὶ τέκνων
chronocrator κοσμοκράτωρ; χρονοκράτωρ
—, be χρονοκρατέω
claim (to house rulership) λόγος
co-house-ruler συνοικοδεσπότης
combination συμπλοκή
combine periods (to same total) συμπάρειμι τοῖς χρόνοις
common return συναποκατάστασις
compute ψηφίζω
computation πῆξις
configuration αἵρεσις; ἀστροθεσία; διάθεμα; διάθεσις; στάσις; σχῆμα
conflict στάσις
conjunction συναφή; συνοδικὴ φάσις; σύνοδος
—, approach to συναφή

—, be in συνάπτω
—, next (following) μέλλουσα σύνοδος
contact κόλλησις; συναφή
—, be about to receive ἐκδέχομαι συναφήν
—, be in κολλάω; συνάπτω
—, leave ἀπορρέω
control διακατέχω; διακρατέω
co-ruler ἐπίκοινος (κύριος); συναιρετίστης
count μερίζω
countenance πρόσωπον
crescent ὀξυγώνιος
crisis; critical time κλιμακτήρ
critical locus κλιμακτηρικὸς τόπος
culminate μεσουρανέω; see also midheaven
— simultaneously with συμμεσουρανέω
— with reference to Lot of Fortune μεσουρανέω τῷ κλήρῳ
culmination μεσουράνησις
—, lower ὑπὸ γῆν
—, simultaneous συμμεσουράνησις
—, upper ὑπέργειον μεσουράνημα
cycle κύκλος

daemon, bad κακοδαιμόνημα; κακὸς δαίμων
—, good ἀγαθὸς δαίμων; ἀγαθοδαιμονέω
death, locus of θανατικὸς τόπος; θάνατος
decan δεκανός; διμ(ερισμός)
degree μοῖρα; μόριον; τριακοντάς
—, reckoned by μοιρικός
depression ταπεινοῦμαι; ταπείνωμα
descending node καταβιβάζων
destined κληρονομέω; ὑπομένω
destroy ἀναιρέω; παραιρέω
destroyer ἀναιρέτης; παραιρέτης
destruction ἀναίρεσις
destructive ἀναιρετικός; παραιρετικός
determine διατεκμαίρομαι
diagram σφαῖρος; σχῆμα
diametrical διαμετρέω; διάμετρος
dim ἀμαυρός
direct motion προσθέτης; προσθετικὸς τοῖς ἀριθμοῖς
disagreement στάσις
distance διάστημα
distant, be ἀφίστημι
division αἵρεσις
dominant aspect καθυπερτέρησις; καθυπερτερέω; κατοπτεύω
downwards κατωφερῶς

eclipse ἐκλειπτικόν
effect; be effective ἀπεργάζομαι; ἀποτελέω; ὑπάρχω; χρηματίζω
elongation ἀπόστασις
epanaphora ἐπαναφορά; ἐπαναφέρομαι; ἐπὶ ἀναφορᾶς
equal degree or position, of ἰσόμοιρον
equator ἰσημερινός
equilateral triangle ἰσοσκελὲς τρίγωνον
equi-zonal ὁμόζωνος
eternal αἰώνιος
exaltation ὕψος; ὑψόω; ὕψωμα; τὸ ὑψωτικόν
— of the nativity ὕψωμα; ὕψωμα κατὰ τὴν γένεσιν; ὕψωμα τῆς γενέσεως
—, sign of ὑψωματικόν (ζῴδιον)
extend φθάνω

faint ἀμαυρός
fall in καταλήγω
favorable position, in οἰκείως
figure εἶδος
finger δάκτυλος
first interval μόνας

GLOSSARY

fixed star ἀπλανὴς ἀστήρ
forecast ἀποτελέω
Fortune, lot of κλῆρος, κλῆρος τύχης
fourfooted (sign) τετράπους
friendship, locus of τόπος περὶ φιλίας
full moon πανσεληνιακὴ φάσις; πανσέληνος
— —, sign of πανσεληνιακὸν ζῴδιον

god, locus of τόπος θεοῦ; τόπος περὶ θεοῦ
good daemon ἀγαθὸς δαίμων
good fortune ἀγαθὴ τύχη
— —, be located in ἀγαθοτυχέω

hem in ἐμπεριέχω; ἐμπεριοχή
hexagon, side of πλευρὰ ἑξάγωνος
horoscope γένεσις, θέμα
Horoscopos ἀνατολικὸν κέντρον; ὡροσκόπος
— —, be in ὡρονομέω; ὡροσκοπέω
hostile (planet) ἐχθρός
hour, seasonal καιρικὴ ὥρα
hourly distinction ὡριαῖα διαστολή
— — mean motion ὡριαῖον μέσον δρόμημα
house οἰκητήριον; οἶκος
house ruler οἰκοδέκτωρ; οἰκοδεσπότης
— —, without ἀνοικοδεσπότης

ideal zodiac νοητὸν ζῳδιακόν
incline λοξόω
inferior aspect, be in καθυπερτεροῦμαι
influence, have ἀποτελέω
inquiry ἐρώτησις
intellect, locus of διανοητικὸς τόπος
intercalary (day) ἐμβολισμός
intercept καταλαμβάνω; μεσεμβολέω; μεσεμβολία
interval διάστασις
— —, first μονάς
— —, second δυάς
— —, third τριάς
invisible; close to the sun ὕπαυγος
itself; its own house ἑαυτός

Jupiter Ζεύς; Φαέθων

king, locus of βασιλικὸς τόπος

latitude πλάτος
lead; forerunner προήγησις
locus τόπος
longitude μῆκος
longitudinal degree τοπικὴ μοῖρα
look upon ὁράω
lot κλῆρος
lots (of Fortune and Daimon) κλῆροι
lower culmination ὑπὸ γῆν; ὑπόγειον
luminary φῶς; φωστήρ

magnitude μέγεθος; ὄγκος
marriage, locus of γαμοστολικός; γαμοστόλος
Mars Ἄρης; Πυρόεις
mean motion, hourly ὡριαῖον μέσον δρόμημα
Mercury Ἑρμῆς; Στίλβων
meridian μεσημβρινόν
middle of ἥμισυς
midheaven κορυφή; μεσουράνημα; μέσσος οὐρανός
minute λεπτόν
misfortune-causing sign μελεοκοπούμενον ζῴδιον
Monday ἡμέρα σελήνης
monomoiria μονομοιρία

morning phase ἑῷος φάσις
motion κινέω; κίνησις
multiply by ποιέω παρά; ψηφίζω παρά

nativity γενέθλη; γένεσις; διάθεμα; θέμα; θεμάτιον
necessity (lot of) ἀνάγκη
new moon day νεομηνία
node σύνδεσμος
— ascending ἀναβαίνων; ἀναβιβάζων
— descending καταβιβάζων
north βορειάς; βορειότερος

operative ἀποτελέω; πρακτικός; χρηματίζω; χρηματιστικός
opposition ἀκρόνυκτος; ἀκρονυχία; ἀκρόνυχος; ἀντίκειμαι; διαμετρέω; διάμετρος; ἐναντίος; ἐναντιοῦμαι; ἐναντίωσις
— to own house ἐναντιοῦμαι ἑαυτῷ
orbit περίπατος
outcome ἀπόλυσις
overtake φθάνω
own οἰκεῖος
— domain, be in ἐπιτόπως; ἰδιοτοπέω; οἰκείως
— house, be in ἰδιοθρονέω

partnership (of triangle rulers) μετοχή
pass by παροδεύω, πάροδος
pass through διοδεύω
passivity πάθος
perceptible αἰσθητός
perigee περίγειος
period of time χρονογραφία
— — life activities πρακτικοὶ χρόνοι
phase φάσις
physical life periods σωματικοὶ χρόνοι
planet ἀστήρ; ἑπτὰ θεοί; πλανώμενος
pole πόλος
portend ἀποτελέω; ἐπιτελέω
portent ἀποτέλεσμα
position ἐποχή; στάσις; see also longitude; latitude
precede (in direction of daily rotation) προαναφέρομαι
preceding προηγούμενος
prestige φαντασία
prognostic ἀποτέλεσμα
progression περίπατος
project rays see radiate
property, locus of τόπος περὶ ἐξουσίας

quartile τετραγωνίζω; τετραγωνικῶς; ἐκ τετραγώνου πλευρᾶς
— —, side of τετράγωνος πλευρά

radiate; project rays ἀκτινοβολέω; ἀκτινοβολία; βάλλω
— between μεσεμβολέω
receive (in house, triangle, etc.) ὑποδέχομαι
receiving (planet), i.e. ruler of receiving house ὑποδεχόμενος
responsible locus αἰτιατικὸς τόπος
retrogradation ἀναποδισμός; ἀφαιρέτης; ἀφαιρετικός
return, common συναποκατάστασις
revolve, turn στρωφάω
rise ἀνατέλλω
— simultaneously with παρανατέλλω; συνανατέλλω
rising ἀνατολή; ἐπιτολή
rising time ἀναφορά; ἀναφορικός
— —, of short ὀλιγανάφορος
— —, of equal ἰσανάφορος
— —, equality ἰσανάφορον
rule (period of life) ἐπικρατέω; κρατέω
ruler of day πολεύων
— — hour διέπων
— — place in zodiac δεσπότης

— — receiving house οἰκοδέκτωρ; ὑποδεχόμενος
— — terms ὁρικὸς κύριος

Saturday ἡμέρα Κρόνου
Saturn Κρόνος; Φαίνων
saturnine κρονικός
seasonal hour καιρικὴ ὥρα
second interval δυάς
sect αἵρεσις
—, of same συναιρετίστης; αἱρετικός
set δύνω; δυσικός; δύσις; δυτικόν
seventh day (of moon; from Venus) ἑβδομαῖα
sextile ἑξάγωνος; ἑξάγωνος πλευρά
share συμμερίζω; συμμετέχω
short rising time ὀλιγανάφορος
side of hexagon ἑξάγωνος πλευρά
sign ζῴδιον; ζῷον
—, according to ζῳδιακῶς
— indicating weakness παθητικὸν ζῴδιον
— of full moon πανσεληνιακὸν ζῴδιον
slow motion, be in ὀλιγοδρομέω
slaves, locus of τόπος περὶ δούλων
solar sign ἡλιακὸν ζῴδιον
solstice ἰσημερία
south νότιος
star ἀστήρ; θεός
start; starting period; first chronocratorship ἄφεσις
starter (first chronocrator) ἀφέτης
starting (i.e. terms of starter) ἀφετικός
stationary στηριγμός; στηρίζω
step βαθμός
subtract ἀναποδίζω
succession (of chronocratorship) ἀντιπαράδοσις
summer solstice ἰσημερία ἐαρινός
superior aspect δεκατεύω; καθυπερτερέω; καθυπερτέρησις

table κανών; ὄργανον
take away παραιροῦμαι
take over (chronocratorship) παραλαμβάνω; παράληψις
terms ὅρια; ὅρος
— ruler of ὁρικὸς κύριος
theme θέμα; θεμάτιον
third (of moon) τριταῖα; τρίτη
thirds λεπτὰ τρίτα
time degree χρόνος; ὡριαῖος χρόνος
— — of daylight ἡμερινοὶ ὡριαῖοι χρόνοι
— — — night νυκτερινοὶ ὡριαῖοι χρόνοι
transfer of power παραδίδωμι; παράδοσις
travel, locus of τόπος περὶ ξένης
triad; third interval τριάς
tropical sign τροπικὸν ζῴδιον
trouble, locus of ἀργὸς τόπος
turning away ἀπόστροφος

unfavorably located ἐκπαραπίπτω; παραπίπτω
upper culmination ὑπέργειον μεσουράνημα
upwards (in direction of daily rotation) ἀνωφερῶς

Venus Ἀφροδίτη; Κύπρις; Φωσφόρος

water clock κλεψύδρα
wax αὐξιφωτέω
weakness, sign indicating παθητικὸν ζῴδιον
weapon (of Centaur) βέλος
winged sign πτερωτὸν ζῴδιον

yield (chronocratorship) παραδίδωμι; παράδοσις

zodiac, ideal νοητὸν ζῳδιακόν
— fixed ἀκίνητον —

CONCORDANCES

A. PAPYRUS PUBLICATIONS

Left column: number of papyrus (not page number) in the original publication;

*: unpublished

Right column: horoscope number (Chapter I)

BGU 957	—9	P. Osl. 6	150	P. S. I. 22	366
Class. Rev. 8,70	316	P. Osl. 163	190		370
P. Aberd. 13	187	P. Osl. 164	177		376,a
P. Aberd. 126	50,1,2,3	P. Osl. 165	394		381,s
P. Cornell Inv. 78 *	282	P. Oxy. 235	15/22	P. S. I. 23	338
P. Fay. 139	161	P. Oxy. 307 *	46		376,b
P. Fouad 6	125	P. Oxy. 585	220	P. S. I. 24	351
P. Fouad 7	293,XII	P. Oxy. 596	139		373
P. Hamb. 96	145		149		381,b
P. Harris 52	171	P. Oxy. 804	—3		385
P. Harris 53	245	P. Oxy. 1476	260	P. S. I. 25	465
P. Iand. 88	320	P. Oxy. 1563	258	P. S. I. 312	345
P. Iand. 89	227	P. Oxy. 1564	283	P. S. I. 764	277
P. Lond. 98	95	P. Oxy. 1565	293,VIII	P. S. I. 765	284
P. Lond. 110	137,b	P. Oxy. 2060	478	P. S. I. 1276	32
P. Lond. 130	81	P. Par. 19	137,a	P. Warren 21	217
P. Mich. 152	184	P. Par. 19 bis	137,c		219,II,1
P. Mich. 153	431	P. Princ. 75	138/161		219,II,12
P. Mich. Inv. 1461 *	208	P. Ryl. 524	250,2		244
P. Mich. Inv. 6329 *	182				

B. VETTIUS VALENS

Left column: page and line of ed. Kroll

Right column: horoscope L-number (Chapter II)

83,10	50	114,6	104,IV	167,11	75
83,27	95,V,18	114,10	108,III	170,15	152
84,3	85,II	114,16	112,VIII	174,8	120,II
84,12	83	128,31	123,VII	210,8	121
84,23	74,XI	129,1	97,II	223,1	120,II
85,1	72	129,8	87,VII	227,27	134,XI
85,9	82	129,15	86,XII	228,15	111,IV
85,19	97,XI	129,18	101,I	228,31	107
85,25	95,V,14	129,24	103	229,6	135,X
86,4	65,X	129,29	115,XII	229,13	112,VII
86,10	105	130,1	65,V	230,7	110,IX
86,17	61,V	130,6	88	230,11	153
86,24	109	130,11	89	230,14	102,XII,4
86,32	62	130,17	91	230,18	120,V
93,21	86,VIII	141,28	75	230,22	122,I,22
93,13	188	142,8	110,III	230,32	114,XI
93,25	78	143,5	61,X	231,5	123,I
94,3	101,III	147,11	114,V	231,32	113,VII
101,22	120,II	147,23	127,XI	233,25	37
111,33	87,I	153,15	75,1	234,7	68
112,5	118	154,1	74,IV	234,33	92,II
113,3	106	154,10	115,II	237,24	104,VII
113,11	85,XI	156,7	75	253,6	132
113,19	116	156,14	135,I	267,28	120,XII
113,25	117,XI	157,15	82	268,7	114,IX
113,32	92,XI	157,20	102,IV,a	268,16	122,XII
114,1	83	157,24	102,IV,b	268,29	118

269,11	117,VI	282,16	110,XII	287,11	118
270,6	74,XI	282,28	113,VII	287,17	127,VII
273,10	173	283,3	129	287,24	122,I,30
273,23	159	283,14	102,XII,14	287,30	133
273,33	162	284,3	105	288,20	142
274,4	122,VI,30	284,12	158	290,6	120,IX
275,12	122,VI,12	285,4	111,IX	318,15	61,X
280,34	124	286,29	114,VII	365,3	431
281,24	134,VI	287,1	120,II	365,29	419
282,3	108,XI				

C. CCAG

Left column: volume; page and line of CCAG
Right column: horoscope L-number (Chapter II)

1;100 (f. 105)	486	5,2;82,30	75	6;66,16	484
1;102 (νς)	474	5,2;85,17	152	6;67 (f. 301)	76
1;103 (νθ)	479	5,2;119,16	132	8,1;146,3	380
1;103 (νη)	475	5,2;120,30	112,VIII	8,1;221,1	428
1;106 (f. 125)	487	5,2;120,33	86,XII	8,1;231,5	516
1;107 (f. 126)	484	5,2;120,35	101,I	8,1;232,1	401
1;171,2 (f. 140)	497	5,2;121,3	103	8,1;235,1	488
2;168,4	122,XII	5,2;121,8	115,XII	8,1;240,17	482
2;168,24	118	5,2;121,13	65,V	8,2;58,3	380
4;106,6	497	5,2;121,17	88	8,2;67,34	380
5,2;44,26	173	5,2;121,22	89	8,2;82 var.	76
5,2;44,38	159	5,2;121,27	91	8,2;84,1	40
5,2;45,8	162	6;14 f. 300/301	479	8,2;85,18	113,IV
5,2;45,16	122,VI,30	6;64,26	478	8,4;221,1	440
5,2;52,24	91	6;65,22	483	8,4;224,21	463
				8,4;236,8	—42
				8,4;236,24	—71

D. OTHER SOURCES

Right column: horoscope number
*: unpublished

Abydos, graffito	353	Dura, Rep. IV, No. 235 *	219,I,c	Manetho III,738—750	L 80
Aelius Aristides	L 117,X	236	219,I,d	Marinus	L 412
Antiochus of Commagene	—61	237	219,I,e	Nimrud Dagh	—61
Bodleian Ostr. 244	207	238	219,I,f	Proclus	L 412
Dura, Rep. II, p. 161	176	239	219,I,g	SEG Nos. 364 to 369	219,I,a to g
Rep. IV, No. 231	p. 54 n. 6	Rep. VI, No. 736	250,1	Usener, Kl. Schr. III	L 621
232	219,I,a,b	Louvre 10390 *	326	Wilcken, Ostr. 1602	207

REFERENCES

Almagest: Claudius Ptolemaeus. *Syntaxis mathematica* ed. J. L. Heiberg; 2 vols., Leipzig, Teubner, 1898, 1903 [= Claudii Ptolemaei opera quae extant omnia, vol. I]. German translation: Karl Manitius, *Des Claudius Ptolemäus Handbuch der Astronomie*. 2 vols. Leipzig, Teubner, 1912, 1913.

Archiv: Archiv für Papyrusforschung.

Aristides: *Aelii Aristides Smyrnaei quae supersunt omnia*, Vol. II. Edidit Bruno Keil, Berlin. Weidmann, 1898.

BGU: *Aegyptische Urkunden aus den koeniglichen Museen zu Berlin*. Herausgegeben von der Generalverwaltung. *Griechische Urkunden*. Berlin, Weidmann, 1895 ff.

Bidez-Cumont, *Mages*: Joseph Bidez – Franz Cumont. *Les mages hellénisés*. 2 vols. Paris, 1938.

Bilabel, *Berichtig.*: Friedrich Bilabel. *Berichtigungsliste der griechischen Papyrusurkunden aus Ägypten* II,1 (1931); II,2 (1932). Continuation of Preisigke, *Berichtig*.

al-Bīrūnī, *Astrol.*: The book of instruction in the elements of the art of astrology by Abu'l-Rayḥān Muḥammad ibn Aḥmad al-Bīrūnī ... The translation facing the text by R. Ramsay Wright. London, Luzac, 1934.

al-Bīrūnī, *India*: *Alberuni's India*. An English edition, with notes and indices by Edward C. Sachau; 2 vols. London, Kegan Paul, 1910.

Blass, *Hyp. or.*[cb]: Fridericus Blass. *Hyperidis orationes quattuor cum ceterarum fragmenta*; ed. altera. Leipzig, Teubner, 1881.

Boer: *See* Paulus Alex. *See* Tetrabiblos.

Boissonade: *See* Proclus.

Boll, *FS*: Franz Boll. *Antike Beobachtungen farbiger Sterne*. Abh. d. Kgl. Bayerischen Akad. d. Wiss., philos.-philol. u. histor. Klasse, 30,1 (1916).

Boll, *SCP*: Franz Boll. *Studien über Claudius Ptolemäus*, reprinted from Jahrbücher f. classische Philologie, Suppl. 21. Leipzig, Teubner. 1894.

Boll, *Sphaera*: Franz Boll. *Sphaera. Neue griechische Texte und Untersuchungen zur Geschichte der Sternbilder*. Leipzig, Teubner, 1903.

Boll [4]: Franz Boll. Beiträge zur Ueberlieferungsgeschichte der griechischen Astrologie und Astronomie. SB d. philos.-philol. u. d. histor. Cl. d. k. bayerischen Akad. d. Wiss. zu München, 1899, 1 p. 77-140.

Bouché-Leclercq, *AG*: A. Bouché-Leclercq. *L'astrologie grecque*. Paris, Leroux, 1899.

Boulanger, *Ael. Ar.*: André Boulanger. *Aelius Aristide et la sophistique dans la province d'Asie au IIe siècle de notre ère*. Paris 1923 (= Bibliothèque des écoles françaises d'Athènes et de Rome. fasc. 26).

Brown, *Tables*: Ernest W. Brown. *Tables of the Motion of the Moon*. Yale University Press, New Haven, 1919.

Browne [1]: Henry Browne. The Egyptian Month Adrian. Journal of Classical and Sacred Philology 2 (1855) p. 44-46.

Burkitt [1]: F. C. Burkitt, Epiphoskein. Journal of Theological Studies 14 (1913) p. 538-546.

Cairo, *Cat. gén. X*: Service des Antiquités de l'Égypte. *Catalogue générale des antiquités égyptiennes du Musée du Caire*. Vol. X. Nos. 10001-10869. *Greek Papyri*, by B. P. Grenfell and A. S. Hunt. Oxford Univ. Press, 1903.

Cat. of Anc. Ms.: *Catalogue of Ancient Manuscripts in the British Museum*. Part I, Greek. London, 1881.

CCAG: *Catalogus Codicum Astrologorum Graecorum*. Bruxelles, Lamertin, 1898 to 1953, 12 vols.

Černý-Parker [1]: J. Černý, P. E. Kahle, R. A. Parker. The Old Coptic Horoscope. *JEA* 43 (1957) p. 86-100.

Champollion-Fig., *Ég.*: Champollion – Figeac. *Égypte ancienne*. L'univers. Histoire et description de tous les peuples. Paris, Firmin Didot, 1840.

Class. Rev.: Classical Review.

Cramer, *ARLP*: Frederick H. Cramer. *Astrology in Roman Law and Politics*. Memoirs of the Amer. Philos. Soc., vol. 37 (1954).

Crum [1]: W. E. Crum. An Egyptian text in Greek characters. *JEA* 28 (1942) p. 20-31.

Cumont, *EA*: Franz Cumont. *L'Égypte des astrologues*, Bruxelles. 1937.

Cumont, *Lux perp.*: F. Cumont, *Lux perpetua*, Paris, Geuthner, 1949.

Cumont [9]: Franz Cumont. Écrits hermétiques. Revue de philologie, de littérature et d'histoire ancienne 42 (1918) p. 63-79, 85-108.

Cumont [10]: Franz Cumont. Les noms des planètes et l'astrolatrie chez les Grecs. L'Antiquité Classique 4 (1935) p. 5-43.

Cumont [11]: Franz Cumont. Astrologues romains et byzantins, École française de Rome. Mélanges d'archéologie et d'histoire 37 (1918-1919) p. 33-54.

Cumont [12]: Franz Cumont, L'astrologue Palchos. Revue de l'instruction publique en Belgique 40 (1897) p. 1-12.

Curtis-Robbins [1]: Heber D. Curtis – Frank E. Robbins, An Ephemeris of 467 A.D. Publications of the Observatory of the University of Michigan, Vol. 6 No. 9.

Daremberg-Saglio, *Dict.*: Ch. Daremberg – Edm. Saglio, *Dictionnaire des antiquités grecques et romaines*, Paris, 1877-1912.

David-van Groningen, *PL*: M. David – B. A. van Groningen, *Papyrologisch Leerboek*, Leiden, Brill, 1940.

David-van Groningen, *PP*: M. David – B. A. van Groningen, *Papyrological Primer*, second (English) edition, Leyden, Brill, 1946.

Delambre, *HAA*: Delambre. *Histoire de l'astronomie ancienne*, 2 vols., Paris, 1817.

Delatte-Stroobant [1]: A. Delatte – P. Stroobant. L'horoscope de Pamprépios, professeur et homme politique de Byzance. Académie Royale de Belgique, Bulletins de la Classe des Lettres et des Sciences Morales et Politiques. 5e série, t. 9 (1923) p. 56-76.

Dittenberger, *OGI*: Wilhelm Dittenberger. *Orientis graeci inscriptiones selectae*. 2 vols. Leipzig, Hirzel, 1903.

Duhem, *SM*: Pierre Duhem, *Le système du monde, histoire des doctrines cosmologique de Platon à Copernic*. 5 vols. Paris, Hermann, 1913-1917.

Dura, *Rep.*: *The Excavations at Dura-Europos, Preliminary Reports*. New Haven, Yale University Press, 1929 ff.

Edelstein, *Asclepius*: Emma J. and Ludwig Edelstein, *Asclepius*. 2 vols. Baltimore, Johns Hopkins Press, 1945.

Engelbrecht, *Heph.*: August Engelbrecht, *Hephaestion von Theben und sein astrologisches Compendium*. Wien, 1887.

Fabricius, *Vita Pr.*: *Procli, Philosophi Platonici, Vita, scriptore Marino Neapolitano*, ed. Fabricius. London, 1703.

Festugière, *Rév.*: A.-J. Festugière. *La révélation d'Hermès Trismégiste I. L'astrologie et les sciences occultes*. 2nd ed. Paris, 1950.

Franz, *CIG 3*: Johann Franz, *Corpus Inscriptionum Graecarum*, vol. 3. Berlin 1853.

Freudenthal [1]: J. Freudenthal. Ueber die Lebenszeit des

Neuplatonikers Proklus, *Rheinisches Museum* NF 43 (1888) p. 486-493.

Gagé [1]: Jean Gagé, L'horoscope de Doura et le culte d'Alexandre sous les Sévères. *Bulletin de la Faculté des Lettres de Strasbourg* 33. No. 3 (1954) p. 151-168.

Garnett [1]: R. Garnett. The date of the Ἀποτελεσματικά of Manetho. *Journal of Philology* 23 (1895) p. 238-240.

Geminus, *Isag.*: Geminus. *Elementa astronomiae*, ed. C. Manitius. Leipzig, Teubner, 1898.

GGA: *Göttingische gelehrte Anzeigen.*

Ginzel, *Chron.*: F. K. Ginzel. *Handbuch der mathematischen und technischen Chronologie*; 3 vols. Leipzig, Hinrichs, 1906-1914.

Ginzel, *Kanon*: F. K. Ginzel. *Spezieller Kanon der Sonnen- und Mondfinsternisse*. Berlin, 1899.

Goell [1]: Theresa Goell. Nimrud Dagh, the tomb of Antiochus I, King of Commagene. *Archaeology* 5 (1952) p. 136-144.

Goodrich [1]: Sidney P. Goodrich. The Dating of a Horoscope. *Archiv* 12 (1937) p. 67 f.

Goodwin [1]: C. W. Goodwin, Sur un horoscope grec contenant les noms de plusieur Decans. *Mélanges Égyptologiques*, 2ᵉ sér. Par F. Chabas. Chalon-sur-Saone, 1864. No. XIII, p. 294-306.

Goodwin [2]: C. W. Goodwin. On an Egyptian text in Greek characters. *Z. f. ägypt. Sprache* 6 (1868) p. 18-24.

Grenfell [1]: B. P. Grenfell. A horoscope of the year 316 A.D. *Class. Rev.* 8 (1894) p. 70 f.

Griffith [1]: F. Ll. Griffith. The Old Coptic horoscope of the Stobart collection. *Z. f. ägypt. Sprache* 38 (1900) p. 71-85.

Griffith [2]: F. Ll. Griffith. The date of the Old Coptic texts and their relation to Christian Coptic. *Z. f. ägypt. Sprache* 39 (1901) p. 78-82.

Grumel, *Chron.*: V. Grumel. Traité d' Études Byzantines 1. *La Chronologie*. Bibliothèque Byzantine, Paris, 1958.

Gundel, *DD*: Wilhelm Gundel. *Dekane und Dekansternbilder. Studien der Bibliothek Warburg* 19. Glückstadt u. Hamburg, 1936.

Gundel, *HT*: Wilhelm Gundel. *Neue astrologische Texte des Hermes Trismegistos*. *Abh. d. Bayerischen Akad. d. Wiss., Philos.-hist. Abt.*, NF 12, 1936.

Halma, *Tab. man.*: N. Halma. *Tables manuelles de Ptolemée et de Théon*, Paris, 1822, 1823, 1825.

Hanell [1]: Krister Hanell. Das Menologium des Liber glossarum. *Bull. de la Soc. Royale des Lettres de Lund*, 1931-1932, p. 7-38.

Heegaard [1]: Poul Heegaard. Three Horoscopes in the Oslo Collection of Papyri. *Symbolae Osloenses*, Fasc. 15-16 (1936) p. 98-100.

Hincks [1]: Edward Hincks, On the various years and months in use among the Egyptians. *Transactions of the Royal Irish Academy*, vol. 24, Polite Literature (1865) p. 25-66.

Hohmann [1]: Franz Hohmann. Zur Chronologie der Papyrusurkunden (Römische Kaiserzeit). Thesis. Greifswald, 1911.

Hombert-Préaux [1]: Marcel Hombert - Claire Préaux. Note sur la durée de la vie dans l'Égypte Gréco-Romain. *Chronique d'Égypte* 20, Nos. 39,40 (1945) p. 139-146.

Honigmann, *SK*: Ernst Honigmann. *Die sieben Klimata*. Heidelberg, 1929.

Hughes [1]: G. R. Hughes. A demotic astrological text. *JNES* 10 (1951) p. 256-264.

Hunt [1]: Arthur S. Hunt. The Warren Magical Papyrus. *Studies presented to F. Ll. Griffith, Egypt Exploration Society*. Humphrey Milford, London, 1932, p. 233-240.

Hunt-Edgar, *Sel. P.*: A. S. Hunt - C. C. Edgar. *Select Papyri*. Vol. I. *Private Affairs*. The Loeb Classical Library, 1932.

Ideler, *Chron.*: Ludwig Ideler. *Handbuch der mathematischen und technischen Chronologie*. Berlin 1825, 1826. (2 vols.).

Jalabert-Mouterde, *IGL SYR*: Louis Jalabert - René Mouterde, *Inscriptions grecques et latines de la Syrie*. Institut Français d'Archéologie de Beyrouth, Paris, Geuthner, 1955 ff.

JAOS: *Journal of the American Oriental Society.*

JCS: *Journal of Cuneiform Studies.*

JEA: *The Journal of Egyptian Archaeology.*

JNES: *Journal for Near Eastern Studies.*

Johnson, *DS*: Jotham Johnson. Dura Studies. Thesis, Univ. of Pennsylvania, Philadelphia, 1932.

Johnson [1]: Jotham Johnson. Tell time by the stars. *Archaeology* 4 (1951) p. 76-82.

Jonckheere [1]: Frans Jonckheere. La durée de la gestation d'après les textes égyptiens. *Chronique d'Égypte* 30 No. 59 (1955) p. 19-45.

Knudtzon-Neugebauer [1]: Erik J. Knudtzon - O. Neugebauer. Zwei astronomische Texte. *Bull. de la Soc. Royale des Lettres de Lund*, 1946-1947, p. 77-88.

Kroll [3]: W. Kroll. Aus der Geschichte der Astrologie. *Neue Jahrbücher für das Klassische Altertum* 7 (1901) p. 559-577.

Kroll [5]: W. Kroll. Astrologisches. *Philologus* 57 (N.S. 11) (1898) p. 123-133; p. 192.

Kubitschek, *Zeitr.*: Wilhelm Kubitschek. *Grundriss der antiken Zeitrechnung. Handbuch d. Altertumswiss.* I,7. München, Beck, 1928.

Kubitschek [1]: Wilhelm Kubitschek. Die Kalenderbücher von Florenz, Rom und Leyden. *Denkschriften d. Kaiserl. Akad. d. Wiss. in Wien, Philos.-hist. Kl.*, 57,3 (1915).

Kugler, *SSB*: Franz Xaver Kugler. *Sternkunde und Sterndienst in Babel*. 2 vols. 1907-1924. 3 Ergänzungshefte, 1913-1935 (No. 3 by Johann Schaumberger). Münster, Aschendorff.

Laudien, *GPO*: Arthur Laudien. *Griechische Papyri aus Oxyrhynchos für den Schulgebrauch ausgewählt*. Berlin, Weidmann, 1912.

Lauth [2]: F. J. Lauth. König Nechepsos, Petosiris, und die Triakontaëteris. *SB d. philos.-philol. u. hist. Cl. d. k. bayer. Akad. d. Wiss. zu München*, 1875 II p. 89-144.

Loth [1]: Otto Loth. Al-Kindī als Astrolog. *Morgenländische Forschungen . . . H. L. Fleischer . . . gewidmet. . . .* Leipzig, Brockhaus, 1875, p. 261-309.

Maass, *CAR*: Ernst Maass. *Commentariorum in Aratum reliquiae*. Berlin, Weidmann, 1898.

Macnaughton, *SEC*: Duncan Macnaughton. *A scheme of Egyptian chronology*. London, Luzac, 1932.

Magie, *Roman Rule*: David Magie. Roman rule in Asia Minor to the end of the third century after Christ. 2 vols., Princeton University Press, Princeton, 1950.

Manetho: *Manethonis Apotelesmaticorum qui feruntur libri VI*, ed. A. Koechly. Leipzig, Teubner, 1858. Also in: *Poetae bucolici et didactici . . . Aratus, Manethonis, Maximi et aliorum astrologica*, ed. A. Koechly, Paris, Firmin-Didot 1862 (reprint 1931).

Manitius: See Almagest.
See Geminus.

Marinus: See Fabricius and Proclus.

Mayser, *Gramm.*: Edwin Mayser. *Grammatik der griechischen Papyri aus der Ptolemäerzeit*. Berlin-Leipzig, Walter de Gruyter. 2 vols. in 5 parts; 123-1934.

Mit. P. Erzh. Rainer: *Mittheilungen aus den Sammlungen der Papyrus Erzherzog Rainer*.

Mond-Myers, *Buch.*: R. Mond - O. H. Myers, *The Bucheum*. Egypt Exploration Society, Mem. 41, London 1934.

Neugebauer [1]: O. Neugebauer. Demotic horoscopes. *JAOS* 63 (1943) p. 115-127.

REFERENCES

Neugebauer [2]: O. Neugebauer. The horoscope of Ceionius Rufius Albinus. *Amer. Jour. Philology* 74 (1953) p. 418-420.

Neugebauer [4]: O. Neugebauer. Egyptian planetary texts. *Trans. Amer. Philos. Soc.*, NS 32 (1942) p. 209-250.

Neugebauer [5]: O. Neugebauer. On some astronomical papyri and related problems of ancient geography. *Trans. Amer. Philos. Soc.*, NS 32 (1942) p. 251-263.

Neugebauer [8]: O. Neugebauer. The astronomical treatise P. Ryl. 27. *Kgl. Danske Videnskabernes Selskab, Hist.-filol. Medd.* 32,2 (1949).

Neugebauer [9]: O. Neugebauer. The Egyptian "Decans." *Vistas in Astronomy*, ed. by Arthur Beer, I p. 47-51. Pergamon Press, London-New York, 1955.

Neugebauer-Sachs [1]: O. Neugebauer - A. Sachs. The "Dodekatemoria" in Babylonian astrology. *Archiv für Orientforschung* 16 (1952) p. 65 f.

Neugebauer-Van Hoesen: O. Neugebauer - H. B. Van Hoesen. Horoscope P. Ryl. 524. *Aegyptus* 32 (1952) p. 333-335.

Neugebauer-Volten [1]: O. Neugebauer - A. Volten. Untersuchungen zur antiken Astronomie IV. Ein demotischer astronomischer Papyrus (Pap. Carlsberg 9). *QS* B4 (1938) p. 383-406.

P. V. Neugebauer, *ACh*: P. V. Neugebauer. *Astronomische Chronologie*. Berlin-Leipzig, Walter de Gruyter, 2 vols., 1929.

P. V. Neugebauer, *TACh*: P. V. Neugebauer. *Tafeln zur astronomischen Chronologie*. Leipzig, Hinrichs; 3 vols., 1912-1922.

P. V. Neugebauer [1]: P. V. Neugebauer. Genäherte Tafeln für Sonne und Planeten. *Astron. Nachrichten* 248, Nr. 5937 (1932) cols. 161-184.

Nicklin [1]: T. Nicklin. A horoscope from Egypt. *Classical Review* 16 (1902) p. 119 f.

Nock-Festugière, *Corpus Herm.*: A. D. Nock - A.-J. Festugière, *Corpus Hermeticum*, Paris 1945-1954, 4 vols.

Noë [1]: Anton Rudolf Noë. Die Proklusbiographie des Marinos. Inaugural-Dissertation, Heidelberg, 1938.

Oertel, *Liturgie*: Friedrich Oertel. *Die Liturgie*. Leipzig, Teubner, 1917.

Oppolzer, *Canon*: Th. Oppolzer. *Canon der Finsternisse. Akad. d. Wiss., Wien Math.-Nat. Cl., Denkschriften* 52 (1887). [Reprinted: Stechert, New York, 1921.]

P. Aberd.: Eric G. Turner. *Catalogue of Greek and Latin Papyri and Ostraca in the Possession of the University of Aberdeen. Aberdeen University Studies* No. 116. Aberdeen, Univ. Press, 1939.

P. Fay.: Bernhard P. Grenfell, Arthur S. Hunt, David G. Hogarth. *Fayûm Towns and their Papyri*. Egypt Exploration Fund, Graeco-Roman Branch, London, 1900.

P. Fouad: A. Bataille, O. Guéraud and others. *Les Papyrus Fouad I.* Nos. 1-89. *Publications de la Société Fouad I de Papyrologie. Textes et Documents III.* Le Caire, 1939.

P. Grenf. II: Bernhard P. Grenfell, Arthur S. Hunt. *Greek Papyri, Series II. New Classical Fragments and other Greek and Latin Papyri.* Oxford, Clarendon Press, 1897.

P. Hamb.: Paul M. Meyer. *Griechische Papyrusurkunden der Hamburger Staats- und Universitätsbibliothek.* Bd. 1, Teubner, Leipzig-Berlin, 1911-1924.

P. Harris: J. Enoch Powell. *The Rendel Harris Papyri of Woodbrooke College, Birmingham.* Cambridge Univ. Press, 1936.

P. Iand.: Josef Sprey. *Papyri Iandanae*, Fasc. 5. Teubner, Leipzig-Berlin, 1931.

P. Lond.: F. G. Kenyon. *Greek Papyri in the British Museum; Catalogue, with Texts. Facsimiles.* London, 1893 [2 vols.].

P. Mich.: John Garrett Winter. *Papyri in the University of Michigan Collection, Miscellaneous Papyri. University of Michigan Studies, Humanistic Series* vol. 40. Michigan Papyri vol. 3. Ann Arbor, Univ. of Michigan Press, 1936.

P. Osl.: S. Eitrem, Leiv Amundsen. *Papyri Osloenses.* Fasc. 1-3. Norske Videnskaps-Akademi i Oslo, 1925-1936.

P. Oxy.: Bernhard P. Grenfell, Arthur S. Hunt and others. *The Oxyrhynchus Papyri.* Egypt Exploration Fund, Graeco-Roman Branch, London 1898 ff.

P. Par.: *Notices et extraits des manuscrits de la Bibliothèque impériale et autres bibliothèques.* Tome 18, IIe partie. *Papyrus Grecs du Louvre et de la Bibliothèque impériale. Publication préparée par Letronne, exécutée par MM. IV. Brunet de Presle et E. Egger.* Paris, 1865 [2 vols., text and plates].

P. Petrie: John P. Mahaffy - J. Gilbart Smyly. *The Petrie Papyri with transcriptions, commentaries and index.* Royal Irish Academy. *Cunningham Memoirs*, Nos. 8, 9, and 11, 1891-1905 (3 vols.).

P. Princ.: Edmund Harris Kase. Princeton University. *Papyri in the Princeton University Collections.* Vol. 2. Princeton Univ. Press, 1936 (= *Studies in Papyrology* No. 1. Edited by Allan Chester Johnson).

P. Reinach: *Papyrus grecs et démotiques recueilles en Égypte et publiés par Théodore Reinach avec le concours de MM. W. Spiegelberg et S. De Ricci.* Paris, Leroux, 1905.

P. Ryl.: C. H. Roberts. *Catalogue of the Greek and Latin Papyri in the John Rylands Library Manchester. Vol. III. Theological and Literary Texts* (Nos. 457-551). Manchester, Univ. Press, 1938.

P. S. I.: G. Vitelli, M. Norsa and others. *Pubblicazioni della Società Italiana per la ricerca dei papiri greci e latini in Egitto. Papiri Greci e Latini.* Firenze, 1912 ff.

P. Tebt.: Bernhard P. Grenfell - Arthur S. Hunt - J. Gilbart Smyly - C. C. Edgar. *The Tebtunis Papyri.* Egypt Exploration Fund, Graeco-Roman Branch. London 1902-1938 (4 vols.).

P. Warren: M. David, B. A. van Groningen, J. C. van Oven. *The Warren Papyri (P. Warren). Papyrologica Lugduno-Batava.* Vol. I. Brill, Leyden, 1941.

Paulus Alexandrinus, *Elementa Apotelesmatica*, ed. Ae. Boer. Leipzig, Teubner, 1958.

Perdrizet-Lefebvre, *GG*: Paul Perdrizet - Gustave Lefebvre. *Les graffites grecs du Memnonion d'Abydos.* Nancy-Paris-Strasbourg, Berger-Levrault, 1919.

Peters-Knobel, *PCS*: Christian Heinrich Friedrich Peters - Edward Ball Knobel. *Ptolemy's Catalogue of Stars.* Carnegie Institution of Washington, Publication No. 86 (1915).

Pinches-Sachs, *LBAT*: Late Babylonian Astronomical and Related Texts, copied by T. G. Pinches and J. N. Strassmaier, prepared for publication by A. J. Sachs. Brown University Press, Providence, R. I., 1955 [= Brown University Studies 18].

Porter-Moss: Bertha Porter - Rosalind L. B. Moss. *Topographical bibliography of ancient Egyptian hieroglyphic texts, reliefs, and paintings.* Oxford, Clarendon Press, 1927 ff.

Preisigke, *Berichtig*: Friedrich Preisigke. *Berichtigungsliste der griechischen Papyrusurkunden aus Ägypten.* Berlin und Leipzig, Ver. Wiss. Verl., 1922. For continuation see Bilabel, *Berichtig*.

Preisigke, *WB*: Friedrich Preisigke. *Wörterbuch der griechischen Papyrusurkunden*, 3 vols., Berlin, 1925-1931; vol. 4,1 (1944) by E. Kiessling.

Proclus: *Opera inedita...*, ed. Victor Cousin, Paris, 1864 [contains, p. 1-66, the Life of Proclus, by Marinus, Greek and Latin tr. by J. F. Boissonade]. See also Steck-Schönberger, *PD*.

Ptolemy: See Almagest.
See Tetrabiblios.

QS: *Quellen und Studien zur Geschichte der Mathematik, Astronomie und Physik.*

RE: *Paulys Real-Encyclopädie der classischen Altertumswissenschaft. Neue Bearbeitung.* 1894 ff.

Rehm [1]: Albert Rehm, ΜΝΗΣΘΗ. *Philologus* 94 (1940-41) p. 1-30.

Rehm, *Parap.*: Albert Rehm. *Parapegmastudien. Abh. d. Bayerischen Akad. d. Wiss., Philos.-hist. Abt.*, N.F. Heft 19 (1941).

Riess [1]: Ernestus Riess. Nechepsonis et Petosiridis fragmenta magica. *Philologus*, Supplementband 6 (1891-1893) p. 325-388.

Roberts, *GLH*: C. H. Roberts. *Greek Literary Hands 350 B.C.-A.D. 400. Oxford Palaeographical Handbooks.* Oxford, Clarendon Press, 1955.

Roscher: W. H. Roscher. *Ausführliches Lexikon der griechischen und römischen Mythologie.* Leipzig, Teubner, 1884-1937. 6 vols.

Sachs [1]: A. Sachs. A classification of the Babylonian astronomical tablets of the Seleucid period. *JCS* 2 (1948) p. 271-290.

Sachs [2]: A. Sachs, Babylonian Horoscopes. *JCS* 6 (1952) p. 49-75.

SB: *Sitzungsberichte*.

Schlachter, *Globus*: Alois Schlachter (ed. Friedrich Gisinger). *Der Globus. Stoicheia* 8 (1927), Leipzig, Teubner.

Schmidt, *EAW*: C. P. Schmidt. *Kulturhistorische Beiträge zur Kenntnis des griechischen und römischen Altertums. Heft II. Die Entstehung der antiken Wasseruhr.* Leipzig, Durr, 1912.

Schott [1]: A. Schott. (Review of) Gundel, Neue astrologische Texte des Hermes Trismegistos. *QS* B 4 (1937) p. 167-178.

Schram, *Taf.*: Robert Schram. *Kalendariographische und chronologische Tafeln.* Leipzig, Hinrichs, 1908.

Schubart, *Pal.*: Wilhelm Schubart. *Griechische Palaeographie. Handbuch d. Altertumswissenschaft* I,4, 1. München, C. H. Beck, 1925.

Schubart, *Pap. Graec. Berol.*: Wilhelm Schubart. *Papyri Graecae Berolinenses.* Bonn, Marcus and Weber, 1911.

SEG: *Supplementum Epigraphicum Graecum*. Brill, Leiden.

Sethe, *Zeitr.*: Kurt Sethe. Die Zeitrechnung der alten Aegypter im Verhältnis zu der der andern Völker. *Nachrichten d. K. Gesellsch. d. Wiss. zu Göttingen, Philol.-histor. Kl.* 1919, p. 287-320; 1920, p. 28-55; p. 97-141.

Seyffarth, *Beitr.*: G. Seyffarth. *Beitraege zur Kenntniss der Literatur, Kunst, Mythologie und Geschichte des alten Aegypten.* Leipzig, Barth, 1826-1833 (5 Hefte).

Smyly [2]: J. Gilbart Smyly. On the fixed Alexandrine year. *Hermathena* 11 (1901) p. 81-88.

Steck-Schönberger, *PD*: Max Steck-Leander Schönberger. *Proklus Diadochus, 410-485, Kommentar zum ersten Buch von Euklids "Elementen."* Halle, 1945.

Stegemann [1]: Victor Stegemann, Horoskopie [Art. in] *Handwörterbuch des deutschen Aberglaubens* IV col. 342-400, Berlin-Leipzig, 1931, 1932.

Stein, *Präf.*: A. Stein. *Die Präfekten von Ägypten in der römischen Kaiserzeit.* Bern, 1950. (Dissertationes Bernenses ser. 1, 1.)

Stein [1]: J. W. Stein. Les Dates de la Naissance et de la Mort du Philosophe Proclus. Specola Astronomica Vaticana. *Miscellanea Astronomica*, vol. 3 (1950) Artic. 99 (p. 71-74).

Stephanus: *See* Usener.

Tannery, *Mém. Sci.*: Paul Tannery, *Mémoires scientifiques*. Paris, 1912 ff.

Taqizadeh [1]: S. H. Taqizadeh, Some Chronological Data relating to the Sasanian Period. *Bull. of the School of Oriental Studies* (University of London) 9 (1937-39) p. 125-139.

Tetrabiblos: *Claudii Ptolemaei opera quae extant omnia* vol. III, 1. *Apotelesmatica*. ed. F. Boll et Ae. Boer. Leipzig, Teubner, 1940.
  Ptolemy, *Tetrabiblos*. Edited and translated by F. E. Robbins. The Loeb Classical Library. 1940.

Thorndike, *Comets*: Lynn Thorndike. *Latin treatises on comets between 1238 and 1368 A.D.* Univ. of Chicago Press, 1950.

Usener, *Kl. Schr.*: Hermann Usener. *Kleine Schriften*. Teubner, Leipzig-Berlin, 1914.

Van der Waerden [2]: B. L. van der Waerden. Egyptian "Eternal Tables." *Koninkl. Nederl. Akad. van Wetensch., Proc.* 50 (1947), p. 536-547, 782-788.

Van Groningen, *PP*: Bernhard Abraham van Groningen - Martin David, *Papyrological Primer*, 2nd ed. Brill, Leyden, 1946.

Vat. gr. 1291: Unpublished MS of the Vatican Library containing tables which are essentially equivalent to Halma, *Tab. man.*

Vettius Valens: *Anthologiarum libri*, ed. Wilhelm Kroll. Berlin, Weidmann, 1908.

Vogt, *GK*: Heinrich Vogt. *Der Kalender des Claudius Ptolemäus* [= *Griechische Kalender V*. Herausgeg. von F. Boll]. *SB Heidelberger Akad. d. Wiss., Philos.-hist. Kl.* 1920, Abh. 15.

Wessely [1]: C. Wessely. Griechische Zauberpapyrus von Paris und London. *Denkschr. d. Kaiserl. Akad. d. Wiss., Phil.-hist. Cl.* 36,2 Wien 1888, p. 27-208.

Wilcken, *Ostr.*: Ulrich Wilcken. *Griechische Ostraca*. Gieseke-Devrient, Leipzig-Berlin, 1899. 2 vols.

Wilcken, *UPZ*: Ulrich Wilcken. *Urkunden der Ptolemäerzeit (ältere Funde).* De Gruyter, Berlin-Leipzig, 1927 ff.

Winter, *LLP*: John Garrett Winter. *Life and Letters in the Papyri.* Ann Arbor, Univ. of Michigan Press, 1933.

Young, *Hierogl.*: *Hieroglyphics collected by The Egyptian Society*, arranged by Thomas Young, London, 1823 [Plates 1 to 40]. *Hieroglyphics continued by the Royal Society of Literature*, arranged by Thomas Young, Vol. II, London, 1828 [plates 41 to 98].

SUBJECT INDEX

Concerning topics not fully covered by the *Glossary of Astrological and Technical Terms* (p. 2 ff.), the list of *Names of Persons and Places* (p. 200), of *Constellations and Star Names* (p. 201) and the *English Glossary* (p. 202) of astrological terms.

The numbers refer to pages.

Abydos, Egypt, 69
Abydos, Hellespont, 142
accuracy of computations, 24, 89, 143, 171, 188
Aelius Aristides, 113
Agathos Daimon, 2, *et passim*
Alexander, era, 152, 166
Alexander Macedonius, 54
Alexandria, 3, 4, 144, 152, 170, 182
Alexandrian calendar, *see* calendar
Almagest, *see* Ptolemy
Anicius Acilius Aginatius Faustus, 147
anomaly, 68, 131, 171
anonymous of the year 379, 187 n. 61
Antares, 153
Antigonus of Nicaea, 79, 80, 90, 108, 115, 149, 176, 186
Antioch, 147, 187
Antiochus I of Commagene, 14, 161
Antiochus, astrologer, 188
Antoninus, 165
aphetes, *see* starter
apogee of Mars, 68
apoklima, 2, *et passim*
Arabic horoscopes, 161, 172
Arabs, 159
archon Nicagoras, 135
Argo, 115
Aristides, 113, 176
Armenia, 159
arrangement of planets, 14, 164
ascendant, *see* Horoscopos
Asclepius, 43, 145, 171
aspects, 2, *et passim*
astrolabe, 158, 160
Athenian calendar, *see* calendar
Athens, 140, 143, 187
Augustus, 3, 165
Avidius Cassius, 120

Babylon, 3, 184, *et passim* (clima of Bab.)
Babylonian astrology, 6, 162
Babylonian astronomy, 3, 12, 14, 25, 26, 170
Babylonian horoscopes, 161, 162, 164
Balbillus, 76, 176
Bartoletti, V., 65
basis, 9, 92, 130, 131
Berossus, 185 n. 19, n. 25
Bes, 69
Beutler, R., 136
biographical events:
  banished, 91, 99, 101, 109, 111, 117, 118
  battles, 119, 159
  danger at sea, 104, 115, 125, 142, 143, 145
  divorce, 106
  escape, 33, 118, 119, 121, 123, 125, 145
  fugitive, 120, 140
  gifts, 80, 97, 108, 111, 126
  imprisoned, 113, 118, 119, 121
  inheritance, 97, 104, 106, 107, 122
  law suits, 32, 103, 104, 106, 108, 111, 113, 117, 118, 120, 122, 125, 126
  riots, 86, 93, 103, 119, 121
  shipwreck, 87, 104, 115, 143, 145
  theft, robbery, 93, 109, 123, 143, 148
  voyage by ship, 104, 115, 125, 142, 143, 144, 146
  women, 32, 33, 107, 108, 113, 117, 118, 121, 126, 127, 142
Bīrūnī, 28, 68 n. 1
bodily defects, *see* injury
Boer, E., 9 n. 25, 10, 155
books and papyrus, as cargo, 145
Bouché-Leclercq, 2, 15, 176, *et passim*
Boulanger, 113
bronze objects, 145
Brown, E. W., Tables, 2
Bulgars, 159
Byzantium, 4, 140, 141, 170 n. 70, 188

Caesarea, 142, 187
calendar, Alexandrian, 3, 17, 18, 25, 46, 50, 51, 173, *et passim*
  Athenian, 135
  Egyptian, 3, 17, 18, 25, 38, 46, 50, 51, 166, 173
  Macedonian, 3, 15, 16, 49, 54, 58, 166
  Roman, 23, 25, 166, 169
camels, 143
Cappadocia, 159
Caracalla, 165
Carinus, 165
Carmody, F., 172
Cassius Dio, 109
Ceionius Rufius Albinus, 161 n. 5
Centaur, 92, 170
centers, 3, 60, 138, *et passim*
Černý, J., 29, 32, 161 n. 3
Chalcidius, 141
Cilicia, 159
Cleomedes, 141
climata, 3, 111, 138, 170, 182, *et passim*
Colchis, 159
collection of horoscopes, 162
Commagene, 14
commanding signs, 5, 28
Commodus, 165
conception, 15, 28, 86, 112, 131, 187
connected, 146 n. 7, 157
Constantius, 165

contact, 5, 146, 157
Coptic, 29, 33, 72
Cornelius Saloninus, 165
coronation horoscopes, 16, 147
Cramer, F. H., 80, 90 n. 2
critical number, 102 n. 1(b)
critical years, 97, 110, 111
Critodemus, 79 n. 1, 82, 83, 84, 97, 102, 176, 185
Cumont, 80, 148, 185, 188, 190, *et passim*

Dating, method of, 1
day, epoch, 18, 25, 167, 169
De planetis, 115 n. 7, 121, 176
death (*see also* length of life):
  in bath, 99, 122
  beheaded, 94, 95, 98
  burned to death, 99, 102
  choked to death, 122
  drowning, 95, 125
  executed, hanged, 96, 109, 141
  killed, 108, 109, 136
  poison, 84, 108
  by snake, 33
  suicide, 96, 99
  wild beasts, 96, 112
death of children, 100, 103, 126, 129, 130, 141, 146
  of concubines, 107, 127
  of father, 108, 109, 115, 116, 128
  of mother, 103, 108, 116
  of relatives, 33
  of slaves, 111
  of wife, 121
decans, 5, 23, 25, 37, 66, 153
decimal fractions of a degree, 25
Delambre, 135
Delatte, A., 140 n. 2, 141
Demetrius, 142 n. 1
demotic horoscopes, 161 n. 33, 168
demotic planetary tables, 166 n. 3, 173, 180
Dendera, 5, 6
depression, 7, 29, 36, *et passim*
destroyer, 76, 78; *cf. also* length of life
diagrams of horoscopes, 18, 49, 54, 55, 58, 76, 80, 90 n. 2, 144 n. 3(a), 144 n. 5(b), 146 n. 2, 147 n. 1, 156, 159 n. 11, 163
diameter (= opposition), 6, 171 n. 77
  of moon, as unit, 26, 171
Diocletian, era, 6, 165
disjunct signs, 13 n. 35
distance, 97
  from M, 134, 150, 152, 158
dodekatemoria, 6, 12, 25, 34, 35, 145, 146, 149

dominant, 3
Domitia Paulina, 90 n. 7a
Doresse, J., 64
Dorotheus, 143, 147 n. 21, 149, 185
double dates, 18, 24, 25, 167
double prerogative, 68 n. 2
dream, 34, 113, 167
Dura, 49, 54, 58, 162
dysis, 7

Eclipse, 136, 148, 171, 174
Edfu, 5, 6, 37
Egypt, 159
Egyptian calendar, see calendar
elongation, maximum, of Venus, 141
epanaphora, 7, 8
ephemeris for 467, 166 n. 29
Epiphanius, 158
equinoctial hours, 169
equinoxes, 4, 24, 25, 171
equi-zonal signs, 142 n. 12
eros, lot of, 9, 45, 67
errors of prediction, 147, 149
Eudaemon, see Valerius Eudaemon
Eudoxus, 15
Eutocius, 172, 176, 188
evening epoch, 18, 25, 169
exaltation, 7, 29, 36, et passim

Fabricius, 135
father, death of, 109, 115, 116, 128
female sign, 9, 19, 23, 29
fictitious data, 83, 87 n. 21, 89, 134, 152
figures, see diagrams
finger, as unit, 25, 171
Firmicus Maternus, 6, 25, 161 n. 5
fixed stars, 26, 153, 171, 187
Flavius Trocondus, 147
forty days of danger, 17, 80, 91, 109
Fotheringham, 52, 56, 60, 164 n. 16
Freudenthal, J., 136

Gagé, 54 n. 10
Galerius, 165
Galienus, 165
Ganymede, 27
geographical latitude, 3, 11
gnomon, horoscopic and solar, 136, 140
Goedicke, H., 168 n. 48
Goell, T., 14
graffiti, 49, 54, 58, 69, 163, 166
Gundel, W., 6, 27, 186

Hadrian, 17, 80, 91, 109, 186
Halley, 114
Handy Tables, 2, 4 n. 5, 9, 10, 134, 188, et passim
Heegaard, P., 48, 50, 52, 73
Hephaestion, 6, 80, 131, 176, 187
Heracleopolis Magna, 16
Heraclides Ponticus, 141
Hermes Trismegistus, 27, 186
Hermogenes, son of Theotimus, 71, 165
Hermopolis, 24
Homer, quoted, 155
horimaia, 146
horoscopes, the 36, 36, 37
horoscopic gnomon, 136, 140

Horoscopos, 7, et passim
hours, 24, 169
houses, 7, 10, 28, 36
Hultsch, 189 n. 76
Humann, 14
Hygeia, 69 n. 2, 145
Hyperberetaios = October, 49, 55
Hyperides, funeral oration, 28, 38

Iberia, 159
illness, 32, 87, 101, 108, 111, 113, 127
 asthma, 90
 breast, chest, throat, 87, 88
 convulsions, 129
 dropsy, 90
 eczema, lichens, mange, 93, 129
 eye, 115, 118, 126
 feet, 103, 118
 fever, 32
 flow of blood, 109
 gout, 94, 95
 leprosy, 93
 lunacy, madness, 103, 104, 127
Illus, 141
imum coeli, 9
India, 159
indictio, 165
injury or bodily defects (see also illness):
 baldness, 115
 broken limbs, 87
 castration, 114
 genitals, groin, 94, 95, 97, 114, 115
 hunchbacked, 108
 short armed, 102
 thigh, 94
 wounded, 119, 127, 129
intervals, 79
Isidoros, son of Pamenes, 69, 165
Islam, horoscope of, 158
islamic astrology, 161, 171, 172, 176
Ismael, 158 n. 1
Italy, 182

Jebb, 114
Johnson, J., 49
Julian the Apostate, 135
Julian of Laodicaea, 188

Kahle, P., 29
Karanis, 50
katarche, 7, 142, 143, 146, 147, 148, 149
Khazars, 159
Khosro I, 16 n. 13, 161
Koch, W., 161 n. 5
Kroll, 80, 82, 91, 111 n. 3, 115, 116 n. 2, 176, et passim

Latin horoscopes, 161 n. 5
latitude of planets, 171
lawsuit, 32, 103, 104, 106, 108, 111, 113, 117, 118, 120, 122, 125, 126
length of daylight, 3, 26
length of life, 77, 78, 86, 87, 88, 89, 92, 100, 102, 105, 109, 110, 112, 120, 123, 126, 129, 130, 136, 138, 174, 177, 178, 179
Lenormant, 114
Leo I, 141 n. 3

Leo III, 159 n. 9
Leontius, 16 n. 13, 141, 147
letter, 149
Libya, 159
linear zigzag functions, 43
lion, tamed, 146
loci, 2, 7, 76, 132, 139, et passim
Loos, 3, 15, 16
Lot of Fortune, 2, 8, 28, 41, 55, 56, 174, et passim
lots, 8, 45, 67, 164, et passim
lunar diameter, 26, 171

Macarius, son of Hermias Gesdos, 72, 166
Macedonian calendar, see calendar
Macrianus, 165
magnitude, stellar, 9, 25, 91, 153, 171
male sign, 9, 19, 29
Manetho, 92, 163, 176
Marcus Aurelius, 165
Marinus, 135
Māshā'allāh, 172, 190
Masson, 114
Maximianus, 165
medical supply, 145
Metonic cycle, 10
midheaven, 8, 9, 78, et passim
midnight epoch, 167, 168, 169 n. 56
monomoiria, 10, 155
months, names, 3, 166
morning epoch, 18, 167, 169
mother, death of, 103, 109, 116
Muḥammad, 158
multiplication table, 66
Munychion, 136 n. 14

Nebuchelus, 54
necessity, lot of, 9, 45, 67
Nechepso, 5, 28, 185
Neugebauer, P. V., Tables, 2, 26, 156
Nicagoras, archon, 135
Nimrud Dagh, 14
nodes of moon, symbols, 159 n. 11
nodes of orbits, 25, 165, 171
non-viable, see stillborn
normal stars, 26
numerals, Greek, 136

Obeying signs, 5
oblique ascension, see rising times
occupation and status:
 consul, 140, 147
 dancer, 121
 distinguished person, 33, 79, 81, 86, 92, 93, 94, 99, 103, 107, 113
 eunuch, 83
 gens, 104
 gladiator, 109
 governorship, 86, 88, 93, 97, 98, 101, 148
 grammarian, 140
 guardsman, 109
 hireling, 33
 night-watchman, 72
 patrician, 141
 pirate, 125
 praefectus augustalis Aegypti, 148
 president, 72
 priesthood, 83, 85, 92, 100, 103, 127

quaestor, 140
robber, 109, 144
seer, 94, 115
slaves, 32, 88, 104, 107, 108, 111, 113, 118, 123, 127, 143
soldier, 93, 113, 118
strategos, 94 n. 1, 98
vagabond, 81, 88
wealthy, 80, 85, 88, 93, 97, 99, 103, 107, 130
Ophiuchus, 145 n. 27, 153, 171
opposition, 10, 171
oracle, 115
ostraca, 52
ostrich feathers, as cargo, 145

P. Anastasi 1073, 38 n. 17
P. Lund inv. 35a, 24
P. Lund inv. 35b, 173
P. Mich. 149, 5, 10 n. 27, 43 n. 8
P. Ryl. 27, 24
P. Tebt. 274, 173
  see also the concordance p. 205
Palchus, 142, 143, 144, 146, 147, 148, 149, 176, 187 n. 61
Palestine, 142 n. 1a, 159, 182, 187
Pamprepius, 140, 187
Panopolis, 141 n. 17, 188
papyrus, as cargo, 145
paradosis, 10
paranatellonta, 43 n. 9, 171
Parker, R. A., 6, 29, 32, 36
Parthian war of Avidius Cassius, 120
Paulus Alexandrinus, 2, 9 n. 25, 10, 35
Pedanius Fuscus, 80, 90 n. 8, 109
Pergamum, 113
perigee, of Mercury. 26, 171
  of sun, 43
period, sidereal, of Saturn, 79
periods, planetary, 10, 101
perpetual tables, 24, 172, 173 n. 82
Persia, 159, 190
phases, 10, 24, 26, 27, 29, 41, 171
Philip, Augustus, 165
Philip, era, 152, 158 n. 1, 166
planetary longitudes, computation of, 2, 105
planetary tables, 173
planets, arrangement, 14, 164
  names, 1, 162
Plato, 113
Pleiades, 95 n. 2
Pliny, 185
Plutarch, 54 n. 4
Posidonius, 185 n. 22
precede, 11
precession, 153, 172, 180, 187
predictions in horoscopes, 101, 122, 145, 162
  errors, 147, 149
pregnancy, 15, 28, 86, 112, 131
prerogative, 68
Probus, 165
Proclus, 135, 176
progression, 132

prosopa, 11, 155
Ptolemaic canon, 24
Ptolemy, 10, 152
  Almagest, 4, 9, 12, 24, 27, 46, 153, 167, 170, 172, et passim
  Tetrabiblos, 12, 138, 139, 155, 188, et passim
Puchstein, 14

Qasr el Banat, 48
Quietus, 165

Retrogradation, 171
Rhetorius, 132, 138, 140, 141, 146, 150, 157, 176, 187, 189
right ascension, 11
rising times, 3, 9, 11, 35, 101, 170, 174, et passim
Roman calendar, see calendar
Rome, 182
Rome, A., 133 n. 22, 150 n. 2
royal canon, 24, 180

Sachs, A., 161 n. 1
Salt collection, 43
Saturn, sidereal period, 79
Scaliger, 6
Schmid, W., 114
Schnabel, 185 n. 19, n. 25
Scythia, 159
seasonal hours, 169
sect, 11
Seleucid era, 166
Septimius Severus, 53, 165
Serapion, 3
Serenus, son of Artemidoros Phanousire, 73, 166
Servianus, 80
Sethe, K., 168
seven climata, 4; see also climata
Severus Alexander, 165
sextile, 91 n. 2, 113
Sirius, 25, 116, 170
Skeat, T. C., 23, 29, 30 n. 2
Smyly, J. G., 168
Smyrna, 113, 144, 187
solar eclipse, 135, 148 n. 9
solar gnomon, 136, 140
solid signs, 12
solstices, 4, 24
Spain, 80, 91 n. 19, 136, 170
sphaera barbarica, 27 n. 15
sphaera recta, 2 n. 2, 4, 10, 11, 12, et passim
starter, 12, 76, 78, 97, 105, 164
stationary points for planets, 28, 41
Stein, J. W., 135
stellar magnitude, 9, 25, 91, 153, 171
step, 12, 152 n. 12, 154
Stephanus, 158, 190
stillborn, 100, 141, 146, 185
Stobart, H., 29, 38
Stobart tables, 166 n. 33, 173
stoicheion, 12, 29
Stroobant, 141
sun, symbol, 1, 163 n. 10

Swallow-Fish, 23, 25, 26
symbols, 163
  for lunar nodes, 1, 159 n. 11
  for νύξ and ἡμέρα, 76, 159 n. 4
  for planets, 1
  for sun, 1, 163 n. 10
  zodiacal signs, 1, 76
synanatellonta, 43, 171
synodic periods, 10
Syria, 159, 182
systems A and B, 3, 12, 174, et passim
syzygies, 174

Tarsus, 148
terms, 12, 20, 29, 36, 41, 67, 82
Tetrabiblos, see Ptolemy
Teucros, 43
Theodorus, augustalian prefect, 148, 171
Theon of Alexandria, 2
  Commentary to the Handy Tables, 154
  Handy Tables, see Handy Tables
Theon of Smyrna, 141
Theophanes, 148 n. 24, 158 n. 1
Theophilus, 185
Thrasyllus, 76
Tiberius, 165
Timocharis, 185
Titus, 24, 165
Trajan, 90, 166
trepidation, 26
triangles, 12, 29
trias, 79
Turks, 159
Turner, E. G., 29
turning away, 13

Usener, 159, 190

Valentinian, 136, 166
Valerian, 165
Valerius Eudaemon, 86 n. 5
van der Waerden, 173 n. 82
Vat. graec. 208, 154 n. 34a
Vat. graec. 1291, 2 n. 2, 4 n. 5, 27, 154 n. 34a, 34b
Verina, 141 n. 3
vernal point, 4, 24, 25, 26, 172, 180, 185
Verus, 165
Vettius Valens, 1, 4, 5, 5 n. 7, 7, 8, 9, 10, 11, 12, 13, 28, 66, 89, 101, 105, 174, 176, 205, et passim
Vogt, H., 10

Washington University, 62 n. 1
water clock, 35, 49, 173
weekdays, 167
Weinstock, S., 143 n. 2
Welles, C. B., 54, 55
Westerman, W. L., 61 n. 1, 86 n. 5

Yielding, 10, 116
Youtie, H., 53

Zero, 156 fig. 20
zodiacal signs, symbols, 1, 76

PLATE 1

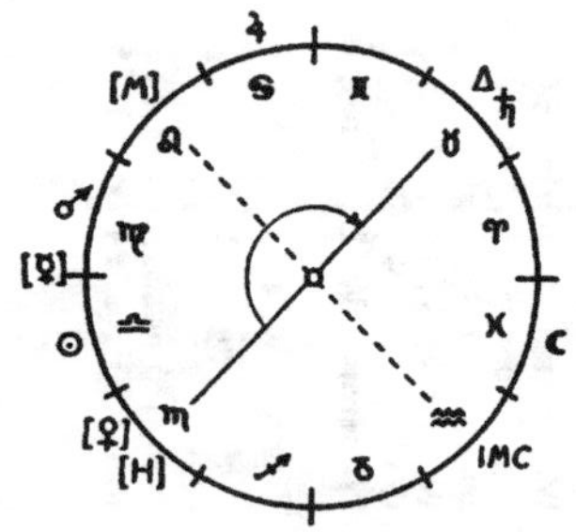

No. −3

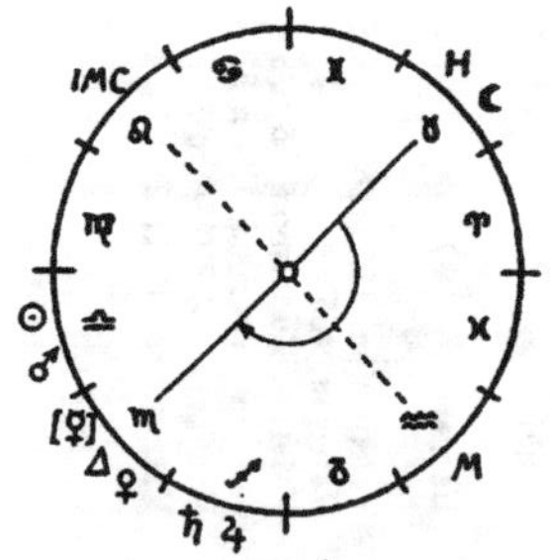

No. 15/22

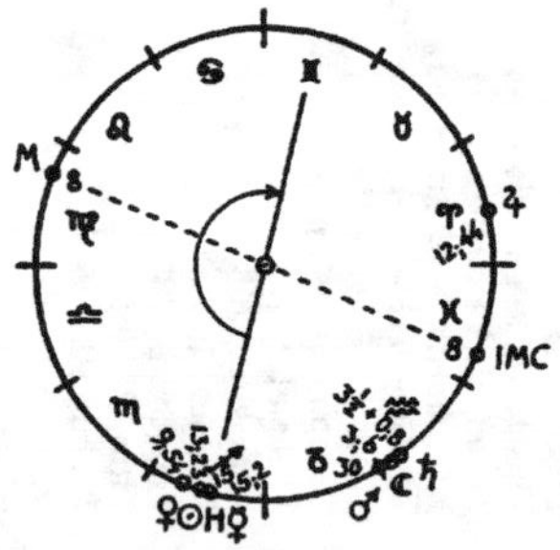

No. 137 a

PLATE 2

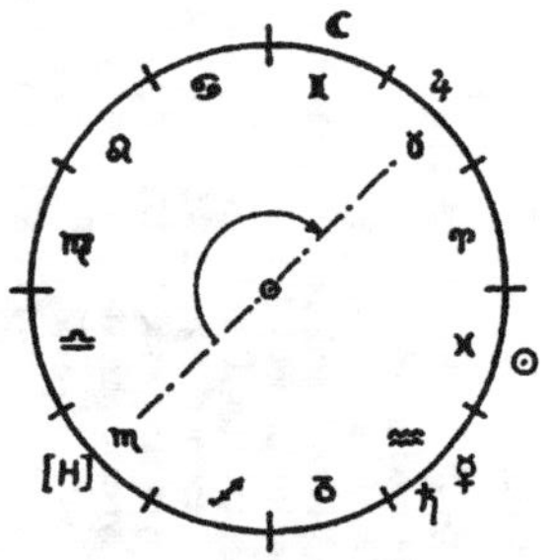

No. 139

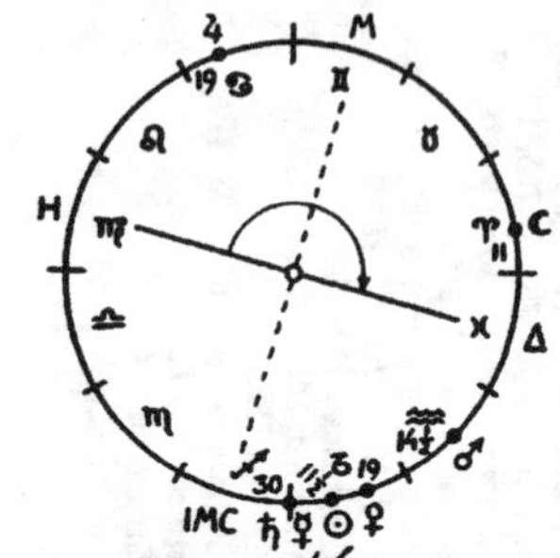

No. 46

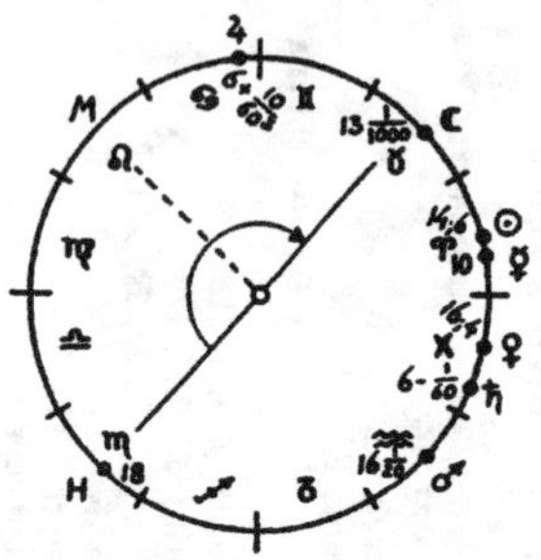

No. 81

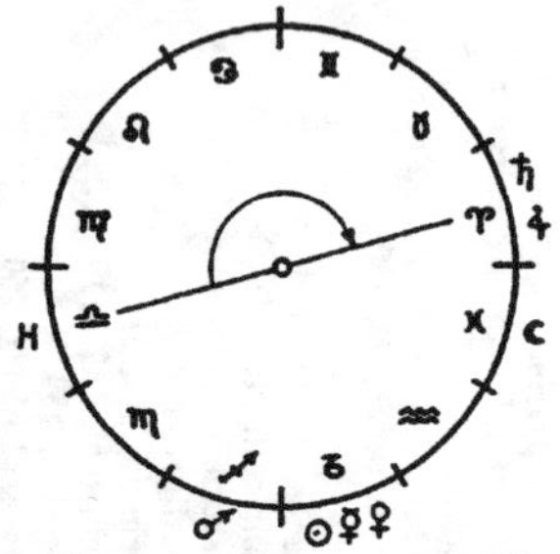

No. 149

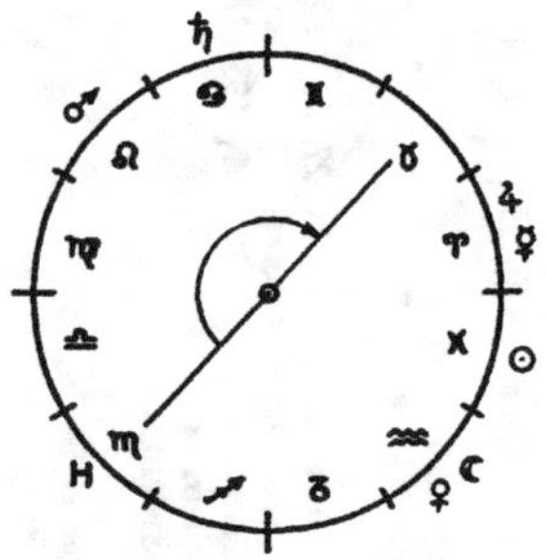

No. 150

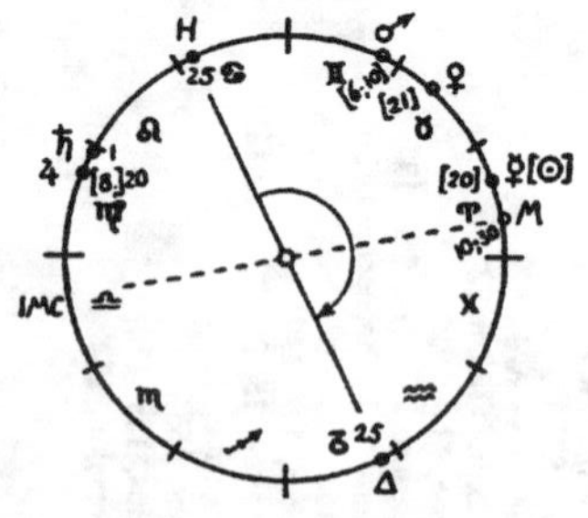

No. 95

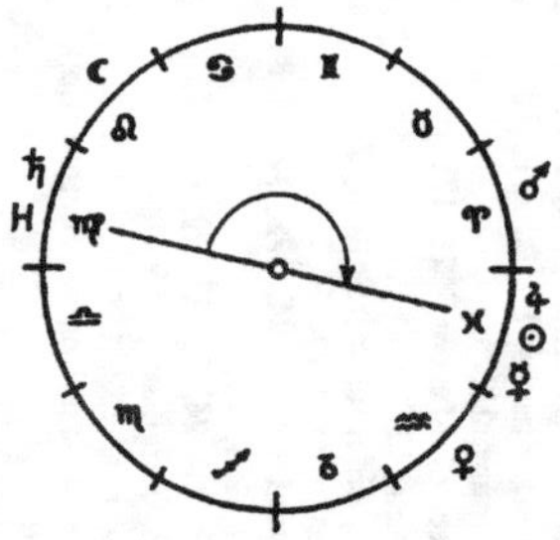

No. 125

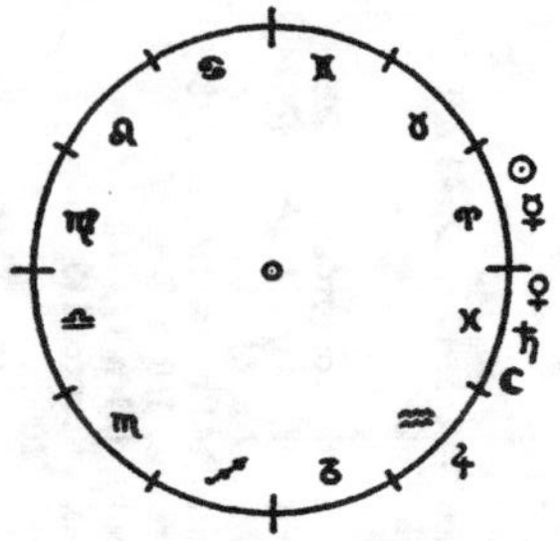

No. 171

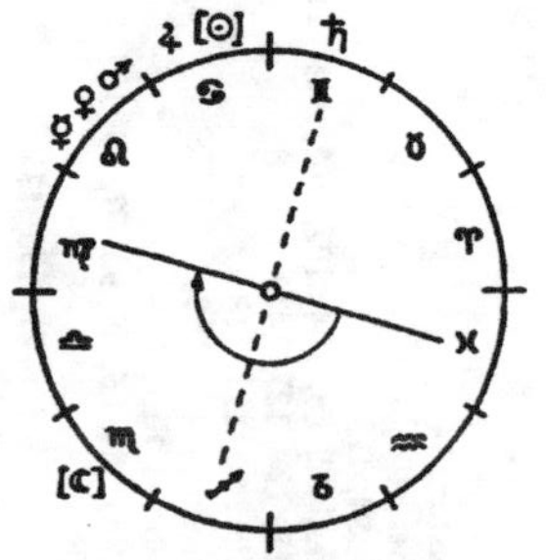

No. 176

214

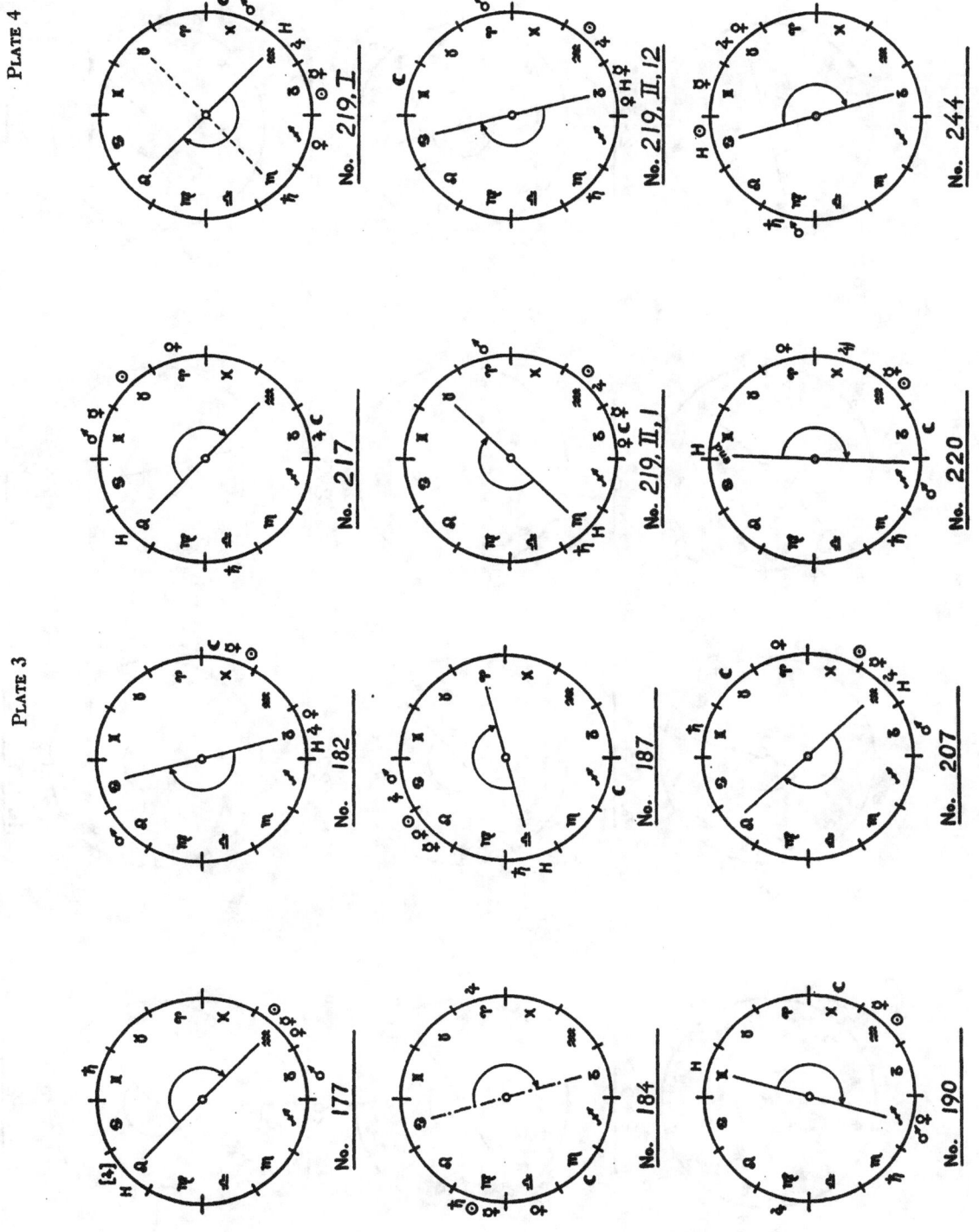

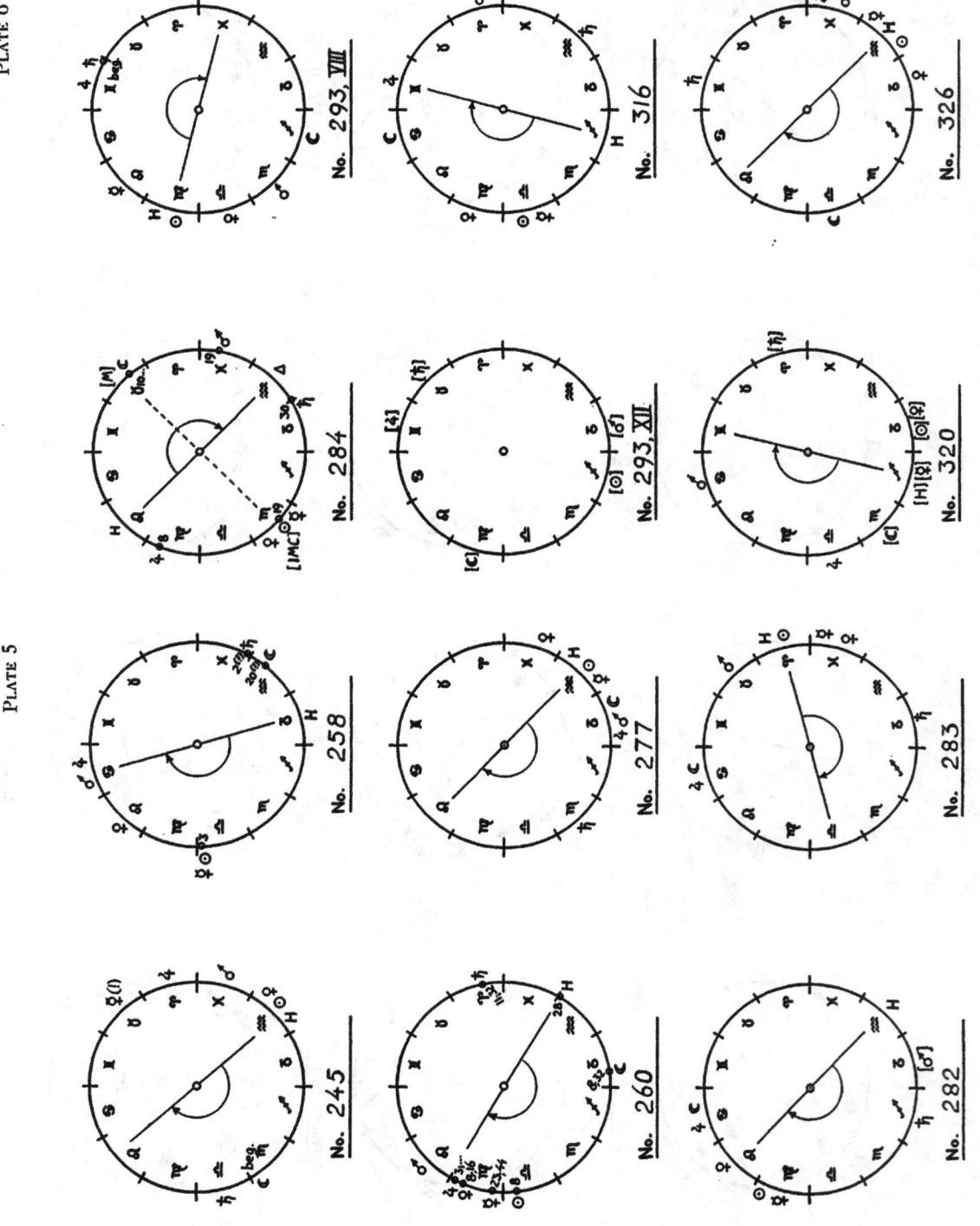

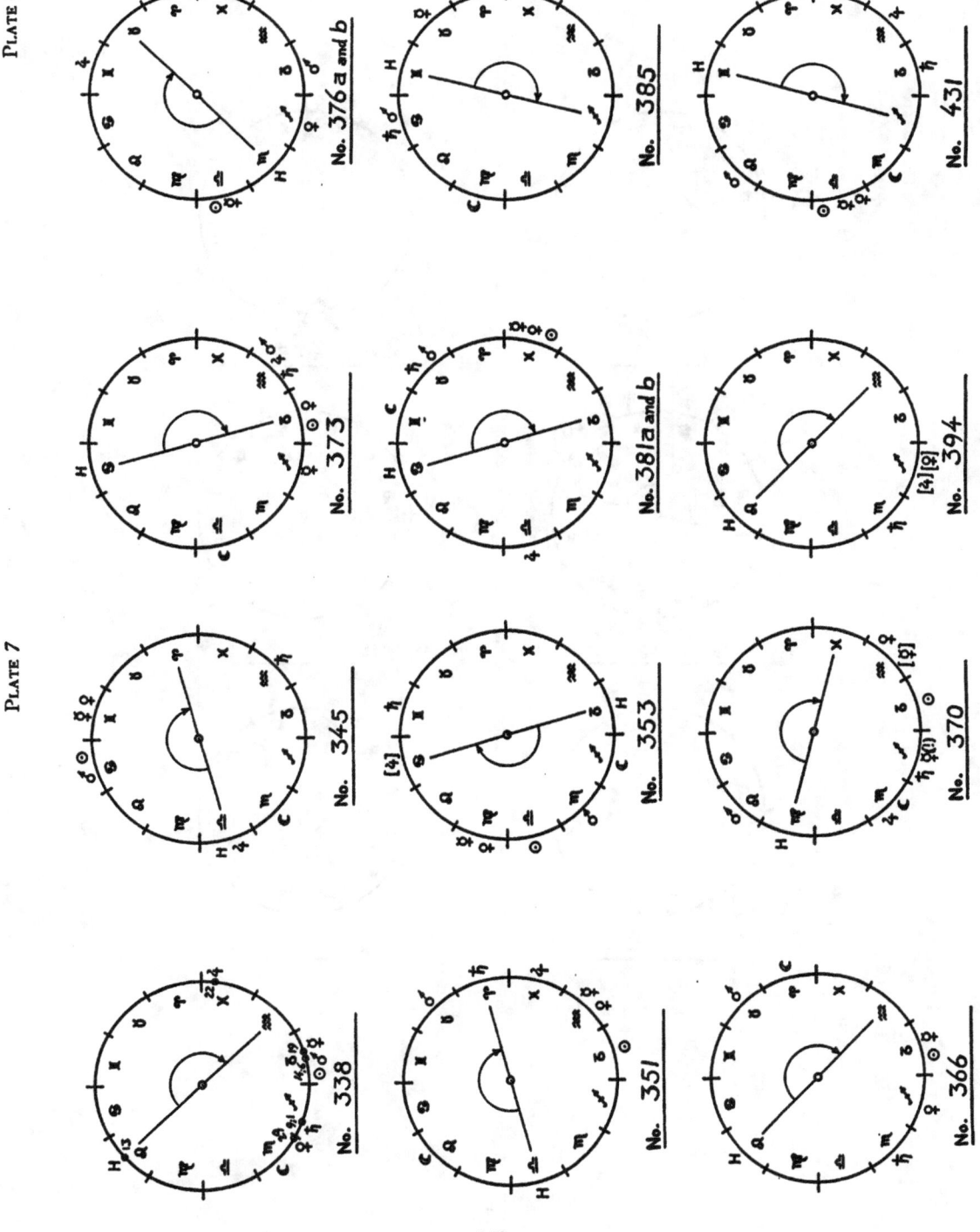

PLATE 9

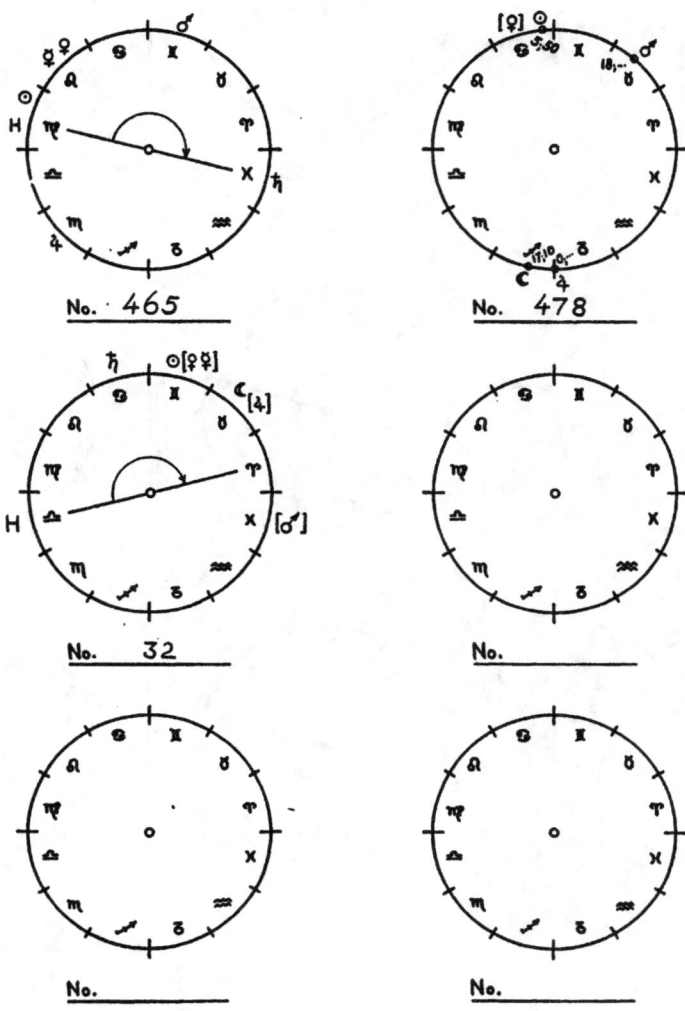

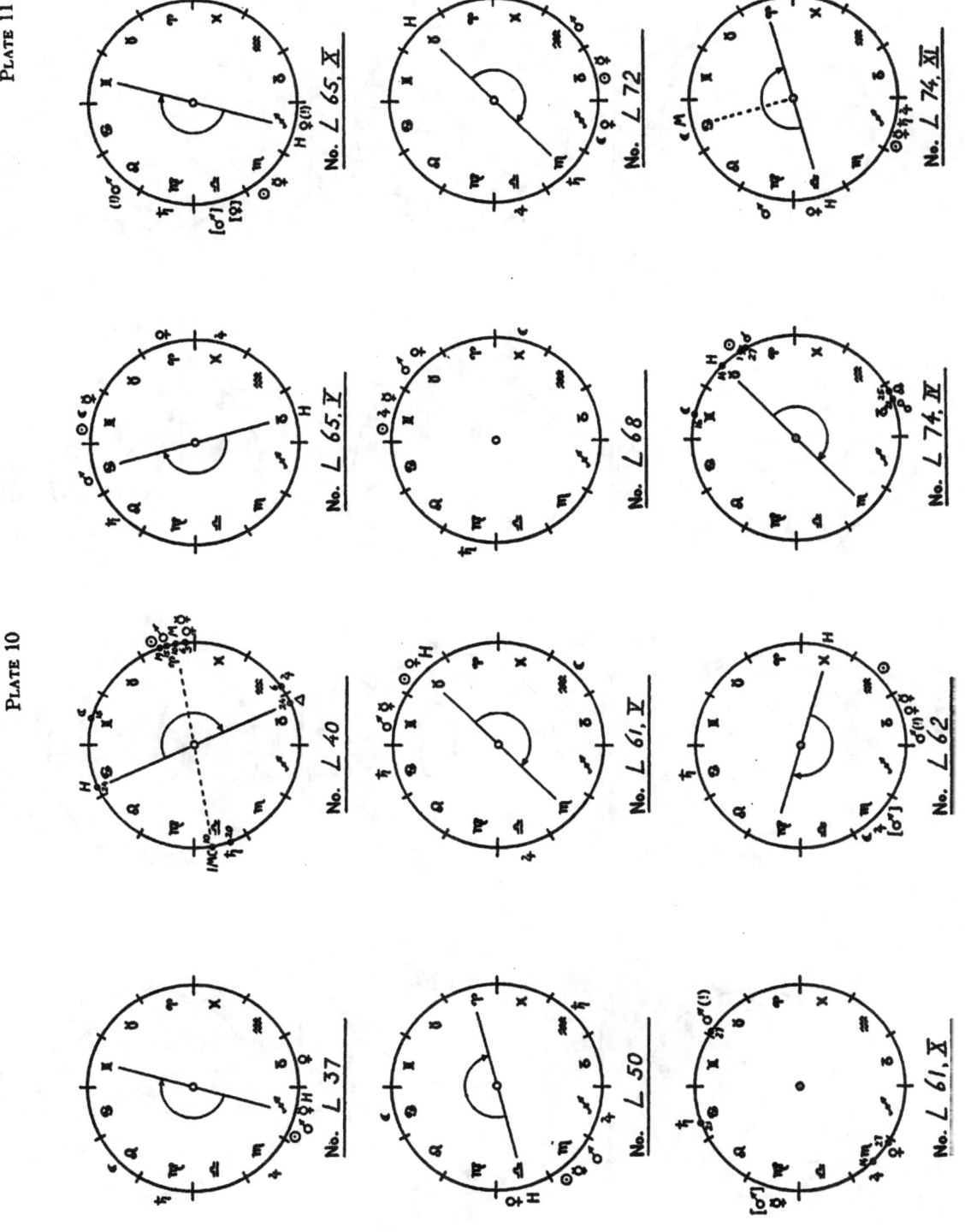

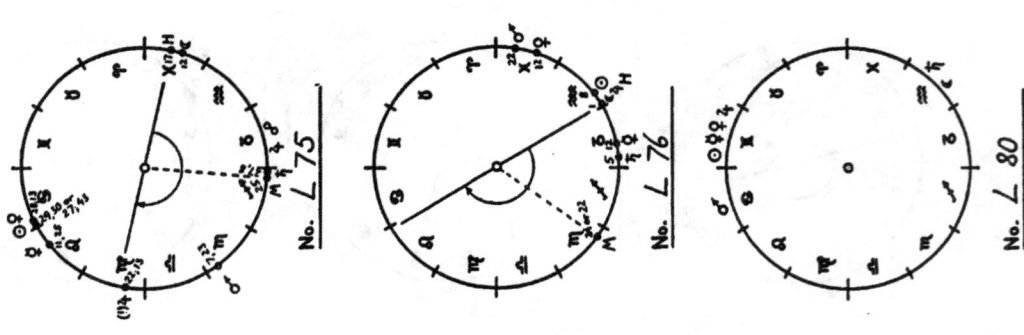

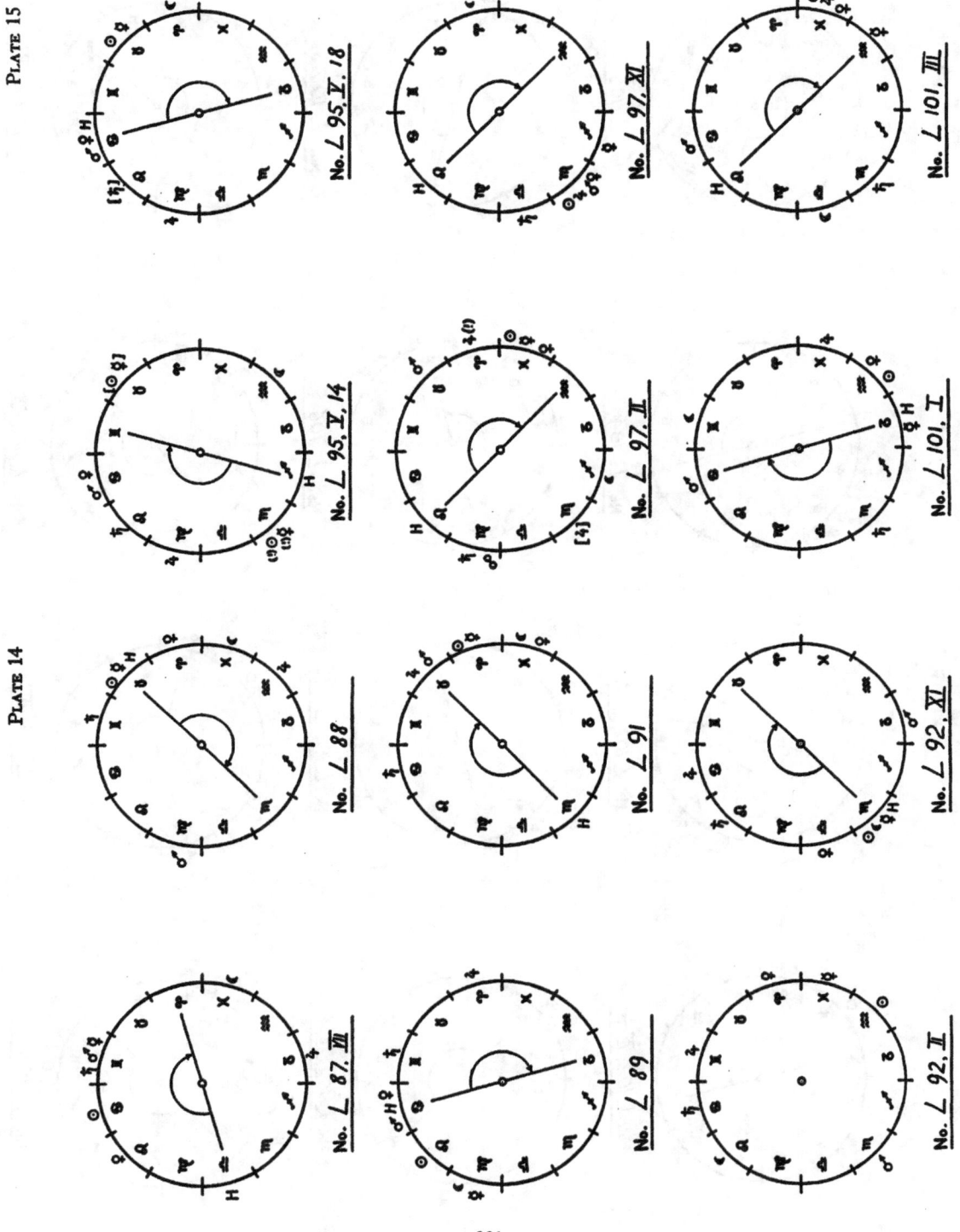

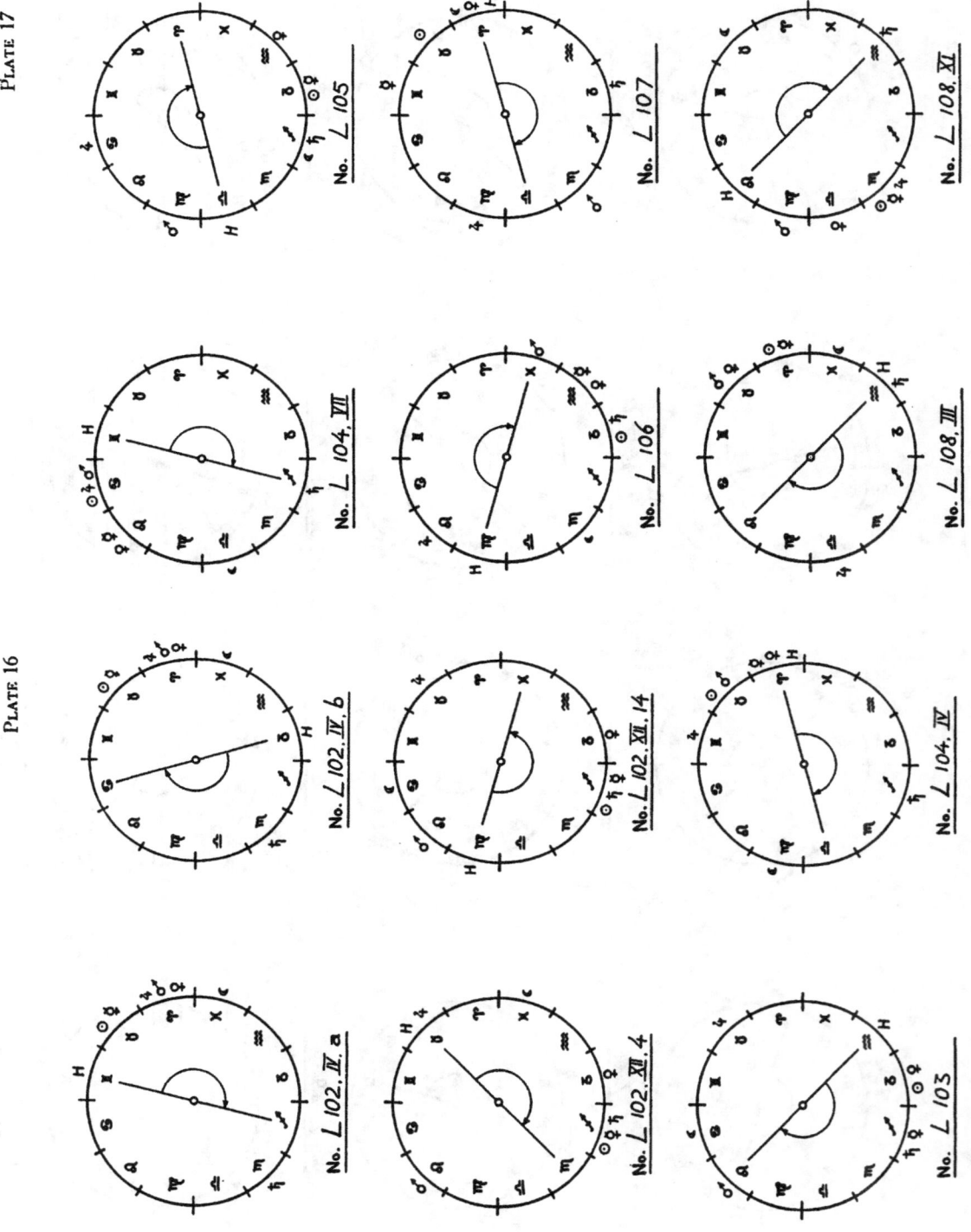

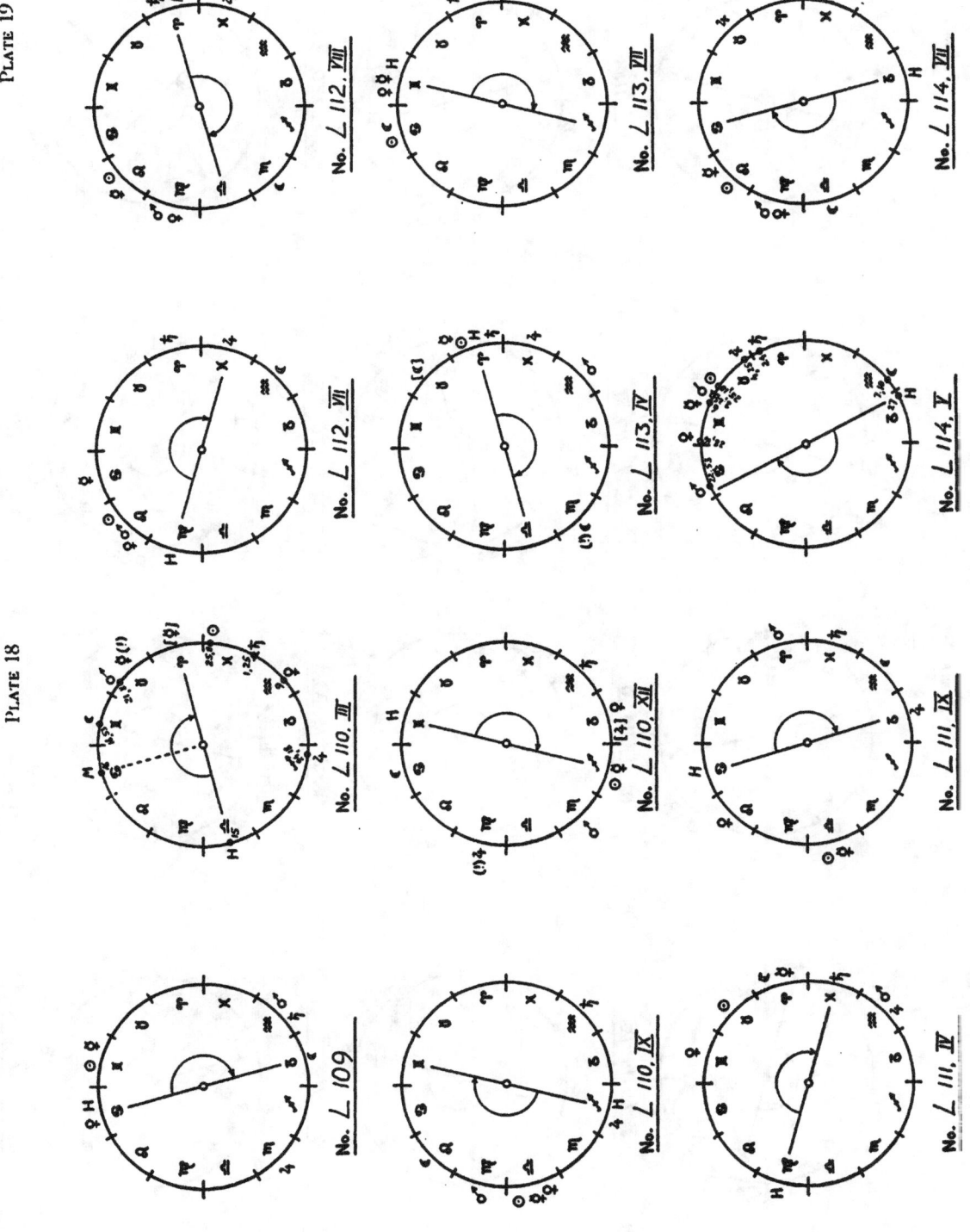

Plate 20

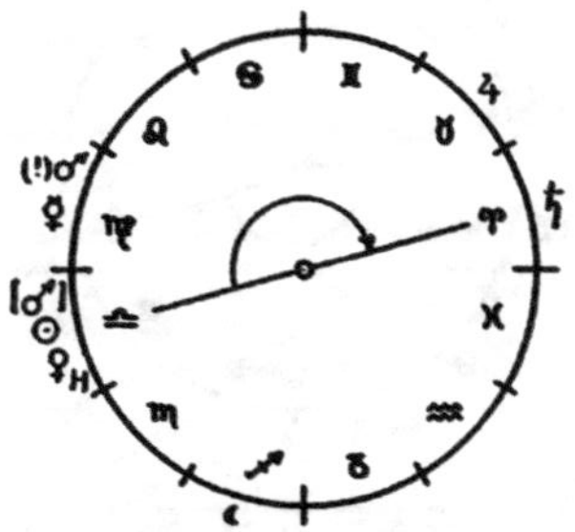
No. ∠ 114, IX

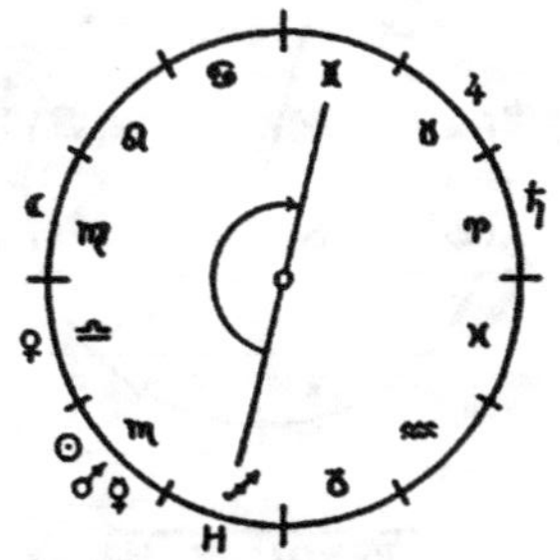
No. ∠ 114, XI

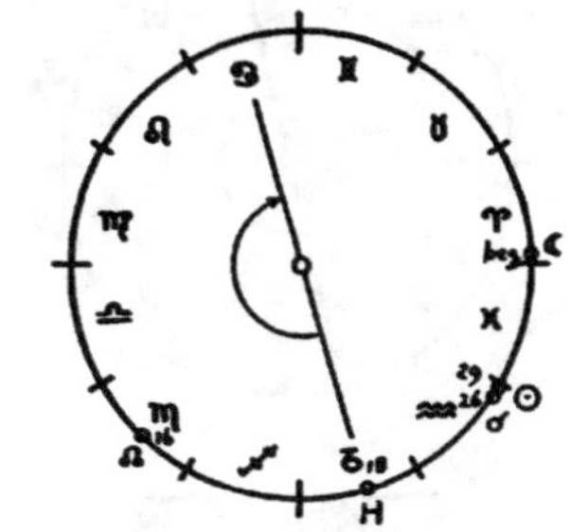
No. ∠ 115, II

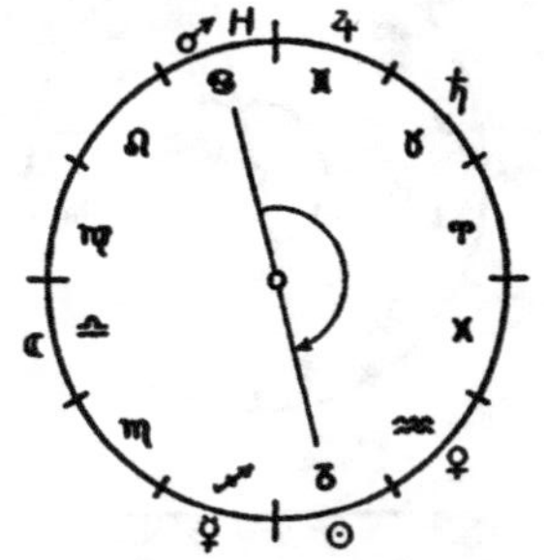
No. ∠ 115, XII

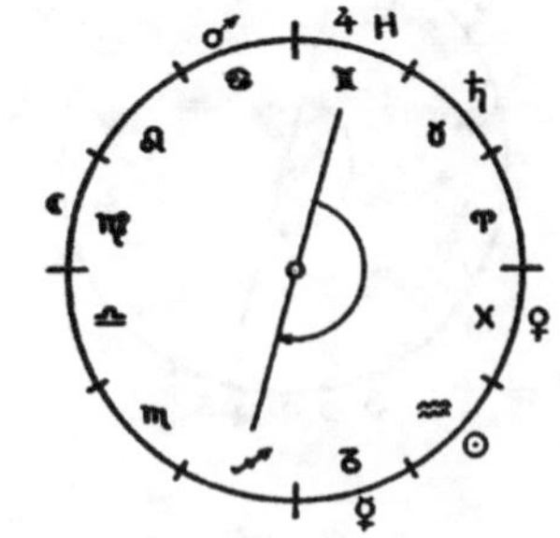
No. ∠ 116

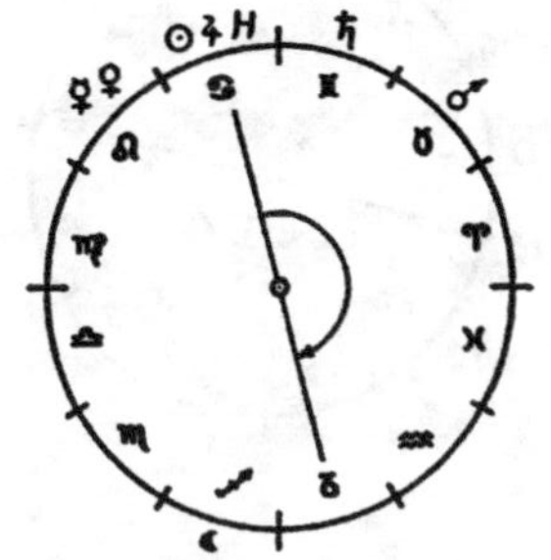
No. ∠ 117, VI

Plate 21

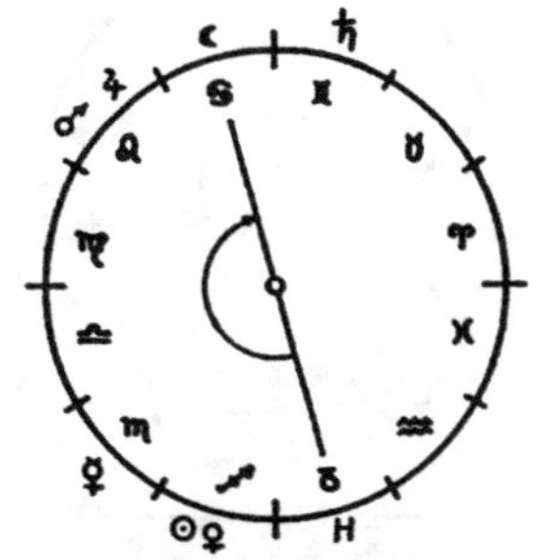
No. ∠ 117, XI

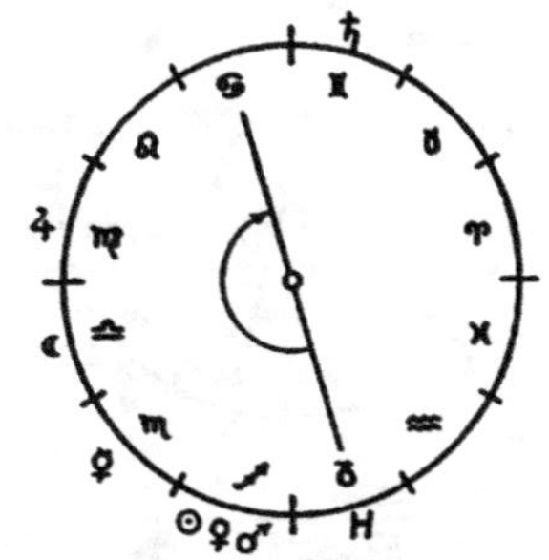
No. ∠ 118

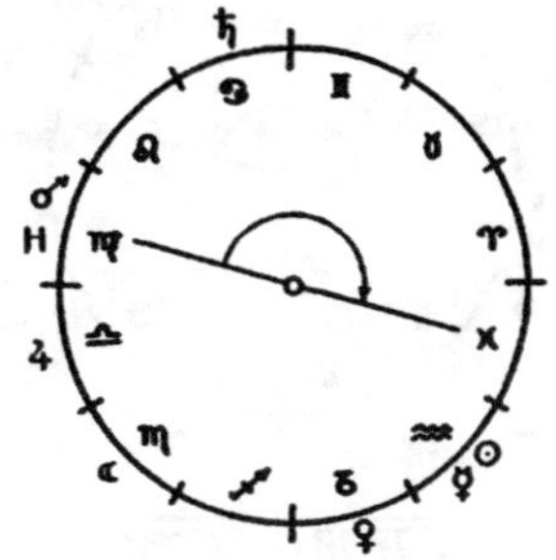
No. ∠ 120, II

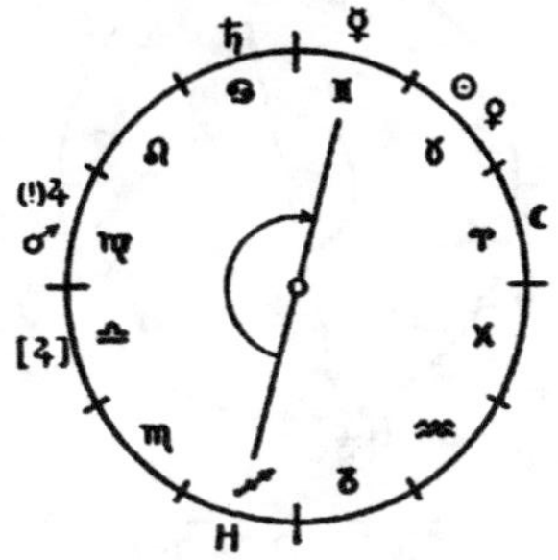
No. ∠ 120, V

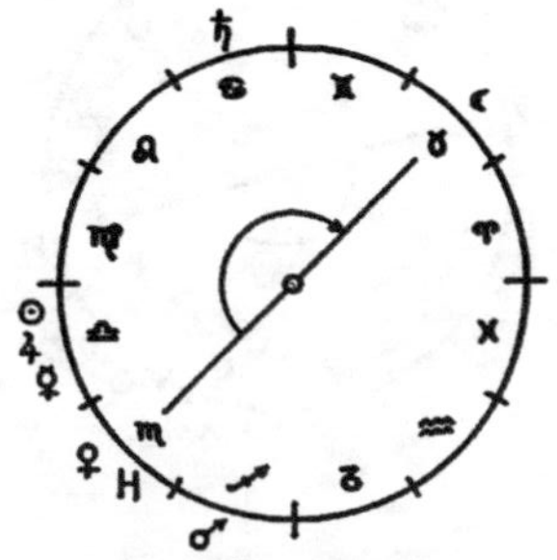
No. ∠ 120, IX

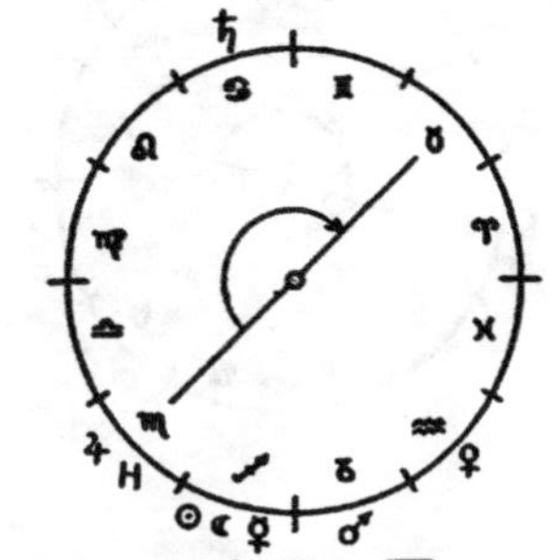
No. ∠ 120, XII

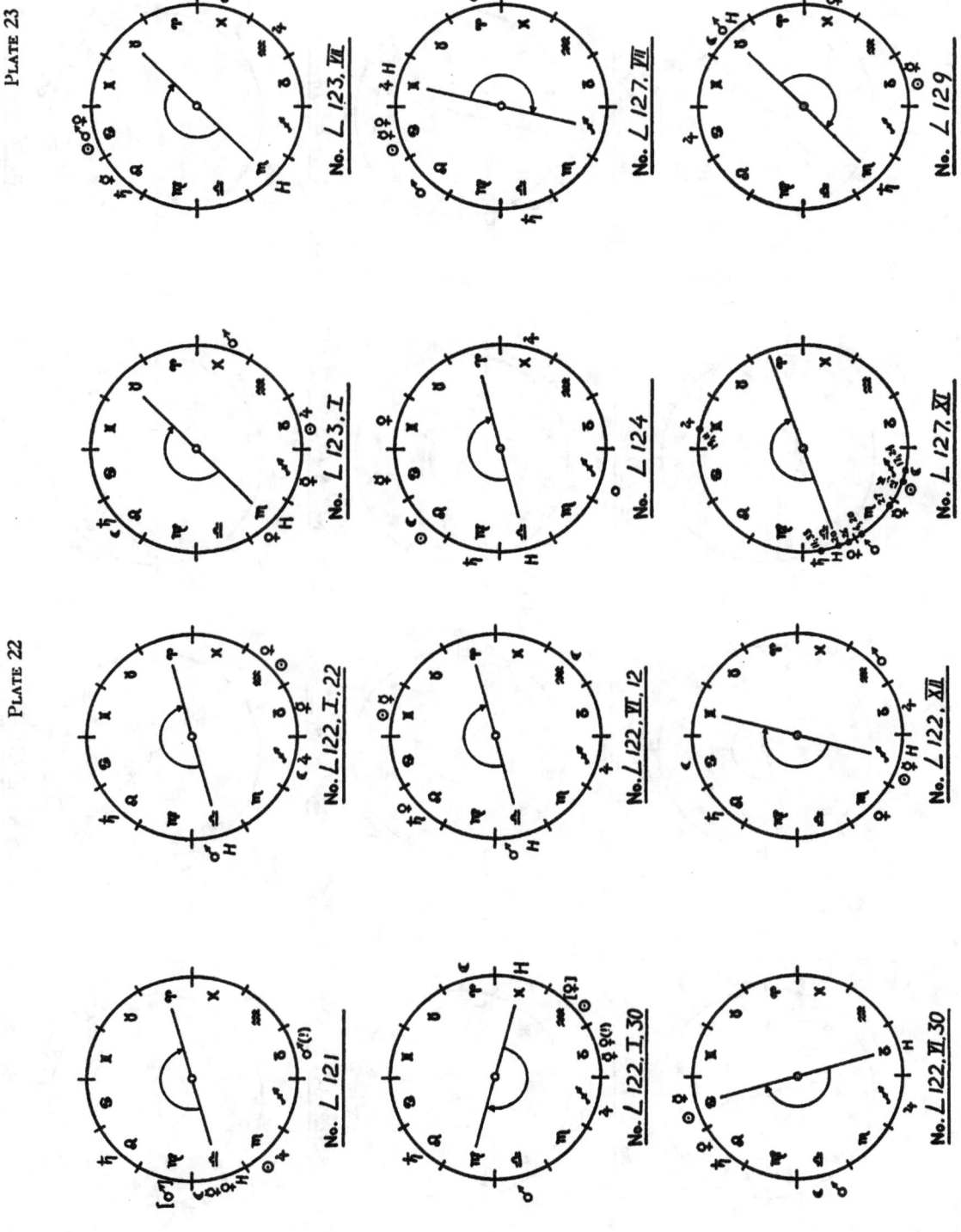

PLATE 25

No. L 152

No. L 158

No. L 162

No. L 142

No. L 153

No. L 159

PLATE 24

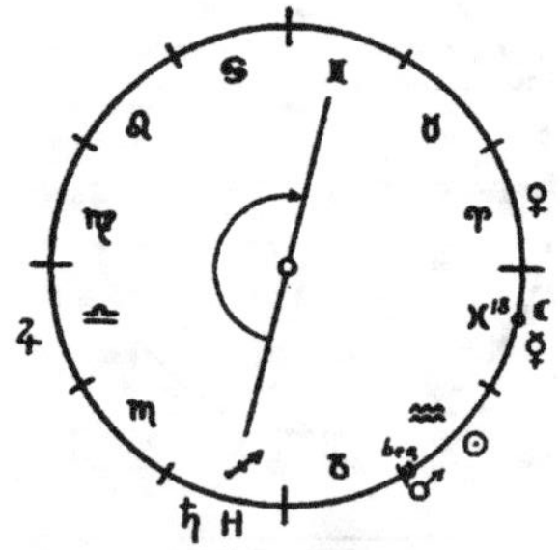

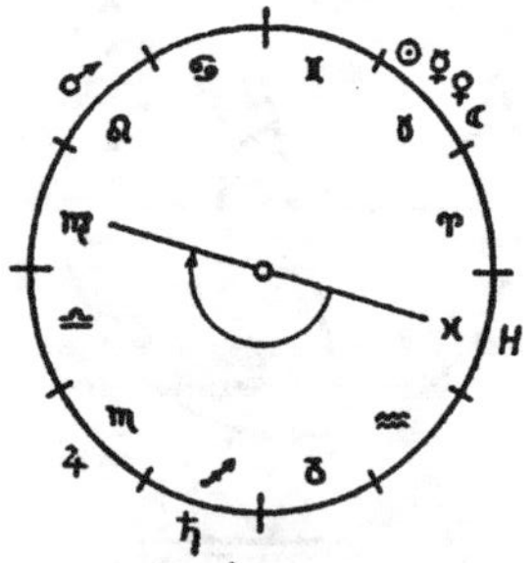

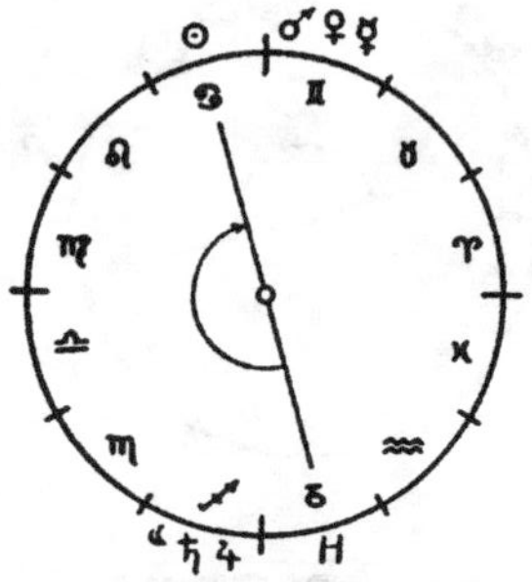

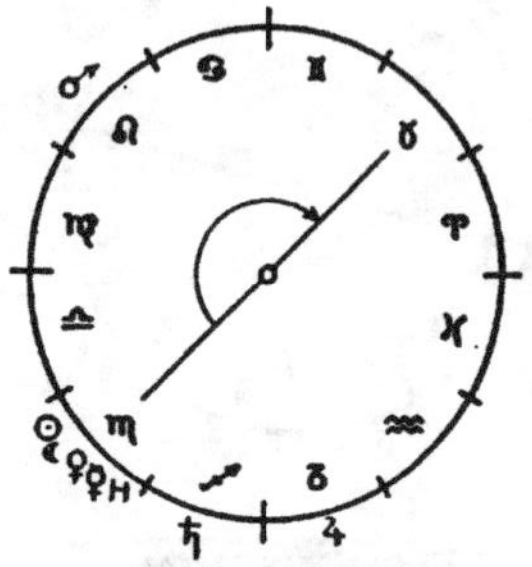

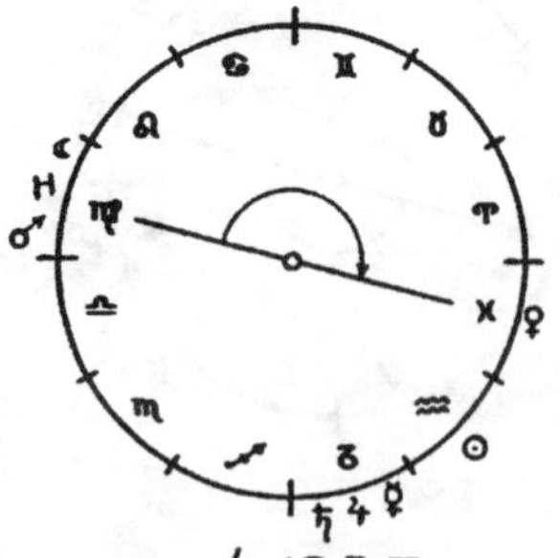

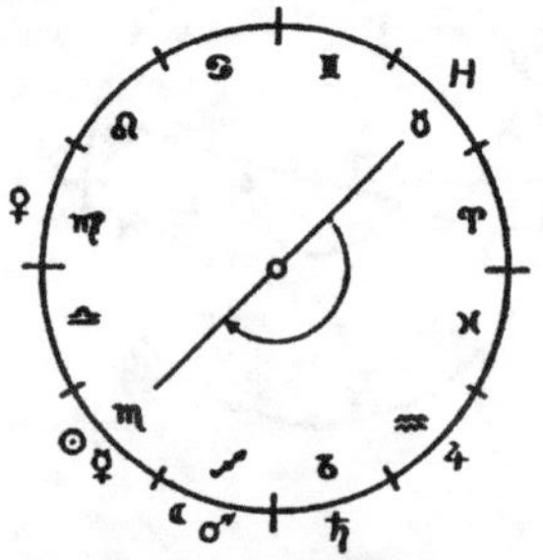

No. L 133

No. L 134, XI

No. L 135, X

No. L 132

No. L 134, II

No. L 135, I

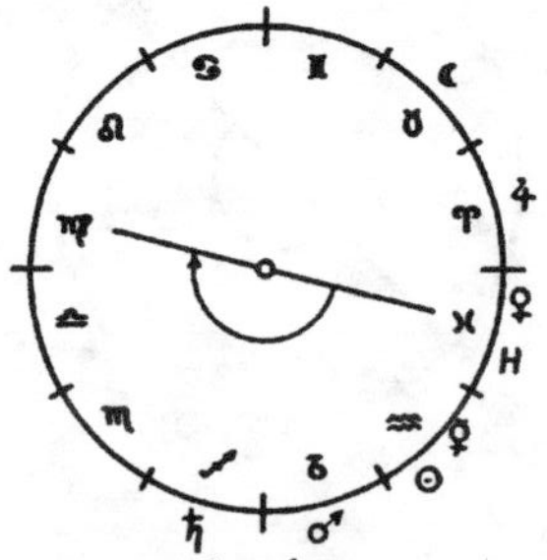

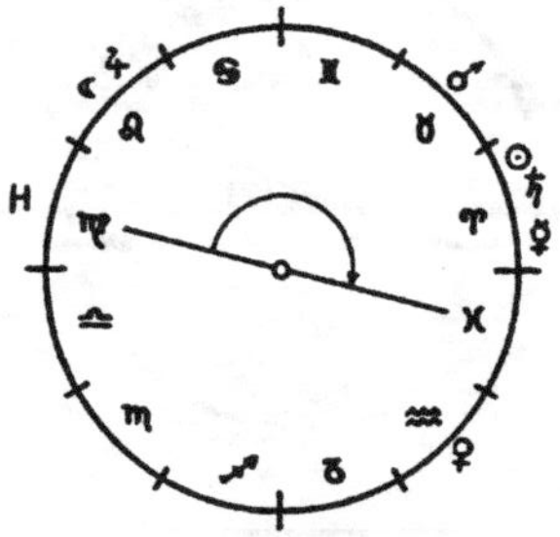

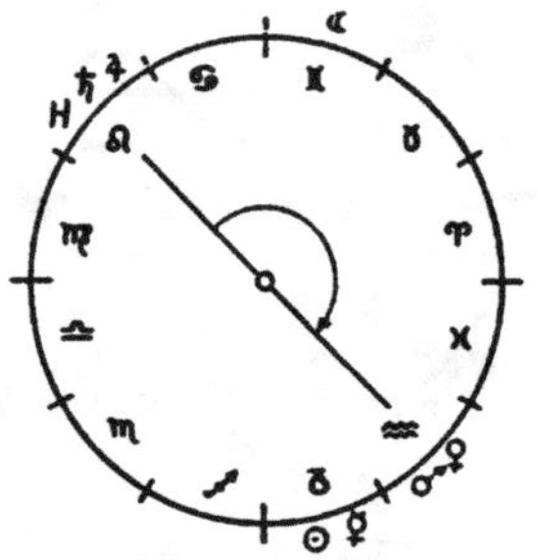

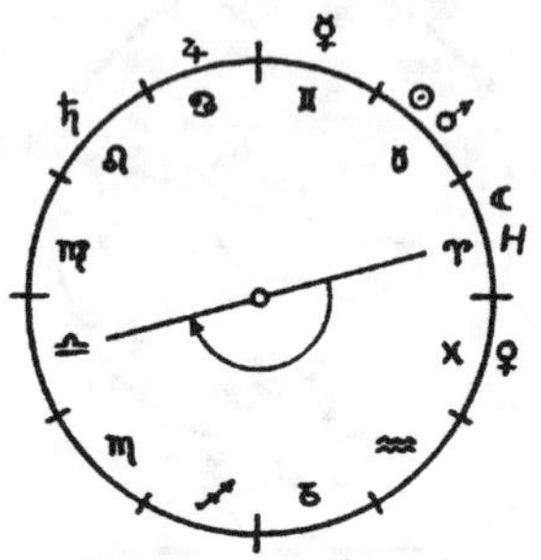

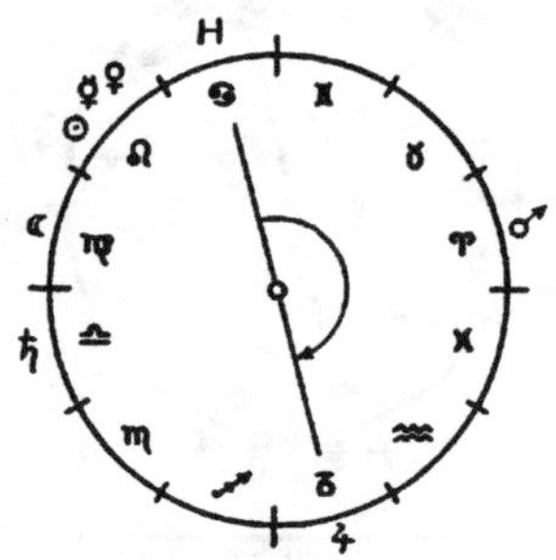

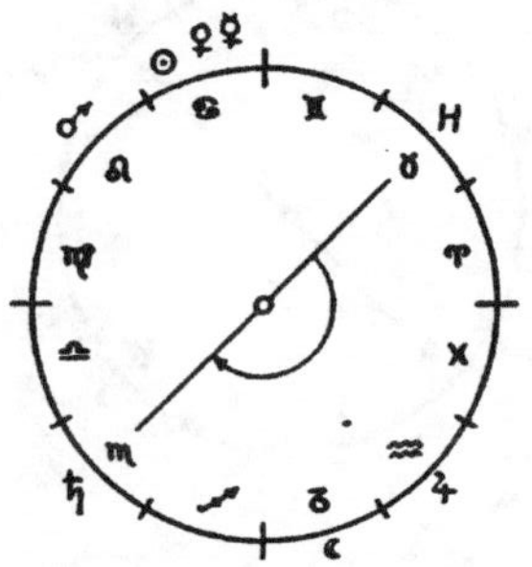

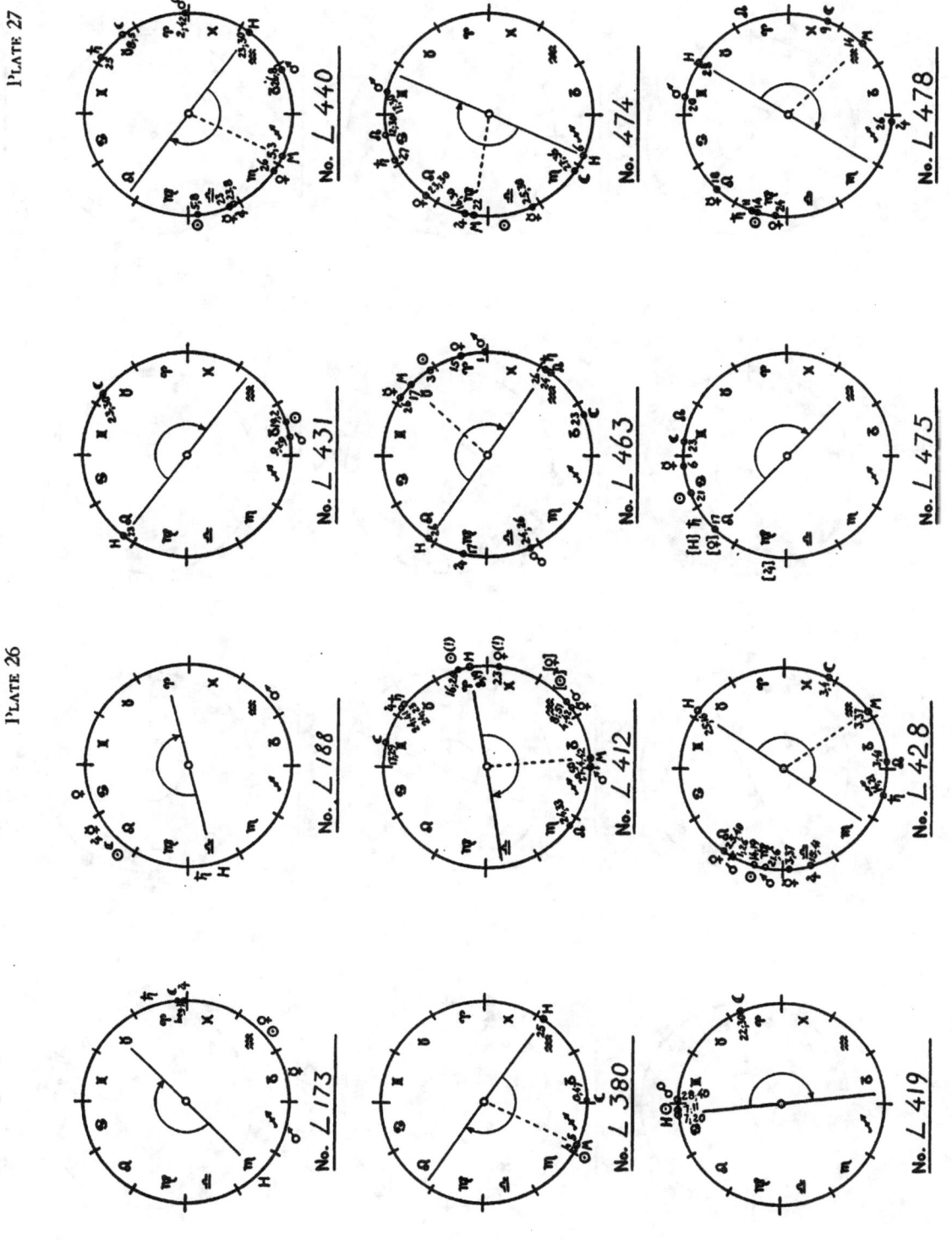

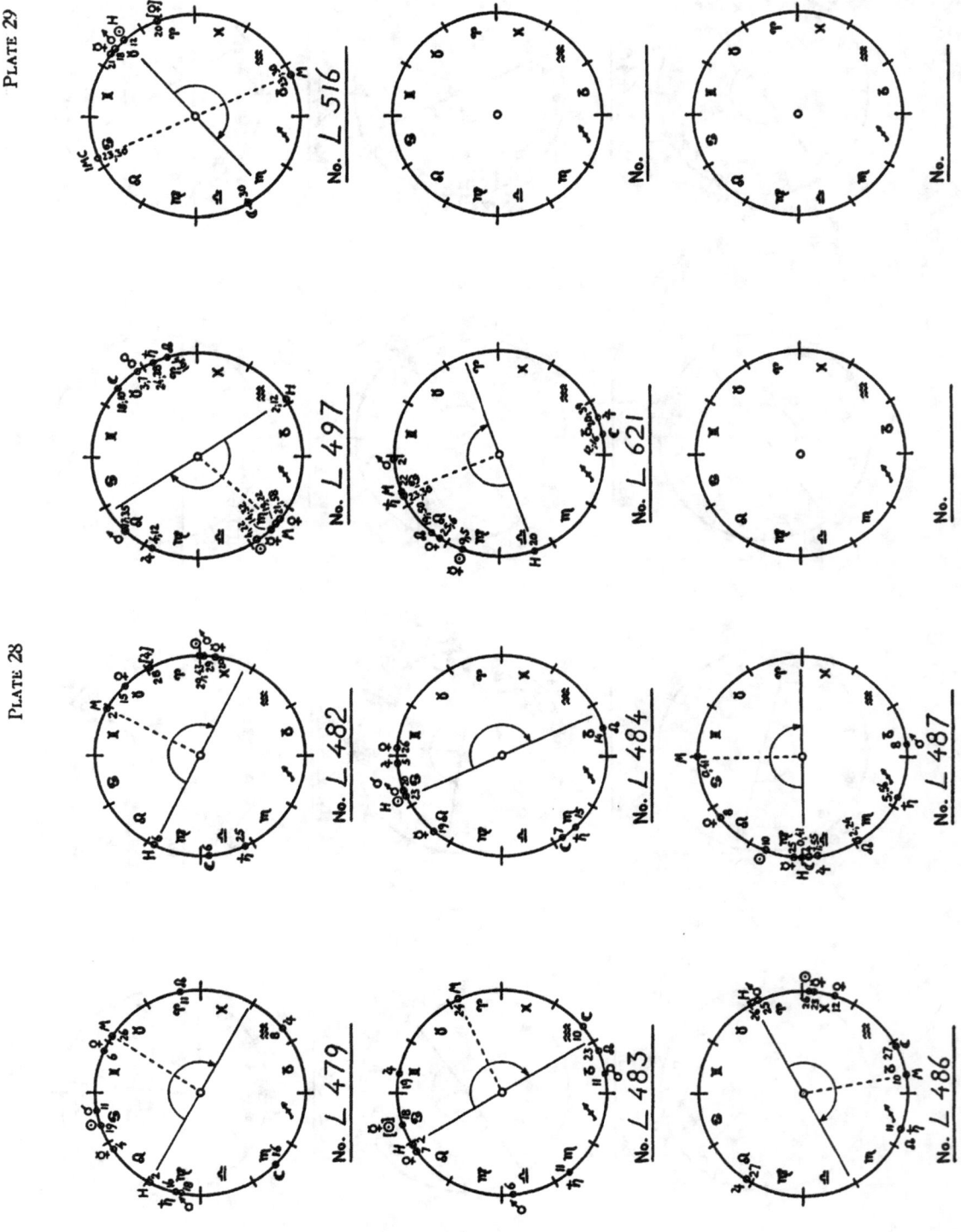

PLATE 30

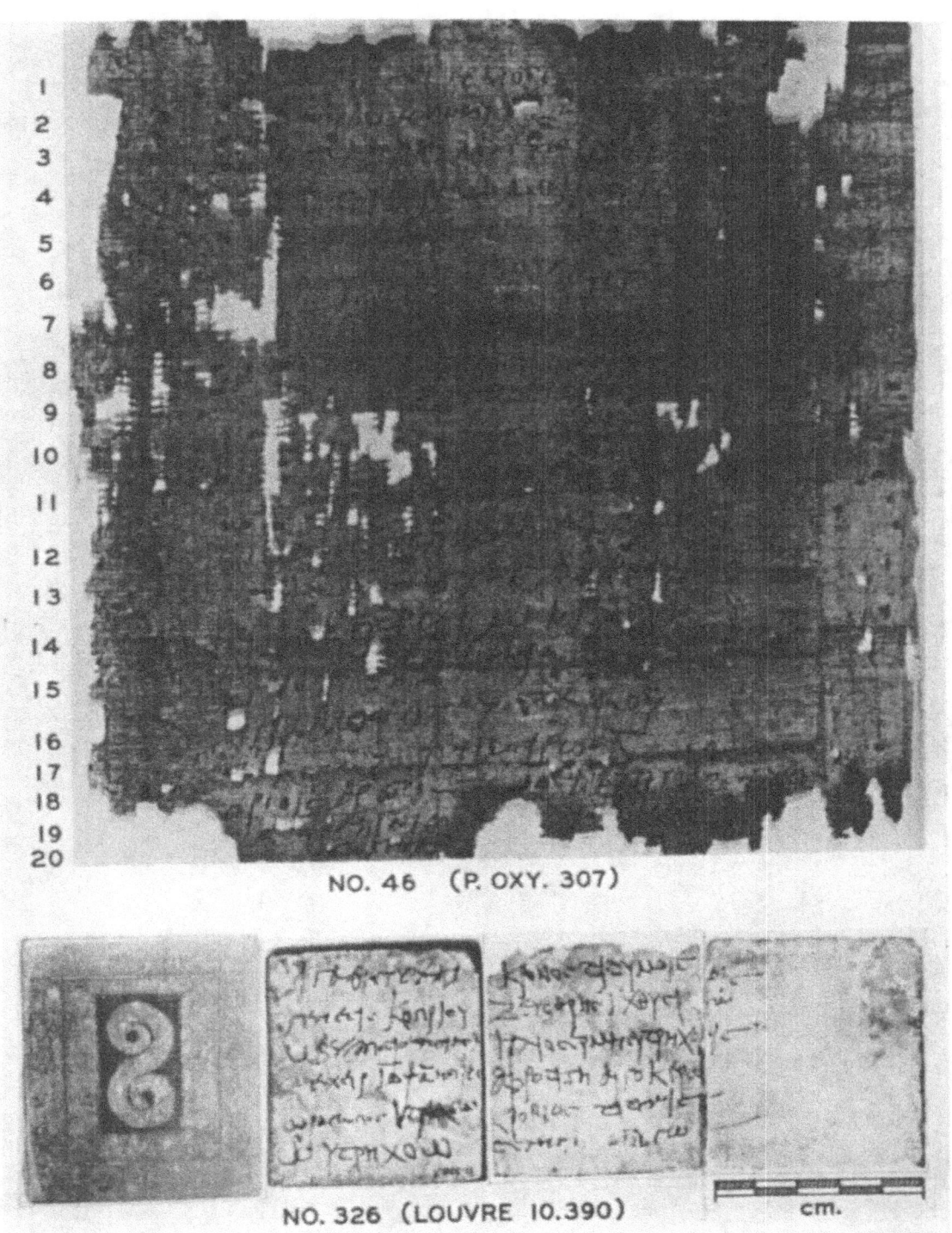

NO. 46 (P. OXY. 307)

NO. 326 (LOUVRE 10.390)

NO. 182
(P. MICH. INV. 6329)

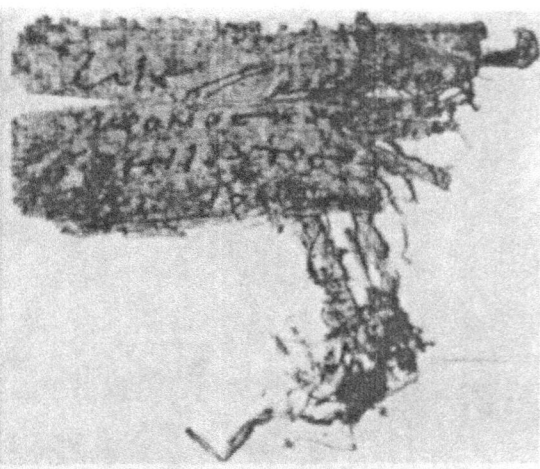

NO. 208
(P. MICH. INV. 1461)

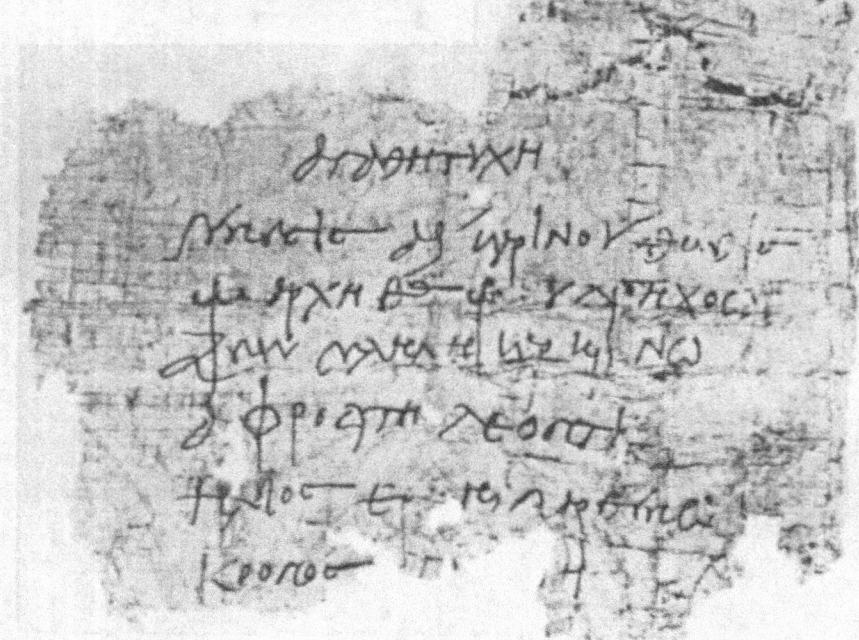

NO. 282
(P. CORNELL INV. 78)

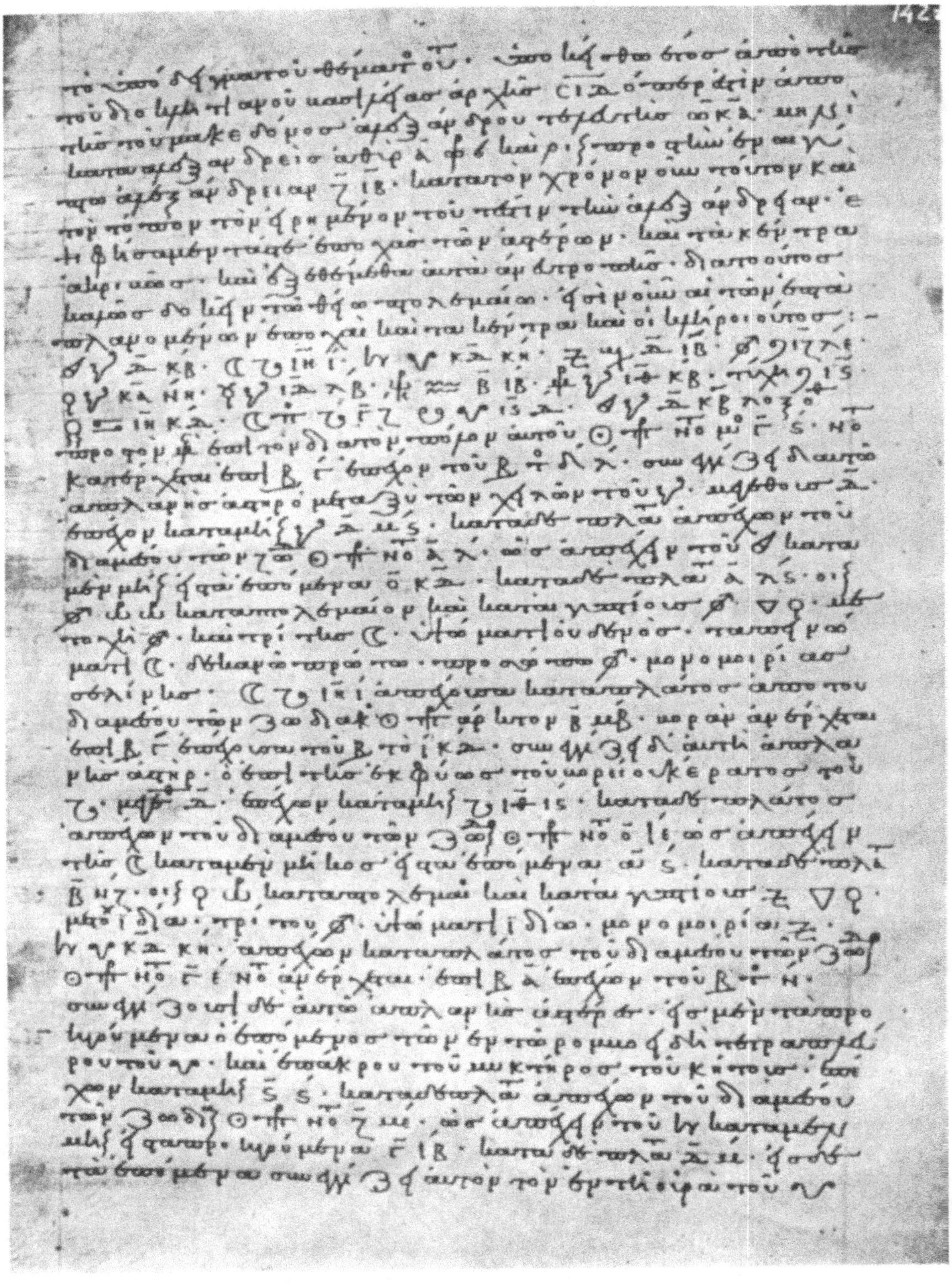

www.ingramcontent.com/pod-product-compliance
Lightning Source LLC
Chambersburg PA
CBHW080848010526
44114CB00018B/2393